William John Macquorn Rankine
Edward Fisher Bamber

A MECHANICAL TEXT-BOOK

OR

INTRODUCTION TO THE STUDY OF MECHANICS AND ENGINEERING

Elibron Classics
www.elibron.com

Elibron Classics series.

ISBN 1-4212-3438-6 (paperback)
ISBN 1-4212-3437-8 (hardcover)

This Elibron Classics Replica Edition is an unabridged facsimile of the edition published in 1873 by Charles Griffin and Company, London.

A

MECHANICAL TEXT-BOOK:

OR,

INTRODUCTION TO THE STUDY OF MECHANICS AND ENGINEERING.

BY

WILLIAM JOHN MACQUORN RANKINE,

CIVIL ENGINEER; LL.D. TRIN. COLL. DUB.; F.R.SS. LOND. AND EDIN.; F.R.S.S.A.;
LATE REGIUS PROFESSOR OF CIVIL ENGINEERING AND MECHANICS IN THE UNIVERSITY OF GLASGOW;

AND

EDWARD FISHER BAMBER, C.E.

With Numerous Diagrams.

LONDON:
CHARLES GRIFFIN AND COMPANY,
STATIONERS' HALL COURT.
1873.

PREFACE.

THIS book is designed as an Introduction to more abstruse works on Engineering and Mechanics, and in particular to those of the late Professor Rankine.

Its study demands only a previous acquaintance with the ordinary Rules of Arithmetic, and with the Elementary Algebraical Notation. A few pages have been devoted to the Differential and Integral Calculus, as these have been used in different parts of the book, their application having been in every instance explained.

Professor Rankine's *Manual of Applied Mechanics* has been taken as the model for this work, the only alteration being the treating of the Theory of *Motion* before that of *Force*, as more in harmony with modern practice, and as proposed by himself for the present purpose.

The general design of the work having been indicated, it only remains for me to explain briefly how my name has been connected with that of Professor Rankine on the Title-page, and also in what condition it was left at the time of his recent lamented death.

I was Professor Rankine's Assistant, and lectured for him during his illness, and it was whilst on a visit which his death suddenly terminated, that the arrangement was made which connected me with him in the task. My duty was simply to assist him in its preparation. On my mentioning to him that the amount of labour I should have to do hardly justified my name appearing with his as joint-author, he replied, that, owing to his state of health, more of the work might devolve upon me than I expected. The issue has proved the correctness of his surmise.

As to the state of the MS. at the time of his death, two hundred pages had been already completed, and the general scope and plan of the work decided upon. I need hardly say that his wishes have been implicitly carried out in every respect, so far as lay in my power. The work has been completed at the request of Professor Rankine's Executrix, and at that of the Publishers, at whose desire also I have undertaken the superintendence of New Editions of his other Scientific Manuals, some of which have already been submitted to the Public.

E. F. B.

GLASGOW, *October, 1873.*

CONTENTS.

MATHEMATICAL INTRODUCTION.

PART I.—PRINCIPLES OF CINEMATICS, OR THE COMPARISON OF MOTIONS.

PART II.—THEORY OF MECHANISM.

CHAPTER I.—DEFINITIONS AND GENERAL PRINCIPLES.

CHAPTER II.—ON ELEMENTARY COMBINATIONS AND TRAINS OF MECHANISM.

CHAPTER III.—DISTRIBUTED FORCES.

SECTION 1.—*Centres of Gravity.*

SECTION 2.—*Of Stress.*

SECTION 3.—*Principles of Hydrostatics and Internal Stress of Solids.*

PART IV.—THEORY OF STRUCTURES.

CHAPTER I.—SUMMARY OF PRINCIPLES OF STABILITY AND STRENGTH.

SECTION 1.—*Of Structures in General.*

SECTION 2.—*Balance and Stability of Frames, Chains, and Blocks.*

CHAPTER II.—PRINCIPLES AND RULES RELATING TO STRENGTH AND STIFFNESS.

SECTION 1.—*Of Strength and Stiffness in General.*

PART V.—PRINCIPLES OF KINETICS.

CHAPTER I.—SUMMARY OF GENERAL PRINCIPLES.—NATURE AND DIVISION OF THE SUBJECT.

CHAPTER 2.—ON UNIFORM MOTION UNDER BALANCED FORCES.

CHAPTER III.—ON THE VARIED TRANSLATION OF POINTS AND RIGID BODIES.

SECTION 1.—*Law of Varied Translation.*

SECTION 2.—*Varied Translation of a System of Bodies.*

Chapter IV.—Rotations of Rigid Bodies.

Chapter V.—Motions of Fluids.

PART VI.—THEORY OF MACHINES.

Chapter I.—Definitions and General Principles.

Chapter II.—Of the Performance of Work by Machines.

MATHEMATICAL INTRODUCTION.

ARITHMETICAL RULES.

For convenience sake the following Arithmetical Rules are here given: they will be referred to hereafter in the designing of toothed gearing, under The Theory of Pure Mechanism, Part II.

Definition.—A *prime number* is one which is only divisible by the number 1.

1. **To find the Prime Factors of a Given Number.**—Try the prime numbers, 2, 3, 5, 7, 11, &c., as divisors in succession, until a prime number has been found to divide the given number without a remainder; then try whether and how many times over the quotient is again divisible by the same prime number, so as to obtain a quotient not divisible again by the same prime number; then try the division of that quotient by the next greater prime number; and so on until a quotient is obtained which is itself a prime number; that is, a number not divisible by any other number except 1. This final quotient and the series of divisors will be the prime factors of the given number. To test the accuracy of the process, multiply all the prime factors together; the product should be the given number.

2. **To find the Greatest Common Measure (otherwise called the greatest common divisor) of Two Numbers.**—Divide the greater number by the less, so as to obtain a quotient, and a remainder less than the divisor; divide the divisor by the remainder as a new divisor; that new divisor by the new remainder; and so on, until a remainder is obtained which divides the previous divisor without a remainder. That last remainder will be the required greatest common measure.

If the last remainder is 1, the two numbers are said to be "prime to each other."

Example.—Required, the greatest common measure of 1420 and 1808.

```
Divisor, 1420) 1808 (1, Quotient.
               1420
Remainder,      388) 1420 (3, Quotient.
                     1164
      Remainder,      256) 388 (1, Quotient.
                           256
           Remainder,      132) 256 (1, Quotient.
                                132
                Remainder,      124) 132 (1, Quotient.
                                     124
                       Remainder,      8) 124 (15, Quotient.
                                          120
                            Remainder,      4) 8 (2, Quotient.
```

The last remainder, 4, is the required greatest common measure.

Definition.—Ratio is the mutual relation of two quantities in respect of magnitude.

3. To reduce the Ratio of Two Numbers to its Least Terms, divide both numbers by their greatest common measure.

For example, $\frac{1808 \div 4}{1420 \div 4} = \frac{452}{355}$.

4. To express the Ratio of Two Numbers in the form of a Continued Fraction.—Let A be the lesser of the two numbers, and B the greater; and let a, b, c, d, &c., be the quotients obtained during the process of finding the greatest common measure of A and B. Then, in the equation

$$\frac{\mathrm{B}}{\mathrm{A}} = a + \cfrac{1}{b + \cfrac{1}{c + \cfrac{1}{d + \&c.,}}}$$

the right-hand side is the continued fraction required.

To save space in printing, a continued fraction is often arranged as follows:—

$$a + \frac{1}{b+}\,\frac{1}{c+}\,\frac{1}{d+}\ \&c.$$

The ratio of two incommensurable quantities is expressed by an endless continued fraction. For example, the ratio of the diagonal to the side of a square is expressed by $1 + \frac{1}{2+}\,\frac{1}{2+}\,\frac{1}{2+}\,\frac{1}{2+}\ \&c.$, without end.

5. To form a series of Approximations to a Given Ratio.—Express the ratio in the form of a continued fraction. Then write the quotients in their order; and in a line below them write $\frac{0}{1}$ to the left of the first quotient, and $\frac{1}{0}$ directly under the first quotient. Then calculate a series of fractions by the following rule:—Multiply the first quotient by the numerator of the fraction that is below it, and add the numerator of the fraction next to the left; the sum will be the numerator of a new fraction: multiply the first quotient by the denominator of the fraction that is below it, and add the denominator of the fraction that is next to the left; the sum will be the denominator of the new fraction; then write that new fraction under the second quotient, and treat the second quotient, the fraction below it, and the fraction next to the left, as before, to find a fraction which is to be written under the third quotient, and so on. For example:

Quotients,....a, b, c, d, &c.

Fractions, $\frac{0}{1}$, $\frac{1}{0}$, $\frac{n}{m}$, $\frac{n'}{m'}$, $\frac{n''}{m''}$;

$$\frac{n}{m}=\frac{0+a}{1+0}=\frac{a}{1};\ \frac{n'}{m'}=\frac{1+bn}{0+bm};\ \frac{n''}{m'}=\frac{n+cn'}{m+cm'};\ \&c.$$

To take a particular case; let the given ratio be as before, $\frac{452}{355}$, then we have the following series:—

Quotients,..................		1	3	1	1	1	15	2	
Fractions,...............	$\frac{0}{1}$	$\frac{1}{0}$	$\frac{1}{1}$	$\frac{4}{3}$	$\frac{5}{4}$	$\frac{9}{7}$	$\frac{14}{11}$	$\frac{219}{172}$	$\frac{452}{355}$
Less or greater than given ratio,.......	L	G	L	G	L	G	L	G	

The fractions in a series formed in the manner just described are called *converging fractions*, and they have the following properties:— *First*, each of them is in its least terms; *secondly*, the difference between any pair of consecutive converging fractions is equal to unity divided by the product of their denominators; for example, $\frac{9}{7}-\frac{5}{4}=\frac{36-35}{7\times 4}=\frac{1}{28}$; $\frac{9}{7}-\frac{14}{11}=\frac{99-98}{7\times 11}=\frac{1}{77}$; *thirdly*, they are alternately less and greater than the given ratio towards which they approximate, as indicated by the letters L and G in the example; and, *fourthly*, the difference between any one of them and the given ratio is less than the difference between that one and the next fraction of the series.

Fractions intermediate between the converging fractions may be found by means of the formula $\frac{hn+kn'}{hm+km'}$; where $\frac{n}{m}$ and $\frac{n'}{m'}$, are any two of the converging fractions, and h and k are any two whole numbers, positive or negative, that are prime to each other.

6. **Logarithms. Definitions.**—The *power* of a number is the product of itself multiplied a certain number of times. The *index* or *exponent* of the power is the small figure placed above the right-hand corner, which denotes the number of times the multiplication takes place. The *Logarithm* of a number to a given base is the index of the power to which the base must be raised to be equal to the given number. That number of which the indices of the powers are the logarithms, is called the *base* of the system. A suffix denotes the base of the logarithm; if $a^x = n$, x is the logarithm of the number n to the base a, or $\log_a n = x$.

Logarithms to the base 10 are called common logarithms.

7. The logarithm of 1 is 0.

8. The common logarithm of 10 is 1, and that of any power of 10 is the index of that power; in other words, it is equal to the number of noughts in the power; thus the common logarithm of 100 is 2; that of 1000, 3; and so on.

9. The common logarithm of ·1 is – 1, and that of any power of ·1 is the index of that power with the negative sign; that is, it is equal to one more than the number of noughts between the decimal point and the figure 1, with the negative sign; for example, the common logarithm of ·01 is – 2; that of ·001, – 3; and so on.

10. The logarithms given in tables, are merely the fractional parts of the logarithms, correct to a certain number of places of decimals, without the integral parts or *indices;* which are supplied in each case according to the following rules:—

The index of the common logarithm of a number not less than 1 is one less than the number of integer places of figures in that number; that is to say, for numbers less than 10 and not less than 1, the index is 0; for numbers less than 100 and not less than 10, the index is 1; for numbers less than 1000, and not less than 100, the index is 2; and so on.

The index of the common logarithm of a decimal fraction less than 1 is *negative,* and is one more than the number of noughts between the decimal point and the significant figures; and the negative sign is usually written above instead of before the index; that is to say, for numbers less than 1 and not less than ·1, the index is $\bar{1}$; for numbers less than ·1 and not less than ·01, the index is $\bar{2}$; and so on.

The fractional part of a common logarithm is always positive, and depends solely upon the series of figures of which the number consists, and not upon the place of the decimal point amongst them.

EXAMPLES.

Number.	Logarithms.
377000	5·57634
37700	4·57634
3770	3·57634
377	2·57634
37·7	1·57634
3·77	0·57634
·377	$\bar{1}$·57634
·0377	$\bar{2}$·57634
·00377	$\bar{3}$·57634

and so on.

11. The logarithm of a product is the sum of the logarithms of its factors.

12. The logarithm of a power is equal to the logarithm of the root multiplied by the index of the power.

13. The logarithm of a quotient is found by subtracting the logarithm of the divisor from the logarithm of the dividend.

14. The logarithm of a root is found by dividing the logarithm of one of its powers by the index of that power.

Note.—In applying these principles to logarithms of numbers less than 1, it is to be observed that negative indices are to be subtracted instead of being added, and added instead of being subtracted.

15. To avoid the inconvenience which attends the use of negative indices to logarithms, it is a very common practice to put, instead of a negative index to the logarithm of a fraction, the *complement* (as it is called) of that index to 10; that is to say, 9 instead of $\bar{1}$, 8 instead of $\bar{2}$, 7 instead of $\bar{3}$, and so on. In such cases, it is always to be understood that each such complementary index has -10 combined with it; and to prevent mistakes, it is useful to prefix $-10+$ to it; for example,

Number.	Logarithm with Negative Index.	Logarithm with Complementary Index.
·377	$\bar{1}$·57634	$-10+9$·57634
·0377	$\bar{2}$·57634	$-10+8$·57634
·00377	$\bar{3}$·57634	$-10+7$·57634

16. To find the fractional part of the common logarithm of a number of five places of figures; take from the table the logarithm corresponding to the first three figures, and the difference between that logarithm and the next greater logarithm in the table; multiply that difference by the two remaining figures of the given number, and divide by 100; the quotient will be a correction, to be added to the logarithm already found.

Example.—Find the common logarithm of 37725.

Log. 377,...........................	57634
Log. 378,...........................	57749
Difference,...........	115
	× 25 ÷ 100
Correction,...........	29
Add log. 377,......................	57634
Log. 37725,.......................	57663 *Answer.*

17. To find the natural number, or *antilogarithm*, corresponding to a common logarithm of five places of decimals, which is not in the table; find the next less, and the next greater logarithm in the table, and take their difference. Opposite the next less logarithm will be the first three figures of the antilogarithm. Subtract the next less logarithm from the given logarithm; annex two noughts to the remainder, and divide by the before-mentioned difference; the quotient will give two additional figures of the required antilogarithm. (The first of those figures may be a nought.)

Example.—Find the antilogarithm of the common logarithm ·57663.

Next less log. in table,...............	57634
Next greater,..........................	57749
Difference,.......................	115
Given logarithm, ...	57663
Subtract log. 377,.....................	57634
Divide by difference,...........	115)2900
Two additional figures,.........	25

so that the answer is 37725.

Note.—The last two rules refer particularly to the tables in Rankine's *Useful Rules and Tables*, but are equally applicable to other tables. For instance, where the logarithm of a number of 5 figures is given in the tables; in these last two rules, for 3 read 5, and for 5 read 7.

TRIGONOMETRICAL RULES.

The following is a summary of the Principles and Chief Rules of Trigonometry:—

Definition.—Every expression which in any way contains a number, or depends for its value upon the value of the number, is said to be a *function* of that number, as $2x$, x^2, log. x, tan x are all functions of x.

18. **Trigonometrical Functions Defined.**—Suppose that A, B, C

stand for the three angles of a right-angled triangle, C being the right angle, and that a, b, c stand for the sides respectively opposite to those angles, c being the hypothenuse; then the various names of trigonometrical functions of the angle A have the following meanings:—

$$\sin A = \frac{a}{c};\ \cos A = \frac{b}{c};$$

$$\text{versin } A = \frac{c-b}{c};\ \text{coversin } A = \frac{c-a}{c};$$

$$\tan A = \frac{a}{b};\ \text{cotan } A = \frac{b}{a};$$

$$\sec A = \frac{c}{b};\ \text{cosec } A = \frac{c}{a}.$$

The *complement* of A means the angle B, such that A + B = a right angle; and the sine of each of those angles is the cosine of the other, and so of the other functions by pairs.

19. **Relations amongst the Trigonometrical Functions of One Angle, A, and of its Supplement, 180° − A:—**

$$\sin A = \sqrt{1-\cos^2 A} = \frac{\tan A}{\sec A} = \frac{1}{\text{cosec } A};$$

$$\cos A = \sqrt{1-\sin^2 A} = \frac{\text{cotan } A}{\text{cosec } A} = \frac{1}{\sec A};$$

$$\text{versin } A = 1 - \cos A;$$

$$\text{coversin } A = 1 - \sin A;$$

$$\tan A = \frac{\sin A}{\cos A} = \frac{1}{\text{cotan } A} = \sin A \cdot \sec A = \sqrt{\sec^2 A - 1};$$

$$\text{cotan } A = \frac{\cos A}{\sin A} = \frac{1}{\tan A} = \cos A \cdot \text{cosec } A = \sqrt{\text{cosec}^2 A - 1};$$

$$\sec A = \frac{1}{\cos A} = \sqrt{1+\tan^2 A};$$

$$\text{cosec } A = \frac{1}{\sin A} = \sqrt{1+\text{cotan}^2 A}.$$

$$\sin(180° - A) = \sin A;$$
$$\cos(180° - A) = -\cos A;$$
$$\text{versin}(180° - A) = 1 + \cos A = 2 - \text{versin } A;$$
$$\text{coversin}(180° - A) = \text{coversin } A;$$
$$\tan(180° - A) = -\tan A;$$
$$\text{cotan}(180° - A) = -\text{cotan } A;$$
$$\sec(180° - A) = -\sec A;$$
$$\text{cosec}(180° - A) = \text{cosec } A.$$

20. **The Circular Measure of an Angle.**—If a right line as radius by revolution about a fixed point at its extremity as centre, traces out an angle from a fixed position, the angle may be measured by the ratio of the arc to the radius; this mode of measurement is called circular measure. The *unit* of circular measure is the angle whose arc is equal to the radius, that is, $360° \div 2\pi = (57° \, 17' \, 45'' = 206265'')$.

To compute sines, &c., approximately by series; reduce the angle to circular measure—that is, to radius-lengths and fractions of a radius-length let it be denoted by A. Then

$$\sin A = A - \frac{A^3}{2.3} + \frac{A^5}{2.3.4.5} - \frac{A^7}{2.3.4.5.6.7} + \&c.$$

$$\cos A = 1 - \frac{A^2}{2} + \frac{A^4}{2.3.4} - \frac{A^6}{2.3.4.5.6} + \&c.$$

21. **Trigonometrical Functions of Two Angles**:—

$$\sin (A \pm B) = \sin A \cos B \pm \cos A \sin B;$$

$$\cos (A \pm B) = \cos A \cos B \mp \sin A \sin B;$$

$$\tan (A \pm B) = \frac{\tan A \pm \tan B}{1 \mp \tan A \tan B}.$$

22. **Formulæ for the Solution of Plane Triangles.**—Let A, B, C be the angles, and a, b, c the sides respectively opposite them.

I. *Relations amongst the Angles*—

$$A + B + C = 180°;$$

or if A and B are given,

$$C = 180° - A - B.$$

II. *When the Angles and One Side are given*, let a be the given side; then the other two sides are

$$b = a \cdot \frac{\sin B}{\sin A}; \quad c = a \cdot \frac{\sin C}{\sin A}.$$

III. *When Two Sides and the Included Angle are given*, let a, b be the given sides, C the given included angle; then

To find the third side. First Method:

$$c = \sqrt{(a^2 + b^2 - 2\,a\,b \cos C)};$$

Second Method: Make $\sin D = \frac{2\sqrt{a\,b}}{a+b} \cdot \cos \frac{C}{2}$; then

$$c = (a+b) \cos D.$$

Third Method: Make $\tan E = \frac{2\sqrt{a\,b}}{a-b} \cdot \sin \frac{C}{2}$; then

$$c = (a-b) \sec E.$$

To find the remaining Angles, A and B.

If the third side has been computed,

$$\sin A = \frac{a}{c} \cdot \sin C; \ \sin B = \frac{b}{c} \cdot \sin C.$$

If the third side has not been computed,

$$\tan \cdot \frac{A + B}{2} = \operatorname{cotan} \frac{C}{2}; \ \tan \frac{A - B}{2} = \frac{a - b}{a + b} \operatorname{cotan} \frac{C}{2};$$

$$A = \frac{A + B}{2} + \frac{A - B}{2}; \ B = \frac{A + B}{2} - \frac{A - B}{2}.$$

IV. *When the Three Sides are given, to find any one of the Angles*, such as C—

$$\cos C = \frac{a^2 + b^2 - c^2}{2ab};$$

or otherwise, let

$$s = \frac{a + b + c}{2}; \text{ then}$$

$$\cos \frac{C}{2} = \sqrt{\frac{s(s - c)}{ab}}; \ \sin \frac{C}{2} = \sqrt{\frac{(s - a)(s - b)}{ab}};$$

$$\operatorname{cotan} \frac{C}{2} = \sqrt{\frac{s(s - c)}{(s - a)(s - b)}}; \ \tan \frac{C}{2} = \sqrt{\frac{(s - a)(s - b)}{s(s - c)}};$$

$$\sin C = \frac{2\sqrt{s(s - a)(s - b)(s - c)}}{ab}.$$

Note.—In all trigonometrical problems, it is to be borne in mind, that small acute angles, and large obtuse angles, are most accurately determined by means of their *sines, tangents*, and *cosecants;* and angles approaching a right angle by their *cosines, cotangents*, and *secants*.

23. **To Solve a Right-angled Triangle.**—Let C denote the right angle; c the hypothenuse; A and B the two oblique angles; a and b the sides respectively opposite them.

Given, the right angle, another angle B, the hypothenuse c. Then

$$A = 90° - B; \ a = c \cdot \cos B; \ b = c \cdot \sin B.$$

Given, the right angle, another angle B, a side a,

$$A = 90° - B; \ b = a \cdot \tan B; \ c = a \cdot \sec B.$$

Given, the right angle, and the sides a, b,

$$\tan A = \frac{a}{b}; \ \tan B = \frac{b}{a}; \ c = \sqrt{a^2 + b^2}.$$

Given, the right angle, the hypothenuse c; a side a,

$$\sin A = \cos B = \frac{a}{c};\ b = \sqrt{c^2 - a^2}.$$

Given the three sides, a, b, c, which fulfilling the equation $c^2 = a^2 + b^2$, the triangle is known to be right-angled at C,

$$\sin A - \frac{a}{c};\ \sin B = \frac{b}{c}.$$

24. To Express the Area of a Plane Triangle in terms of its Sides and Angles.

Given, one side, c, and the angles.

$$\text{Area} - \frac{c^2}{2} \cdot \frac{\sin A \sin B}{\sin C}.$$

Given, two sides, b, c, and the included angle A.

$$\text{Area} = \frac{b\,c \cdot \sin A}{2}.$$

Given, the three sides a, b, c. Let $\frac{a+b+c}{2} = s$; then

$$\text{Area} = \sqrt{\left\{s\,(s-a)\,(s-b)\,s - c)\right\}}.$$

RULES OF THE DIFFERENTIAL AND INTEGRAL CALCULUS.

25. *Definitions.*—A *function* has already been defined. When a function of one quantity is assumed equal to another quantity, both quantities are called *variables*, the one upon whose assumed value the other depends being called the *independent* variable, while the other, whose value depends upon it, is called the *dependent* variable. The expression $y = \phi\, x$ for instance denotes that the *dependent variable* y, depends for its value upon the *independent variable* x, the expression denoting that y is a *function* of x.

A quantity, x, may be assumed to be made up of an infinite number of infinitesimal parts, dx, this expression meaning simply one of the small infinitesimal differences of which x is made up, *i.e.*, $x - n \cdot dx$, where n is assumed to increase without limit, and dx to diminish without limit: this process of considering a quantity to be diminished without limit is called *differentiation.*

The quotient, if it has a limit formed by taking the difference of the function of a quantity, and the function of that quantity with a small increment, and dividing by the increment, is termed the *differential coefficient* of the function, with regard to the quantity $\frac{\phi\,(x+dx) - \phi\, x}{dx}$ is the differential coefficient of $\phi\, x$ with respect to x, this is generally written $\phi'\, x$; or otherwise the small increment or decrement of the dependent variable divided by that of the independent variable, the former being a function of the latter, is called the differential coefficient, thus $\frac{dy}{dx}$ is the differential coefficient of y with respect to x, it being always borne in mind that $\frac{dy}{dx}$ is *one* quantity, which cannot be divided into a numerator dy, and a denominator dx.

26. **Rules for finding differential coefficients,—**

If $y = C$ (a constant); $\frac{dy}{dx} = 0$.

The Differential Coefficient of the sum of functions is equal to the sum of the differential coefficients of the functions, or if $v = w + y + z$ where all of these quantities are functions of x, then $\frac{dv}{dx} = \frac{dw}{dx} + \frac{dy}{dx} + \frac{dz}{dx}$.

In the same way to find the differential coefficient of the difference, product, and quotient of functions of quantities.

If $v = y - z$, then $\frac{dv}{dx} = \frac{dy}{dx} - \frac{dz}{dx}$, where v, y, z, are functions of x.

If $v = w\,y\,z$, then $\frac{dv}{dx} = \frac{dw}{dx} \cdot y \cdot z + \frac{dy}{dx} \cdot w \cdot z + \frac{dz}{dx} \cdot w \cdot y$, where v, w, y, z, are functions of x.

If $v = \frac{y}{z}$, then $\frac{dv}{dx} = \frac{z \cdot \frac{dy}{dx} - y \frac{dz}{dx}}{z^2}$, where v, y, z, are functions of x.

If $\phi x = nx$, $\phi' x = n$; or otherwise let $\phi x = y = nx$, then $\frac{dy}{dx} = n$, thus if $\phi x = 7x$, $\phi' x = 7$.

If $\phi x = x^n$, $\phi' x = nx^{n-1}$, thus if $\phi x = x^7$, $\phi' x = 7x^6$.

If $\phi x = \log_a x$, $\phi' x = \frac{1}{x \log_\epsilon a}$, thus if $\phi x = \log_{10}\, x$, $\phi' x = \frac{1}{x \log_\epsilon 10} = \frac{1}{2{\cdot}30258x} = \frac{{\cdot}43429}{x}$

If $\phi x = \log_\epsilon x$, $\phi' x = \frac{1}{x}$.

If $\phi x = a^x$, $\phi' x = a^x \log_\epsilon a$.

If $\phi x = \epsilon^x$, $\phi' x = \epsilon^x$.

If $\phi x = \sin x$, $\phi' x = \cos x$.

If $\phi x = \cos x$, $\phi' x = -\sin x$.

If $\phi x = \tan x$, $\phi' x = \frac{1}{\cos^2 x}$.

Definition.—By sine $^{-1} x$ is meant the angle whose sine is x, thus if $x = \text{sine}\ y$, $y = \text{sine}^{-1} x$.

$$\text{If } \phi x = \sin^{-1} x,\ \phi' x = \frac{1}{\sqrt{(1-x^2)}}$$

$$\phi x = \cos^{-1} x,\ \phi' x = -\frac{1}{\sqrt{(1-x^2)}}$$

$$\phi x = \tan^{-1} x,\ \phi' x = \frac{1}{1+x^2}.$$

If $v = \phi x$, and $\psi v = \chi x$, then $\chi' x = \psi' v \cdot \phi' x$, or otherwise;

$$\text{If } v = \phi x,\ y = \chi x,\ y = \psi v \text{ then } \frac{dy}{dx} = \frac{dy}{dv} \cdot \frac{dv}{dx}.$$

If $\phi x = \log. \sin x$, then $\phi' x = \frac{1}{\sin x} \cdot \cos x$; here *log. sin x* is first differentiated with respect to *sin x*, and then *sin x* with respect to x.

Definition.—The differential coefficient of the differential coefficient is called the *second* differential coefficient; the differential coefficient of the second is called the *third;* and so on :—

Thus let $\phi x = x^n$, then $\phi' x = n x^{n-1}$, $\phi'' x = n(n-1)x^{n-2}$, $\phi''' x = n(n-1)(n-2)x^{n-3}$; where $\phi'' x$ stands for the second differential coefficient, $\phi''' x$ for the third differential coefficient. So let $\phi x = \sin x$, then $\phi' x = \cos x$, $\phi'' x = -\sin x$, $\phi''' x = -\cos x$. This process is called successive differentiation.

Another mode of representing this is the following :— $\frac{dy}{dx}$ is the differential coefficient of y with respect to x; the second differential coefficient $\frac{d\frac{dy}{dx}}{dx}$ is represented by $\frac{d^2 y}{dx^2}$, where $\frac{d^2 y}{dx^2}$ is a quantity which cannot be divided into a numerator $d^2 y$ and denominator dx^2, as already stated of the quantity $\frac{dy}{dx}$.

27. The following is an illustration of the application of the differential calculus to Geometry. In fig. 1, let O X, O Y be two axes of co-ordinates at right angles to each other. Let A B be a curve, whose equation is represented by $y = \phi x$, *i.e.*, each point of the curve has ordinates and abscissæ bearing to each other the ratio represented by the equation, and all the points in the curve are known when their ordinates and abscissæ (co-ordinates) are known. Let O C $= a$ be a particular value of x, and A C $= b = \phi a$ be a particular value of y;

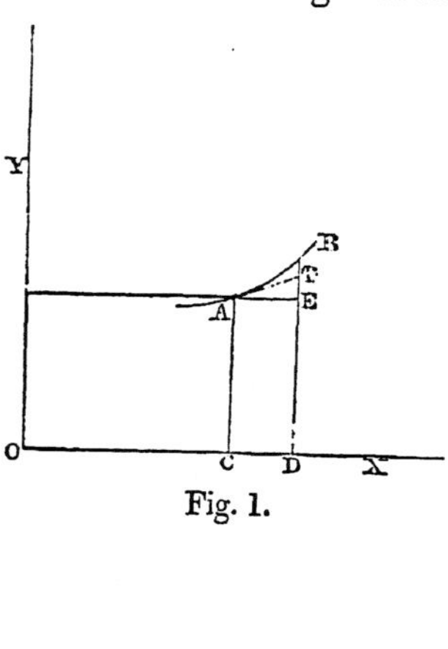

Fig. 1.

also let A E be a small increment of $x = \Delta x$, and E B a corresponding increment of $y = \Delta y$. The trigonometrical tangent of the angle E A B $= \frac{\Delta y}{\Delta x}$. Let there be a line, A T, lying between A B and A E, whose tangent is $\frac{d\,y}{d\,x} = \varphi' x$. Then as B approaches towards A, the line A B will ultimately be the line A T; that is, its limiting value as soon as A and B coincide, will be A T. Hence if x be an abscissa of a curve, and $\varphi x = y$ be an ordinate, the differential coefficient $\frac{d\,y}{d\,x} = \varphi' x$ is the trigonometrical tangent of the angle which the geometrical tangent of the curve makes with the axis of x, at the point where the abscissa is x.

The differential calculus will occasionally be applied in an elementary manner to portions of the following work, as, for example, in treating of the Varied Motion of Points (Part I., Section 3).

28. The integral calculus is the inverse of the differential; it determines the whole magnitude of a quantity of which the differentials are given.

If a number of points be taken in a curve, and chords drawn joining the points, and also tangents drawn through the points intersecting each other, the sum of the one will be less than, and that of the other (intersected portions) will be greater than the length of the curve; if the chords, and tangents are increased in number, they will approximate to the length of the curve. The integral calculus is used for finding the exact length of the curve. A mechanical illustration is the computation of the space passed over by a point having varied motion.

29. **Approximate Computation of Integrals.**—The present article is intended to afford to those who have not made that branch of mathematics which treats of the process of *integration* a special study, some elementary information respecting it.

The meaning of the symbol of an integral, viz.:—

$$\int u\,d\,x,$$

is of the following kind:—

In fig. 2, let A C B D be a plane area, of which one boundary, A B, is a portion of an axis of abscissæ O X,—the opposite boundary, C D, a curve of any figure,—and the remaining boundaries A C, B D, ordinates perpendicular to O X, whose respective abscissæ, or distances from the origin O, are

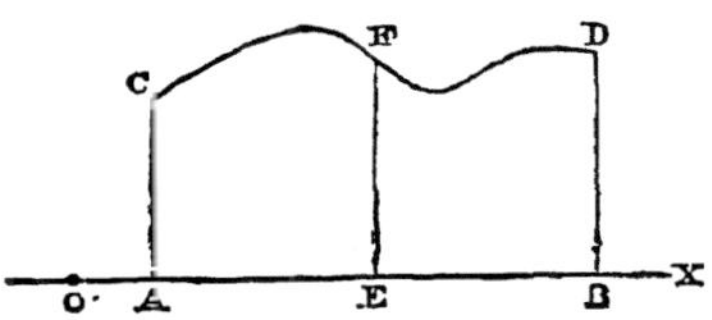

Fig. 2.

$$\overline{OA} = a; \ \overline{OB} = b.$$

Let $\overline{EF} = u$ be any ordinate whatsoever of the curve C D, and $\overline{OE} = x$ the corresponding abscissa. Then the integral denoted by the symbol,

$$\int_a^b u \, d \, x,$$

means, *the area of the figure* A C B D. The abscissæ a and b which are the least and greatest values of x, and which indicate the longitudinal extent of the area, are called the *limits of integration;* but when the extent of the area is otherwise indicated, the symbols of those limits are sometimes omitted.

When the relation between u and x is expressed by any ordinary algebraical equation, the value of the integral for a given pair of values of its limits can generally be found by means of formulæ which are contained in works on the Integral Calculus, or by means of mathematical tables.

Cases may arise, however, in which u cannot be so expressed in terms of x; and then approximate methods must be employed. Those approximate methods, of which two are here described, are founded upon the division of the area to be measured into bands by parallel and equi-distant ordinates, the approximate computation of the areas of those bands, and the adding of them together; and the more minute that division is, the more near is the result to the truth.

First Approximation.

Divide the area A C D B, as in fig. 3, into any convenient number of bands by parallel ordinates, whose uniform distance apart is Δx; so that if n be the number of bands, $n + 1$ will be the number of ordinates, and

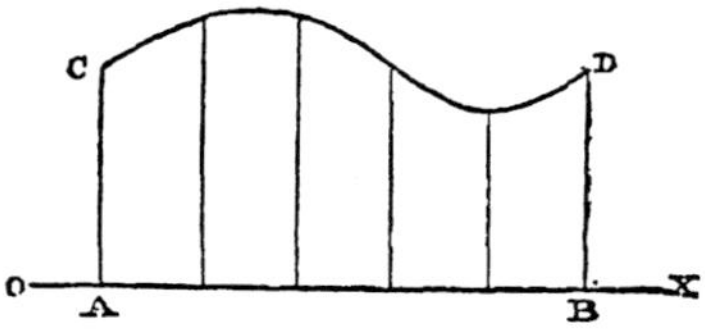

Fig. 3.

$$b - a = n \Delta x,$$

the length of the figure.

Let u', u'', denote the two ordinates which bound one of the bands; then the area of that band is

$$\frac{u' + u''}{2} \cdot \Delta x, \textit{ nearly;}$$

and consequently, adding together the approximate areas of all the bands,—denoting the extreme ordinates as follows,—

$$\overline{AC} = u_a; \ \overline{BD} = u_b;$$

and the intermediate ordinates by u_i, we find for the approximate value of the integral—

$$\int_a^b u\,dx = \left(\frac{u_a}{2} + \frac{u_b}{2} + \Sigma \cdot u_i\right) \Delta x, \dots\dots\dots\dots\dots\dots (1.)$$

Second Approximation.

Divide the area A C D B, as in fig. 4, into an *even* number of bands, by parallel ordinates, whose uniform distance apart is Δx. The ordinates are marked alternately by plain lines and by dotted lines, so as to arrange the bands in pairs. Considering any one pair of bands, such as E F H G, and assuming that the curve F H is nearly a parabola, it appears from the properties of that curve, that the area of that pair of bands is

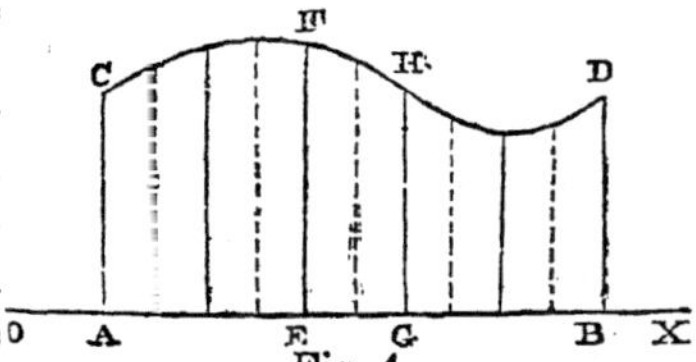

Fig. 4.

$$\frac{(u' + 4\,u'' + u''')\,\Delta x}{3}, \textit{ nearly};$$

in which u' and u''' denote the plain ordinates $\overline{EF}$ and $\overline{GH}$, and u'' the intermediate dotted ordinate; and consequently, adding together the approximate areas of all the pairs of bands, we find, for the approximate value of the integral—

$$\int_a^b u\,dx = \Big(u_a + u_b + 2\,\Sigma \cdot u_i \text{ (plain)}$$

$$+ 4\,\Sigma \cdot u_i \text{ (dotted)}\Big)\frac{\Delta x}{3}, \dots\dots\dots\dots\dots\dots (2.)$$

It is obvious, that if the values of the ordinates u required in these computations can be calculated, it is unnecessary to draw the figure to a scale, although a sketch of it may be useful to assist the memory.

When the symbol of integration is repeated, so as to make a *double integral*, such as

$$\int\int u \cdot dx\,dy,$$

or a *triple integral*, such as

$$\int\int\int u \cdot dx\,dy\,dz,$$

it is to be understood as follows:—

Let

$$v = \int u \cdot dx$$

be the value of this single integral for a given value of y. Construct a curve whose abscissæ are the various values of y within the prescribed limits, and its ordinates the corresponding values of v. Then the area of that curve is denoted by

$$\int v \cdot dy = \int \int u \cdot dx\, dy.$$

Next, let

$$t = \int v \cdot dy$$

be the value of this double integral for a given value of z. Construct a curve whose abscissæ are the various values of z within the prescribed limits, and its ordinates the corresponding values of t. Then the area of that curve is denoted by

$$\int t \cdot dz = \int \int v \cdot dy\, dz = \int \int \int u \cdot dx\, dy\, dz;$$

and so on for any number of successive integrations.

RULES FOR THE MENSURATION OF FIGURES AND FINDING OF CENTRES OF MAGNITUDE.

SECTION 1.—AREAS OF PLANE SURFACES.

30. **Parallelogram.** *Rule A.*—Multiply the length of one of the sides by the perpendicular distance between that side and the opposite side.

Rule B.—Multiply together the lengths of two adjacent sides and the sine of the angle which they make with each other. (When the parallelogram is right-angled, that sine is = 1.)

31. **Trapezoid** (or four-sided figure bounded by a pair of parallel straight lines, and a pair of straight lines not parallel). Multiply the half sum of the two parallel sides by the perpendicular distance between them.

32. **Triangle.** *Rule A.*—Multiply the length of any one of the sides by one-half of its perpendicular distance from the opposite angle.

Rule B.—Multiply one-half of the product of any two of the sides by the sine of the angle between them.

Rule C.—Multiply together the following four quantities: the half sum of the three sides, and the three remainders left after subtracting each of the three sides from that half sum; extract the square root of the quotient; that root will be the area required.

Note.—*Any Polygon* may be measured by dividing it into triangles, measuring those triangles, and adding their areas together.

33. **Parabolic Figures of the Third Degree.**—The parabolic

figures to which the following rules apply are of the following kind (see figs. 5 and 6.) One boundary is a straight line, A X, called the *base* or *axis;* two other boundaries are either points in that line, or straight lines at right angles to it, such as A B and X C, called *ordinates;* and the fourth boundary is a curve, B C, of the *parabolic class*, and of the *third degree;* that is, a curve whose *ordinate* (or perpendicular distance from the base A X) at any point is expressed by what is called an *algebraical function of the third degree* of the *abscissa* (or distance of that ordinate from a fixed point in the base). An algebraical function of the third degree of a quantity consists of terms not exceeding four in number, of which one may be constant, and the rest must be proportional to powers of that quantity not higher than the cube.

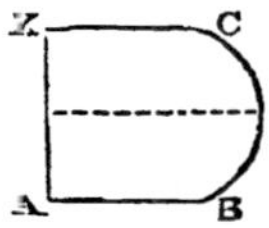

Fig. 5.

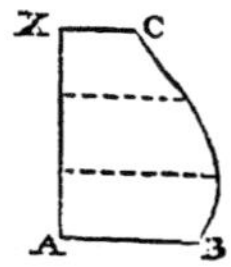

Fig. 6.

Rule A.—Divide the base, as in fig. 5, into two equal parts or intervals; measure the endmost ordinates, A B and X C, and the middle ordinate (which is dotted in the figure) at the point of divison; add together the endmost ordinates and *four times* the middle ordinate, and divide the sum by *six;* the quotient will be the *mean breadth* of the figure, which, being multiplied by the length of the base, A X, will give the area.

Rule B.—Divide the base, as in fig. 6, into three equal intervals; measure the endmost ordinates, A B and X C, and the two intermediate ordinates (which are dotted) at the points of division; add together the endmost ordinates and three times each of the intermediate ordinates; divide the sum by *eight;* the quotient will be the *mean breadth* of the figure, which, being multiplied by the length of the base, A X, will give the area.

In applying either of those rules to figures whose curved boundaries meet the base at one or both ends, the ordinate at each such point of meeting is to be made = 0.

34. Any Plane Area.—Draw an axis or base-line, A X, in a convenient position. The most convenient position is usually parallel to the greatest length of the area to be measured. Divide the length of the figure into a convenient number of equal intervals, and measure breadths in a direction perpendicular to the axis at the two ends of that length, and at the points of division, which breadths will, of course, be one more in number than the intervals. (For example, in fig. 7, the length of the figure is divided into ten equal intervals, and eleven breadths are measured at b_0, b_1, &c.) Then the following rules are exact, if the sides of the figures are bounded by straight lines, and by

Fig. 7.

parabolic curves not exceeding the third degree, and are approximate for boundaries of any other figures.

Rule A.—("*Simpson's First Rule,*" to be used when the number of intervals is even.)—Add together the two endmost breadths, *twice* every second intermediate breadth, and *four times* each of the remaining intermediate breadths; multiply the sum by the common interval between the breadths, and divide by 3; the result will be the area required.

For two intervals the multipliers for the breadths are 1, 4, 1 (as in Rule A of the preceding Article; for four intervals, 1, 4, 2, 4, 1; for six intervals, 1, 4, 2, 4, 2, 4, 1; and so on. These are called "Simpson's Multipliers."

Example.—Length, 120 feet, divided into six intervals of 20 feet each.

Breadths in Feet and Decimals.	Simpson's Multipliers.	Products.
17·28	1	17·28
16·40	4	65·60
14·08	2	28·16
10·80	4	43·20
7·04	2	14·08
3·28	4	13·12
0	1	0·00
	Sum,	181·44
	× Common interval,	20 feet.
	÷ 3)	3628·8
	Area required,	1209·6 square feet.

Rule B.—("*Simpson's Second Rule*" to be used when the number of intervals is a multiple of 3.—Add together the two endmost breadths, *twice* every third intermediate breadth, and *thrice* each of the remaining intermediate breadths; multiply the sum by the common interval between the breadths, and by 3; divide the product by 8; the result will be the area required.

"Simpson's Multipliers" in this case are, for three intervals, 1, 3, 3, 1; for six intervals, 1, 3, 3, 2, 3, 3, 1; for nine intervals, 1, 3, 3, 2, 3, 3, 2, 3, 3, 1; and so on.

Example.—Length, 120 feet, divided into six intervals of 20 feet each.

Breadths in Feet and Decimals.	Simpson's Multipliers.	Products.
17·28	1	17·28
16·40	3	49·20
14·08	3	42·24
10.80	2	21·60
7·04	3	21·12
3·28	3	9·84
0	1	0·00
	Sum,	161·28
	× Common interval,	20 feet.
		3225·6
		× 3
	÷ 8)	9676·8
	Area required,	1209·6 square feet.

Remarks.—The preceding examples are taken from a parabolic figure of the third degree, for which both Simpson's Rules are exact; and the results of using them agree together precisely. For other figures, for which the rules are approximate only, the first rule is in general somewhat more accurate than the second, and is therefore to be used unless there is some special reason for preferring the second.

The probable extent of error in applying Simpson's First Rule to a given figure is, in most cases, nearly proportional to the fourth power of the length of an interval.

The errors are greatest where the boundaries of the figure are most curved, and where they are nearly perpendicular to the axis. In such positions of a figure the errors may be diminished by subdividing the axis into smaller intervals.

Rule C.—("*Merrifield's Trapezoidal Rule,*" for calculating separately the areas of the parts into which a figure is subdivided by its equidistant ordinates or breadths.)—Write down the breadths in their order. Then take the *differences* of the successive breadths, distinguishing them into positive and negative, according as the breadths are increasing or diminishing, and write them opposite the intervals between the breadths. Then take the differences of those differences, or *second differences,* and write them opposite the intervals between the first differences, distinguishing them into positive and negative, according to the following principles:—

First Differences.		Second Difference.
Positive increasing, or Negative diminishing,	}	Positive.
Negative increasing, or Positive diminishing,	}	Negative.

In the column of second differences there will now be two blanks opposite the two endmost breadths; those blanks are to be filled up with numbers each forming an arithmetical progression with the two adjoining second differences, if these are unequal, or equal to them, if they are equal.

Divide each second difference by 12; this gives a *correction*, which is to be *subtracted* from the breadth opposite it if the second difference is *positive*, and *added* to that breadth if the second difference is *negative*.

Then to find the area of the division of the figure contained between a given pair of ordinates or breadths; *multiply the half sum of the corrected breadths by the interval between them.*

The area of the whole figure may be formed either by adding together the areas of all its divisions, or by adding together the halves of the endmost corrected breadths, and the whole of the intermediate breadths, and multiplying the sum by the common interval.

Example.—Length, 120 feet, divided into six intervals of 20 feet each.

Breadths in Feet and Decimals.	First Differences.	Second Differences.	Corrections	Corrected Breadths. Feet	Areas of Divisions. Sq. Feet.
17·28		(−1·92)	+0·16	17·44	
	−0·88				339·6
16·40		−1·44	+0·12	16·52	
	−2·32				306·8
14·08		−0·96	+0·08	14·16	
	−3·28				250·0
10·80		−0·48	+0·04	10·84	
	−3·76				178·8
7·04		0	0	7·04	
	−3·76				102·8
3·28		+0·48	−0·04	3·24	
	−3.28				31·6
0		(+0·96)	−0·08	−0·08	
				Total area, square feet,	1209·6

The second differences enclosed in parentheses at the top and bottom of the column are those filled in by making them form an arithmetical progression with the second differences adjoining them.

The last corrected breadth in the present example is negative, and is therefore subtracted instead of added in the ensuing computation.

Rule D.—(" *Common Trapezoidal Rule,*" to be used when a rough approximation is sufficient.) Add together the halves of the endmost breadths, and the whole of the intermediate breadths, and multiply the sum by the common interval.

Example.—The same as before.

	Feet.	
Half breadth at one end, 17·28 ÷ 2 -	8·64	
Intermediate breadths,	16·40	
	14·08	
	10·80	
	7·04	
	3·28	
Half breadth at the other end, . .	0	
	60·24	
× Common interval, .	20	
Approximate area, ...	1204·8	square feet.
True area as before computed, . .	1209·6	
Error,	−4·8	square feet.

35. Circle.—The area of a circle is equal to its circumference multiplied by one-fourth of its diameter, and therefore to the square of the diameter multiplied by one-fourth of the ratio of the circumference to the diameter. The ratio of the area of a circle to the square of its diameter $\left(\text{which ratio is denoted by the symbol } \frac{\pi}{4}\right)$ is *incommensurable;* that is, not expressible exactly in figures; but it can be found approximately, to any required degree of precision. Its value has been computed to 250 places of decimals; but the following approximations are close enough for most purposes, scientific or practical:—

Approximate Values of $\frac{\pi}{4}$.	Errors in Fractions of the Circle, about
·7853981634 −	+ one-300,000,000,000th.
·785398 + ...	− one-5,000,000th.
·7854 −	+ one-400,000th.
$\frac{355}{4 \times 113}$ −	+ one-13,000,000th.
$\frac{11}{14}$ −	+ one-2,500th.

The *diameter of a circle equal in area to a given square* is very nearly 1·12838 × the side of the square. The following table gives examples of this:—

TABLE—MULTIPLIERS FOR CONVERTING

	Sides of Squares into Diameters of Equal Circles.	Diameters of Circles into sides of Equal Squares.	
1	1·12838	0·88623	1
2	2·25676	1·77245	2
3	3·38514	2·65868	3
4	4·51352	3·54491	4
5	5·64190	4·43113	5
6	6·77028	5·31736	6
7	7·89866	6·20359	7
8	9·02704	7·08981	8
9	10·15542	7·97604	9
10	11·28380	8·86227	10

36. The area of a **Circular Sector** (O A C B, fig. 8) is the same fraction of the whole circle that the angle A O B of the sector is of a whole revolution. In other words, multiply *half the square of the radius*, or *one-eighth of the square of the diameter*, by the circular measure (to radius unity) of the angle A O B; the product will be the area of the sector.

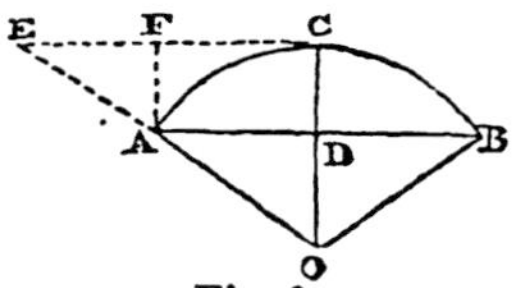

Fig. 8.

SECTION 2.—VOLUMES OF SOLID FIGURES.

37. **To Measure the Volume of any Solid.**—*Method I. By layers.*—Choose a straight axis in any convenient position. (The most convenient is usually parallel to the greatest length of the solid.) Divide the whole length of the solid, as marked on the axis, into a convenient number of equal intervals, and measure the sectional area of the solid upon a series of planes crossing the axis at right angles at the two ends and at the points of division. Then treat those areas as if they were the breadths of a plane figure, applying to them Rule A, B, or C of Article 34, page 17; and the result of the calculation will be the volume required. If Rule C is used, the volume will be obtained in separate layers.

Method II. By prisms or columns ("Wooley's Rule").—Assume a plane in a convenient position as a base, divide it into a network of equal rectangular divisions, and conceive the solid to be built of a set of rectangular prismatic columns, having those rectangular divisions for their sectional areas. Measure the thickness of the solid at the *centre* and at the *middle of each of the sides* of each of those rectangular columns; add together the doubles of all the thicknesses before-mentioned, which are in the interior of the solid, and the simple thicknesses which are at its boundaries; divide the sum by *six*, and multiply by the area of one rectangular division of the base.

SECTION 3.—LENGTHS OF CURVED LINES.

38. **To Calculate the Lengths of Circular Arcs.**—When the proportion of the arc to an entire circumference is given, the length of the arc, in terms of the radius, is to be calculated by multiplying that proportion by the well-known approximate value of the ratio of the circumference of a circle to its radius: viz., $\frac{\text{circumference}}{\text{radius}} = \frac{710}{113}$ nearly, $= 6{\cdot}283185$ nearly: the above ratio is commonly denoted by the symbol 2π; the reciprocal of the above ratio is very nearly $\frac{113}{710} = 0{\cdot}159155$ nearly; but it is often much more convenient in practice to proceed by drawing; and then the following rules are the most accurate yet known :—*

I. (Fig. 9). *To draw a straight line approximately equal to a given circular arc,* A B. Draw the straight chord B A; produce A to C, making $AC = \frac{1}{2} BA$; about C, with the radius $CB = \frac{3}{2} BA$, draw a circle; then draw the straight line A D, touching the given arc in A, and meeting the last-mentioned circle in D; A D will be the straight line required.

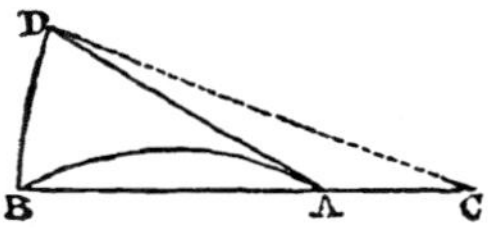

Fig. 9.

The error of this rule consists in the straight line being a little shorter than the arc: in fractions of the length of the arc, it is about $\frac{1}{1000}$ for an arc equal in length to its own radius; and it varies as the fourth power of the angle subtended by the arc; so that it may be diminished to any required extent by subdividing the arc to be measured by means of bisections. For example, in drawing a straight line approximately equal to an arc subtending 60°, the error is about $\frac{1}{900}$ of the length of the arc; divide the arc into two arcs, each subtending 30°; draw a straight line approximately equal to one of these, and double it; the error will be reduced to *one-sixteenth* of its former amount; that is, to about $\frac{1}{14400}$ of the length of the arc. The greatest angular extent of the arcs to which the rule is applied should be limited in each case according to the degree of precision required in the drawing.

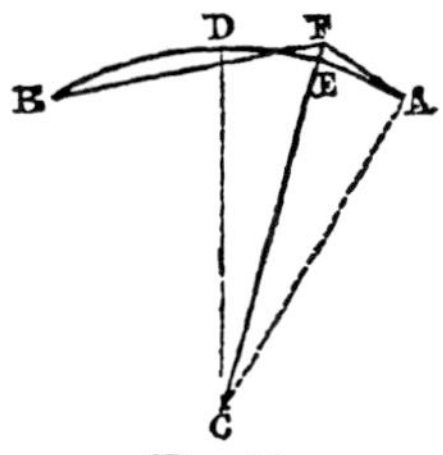

Fig. 10.

II. (Fig. 10). *To draw a straight line approximately equal to a given circular arc,* A B. (Another Method.) Let C be the centre of the arc. Bisect the arc A B in D, and the arc A D in E; draw the straight secant

* These rules are extracted from Papers read to the British Association in 1867, and published in the *Philosophical Magazine* for September and October of that year.

C E F, and the straight tangent A F, meeting each other in F; draw the straight line F B; then a straight line of the length A F + F B will be approximately equal in length to the arc A B.

The error of this rule, in fractions of the length of the arc, is just one-fourth of the error of Rule I., but in the contrary direction; and it varies as the fourth power of the angle subtended by the arc.

III. *To lay off upon a given circle an arc approximately equal in length to a given straight line.* In fig. 11, let A D be part of the circumference of the given circle, A one end of the required arc, and A B a straight line of the given length, drawn so as to touch the circle at the point A. In A B take A C $= \frac{1}{4}$ A B, and about C, with the radius C B $= \frac{3}{4}$ A B draw a circular arc B D, meeting the given circle in D. A D will be the arc required.

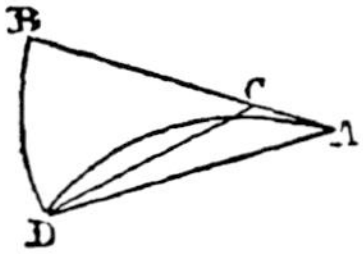

Fig. 11.

The error of this rule, in fractions of the given length, is the same as that of Rule I., and follows the same law.

IV. (Fig. 11.) *To draw a circular arc which shall be approximately equal in length to the straight line* A B, *shall with one of its ends touch that straight line at* A, *and shall subtend a given angle.* In A B take A C $= \frac{1}{4}$ A B; and about C, with the radius C B $= \frac{3}{4}$ A B, draw a circle, B D. Draw the straight line A D, making the angle B A D = one-half of the given angle, and meeting the circle B D in E. Then D will be the other end of the required arc, which may be drawn by well-known rules.

The error of this rule, in fractions of the given length, is the same with that of Rules I. and III., and follows the same law.

V. *To divide a circular arc, approximately, into any required number of equal parts.* By Rule I. or II., draw a straight line approximately equal in length to the given arc; divide that straight line into the required number of equal parts, and then lay off upon the given arc, by Rule III., an arc approximately equal in length to one of the parts of the straight line.

Rule V. becomes unnecessary when the number of parts is 2, 4, 8, or any other power of 2; for then the required division can be performed *exactly* by plane geometry.

VI. *To divide the whole circumference of a circle approximately into any required number of equal arcs.* When the required number of equal arcs is any one of the following numbers, the division can be made exactly by plane geometry, and the present rule is not needed:—any power of 2; 3; 3 × any power of 2; 5; 5 × any power of 2; 15; 15 × any power of 2.* In other cases

* It may be convenient here to state the methods of subdividing arcs and whole circles by plane geometry. (1.) *To bisect any circular arc* On the chord of the arc as a base, construct any convenient isosceles triangle, with the summit pointing away from the centre of the arc; a straight line from

proceed as follows:—Divide the circumference exactly, by plane geometry, into such a number of equal arcs as may be required, in order to give sufficient precision to the approximative part of the process. Let the number of equal arcs in that preliminary division be called n. Divide one of them, by means of Rule V., into the required number of equal parts; n times one of those parts will be one of the required equal arcs into which the whole circumference is to be divided.

Rules I., III., and V., are applicable to arcs of other curves besides the circle, provided the changes of curvature in such arcs are small and gradual.

39. **To Measure the Length of any Curve.**—Divide it into short arcs, and measure each of them by Rule I. of Article 38, page 23.

SECTION 4.—GEOMETRICAL CENTRES AND MOMENTS.

40. **Centre of Magnitude—General Principles.**—By the *magnitude* of a figure is to be understood its length, area, or volume, according as it is a line, a surface, or a solid.

The *centre of magnitude* of a figure is a point such that, if the figure be divided in any way into equal parts, the distance of the centre of magnitude of the whole figure from any given plane is the mean of the distances of the centres of magnitude of the several equal parts from that plane.

The *geometrical moment* of any figure relatively to a given plane is the product of its magnitude into the perpendicular distance of its centre from that plane.

I. *Symmetrical figure.*—If a plane divides a figure into two symmetrical halves, the centre of magnitude of the figure is in that plane; if the figure is symmetrically divided in the like manner by two planes, the centre of magnitude is in the line where those planes cut each other; if the figure is symmetrically divided by three planes, the centre of magnitude is their point of intersection; and if a figure has a *centre of figure* (for example, a circle, a sphere,

the centre of the arc to that summit will bisect the arc. (2.) *To mark the sixth part of the circumference of a circle.* Lay off a chord equal to the radius. (3.) *To mark the tenth part of the circumference of a circle.* In fig. 12, draw the straight line A B = the radius of the circle; and perpendicular to A B, draw B C = ½ A B. Join A C, and from it cut off C D = C B. A D will be the chord of one-tenth part of the circumference of the circle. (4.) *For the fifteenth part,* take the difference between one-sixth and one-tenth. It may be added that Gauss discovered a method of dividing the circumference of a circle by geometry exactly, when the number of equal parts is any prime number that is equal to 1 + a power of 2; such as $1+2^4=17$; $1+2^8=257$, &c.; but the method is too laborious for use in designing mechanism.

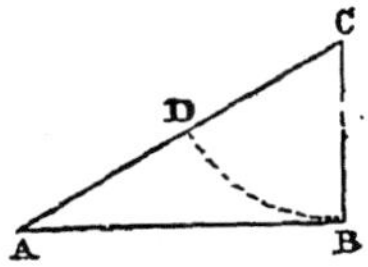

Fig. 12.

an ellipse, an ellipsoid, a parallelogram, &c.), that point is its centre of magnitude.

II. *Compound figure.*—To find the perpendicular distance from a given plane of the centre of a compound figure made up of parts whose centres are known. Multiply the magnitude of each part by the perpendicular distance of its centre from the given plane; distinguish the products (or *geometrical moments*) into positive or negative, according as the centres of the parts lie to one side or to the other of the plane; add together, separately, the positive moments and the negative moments: take the difference of the two sums, and call it positive or negative according as the positive or negative sum is the greater; this is the *resultant moment* of the compound figure relatively to the given plane; and its being positive or negative shews at which side of the plane the required centres lies. Divide the resultant moment by the magnitude of the compound figure; the quotient will be the distance required.

The centre of a figure in three dimensions is determined by finding its distances from three planes that are not parallel to each other. The best position for those planes is perpendicular to each other; for example, one horizontal, and the other two cutting each other at right angles in a vertical line. To determine the centre of a plane figure, its distances from two planes perpendicular to the plane of the figure are sufficient.

41. **Centre of a Plane Area.**—To find, approximately, the centre of any plane area.

Rule A.—Let the plane area be that represented in fig. 7 (of Article 34, page 17). Draw an axis, A X, in a convenient position, divide it into equal intervals, measure breadths at the ends and at the points of division, and calculate the area, as in Article 34.

Then multiply each breadth by its distance from one end of the axis (as A); consider the products as if they were the breadths of a new figure, and proceed by the rules of Article 34 to calculate the area of that new figure. The result of the operation will be the *geometrical moment* of the original figure relatively to a plane perpendicular to A X at the point A.

Divide the *moment* by the *area* of the original figure; the quotient will be the distance of the centre required from the plane perpendicular to A X at A.

Draw a second axis intersecting A X (the most convenient position being in general perpendicular to A X), and by a similar process find the distance of the centre from a plane perpendicular to the second axis at one of its ends; the centre will then be completely determined.

Rule B.—If convenient, the distance of the required centre from a plane cutting an axis at one of the intermediate points of division, instead of at one of its ends, may be computed as follows:—Take separately the moments of the two parts into which that plane divides the figure; the required centre will lie in the part which has the greater moment. Subtract the less moment from the greater; the remainder will be the *resultant moment* of the whole figure, which being divided by the whole area, the quotient will be the distance of the required centre from the plane of division.

Remark.—When the resultant moment is $=0$, the centre is in the plane of division.

Rule C.—To find the perpendicular distance of the centre from the axis A X. Multiply each breadth by the distance of the middle point of that breadth from the axis, and by the proper "Simpson's Multiplier," Article 34, page 18; distinguish the products into right-handed and left-handed, according as the middle points of the breadths lie to the right or left of the axis; take separately the sum of the right-handed products and the sum of the left-handed products; the required centre will lie to that side of the axis for which the sum is the greater; subtract the less sum from the greater, and multiply the remainder by $\frac{1}{3}$ of the common interval if Simpson's first rule is used, or by $\frac{3}{8}$ of the common interval if Simpson's second rule is used; the product will be the *resultant moment* relatively to the axis A X, which being divided by the area, the quotient will be the required distance of the centre from that axis.*

42. **Centre of a Volume.**—To find the perpendicular distance of the centre of magnitude of any solid figure from a plane perpendicular to a given axis at a given point, proceed as in Rule A of the preceding Article to find the moment relatively to the plane, substituting *sectional areas* for *breadths;* then divide the moment by the volume (as found by Article 37); the quotient will be the required distance.

To determine the centre completely, find its distances from three planes, no two of which are parallel. In general it is best that those planes should be perpendicular to each other.

43. **Centre of Magnitude of a Curved Line.**—*Rule A.—To find approximately the centre of magnitude of a very flat curved line.*—In fig. 13, let A D B be the arc. Draw the straight chord A B, which bisect in C; draw C D (the *deflection* of the arc) perpendicular to A B; from D lay off D E $=\frac{1}{3}$ C D; E will be very nearly the centre required.

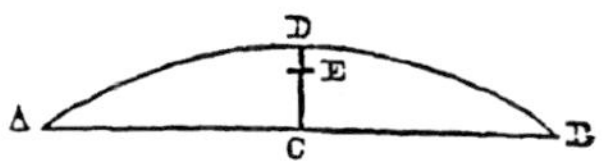

Fig. 13.

* The rules of this Article are expressed in symbols, as follows:—Let x and y be the perpendicular distances of any point in the plane area from two

This process is exact for a cycloidal arc whose chord, A B, is parallel to the base of the cycloid. For other curves it is approximate. For example, in the case of a circular arc, it gives D E too small; the error, for an arc subtending 60°, being about $\frac{1}{330}$ of the deflection, and its proportion to the deflection varying nearly as the square of the angular extent of the arc.

Rule B.—*When the curved line is not very flat,* divide it into very flat arcs; find their several centres of magnitude by Rule A, and measure their lengths; then treat the whole curve as a compound figure, agreeably to Rule II. of Article 40, page 26.

44. **Special Figures.**—I. *Triangle* (fig. 14).—From any two of the angles draw straight lines to the middle points of the opposite sides; these lines will cut each other in the centre required;—or otherwise,—from any one of the angles draw a straight line to the middle of the opposite side, and cut off one-third part from that line commencing at the side.

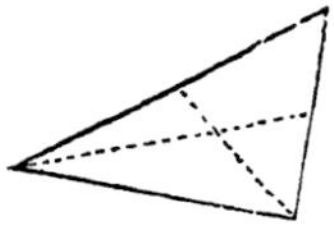

Fig. 14.

II. *Quadrilateral* (fig. 15).—Draw the two diagonals A C and B D, cutting each other in E. If the quadrilateral is a parallelogram, E will divide each diagonal into two equal parts, and will itself be the centre. If not, one or both of the diagonals will be divided into unequal parts by the point E. Let B D be a diagonal that is unequally divided. From D lay off D F in that diagonal = B E. Then the centre of the triangle F A C, found as in the preceding rule, will be the centre required.

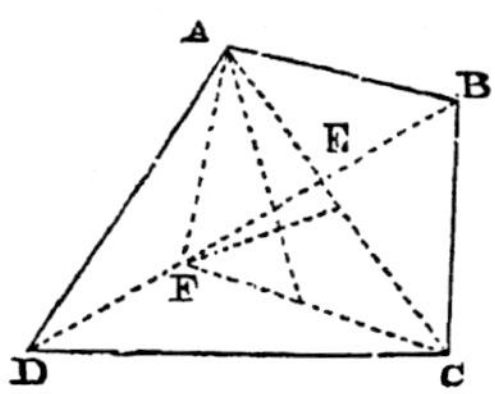

Fig. 15.

III. *Plane polygon.*—Divide it into triangles; find their centres, and measure their areas; then treat the polygon as a compound figure made up of the triangles, by Rule II. of Article 40, page 26.

IV. *Prism or cylinder with plane parallel ends.*—Find the centres of the ends; a straight line joining them will be the axis of the prism or cylinder, and the middle point of that line will be the centre required.

planes perpendicular to the area and to each other, and x_0 and y_0 the perpendicular distances of the centre of magnitude of the area from the same planes; then

$$x_0 = \frac{\int\int x\,dx\,dy}{\int\int dx\,dy};\quad y_0 = \frac{\int\int y\,dx\,dy}{\int\int dx\,dy}.$$

See Article 29, page 16.

V. *Tetrahedron, or triangular pyramid* (fig. 16).—Bisect any two opposite edges, as A D and B C, in E and F; join E F, and bisect it in G; this point will be the centre required.

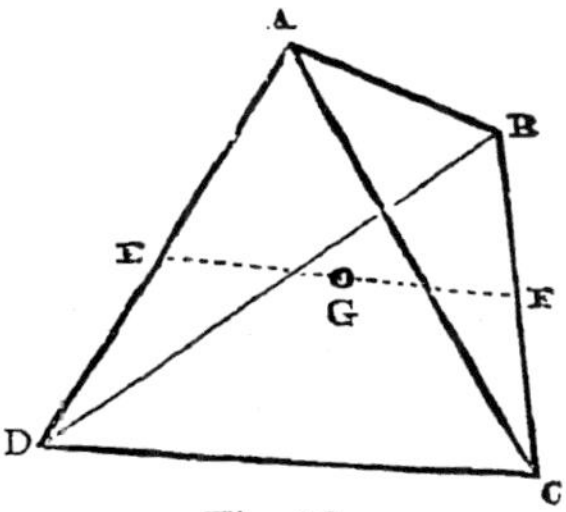

Fig. 16.

VI. *Any pyramid or cone with a plane base.*—Find the centre of the base, from which draw a straight line to the summit; this will be the axis of the pyramid or cone. From the axis cut off one-fourth of its length, beginning at the base; this will give the centre required.

VII. *Any polyhedron or plane-faced solid.*—Divide it into pyramids; find their centres and measure their volumes; then treat the whole solid as a compound figure by Rule II. of Article 22.

VIII. *Circular arc.*—In fig. 17, let A B be the arc, and C the the centre of the circle of which it is part. Bisect the arc in D, and join C D and A B. Multiply the radius C D by the chord A B, and divide by the length of the arc A D B; lay off the quotient C E upon C D; E will be the centre of magnitude of the arc.

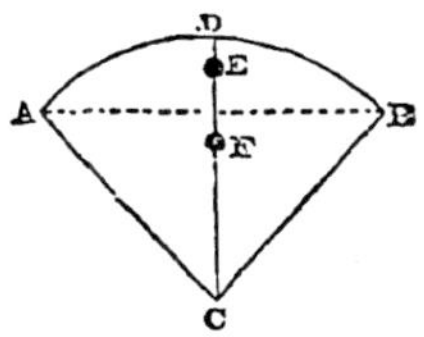

Fig. 17.

IX. *Circular sector*, C A D B, fig. 17.—Find C E as in the preceding rule, and make C F = $\frac{2}{3}$ C E; F will be the centre required.

X. *Sector of a flat ring.*—Let r be the external and r' the internal radius of the ring. Draw a circular arc of the same angular extent with the sector, and of the radius $\frac{2}{3} \cdot \frac{r^3 - r'^3}{r^2 - r'^2}$, and find its centre of magnitude by Rule VIII.

ELEMENTARY MECHANICAL NOTIONS.

Definition of General Terms and Division of the Subject.

45. **Mechanics** is the science of rest, motion, and force.

The *laws*, or *first principles* of mechanics, are the same for all bodies, celestial and terrestrial, natural and artificial.

The *methods of applying* the principles of mechanics to particular cases are more or less different, according to the circumstances of the case. Hence arise branches in the science of mechanics.

46. **Matter** (considered mechanically) is that which fills space.

47. **Bodies** are limited portions of matter. Bodies exist in three conditions—the solid, the liquid, and the gaseous. Solid bodies tend to preserve a definite size and shape. Liquid bodies tend to preserve a definite size only. Gaseous bodies tend to expand indefinitely. Bodies also exist in conditions intermediate between the solid and liquid, and possibly also between the liquid and the gaseous.

48. **A Material or Physical Volume** is the space occupied by a body or by a part of a body.

49. **A Material or Physical Surface** is the boundary of a body, or between two parts of a body.

50. **Line, Point, Physical Point, Measure of Length.**—In mechanics, as in geometry, a Line is the boundary of a surface, or between two parts of a surface; and a Point is the boundary of a line, or between two parts of a line; but the term "*Physical Point*" is sometimes used by mechanical writers to denote an *immeasurably small body*—a sense inconsistent with the strict meaning of the word "point;" but still not leading to error, so long as it is rightly understood.

In *measuring* the dimensions of bodies, the standard British unit of length is the *yard*, being the length at the temperature of 62° Fahrenheit, and at the mean atmospheric pressure, between the two ends of a certain bar which is kept in the office of the Exchequer, at Westminster.

In computations respecting motion and force, and in expressing the dimensions of large structures, the unit of length commonly employed in Britain is the *foot*, being one-third of the yard.

In expressing the dimensions of machinery, the unit of length commonly employed in Britain is the *inch*, being one-thirty-sixth part of the yard. Fractions of an inch are very commonly stated by mechanics and other artificers in halves, quarters, eighths, sixteenths, and thirty-second parts; but according to a resolution of the Institution of Mechanical Engineers, passed at the meeting held at Manchester in June, 1857, the practice has been introduced of expressing fractions of an inch in decimals.

The French unit of length is the mètre, being about $\frac{1}{40000000}$ of the earth's circumference, measured round the poles.

51. **Rest** is the relation between two points, when the straight line joining them does not change in length nor in direction.

A body is at rest relatively to a point, when every point in the body is at rest relatively to the first mentioned point.

52. **Motion** is the relation between two points when the straight line joining them changes in length, or in direction, or in both.

A body moves relatively to a point when any point in the body moves relatively to the first mentioned point.

53. **Fixed Point.**—When a single point is spoken of as having motion or rest, some other point, either actual or ideal, is always either expressed or understood, *relatively* to which the motion or rest of the first point takes place. Such a point is called a *fixed* point.

So far as the phenomena of motion alone indicate, the choice of a fixed point with which to compare the positions of other points appears to be arbitrary, and a matter of convenience alone; but when the laws of force, as affecting motion, come to be considered, it will be seen that there are reasons for calling certain points fixed, in preference to others.

In the mechanics of the solar system, the fixed point is what is called the *common centre of mass* of the bodies composing that system. In applied mechanics, the fixed point is either a point which is at rest relatively to the earth, or (if the structure or machine under consideration be movable from place to place on the earth), a point which is at rest relatively to the structure, or to the frame of the machine, as the case may be.

Points, lines, surfaces, and volumes, which are at rest relatively to a fixed point, are fixed.

54. **Cinematics.**—The comparision of motions with each other, without reference to their causes, is the subject of a branch of geometry called "*Cinematics*."

55. **Force** is an action between two bodies, either causing or tending to cause change in their relative rest or motion.

The notion of force is first obtained directly by sensation; for the forces exerted by the voluntary muscles can be felt. The existence of forces other than muscular tension is inferred from their effects.

56. **Equilibrium** or **Balance** is the condition of two or more forces which are so opposed that their combined action on a body produces no change in its rest or motion.

The notion of balance is first obtained by sensation; for the forces exerted by voluntary muscles can be felt to balance sometimes each other, and sometimes external pressures.

57. **Dynamics—Statics and Kinetics.**—Forces may take effect, either by balancing other forces, or by producing change of motion. The former of those effects is the subject of *Statics;* the latter that of *Kinetics,* and the Science which treats of both is by modern practice entitled *Dynamics;* these, together with *Cinematics,* already defined, form the three great divisions of pure, abstract, or general mechanics.

58. **Structures and Machines.**—The works of human art to which the science of applied mechanics relates, are divided into two classes, according as the parts of which they consist are intended to rest or to move relatively to each other. In the former case they are called *Structures;* in the latter, *Machines.* Structures are subjects of Statics alone; Machines, when the motions of their parts are considered alone, are subjects of Cinematics; when the forces acting on and between their parts are also considered, machines are subjects of Dynamics.

PART I.

PRINCIPLES OF CINEMATICS, OR THE COMPARISON OF MOTIONS.

59. **Division of the Subject.**—The Science of Cinematics, and the fundamental notions of rest and motion to which it relates, having already been defined among the ELEMENTARY MECHANICAL NOTIONS, Articles 51, 52, 53, 54, it remains to be stated, that the principles of Cinematics, or the comparison of motions, will be divided and arranged in the present part of this treatise in the following manner :—

I. Motions of Points.
II. " Rigid Bodies or Systems.
III. " Pliable Bodies and Fluids.

CHAPTER I.

MOTIONS OF POINTS.

SECTION 1.—MOTIONS OF A PAIR OF POINTS.

60. **Fixed and Nearly Fixed Directions.**—From the definition of motion given in Article 52, it follows, that in order to determine the relative motion of a pair of points, which consists in the change of length and direction of the straight line joining them, that line must be compared, at the beginning and end of the motion considered, with some fixed or standard length, and with at least two fixed directions. Standard lengths have already been considered in Article 50.

An *absolutely fixed direction* may be ascertained by means whose principles cannot be demonstrated until the subject of kinetics is considered. For the present it is sufficient to state, that when a solid body rotates free from the influence of any external force tending to change its rotation, there is an absolutely fixed direction called that of the *axis of angular momentum*, which bears certain relations to the successive positions of the body.

A *nearly fixed direction* is that of a straight line joining a pair of points in two bodies whose distance from each other is very great, such as the earth and a fixed star.

A *line fixed relatively to the earth* changes its absolute direction (unless parallel to the earth's axis in a manner depending on the earth's rotation, and returns periodically to its original absolute direction at the end of each *sidereal day* of 86,164 seconds. This rate of change of direction is so slow compared with that which takes place in almost all pieces of mechanism to which cinematical and kinetic principles are applied, that in almost all questions of applied mechanics, directions fixed relatively to the earth may be treated as sufficiently nearly fixed for practical purposes.

When the motions of pieces of mechanism relatively to each other, or to the frame by which they are carried, are under consideration, directions fixed relatively to the frame, or to one of the pieces of the machine, may be considered provisionally as fixed for the purposes of the particular question.

POSTULATE.—Let it be granted that a line may represent a motion, where the term motion is employed to represent the path of motion, the direction and the velocity or length of motion. This is a self-evidently possible problem, for a line may be drawn to represent any path, in any direction to represent any direction of motion, and of any length to represent any length of motion, or velocity, limited always by the space within which motions can take place or lines be drawn.

61. **Motion of a Pair of Points.**—In fig. 18, let $A_1 B_1$ represent the relative situation of a pair of points at one instant, and $A_2 B_2$ the relative situation of the same pair of points at a later instant. Then the change of the straight line $\overline{A B}$ between those points, from the length and direction represented by $\overline{A_1 B_1}$ to the length and direction represented by $\overline{A_2 B_2}$, constitutes the *relative motion* of the pair of points A B, during the interval between the two instants of time considered.

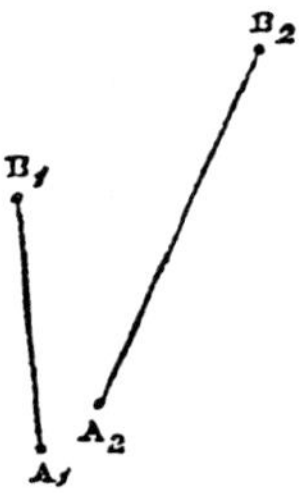

Fig. 18.

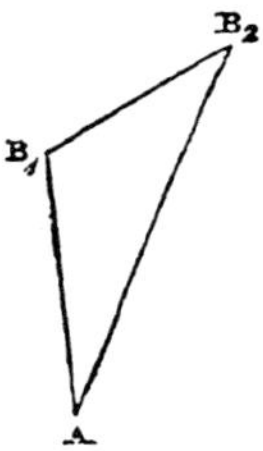

Fig. 19.

Fig. 20.

To represent that relative motion by one line, let there be drawn, from one point A, fig. 19, a pair of lines, $\overline{A B_1}$, $\overline{A B_2}$, equal and parallel to $\overline{A_1 B_1}$, $\overline{A_2 B_2}$, of fig. 18; then A represents one of the pair of points whose relative motion is under consideration, and B_1, B_2, represent the two successive positions of the other point B relatively to A; and the line $\overline{B_1 B_2}$ represents *the motion of* B *relatively to* A, which, for the purposes of the representation, is assumed to be fixed.

Or otherwise, as in fig. 20, from a single point B let there be drawn a pair of lines, $\overline{B A_1}$, $B A_2$, equal and parallel to $\overline{A_1 B_1}$, $\overline{A_2 B_2}$, of fig. 18; then $A_1 A_2$, represent the two successive positions of A relatively to B; and the line $\overline{A_1 A_2}$, equal and parallel to $\overline{B_1 B_2}$ of fig. 19, but *pointing in the contrary direction*, represents *the motion of* A *relatively to* B.

62. **Fixed Point and Moving Point.**—In fig. 19, A is treated as the fixed point, and B as the moving point; and in fig. 20, B is treated as the fixed point, and A as the moving point; and these are simply two different methods of representing to the mind the same relation between the points A and B (see Article 53).

63. **Component and Resultant Motions.**—Let O be a point assumed as fixed, and A and B two successive positions of a second point relatively to O. In order to express mathematically the amount and direction of $\overline{A B}$, the motion of the second point relatively to O, that line may be compared with three *axes*, or lines in fixed directions, traversing the fixed point O, such as O X, O Y, O Z.

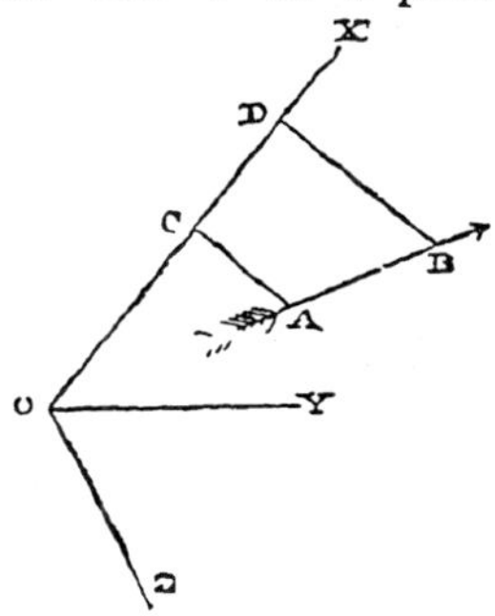

Fig. 21.

Through A and B draw straight lines A C, B D, parallel to the plane of O Y and O Z, and cutting the axis O X in C and D. Then $\overline{C D}$ is said to be the *component* of the motion of the second point relatively to O, *along*, o *in the direction of* the axis O X; and by a similar process are found the components of the motion A B along O Y and O Z. The entire motion $\overline{A B}$ is said to be the *resultant* of these components, and is evidently the diagonal of a parallelopiped of which the components are the sides.

The three axis are usually taken at right angles to each other; in which case A C and B D are perpendiculars let fall from A and B upon O X; and if α be the angle made by the direction of the motion $\overline{A B}$ with O X,

$$\overline{C D} = \overline{A B} \cdot \cos \alpha.$$

64. **The Measurement of Time** is effected by comparing the events, and especially the motions, which take place in intervals of time.

Equal times are the times occupied by the same body, or by equal and similar bodies, under precisely similar circumstances, in performing equal and similar motions. The *standard unit of time* is the period of the earth's rotation, or *sidereal day*, which has been proved by Laplace, from the records of celestial phenomena, not to have changed by so much as one *eight-millionth* part of its length in the course of the last two thousand years.

A subordinate unit is the *second*, being the time of one swing of a pendulum, so adjusted as to make 86,400 oscillations in 1·00273791 of a sidereal day; so that a sidereal day is 86164·09 seconds.

The length of a solar day is variable; but the *mean solar day*, being the exact mean of all its different lengths, is the period already mentioned of 1·00273791 of a sidereal day, or 86,400 seconds. The divisions of the mean solar day into 24 hours, of each hour into 60 minutes, and of each minute into 60 seconds, are familiar to all.

Fractions of a second are measured by the oscillations of small pendulums, or of springs, or by the rotations of bodies so contrived as to rotate through equal angles in equal times.

65. **Velocity** is the ratio of the number of units of length described by a point in its motion relatively to another point, to the number of units of time in the interval occupied in describing the length in question; and if that ratio is the same, whether it be computed for a longer or a shorter, an earlier or a later, part of the motion, the velocity is said to be UNIFORM. Velocity is expressed in *units of distance per unit of time*. For different purposes, there are employed various units of velocity, some of which, together with their proportions to each other, are given in the following table:—

Comparison of Different Measures of Velocity.

	Miles per hour.	Feet per second.	Feet per minute	Feet per hour
	1	= $1\cdot4\dot{6}$	= 88	= 5280
	$0\cdot68\dot{1}\dot{8}$	= 1	= 60	= 3600
	$0\cdot011\dot{3}\dot{6}$	= $0\cdot01\dot{6}$	= 1	= 60
	$0\cdot00018\dot{9}\dot{3}$	= $0\cdot0002\dot{7}$	= $0\cdot01\dot{6}$	= 1
1 nautical mile per hour, or "knot,"......	= 1·1507	= 1·6877	= 101·262	= 6075·74

In treating of the general principles of mechanics, the *foot per second* is the unit of velocity commonly employed in Britain. The units of time being the same in all civilized countries, the proportions amongst their units of velocity are the same with those amongst their linear measures.

Component and resultant velocities are the velocities of component and resultant motions, and are related to each other in the same way with those motions, which have already been treated of in Article 63.

66. **Uniform Motion** consists in the combination of uniform velocity with uniform direction; that is, with motion along a straight line whose direction is fixed.

SECTION 2.—UNIFORM MOTION OF SEVERAL POINTS.

67. **Motion of Three Points.**—THEOREM. *The relative motions of three points in a given interval of time are represented in direction and magnitude by the three sides of a triangle.* Let O, A, B, denote the three points. Any one of them may be taken as a fixed point; let O be so chosen; and let O X, O Y, O Z, fig. 22, be axes traversing it in fixed directions. Let A_1 and B_1 be the positions of A and B relatively to O at the beginning of the given interval of time, and A_2 and B_2 their positions at the end of that interval. Then $\overline{A_1 A_2}$ and $\overline{B_1 B_2}$ are the respective motions of A and B relatively to O. Complete the parallelogram $A_1 B_1 b A_2$; then because $\overline{A_2 b}$ is parallel and equal to $A_1 B_1$, b is the position which B would have at the end of the interval, *if it had no motion relatively to* A; but B_2 is the actual position of B at the end of the interval; therefore, $\overline{b B_2}$ is the motion of B relatively to A. Then in the triangle $B_1 b B_2$,

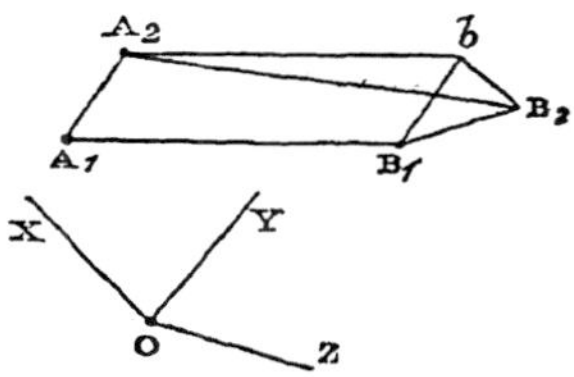

Fig. 22.

$\overline{B_1 b} = \overline{A_1 A_2}$ is the motion of A relatively to O,

$\overline{b B_2}$ is the motion of B relatively to A,

$\overline{B_1 B_2}$ is the motion of B relatively to O;

so that those three motions are represented by the three sides of a triangle.—Q. E. D.

This Theorem might be otherwise expressed by saying, *that if three moving points be considered in any order, the motion of the third relatively to the first is the resultant of the motion of the third relatively to the second, and of the motion of the second relatively to the first;* the word "*resultant*" being understood as already explained in Article 63.

68. **Motions of a Series of Points.**—COROLLARY. *If a series of points be considered in any order, and the motion of each point determined relatively to that which precedes it in the series, and if the relative motion of the last point and the first point be also determined, then will those motions be represented by the sides of a closed*

polygon. Let O be the first point, A, B, C, &c., successive points following it, M the last point but one, and N the last point; and, for brevity's sake, let the relative motion of two points, such as B and C, be denoted thus (B, C). Then by the Theorem of Article 67, (O, A), (A, B), and (O, B) are the three sides of a triangle; also (O, B), (B, C), and (O, C), are the three sides of a triangle; therefore (O, A), (A, B), (B, C), and (O, C), are the four sides of a quadrilateral; and by continuing the same process, it is shewn, that how great soever the number of points, (O, N), is the closing side of a polygon, of which (O, A), (A, B , (B, C), (C, D), &c, (M, N) are the other sides.—Q. E. D. In other words, *the motion of the last point relatively to the first is the resultant of the motions of each point of the series relatively to that preceding it.*

69. **The Parallelopiped of Motions.**—In fig. 23, let there be four points, O, A, B, C, of which one, O, is assumed as fixed, and is traversed by three axes in fixed directions, O X, O Y, O Z. In a given interval of time, let A have the motion $\overline{A_1 A_2}$ along or parallel to O X; let B have, in the same interval, the motion $\overline{b B_2}$ parallel to O Y, and relatively to A; then $\overline{B_1 B_2}$, the diagonal of the parallelogram whose sides are $\overline{B_1 b} = \overline{A_1 A_2}$ and $\overline{b B_2}$, is the motion of B relatively to O. Let C have, relatively to B, the motion $\overline{c C_2}$ parallel to O Z; then $\overline{C_1 C_2}$, the diagonal of the parallelopiped whose edges are $\overline{A_1 A_2}$, $\overline{b B_2}$, and $\overline{c C_2}$, is the motion of C relatively to O, being the resultant of the motions represented by those three edges. This is a *mechanical* explanation of the composition of motions, leading to results corresponding with the *geometrical* explanation of Article 63.

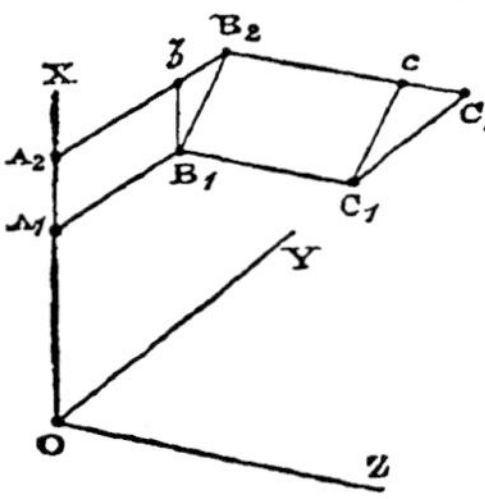

Fig. 23.

70. **Comparative Motion** is the relation which exists between the simultaneous motions of two points relatively to a third, which is assumed as fixed. The comparative motion of two points is expressed, in the most general case, by means of four quantities, viz. :—

(1.) The *velocity ratio,** or the proportion which their velocities bear to each other, that is, the proportion borne to each other by the distances moved through by the two points in the same interval of time.

(2.) (3.) (4.) The *directional relation,** which is the relation between the directions in which the two points are moving at the same instant, and which requires, for its complete expression, three

* These terms are adopted from Prof. Willis's work on Mechanism.

angles. Those three angles may be measured in different ways, and one of those ways is the following :—

(2.) The angle made by the directions of the compared motions with each other.

(3.) The angle made by a plane parallel to those two directions with a fixed plane.

(4.) The angle made by the intersection of those two planes with a fixed direction in the fixed plane.

Thus, the comparative motion of two points relatively to a third, is expressed by means of one of those groups of four elements which Sir William Rowan Hamilton has called "*quaternions.*" In most of the practical applications of cinematics, the motions to be compared are limited by conditions which render the comparision more simple than it is in the general case just described. In machines, for example, the motion of each point is limited to two directions, forward or backward in a fixed path; so that the comparative motion of two points is sufficiently expressed by means of the velocity ratio, together with a directional relation expressed by + or −, according as the motions at the instant in question are similar or contrary.

Section 3.—Varied Motion of Points.

71. **Velocity and Direction of Varied Motion.**—The motion of one point relatively to another may be varied, either by change of velocity, or by change of direction, or by both combined, which last case will now be considered, as being the most general.

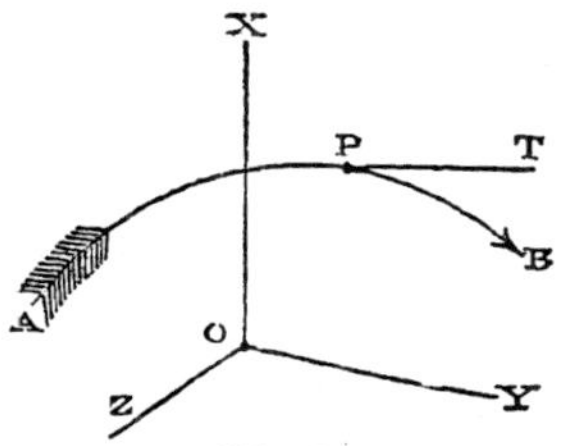

Fig. 24.

In fig. 24, let O represent a point assumed as fixed, O X, O Y, O Z, fixed directions, and A B part of the *path* or orbit traced by a second point in its varied motion relatively to O. At the instant when the second point reaches a given position, such as P, in its path, the *direction* of its motion is obviously that of $\overline{\text{P T}}$, a tangent to the path at P.

To find the velocity at the instant of passing P, let Δt denote an interval of time which includes that instant, and Δs the distance traced in that interval. Then

$$\frac{\Delta s}{\Delta t}$$

is an *approximation* to the velocity at the instant in question, which will approach continually nearer and nearer to the exact velocity as the interval Δt and the distance Δs are made shorter and shorter; and the *limit* towards which $\frac{\Delta s}{\Delta t}$ converges, as Δs and Δt are indefinitely diminished, and which is denoted by

$$v = \frac{d s}{d t}, \ldots\ldots\ldots\ldots\ldots\ldots\ldots\ldots\ldots (1.)$$

is the *exact* velocity at the instant of passing P. In the language of the differential calculus, the space is a function of the time and the velocity is the differential coefficient of the space with respect to the time, thus $s = \phi\, t$ and $\frac{d s}{d t} = \phi' t = v$. It will be seen hereafter that, the velocity v) itself is a function of the time t). This is the process called "*differentiation.*"

Should the velocity at each instant of time be known, then the distance $s_1 - s_0$, described during an interval of time $t_1 - t_0$, is found by *integration* (see Article 29), as follows:—

$$s_1 - s_0 = \int_{t}^{t_1} v\, d\, t. \ldots\ldots\ldots\ldots\ldots\ldots\ldots\ldots\ldots (2.)$$

72. **Components of Varied Motion.**—All the propositions of the two preceding sections, respecting the composition and resolution of motions, are applicable to the velocities of varied motions at a given instant, each such velocity being represented by a line, such as $\overline{\text{P T}}$, in the direction of the tangent to the path of the point which moves with that velocity, at the instant in question. For example, if the axes O X, O Y, O Z, are at right angles to each other, and if the tangent $\overline{\text{P T}}$ makes with their directions respectively the angles α, β, γ then the three rectangular components of the velocity of the point parallel to those three axes are

$$v \cos \alpha; \; v \cos \beta; \; v \cos \gamma.$$

Let x, y, z, be the co-ordinates of any point, such as P, in the path A P B, as referred to the three given axes. If a point p be assumed indefinitely near to the point P, its co-ordinates will be $x + d x$, $y + d y$, $z + d z$, and if $d s$ have the already assumed value, $d x$, $d y$, $d z$, will be its projections on the three axes; that is, the lengths bounded by perpendiculars let fall from the extremities of $d s$ on the three respective axes. Then it is well known that

$$\cos \alpha = \frac{d x}{d s}; \; \cos \beta = \frac{d y}{d s}; \; \cos \gamma = \frac{d z}{d s};$$

and consequently the three components of the velocity $v \left(= \frac{d s}{d t}\right)$ are

$$v \cos \alpha = \frac{d x}{d t}; \; v \cos \beta = \frac{d y}{d t}; \; v \cos \gamma = \frac{d z}{d t}; \ldots\ldots (3.)$$

now by the Geometry of three dimensions

$$\cos^2 \alpha + \cos^2 \beta + \cos^2 \gamma = 1.$$

and hence these are related to their resultant by the equation

$$\left(\frac{dx}{dt}\right)^2 + \left(\frac{dy}{dt}\right)^2 + \left(\frac{dz}{dt}\right)^2 = v^2 \ldots\ldots\ldots\ldots (4.)$$

73. **Uniformly-Varied Velocity.**—Let the velocity of a point either increase or diminish at an uniform rate; so that if t represents the time elapsed from a fixed instant when the velocity was v_0, the velocity at the end of that time shall be

$$v = v_0 + a t; \ldots\ldots\ldots\ldots\ldots\ldots (1.)$$

a being a constant quantity, which is the *rate of variation* of the velocity, and is called *acceleration* when positive, and *retardation* when negative. Then the *mean* velocity during the time t is

$$\frac{v_0 + v}{2} = \frac{v_0 + v_0 + a t}{2} = v_0 + \frac{a t}{2} \ldots\ldots\ldots (2.)$$

and the distance described is

$$s = v_0 t + \frac{a t^2}{2} \ldots\ldots\ldots\ldots\ldots\ldots (3.)$$

If there be no initial velocity, that is, if the body start from a state of rest, then $v = a t$ and $s = \frac{a t^2}{2}$, and these equations are illustrations of the use of the differential calculus; for first differentiate s with respect to t in the equation $s = \frac{a t^2}{2}$, and there is obtained $\frac{ds}{dt}(= v) = \frac{a \cdot 2t}{2} = a t$, which is the first equation, then differentiate $v = a t$, and there is obtained $\frac{dv}{dt} = a$. To find the velocity of a point, whose velocity is uniformly varied, at a given instant, and the rate of variation of that velocity, let the distances, Δs_1, Δs_2, described in two equal intervals of time, each equal to Δt, before and after the instant in question, be observed. Then the velocity at the instant between those intervals is

$$v = \frac{\Delta s_1 + \Delta s_2}{2 \Delta t} \ldots\ldots\ldots\ldots\ldots\ldots (4.)$$

and its rate of variation is

$$a = \frac{\Delta v}{\Delta t} = \frac{\Delta s_2 - \Delta s_1}{(\Delta t)^2} \ldots\ldots\ldots\ldots\ldots (5.)$$

where the variation of velocity $=\frac{\Delta s_2 - \Delta s_1}{\Delta t}$, and the rate of variation being either acceleration or retardation, as the velocity of the point is being increased or diminished, is that quantity divided by Δt.

74. **Graphical Representation of Motions.**—Since in uniform motion the space is equal to the product of the velocity and time, and since in geometry a rectangular area is the product of a base line and perpendicular, an uniform motion may be represented by a rectangular area, as in fig. 25, where A B represents a certain number of *units* of time, and A C a certain number of units of velocity per *unit* of time. It will be noticed that in uniform motion, the velocity or number of units of velocity at each unit of time is the same, as at A, B, E. Varied motion and uniformly varied motion may also be graphically represented: in the first, the line C D will be a curve; and in the second, the line C D will form a constant angle with A B; hence in varied motion any ordinate, E F, depends upon the abscissa A E, and the mean velocity is the mean ordinate of a figure so formed, or is the quotient of the area (space) divided by the base (time), whereas in uniformly-varied motion, the space described depends upon the initial and final velocities alone, and not upon the intermediate velocities. Fig. 26 represents varied motion where the velocity at each point is represented by the ordinate at that point, and the mean velocity is equal to the area of the figure divided by the base A B. Fig. 27 represents uniformly-varied motion, and it is evident that, in order to estimate the area of the figure A B C D, that is, the space, it is only necessary to consider the initial and final velocities. In these figures, if the velocity be null at any point, there will be no ordinate at that point: if the direction of motion change, this will be represented by a change of sign of the ordinate or velocity.

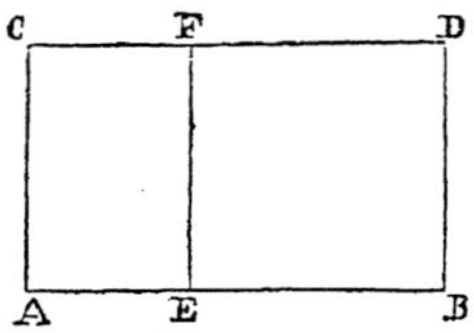

Fig. 25.

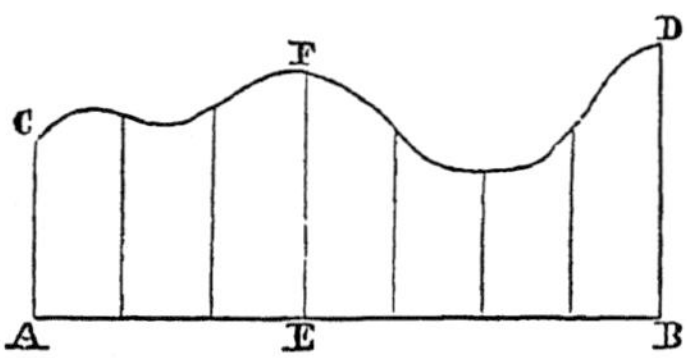

Fig. 26.

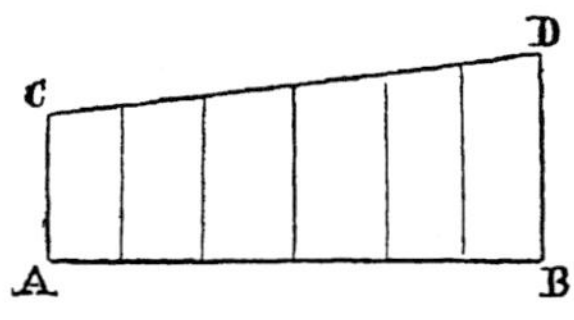

Fig. 27.

There is another method of graphically representing the motion of a point: in this the abscissæ represent the time, and the ordinates at each point the space passed over in the corresponding number of units of time, or the distance of the point from a certain datum point. In this case the space described in any number of units of time is equal to the difference of the lengths of the ordinates at the corresponding intervals, and the velocity is proportional to the quotient of the difference of the ordinates divided by the difference of the abscissæ.

75. **Varied Rate of Variation of Velocity.**—When the velocity of a point is neither constant nor uniformly-varied, its rate of variation may still be found by applying to the velocity the same operation of *differentiation*, which, in Article 73, was applied to the distance described in order to find the velocity. The result of this operation is expressed by the symbols,

$$a=\frac{d\,v}{d\,t}=\frac{d\cdot\frac{d\,s}{d\,t}}{d\,t}=\frac{d^2 s}{d\,t^2};$$

and is the limit to which the quantity obtained by means of the formula 5 of Article 73 continually approximates, as the interval denoted by Δt is indefinitely diminished. In the fraction $\frac{d\cdot\frac{d\,s}{d\,t}}{d\,t}$, $d\,s$ is the limit of the difference of either of the spaces Δs in equation (5), Article 73, and $d\cdot d\,s$, is the limit of the difference of that difference, viz., $\Delta s_2-\Delta s_1$; that is, d in this fraction is represented by the minus sign $(-)$ in the other, and $d\,s$ by the limit of either of the quantities Δs_1, Δs_2. Here in the language of the differential calculus, the velocity (v) is a function of the time (t), and the acceleration (a) is the differential coefficient of the velocity with respect to the time, thus $v=\phi\,t$ and $a=\phi'\,t$, or $=\frac{d\,v}{d\,t}$. Also the velocity, v, being the differential coefficient of the space with respect to the time, see Article 71; the acceleration a is the 2nd differential coefficient of the space with respect to the time, or v being $\psi' t$, $a=\psi''\,t$.

76. **Combination of Uniform and Uniformly Accelerated Motion.**—Assume a pair of rectangular axes of co-ordinates. Let the uniform motion be represented by abscissæ along O X, and the uniformly accelerated motion by ordinates parallel to O Y; let O B $(=x)=v\,t$, represent the space described in the time t with the velocity v, and let O C $(=y)=\frac{a\,t^2}{2}$, represent the space de-

scribed with a uniform rate of acceleration, a, in the same time t, see Article 73, then $x^2 = v^2 t^2$ and $y = \frac{a t^2}{2}$, $\therefore\ x^2 = y \frac{2 v^2}{a}$, where the square of any abscissa bears a constant ratio to the corresponding ordinate, and the path of the point is known by Conic Sections to be a Parabola. The same follows for any axes of co-ordinates; but if the direction of the uniformly accelerated motion be that of the uniform motion or directly opposed to it, the resultant direction will be the same as that of either motion, or will be that of the greater component.

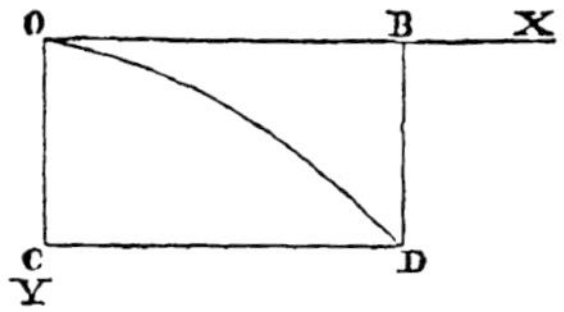

Fig. 23.

77. **Uniform Deviation** is the change of motion of a point which moves with uniform velocity in a circular path. The *rate* at which uniform deviation takes place is determined in the following manner:—

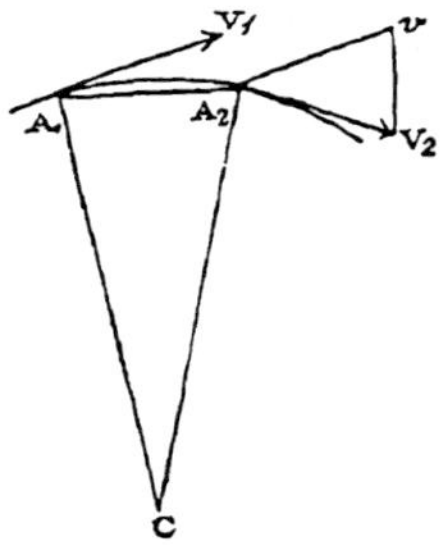

Fig. 29.

Let C, fig. 29, be the centre of the circular path described by a point A with an uniform velocity v, and let the radius $\overline{CA}$ be denoted by r. At the beginning and end of an interval of time Δt, let A_1 and A_2 be the positions of the moving point. Then

$$\text{the arc } A_1 A_2 = v \Delta t;$$

$$\text{the chord } A_1 A_2 = v \Delta t \cdot \frac{\text{chord}}{\text{arc}}$$

The velocities at A_1 and A_2 are represented by the equal lines $\overline{A_1 V_1} = \overline{A_2 V_2} = v$, touching the circle at A_1 and A_2 respectively. From A_2 draw $\overline{A_2 v}$ equal and parallel to $\overline{A_1 V_1}$, and join $\overline{V_2 v}$. Then the velocity $\overline{A_2 V_2}$ may be considered as compounded of $\overline{A_2 v}$ and $\overline{v V_2}$; so that $\overline{v V_2}$ is the *deviation* of the motion during the interval Δt; and because the isosceles triangles $A_2 v V_2$, $C A_1 A_2$, are similar:—

$$\overline{v V_2} = \frac{\overline{A_2 V_2} \cdot \overline{A_1 A_2}}{C A} = \frac{v^2 \cdot \Delta t}{r} \cdot \frac{\text{chord}}{\text{arc}},$$

deduced by substituting the value of $\overline{A_1 A_2}$ already found; and the *approximate rate* of that deviation being the deviation divided by the interval of time in which it occurs, is

$$\frac{v^2}{r} \cdot \frac{\text{chord}}{\text{arc}};$$

but the deviation does not take place by instantaneous changes of velocity, but by insensible degrees; so that the true rate of deviation is to be found by finding the limit to which the approximate rate continually approaches as the interval Δt is diminished indefinitely. Now the factor $\frac{v^2}{r}$ remains unaltered by that diminution; and the ratio of the chord to the arc approximates continually to equality; so that the limit in question, or *true rate of deviation*, is expressed by

$$\frac{v^2}{r} \ldots\ldots\ldots\ldots\ldots\ldots\ldots\ldots\ldots\ldots (1.)$$

78. **Varying Deviation.**—When a point moves with a varying velocity, or in a curve not circular, or has both these variations of motion combined, the *rate of deviation* at a given instant is still represented by Equation 1 of Article 77, provided v be taken to denote the velocity, and r the radius of curvature of the path, of the point at the instant in question.

79. **The Resultant Rate of Variation** of the motion of a point is found by considering the rate of variation of velocity and the rate of deviation as represented by two lines, the former in the direction of a tangent to the path of the point, and the latter in the direction of the radius of curvature at the instant in question, and taking the diagonal of the rectangle of which those two lines are the sides, which has the following value:—

$$\sqrt{\left(\frac{d\,v}{d\,t}\right)^2 + \frac{v^4}{r^2}} = \sqrt{\left\{\left(\frac{d^2\,s}{d\,t^2}\right)^2 + \frac{1}{r^2}\left(\frac{d\,s}{d\,t}\right)^4\right\}} \ldots (1.)$$

the first term of the quantity under the first radical is the square of $\frac{d\,v}{d\,t}$ in Article 73, and the second the square of $\frac{v^2}{r}$, Equation (1), Article 77.

80. **The Rates of Variation of the Component Velocities** of a point parallel to three rectangular axes, are represented as follows:—

$$\frac{d^2\,x}{d\,t^2};\ \frac{d^2\,y}{d\,t^2};\ \frac{d^2\,z}{d\,t^2};\ldots\ldots\ldots\ldots\ldots\ldots\ldots (1.)$$

and if a rectangular parallelopiped be constructed, of which the edges represent these quantities, its diagonal, whose length is

$$\sqrt{\left\{\left(\frac{d^2\,x}{d\,t^2}\right)^2 + \left(\frac{d^2\,y}{d\,t^2}\right)^2 + \left(\frac{d^2\,z}{d\,t^2}\right)^2\right\}} \ldots\ldots\ldots\ldots (2.)$$

will represent the *resultant rate of variation*, already given in another form in Equation 1 of Article 79.

81. **The Comparison of the Varied Motions** of a pair of points

relatively to a third point assumed as fixed, is made by finding the ratio of their velocities, and the directional relation of the tangents of their paths at the same instant, in the manner already described in Article 70, as applied to uniform motions. It is evident that the comparative motions of a pair of points may be so regulated as to be constant, although the motion of each point is varied, provided the variations take place for both points at the same instant, and at rates proportional to their velocities.

CHAPTER II.

MOTIONS OF RIGID BODIES.

SECTION 1.—RIGID BODIES, AND THEIR TRANSLATION.

82. The term **Rigid Body** is to be understood to denote a body, or an assemblage of bodies, or a system of points, whose figure undergoes no alteration during the motion which is under consideration.

83. **Translation or Shifting** is the motion of a rigid body relatively to a fixed point, when the points of the rigid body have no motion relatively to each other; that is to say, when they all move with the same velocity and in the same direction at the same instant, so that no line in the rigid body changes its direction.

It is obvious that if three points in the rigid body, not in the same straight line, move in parallel directions with equal velocities at each instant, the body must have a motion of translation.

The paths of the different points of the body, provided they are all equal and similar, and at each instant parallel, may have any figure whatsoever.

SECTION 2.—SIMPLE ROTATION.

84. **Rotation or Turning** is the motion of a rigid body when lines in it change their direction. Any point in or rigidly attached to the body may be assumed as a fixed point to which to refer the motions of the other points. Such a point is called a *centre of rotation.*

85. **Axis of Rotation.**—THEOREM. *In every possible change of position of a rigid body, relatively to a fixed centre, there is a line traversing that centre whose direction is not changed.* In fig. 30, let O be the centre of rotation, and let A and B denote any two other points in the body, whose situations relatively to O are, before the turning, A_1, B_1, and after the turning, A_2, B_2. Join $\overline{A_1 A_2}$, $\overline{B_1 B_2}$, forming the isosceles triangles $O A_1 A_2$, $O B_1 B_2$. Bisect the bases of those triangles in C

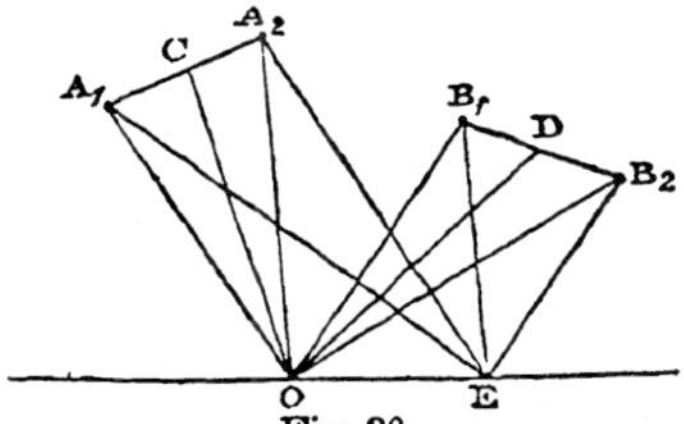

Fig. 30.

and D respectively, and through the points of bisection draw two planes perpendicular to the respective bases, intersecting each other in the straight line $\overline{OE}$, which must traverse O. Let E be any point in the line O E; then E A_1 A_2, and E B_1 B_2, are isosceles triangles; and E is at the same distance from O, A, and B, before and after the turning; therefore E is one and the same point in the body, whose place is unchanged by the turning; and this demonstration applies to every point in the straight line $\overline{OE}$; therefore that line is unchanged in direction.—Q. E. D.

In fig. 31, the same construction and reasoning being applied, the point E being supposed vertically above or below the point O, it is evident that the planes through O D, and O C intersect, and the axis will be represented by a straight line perpendicular to the plane of the paper through O and E.

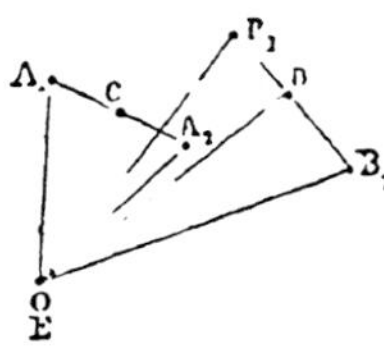

Fig. 31.

COROLLARY. It is evident that every line in the body, parallel to the axis, has its direction unchanged.

86. **The Plane of Rotation** is any plane perpendicular to the axis, such as any plane parallel to the plane of the paper, in fig. 31. **The Angle of Rotation,** or angular motion, is the angle made by the two directions, before and after the turning, of a line perpendicular to the axis, as A_1 O A_2, or B_1 O B_2, in fig. 31.

87. **The Angular Velocity** of a turning body is the ratio of the angle of rotation, expressed in terms of radius, to the number of units of time in the interval of time occupied by the angular motion. Speed of turning is sometimes expressed also by the number of turns or fractions of a turn in a given time. The relation between these two modes of expression is the following:—Let a be the angular velocity, as above defined, and T the turns in the same unit of time; then

$$T = \frac{a}{2\pi};$$

$$a = 2\pi T;$$

$$\left(2\pi = 6{\cdot}2831852 = \frac{710}{113}.\right)^*$$

88. **Uniform Rotation** consists in uniformity of the angular

* The value of π may be easily remembered by taking the first three odd numbers twice each, and placing the six in a row, using the first three as the denominator, and the last three as the numerator of a fraction: we thus obtain 113 | 355 $=\frac{355}{113}$; this is a nearer approximation than 3·14159, and is generally much more easily employed in calculation.

velocity of the turning body, and constancy of the direction of its axis of rotation.

89. **Rotation common to all Parts of Body.**—Since the angular motion of rotation consists in the change of direction of a line in a plane of rotation, and since that change of direction is the same how short soever the line may be, it is evident that the condition of rotation, like that of translation, is common to every particle, how small soever, of the turning rigid body, and that the angular velocity of turning of each particle, how small soever, is the same with that of the entire body. This is otherwise evident by considering, that each part into which a rigid body can be divided turns completely about in the same time with every other part, and with the entire body.

90. **Right and Left-Handed Rotation.**—The *direction* of rotation round a given axis is distinguished in an arbitrary manner into *right-handed* and *left-handed.* One end of the axis is chosen, as that from which an observer is supposed to look along the direction of the axis towards the rotating body. Then if the body seems to the observer to turn in the same direction in which the sun seems to revolve to an observer north of the tropics, or in that in which the hands of a watch or clock revolve, the rotation is said to be *right-handed;* if in the contrary direction, *left-handed:* and it is usual to consider the angular velocity of right-handed rotation to be positive, and that of left-handed rotation to be negative; but this is a matter of convenience. It is obvious that the same rotation which seems right-handed when looked at from one end of the axis, seems left-handed when looked at from the other end.

91. **Relative Motion of a Pair of Points in a Rotating Body.**—Let O and A denote any two points in a rotating body; and considering O as fixed, let it be required to determine the motion of A relatively to an axis of rotation drawn through O. On that axis let fall a perpendicular from A; let r be the length of that perpendicular. Then the motion of A relatively to the axis traversing O is one of *revolution,* or *translation in a circular path of the radius r;* the centre of that circular path being at the point where the perpendicular from A meets the axis. If a be the angular velocity of the body, that is, the velocity of a point situate at the distance unity from the axis of rotation, then the *velocity* of A relatively to the axis traversing O is

$$v = a\, r; \qquad \text{(1.)}$$

and the *direction* of that velocity is at each instant perpendicular to the plane drawn through A and the axis. The *rate of deviation* of A in its motion relatively to the given axis is

$$\frac{v^2}{r} = a^2\, r, \qquad \text{(2.)}$$

in which the first expression is that already found in Article 77, and the second is deduced from the first by the aid of Equation 1 of this Article. It is evident that for a given rotation the motion of O relatively to an axis of rotation traversing A is exactly the same with that of A relatively to a parallel axis traversing O; for it depends solely on the angular velocity a, the perpendicular distance r of the moving point from the axis, and the direction of the axis; all which are the same in either case.

r is called the *radius-vector* of the moving point.

92. **Cylindrical Surface of Equal Velocities.**—If a cylindrical surface of circular cross section be described about an axis of rotation, all the points in that surface have equal velocities relatively to the axis, and the direction of motion of each point in the cylindrical surface relatively to the axis is a tangent to the surface in a plane perpendicular to the axis.

93. **Comparative Motions of Two Points relatively to an Axis.**—Let O, A, B, denote three points in a rotating rigid body; let O be considered as fixed, and let an axis of rotation be drawn through it. Then the *comparative* motions of A and B relatively to that axis are expressed as follows:—*The velocity-ratio is that of the radii-vectores of the points, and the directional relation consists in the angle between their directions of motion being the same with that between their radii-vectores.* Or symbolically: Let r_1, r_2, be the perpendicular distances of A and B from the axis traversing O, and v_1 and v_2 their velocities; then

$$\frac{v_2}{v_1} = \frac{r_2}{r_1}; \text{ and } \widehat{v_1 v_2} = \widehat{r_1 r_2}.$$

94. **Components of Velocity of a Point in a Rotating Body.**—The component parallel to an axis of rotation, of the velocity of a point in a rotating body relatively to that axis, is null. That velocity may be resolved into components in the plane of rotation. Thus let O, in fig. 32, represent an axis of rotation of a body whose plane of rotation is that of the figure; and let A be any point in the body whose radius-vector is $\overline{OA} = r$. The velocity of that point being $v = a r$ (a representing the velocity of a point situated at the distance unity from the axis of rotation), let that velocity be represented by the line $\overline{AV}$ perpendicular to O A. Let B A be any direction in the plane of rotation, along which it is desired to find the component of the velocity of A; and let $\angle$ V A U $= \theta$ be the angle made by that line with A V. From V let fall V U perpendicular

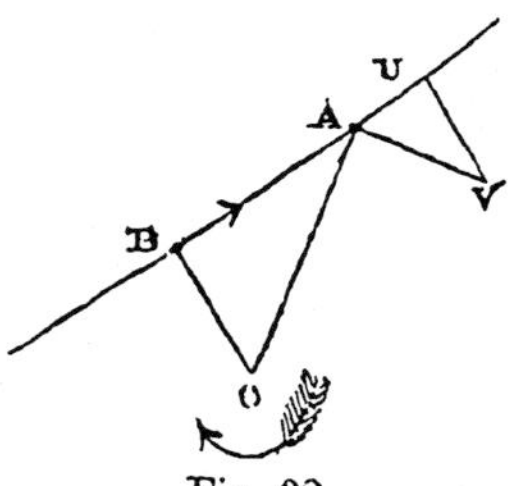

Fig. 32.

to B A; then $\overline{AU}$ represents the component in question; and denoting it by u,

$$u = v \cdot \cos \theta = a r \cdot \cos \theta \ldots\ldots\ldots\ldots\ldots\ldots (1.)$$

From O let fall O B perpendicular to B A. Then $\angle$ A O B $=$ $\angle$ V A U $= \theta$; and the right-angled triangles O B A and A U V are similar; so that

$$\overline{AV} : \overline{AU} :: \overline{OA} : \overline{OB} = r \cos \theta \ldots\ldots\ldots\ldots (2.)$$

Now the *entire* velocity of B relatively to the axis O is

$$a r \cos \theta = u, \ldots\ldots\ldots\ldots\ldots\ldots\ldots\ldots (3.)$$

so that *the component, along a given straight line in the plane of rotation, of the velocity of any point in that line, is equal to the velocity of the point where a perpendicular from the axis meets that line.*

SECTION 3.—COMBINED ROTATIONS AND TRANSLATIONS.

95. **Property of all Motions of Rigid Bodies.**—The foregoing proposition may be regarded as a particular case of the following, which is true of all motions of a rigid body.

The components, along a given straight line in a rigid body, of the velocities of the points in that line relatively to any point, whether in or attached to the body or otherwise, are all equal to each other; for otherwise, the distances between points in the given straight line must alter, which is inconsistent with the idea of rigidity.

96. **Helical Motion.**—Rotation is the only movement which a rigid body as a whole can have relatively to a point belonging to it or attached to it. But if the motion of the body be determined relatively to a point not attached to it, a translation may be combined with the rotation. When that translation takes place in the direction of the axis of rotation, the motion of the rigid body is said to be *helical*, or *screw-like*, because each point in the rigid body describes a helix or screw, or a part of a helix or screw.

Let v_1 denote the velocity of translation, parallel to the axis of rotation, which is common to all points of the body; this is called the *velocity of advance.* The advance during one complete turn of the rotating body is the *pitch* of each of the helical or screw-like paths described by its particles; that is, the distance, in a direction parallel to the axis, between one turn of each such helix and the next; and a being the angular velocity, so that $\frac{2\pi}{a}$ is the time of one turn (2π being the space traversed in one turn by a point at the distance unity from the axis), the value of the pitch (or the space passed over, which is equal to the product of the velocity and time) is

$$p = \frac{2\pi v_1}{a}; \text{ whence } v_1 = \frac{a p}{2\pi} \dots\dots\dots\dots (1.)$$

Let r, as before, be the radius-vector of any point in the body, and let

$$v_2 = a r \dots\dots\dots\dots (2.)$$

denote its *velocity of revolution*, or velocity relatively to the axis, due to the rotation alone. Then the *resultant* velocity of that point is

$$v = \sqrt{v_1^2 + v_2^2} = a \cdot \sqrt{\left\{ \frac{p^2}{4\pi^2} + r^2 \right\}} \dots\dots\dots (3.)$$

The *inclination* of the helix described by that point to the *plane of rotation* is given by the equation

$$i = \text{arc} \cdot \tan \cdot \frac{v_1}{v_2} = \text{arc} \cdot \tan \cdot \frac{p}{2\pi r}; \dots\dots\dots\dots (4.)$$

that is, an angle whose tangent is equal to v_1 divided by v_2, or to p divided by $2\pi r$, the tangent of that angle being the ratio of the pitch to the circumference of the circle described by the point relatively to the axis of rotation.

97. **Problem.—To find the Motion of a Rigid Body from the Motions of Three of its Points.**—Let A, B, C, fig. 33, be three points in a rigid body, and at a given instant let them have motions relatively to a point independent of the body, which motions are represented in velocity and direction by the three lines $A V_a$, $B V_b$, $C V_c$. It is required to find the motion of the entire rigid body relatively to the same fixed point.

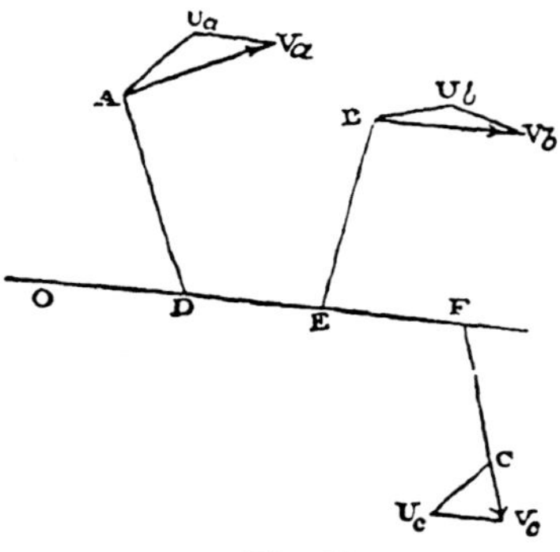

Fig. 33.

Through any point o, fig. 34, draw three lines oa, ob, oc, equal and parallel to the three lines $\overline{A V_a}$, $\overline{B V_b}$, $C V_c$. Through a, b, and c, draw a plane $a\,b\,c$, on which let fall a perpendicular $o\,n$ from o. Then $o\,n$ represents a component, which is common to the velocities of all the three points A, B, C, and must therefore be common to all the points in the body; that is, it is a *velocity of translation*.

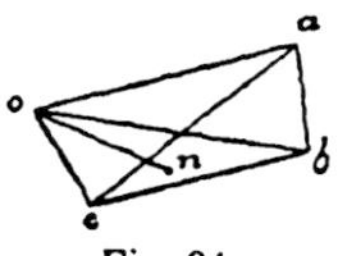

Fig. 34.

From the points V_a, V_b, V_c, draw lines $\overline{V_a U_a}$, $\overline{V_b U_b}$, $\overline{V_c U_c}$, equal and parallel to $o\,n$, but opposite in direction to it; and join

A U_a, B U_b, C U , which will all be parallel to the same plane; that is, to the plane *a b c*. The last three lines will represent the component velocities which, along with the common vel city of translation parallel to *o n*, make up the resultant velocities of the three points. Through the point A draw a plane perpendicular to the comp nent of its motion, which is parallel to *a b c*; that is, to A U_a, and through B lraw a plane perpendicular to B U_b. These two planes will interse t each other in a line O D E, which will be parallel to *o n*. The perpendicular distances of that line from the points A B being unchanged by the motion, it represents one and the same line in or attached to the rigid b dy, and it is therefore the axis of rotation. A plane drawn through the third point, C, perpendicular to C U_c, will cut the other two planes in the same axis: the three revolving c mponent velocities

$$A U_a, B U_b, C U_c,$$

will be respectively proportional to the perpendicular distances, or *radii vect res*, A D, B E, C F,

f the three points from that axis; and the angular velocity will be equal to each of the three quotients made by dividing the revolving component velocities of the p ints by their respective radii vectores. This rotation, combined with a translation parallel to the axis, with a velocity represented by *o n*, constitutes a *helical* or *screw-like* motion, being the required motion of the rigid body.—Q. E. I.

98. Special Cases of the preceding problem occur, in which either a m re simple method of solution is sufficient, or the general method fails, and a special method has to be employed.

I. *When the m ti ns of the points of the b dy are known to be all parallel t one plane*, it is sufficient to know the motions of two points, such as A, B, fig. 35. Let A O, B O, be two planes traversing A and B, and perpendicular to the respective directions of the simultaneous velocities of those points; if those planes cut each other, the entire moti n is a rotation; the line of intersection of the planes O, being the axis of rotation, and the angular velocity, are found as in the last Article. If the two planes are parallel, the motion is a translation.

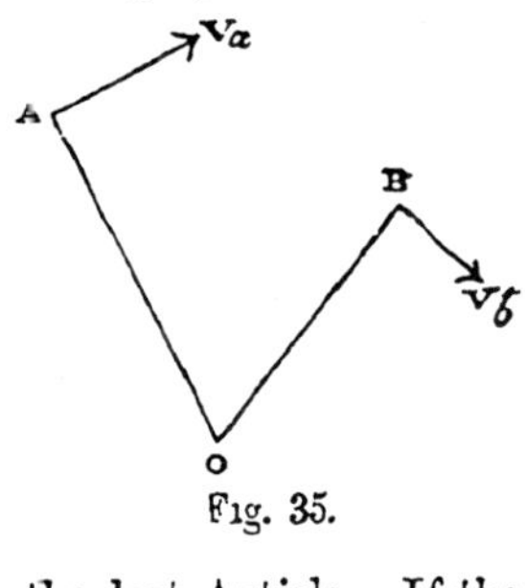

Fig. 35.

II. *If three points in the same plane have parallel motions oblique to the plane*, the motion is a translation.

III. *If three points in the same plane move perpendicularly to the plane*, as A B C, fig. 35 *a*, then if their velocities are equal, the motion is a translation; and if their velocities are unequal, the motion is a rotation about the axis which is the intersection of the plane of the three points with the plane drawn through the extre-

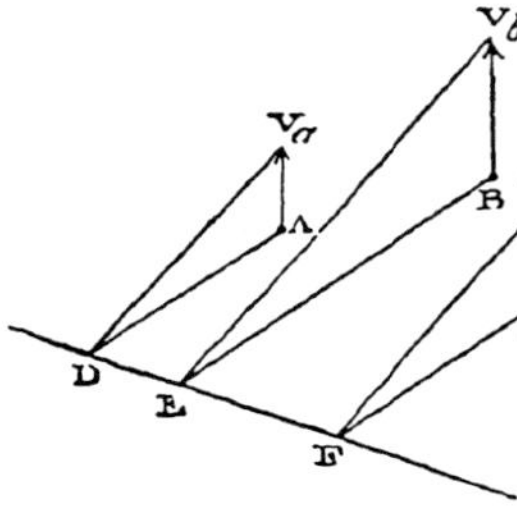

Fig. 35 *a.*

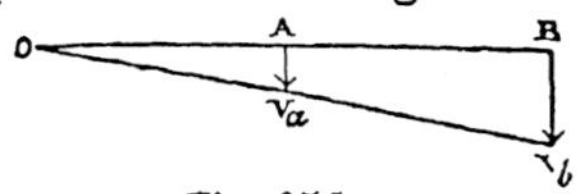

Fig. 35 *b.*

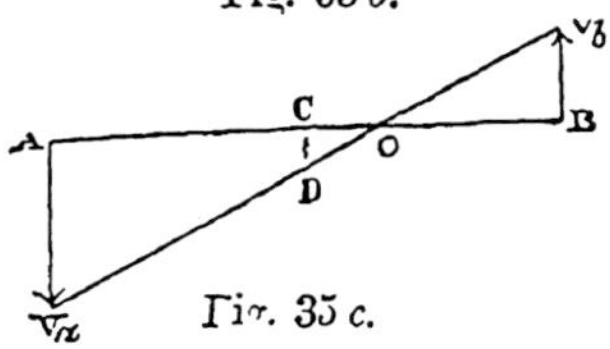

Fig. 35 *c.*

mities of the three lines which represent their velocities viz., through the points, V_a, V_b, V_c; the angular velocity being found as in Article 97.

If the plane of rotation is known, then the simultaneous velocities of two points, as A and B in figs. 35 *b* and 35 *c*, are sufficient to determine the axis O.

99. **Rotation Combined with Translation in the Same Plane.**—

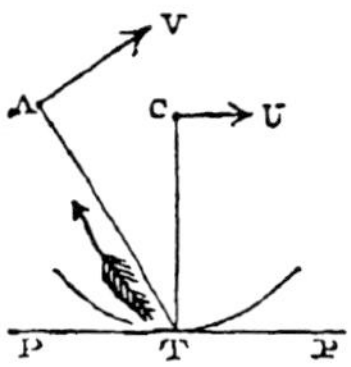

Fig. 36.

Let a body rotate about an axis C (fig. 36), fixed relatively to the body, with an angular velocity *a*, and at the same time let that axis have a motion of translation in a straight path perpendicular to the direction of the axis, with the velocity *u*, represented by the line C U. It is required to find the velocity and direction of motion of any point in the body. From the moving axis draw a straight line C T perpendicular to that axis and to C U, and in that direction into which the rotation (as represented by the feathered arrow) tends to turn C U, and make

$$CT = \frac{u}{a} \quad \text{.............................} \quad (1.)$$

Then the point T has, in virtue of *translation along with* the axis C, a *forward* motion with the velocity u; and in virtue of *rotation about* that axis, it has a *backward* motion with the velocity

$$a \cdot CT = u,$$

equal and opposite to the former; and its resultant velocity is 0. Hence every point in the body, which comes in succession into the position T, situated at the distance $\frac{u}{a}$ from the axis C in the direction above described, is *at rest at the instant of its arriving at that position;* that is, it has just ceased to move in one direction, and is about to move in another direction; and this is true of every point which arrives at *a line traversing* T *parallel to* C. Consequently the resultant motion of the body, at any given instant, is the same as if it were rotating about the line which *at the instant in question* occupies the position T, parallel to C, at the distance $\frac{u}{a}$; and that line is called THE INSTANTANEOUS AXIS. To find the motion of any point A in the body at a given instant, let fall the perpendicular A T from that point on the instantaneous axis; then the motion of A is in the direction A V perpendicular to the plane of the instantaneous axis and of the *instantaneous radius vector* A T, and the velocity of that motion is

$$v = a \cdot AT \quad \ldots\ldots\ldots\ldots\ldots\ldots (2.)$$

100. **Rolling Cylinder; Trochoid.**—Every straight line parallel to the moving axis C, in a cylindrical surface described about C with the radius $\frac{u}{a}$, becomes in turn the instantaneous axis. Hence the motion of the body is the same with that produced by the rolling of such a cylindrical surface on a plane P T P parallel to C and to C U, at the distance $\frac{u}{a}$.

The path described by any point in the body, such as A, which is not in the moving axis C, is a curve well known by the name of *trochoid*. The particular form of trochoid called the *cycloid*, is described by each of the points in the rolling cylindrical surface; being such a curve as is described by a nail in the tyre of a revolving wheel.

101. **Plane Rolling on Cylinder; Spiral Paths.**—Another mode of representing the combination of rotation with translation in the same plane as follows:—Let O, fig. 37, be an axis assumed as fixed, about which let the plane O C containing the axis O rotate right-handedly, in the figure), with the angular velocity a. Let a rigid body have, *relatively to the rotating plane*, and in a direction perpendicular to it, a translation with the velocity u. In the plane O C, and at right angles to the axis O, take $OT = \frac{u}{a}$, in such a direction that the velocity

$$u = a \cdot OT,$$

which the point T *in the rotating plane* has at a given instant, shall be in the contrary direction to the equal velocity of translation u, which the rigid body has relatively to the rotating

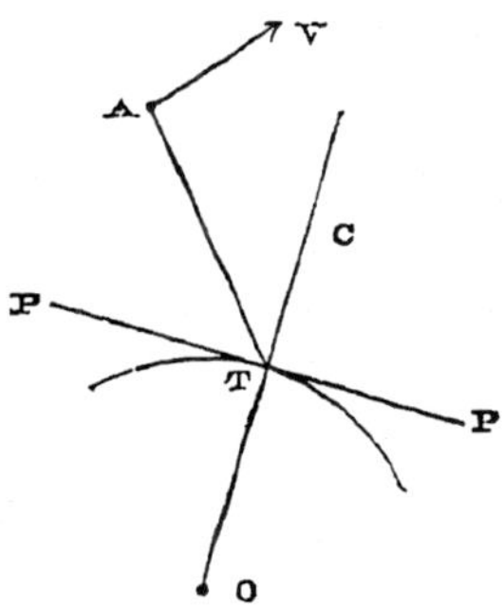

Fig. 37.

plane. Then each point *in the rigid body* which arrives at the position T, or at any position in a line traversing T parallel to the fixed axis O, is at rest *at the instant* of its occupying that position; therefore the line traversing T parallel to the fixed axis O is *the instantaneous axis;* the motion at a given instant of any point in the rigid body, such as A, is at right angles to the radius-vector A T drawn perpendicular to the instantaneous axis; and the velocity of that motion is given by the equation

$$v = a \cdot \mathrm{A\,T}.$$

All the lines in the rigid body which successively occupy the position of instantaneous axis are situated in a plane of that body, P T P, perpendicular to O C; and all the positions of the instantaneous axis are situated in a cylinder described about O with the radius $\overline{\mathrm{O\,T}}$; so that the motion of the rigid body is such as is produced by the *rolling of the plane* P P *on the cylinder whose radius is* $\mathrm{O\,T} = \frac{u}{a}$. Each point in the rigid body, such as A, describes a plane spiral about the fixed axis O. For each point in the *rolling plane,* P P, that spiral is the involute of the circle whose radius is O T. The simplest method of understanding the nature of this curve, is to wrap a cord round the perimeter of a cylinder, placed on a sheet of paper, to attach a tracing point to any point in the cord in juxtaposition with the cylinder, and then to unwrap the cord from the cylinder, keeping the cord always in the same plane parallel to the plane of the paper; the tracing point will trace the involute of a circle on the sheet of paper. For each point whose path of motion traverses the fixed axis O; that is, for each point in a plane of the rigid body traversing O parallel to P P, the spiral is Archimedean, having a radius-vector increasing by the length u for each angle a through which it rotates; this spiral is traced by a point moving uniformly from the centre along the radius, while the radius itself revolves.

102. **Combined Parallel Rotations.** In figs. 38, 39, and 40, let O be an axis assumed as fixed, and O C a plane traversing that axis, and rotating about it with the angular velocity a. Let C be an axis in that plane, parallel to the fixed axis O; and about the moving axis C let a rigid body rotate with the angular velocity b *relatively to the plane* O C; and let the directions of the rotations a and b be distinguished by positive and negative signs. The body is said to have the rotations about the parallel axes O and C *combined* or *compounded,* and it is required to find the result of that combination of parallel rotations.

Fig. 38 respresents the case in which a and b are similar in direction; fig. 39, that in which a and b are in opposite directions, and b is the greater; and fig. 40, that in which a and b are in opposite directions, and a is the greater.

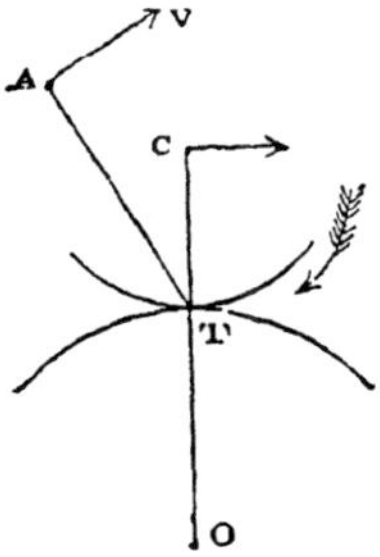

Fig. 38.

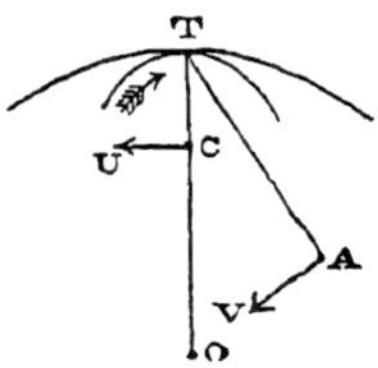

Fig. 39.

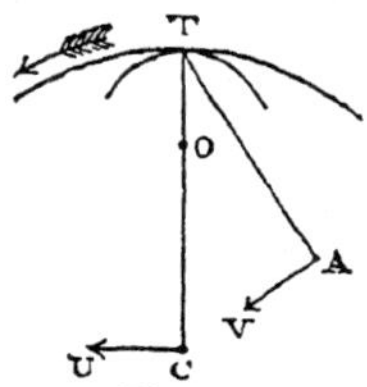

Fig. 40.

Let a common perpendicular O C to the fixed and moving axes be intersected in T by a straight line parallel to both these axes, in such a manner that the distances of T from the fixed and moving axes respectively shall be inversely proportional to the angular velocities of the component rotations about them, as is expressed by the following proportion:

$$a : b :: \mathrm{CT} : \mathrm{OT} \ldots\ldots\ldots\ldots \quad 1.$$

When a and b are similar in direction, let T fall between O and C, as in fig. 38; when they are contrary, beyond, as in figs. 39 and 40. Then the velocity of the line T *of the plane* O C is $a \cdot \mathrm{OT}$; and the velocity of the line T *of the rigid body, relatively to the plane* O C, is $b \cdot \mathrm{CT}$, equal in amount and contrary in direction to the former; therefore each line of the rigid body which arrives at the position T is at rest at the instant of its occupying that position, and is then *the instantaneous axis.* The *resultant angular velocity* is given by the equation

$$c = a + b; \ldots\ldots\ldots\ldots (2.)$$

regard being had to the directions or signs of a and b; that is to say, if we now take a and b to represent *arithmetical* magnitudes, and affix explicit signs to denote their directions, the direction of c will be the same with that of the greater; the case of fig. 38 will be represented by Equation 2, already given; and those of figs. 39 and 40 respectively by

$$c = b - a; \ c = a - b \ldots\ldots\ldots\ldots (2\ \mathrm{A.})$$

The relative proportions of a, b, and c, and of the distances between the fixed, moving, and instantaneous axes, are given by the equation

$$a : b : c :: \overline{\mathrm{CT}} : \mathrm{OT} : \mathrm{OC} \ldots\ldots\ldots\ldots (3.)$$

The motion of any point, such as A, in the rigid body, is at each instant at right angles to the radius-ve tor A T drawn from the point perpendicular to the instantaneous axis; and the velocity of that motion is

$$v \quad c \cdot \mathrm{A\,T} \ldots\ldots\ldots\ldots\ldots\ldots\ldots\ldots \text{(4.)}$$

103. **Cylinder Rolling on Cylinder; Epitrochoids.**—All the lines in the rigid body which successively occupy the position of instantaneous axis are situated in a cylindrical suıface described about C with the radius C T; and all the p sitions of the instantaneous axis are contained in a cylindıical surface described about O with the radius O T; therefore the resultant motion of the rigi l body is that which is produced by r lling the former cylinder, attached to the body, on the latter cylınder, considered as fixed.

In fig. 38, a convex cylinder rolls on a c nvex cylinder; in fig. 39, a smaller convex cylinder r lls in a larger concave cylinder; in fig. 40, a larger c ncave cylinder rolls on a smaller convex cylinder.

Each point in the ıolling rigid b dy traces, relatively to the fixed axis, a curve of the kind called *epitroch ids*. The ep itrochoid traced by a point in the surface of the rolling cylınder is an *epicycloid*.

In certain cases, the epitr choids become curves of a more simple class. For example, each point in the *m ving axis* C traces a circle.

When a cylinder, as in fig. 39, r lls within a concave cylinder of *double its radius*, each p int in the surface of the rolling cylinder moves backwaı ds and foı wards in a straight line, being a diameter of the fixed cylin ler; each point in the axis of the ıolling cylinder traces a circle of the same ıadius with that cylinder, and each other point in or attached to the r lling cylinder traces an ellipse of greater or less eccentricity, having its centre in the fixed axis O.

In the examples shewn in figs. 41, 42, and 43 the ratio of the rolling-circle to the base circle* is $\frac{1}{3}$, so that the epitr cl ids are three-lobed. Each figure shews an external and an internal epitrochoid, traced by rolling the rolling-circle outside and inside the base circle respectively. The centres of the base circles are marked A; those of the external rolling circles, B; those of the internal r lling-circles, *b*; and the tracing points of the external and internal rolling circles are marked C and *c* respectively.

In fig. 41 the tracing points are in the circumferences of the rolling circles; and the cuı ves traced are ep icycloids, distinguished by having *cusps* at the points where the tracing-point coincides with the base-circle. In fig. 42 the tracing points are inside the rolling-circles; and the curves traced are *pr late epitrochoids*, dis tinguished by their wave like form. In fig. 43 the tracing-points

* The fixed cırcle is called a base circle.

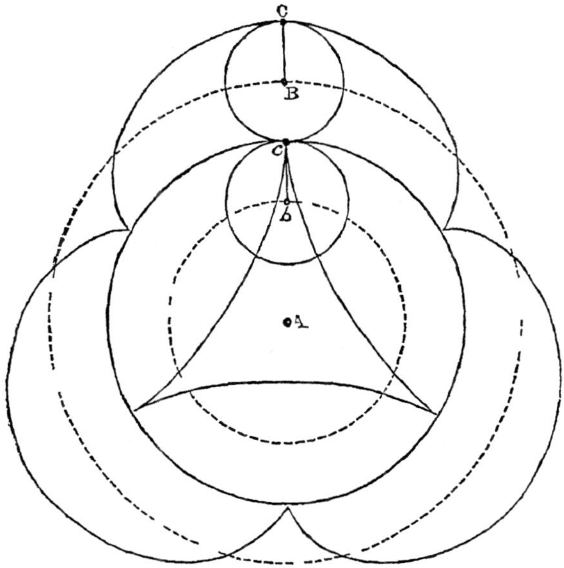

Fig, 41.

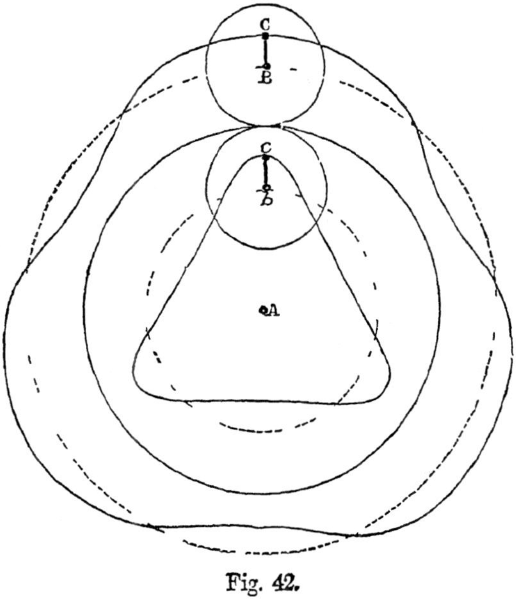

Fig. 42.

are outside the rolling-circles; and the curves traced are *curtate epitrochoids,* distinguished by their looped form.

An important property of curves traced by rolling is that at

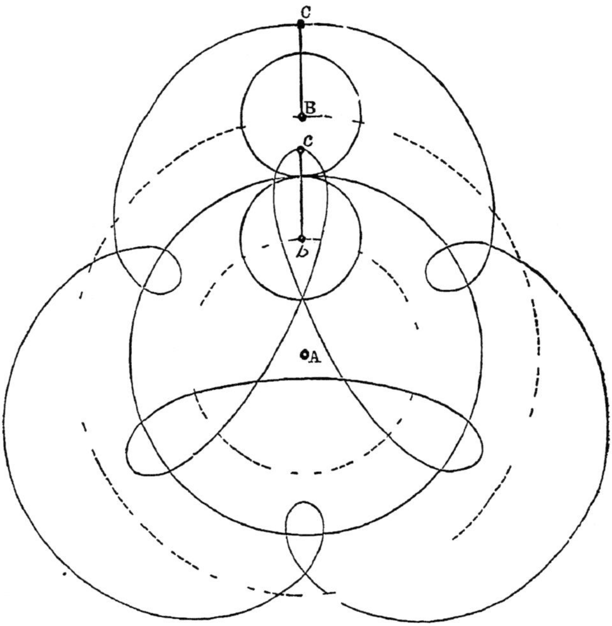

Fig. 43.

every instant the straight line j ining the tracing-point and the pitch-point, or point of c ntact of the rolling curve and base-curve, is normal to the traced curve at the tracing point.

The distance B C or $b\,c$ may in each case be called the *tracing-arm.*

In mechanism f r the tracing of epitrochoids (used chiefly in ornamental turning), the r lling and base-circles are the pitch-circles of a pair of spur wheels, made with great accuracy.

Elliptic paths traced by rolling form a particular case of internal epitrochoids. In fig. 44 is represented a rolling circle, which rolls inside a base-circle of exactly twice its radius. Then considering a quarter of a revolution at a time), while the centre of the rolling-circle traces a quadrant, B b, of an equal circle about A, a point D in the circumference of the rolling-circle traces a straight line traversing A, and a point C, inside the rolling circle, traces a quadrant, C c, of an ellipse whose semiaxes are A C A B + B C, and A c C D A B − B C; also a p int C′ outside the rolling-circle, but rigidly attached to it, traces a quadrant, C′ c′, of an ellipse whose semiaxes are A C B C + A B, and A c′ C′ D B C − A B. The former may be called an *internal*, and the latter an *external*, ellipse. The proportions of the axes of either of them

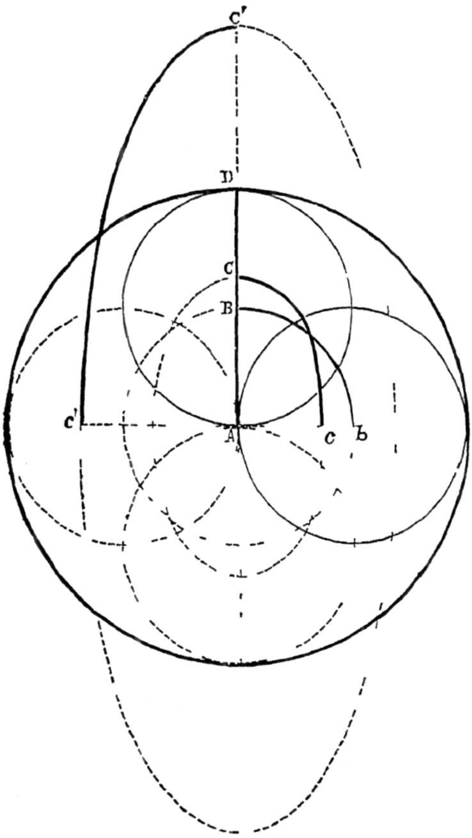

Fig. 44.

may be indefinitely varied by adjusting the position of the tracing point; but in every internal ellipse the sum, and in every external ellipse the difference, of the semiaxes is equal to the diameter of the rolling-circle; that is, to the radius of the base-circle.

This is the principle of the mechanism commonly used for turning ellipses.

It is evident that by having a number of tracing-points carried by one rolling-circle, several ellipses differently proportioned and in different positions may be traced at the same time.

104. Equal and Opposite Parallel Rotations Combined.—Let a plane O C rotate with an angular vel city a about an axis O c n-tained in the plane, and let a rigid body rotate about the axis C in that plane parallel to O, with an angular velocity $-a$, equal and opposite to that of the plane. Then the angular velocity of the rigid body is nothing; that is, its motion is one of *translati n* only, all its points moving in equal circles of the radius O C, with the velocity $a \cdot$ O C. This case is not capable of being represented by a rolling acti n.

105. Rotations about Intersecting Axes Combined. In fig. 45, let O A be an axis assumed as fixed; and about it let the plane A O C rotate with the angular velocity a. Let O C be an axis in the rotating plane; and about that axis let a rigid body rotate with the angular velocity b relatively to the rotating plane.

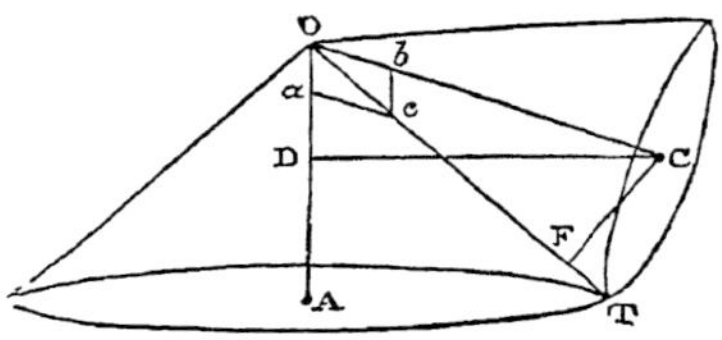

Fig. 45.

Because the point O in the rigid body is fixed, the instantaneous axis must traverse that point. The direction of that axis is determined, as before, by considering that each point which arrives at that line must have, in virtue of the rotation ab ut O C, a velocity relatively to the rotating plane, equal and directly opposed to that which the coincident point of the rotating plane has. Hence it follows, that the ratio of the perpendicular distances of each point in the instantaneous axis from the fixed and moving axes respe tively—that is, the ratio of the sines of the angles which the instantaneous axis makes with the fixed and moving axes must be the reciprocal of the ratio of the comp nent angular velocities about those axes; or symbolically, if O T be the instantane us axis,

$$\sin \text{A O T} : \sin \text{C O T} : : b : a \text{.................(1.}$$

This determines the direction of the instantaneous axis, which may also be found by graphic construction as follows: On O A take O a proportional to a; and on O C take O b proportional to b. Let those lines be taken in such directions, that to an observer looking from their extremities towards O, the component rotations seem both right-handed. Complete the parallelogram O b c a; the diagonal O c will represent the direction of the instantaneous axis.

The resultant angular velocity about this instantaneous axis is found by considering, that if C be any point in the moving axis, the linear velocity of that point must be the same, whether computed from the angular velocity a of the rotating plane about the fixed axis O A, or from the resultant angular velocity c of the rigid body about the instantaneous axis. That is so say, let C D, C E, be perpendiculars from C upon O A, O T, respectively; then

$$a \cdot \mathrm{CD} \quad c \cdot \mathrm{CE};$$

but C D : C E : : sin ∠ A O C : sin ∠ C O T; and therefore

$$\sin \angle \mathrm{COT} : \sin \angle \mathrm{AOC} :: a : c;$$

and, combining this proportion with that given in Equation 1, we obtain the following proportional equation:—

$$\left. \begin{array}{ccccccc} & \sin \angle \mathrm{COT} & : & \sin \angle \mathrm{AOT} & : & \sin \angle \mathrm{AOC} \\ :: & a & : & b & : & c \\ :: & \mathrm{O}a & : & \mathrm{O}b & : & \mathrm{O}c \end{array} \right\} \ldots\ldots (2)$$

that is to say, *the angular velocities of the component and resultant rotations are each proportional to the sine of the angle between the axes of the other two;* and *the diagonal of the parallelogram* O b c a *represents both the direction of the instantaneous axis and the angular velocity about that axis.*

106. **Rolling Cones.**—All the lines which successively come into the position of instantaneous axis are situated in the surface of a cone described by the revolution of O T about O C; and all the positions of the instantaneous axis lie in the surface of a cone described by the revolution of O T about O A. Therefore the motion of the rigid body is such as would be produced by the rolling of the former of these cones upon the latter.

It is to be understood, that either of the cones may become a flat disc, or may be hollow, and touched internally by the other. For example, should ∠ A O T become a right angle, the fixed cone would become a flat disc; and should ∠ A O T become obtuse, that cone would be hollow, and would be touched internally by the rolling cone; and similar changes may be made in the rolling cone.

The path described by a point in or attached to the rolling cone is a *spherical epitrochoid;* but for the purposes of the present treatise, it is unnecessary to enter into details respecting the properties of that class of curves.

107. **Comparative Motions in Compound Rotations.** The velocity ratio of two points in a rotating rigid body at any instant, is that of their perpendicular distances from its instantaneous axis; and the angle between the directions of motion of the two points is equal to that between the two planes which traverse the points and the instantaneous axis.

Section 4.—Varied Rotation.

108. **Variation of Angular Velocity** is measured like variation of linear velocity, by comparing the change which takes place in

the angular velocity of a rotating body, Δa, during a given interval of time, with the length of that interval, Δt, and the *rate of variation* is the value towards which the ratio of the change of angular velocity to the interval of time, $\frac{\Delta a}{\Delta t}$, converges, as the length of the interval is indefinitely diminished; being represented by $\frac{d a}{d t}$, and found by the operation of differentiation.

109. **Components of Varied Rotation.** The most convenient way, in most cases, of expressing the mode of variation of a rotatory motion, is to resolve the angular velocity at each instant into three component angular velocities about three rectangular axes fixed in direction. The values of these components, at any instant shew at once the resultant angular velocity and the direction of the instantaneous axis. For example, let a_x, a_y, a_z, be the rectangular components of the angular velocity of a rigid body at a given instant,—

rotation about x from y towards z,
about y from z towards x,
and about z from x towards y,

being considered as positive; then

$$a = \sqrt{a_x^2 + a_y^2 + a_z^2} \quad \ldots\ldots\ldots\ldots\ldots\ldots(1.)$$

is the resultant angular velocity, and

$$\cos \alpha = \frac{a_x}{a}; \ \cos \beta = \frac{a_y}{a}; \ \cos \gamma = \frac{a_z}{a}; \ldots\ldots\ldots\ldots(2.$$

are the cosines of the angles which the instantaneous axis makes with the axis of x, y, and z respectively.

CHAPTER III.

MOTIONS OF PLIABLE BODIES, AND OF FLUIDS.

110. **Division of the Subject.**—The subject of the present chapter will be considered under the following branches:—

I. The Motions of Flexible Cords.
II. The Motions of Fluids not altering in Volume.
III. The Motions of Fluids altering in Volume.

SECTION 1.—MOTIONS OF FLEXIBLE CORDS.

111. **General Principles.**—As those relative motions of the points of a cord which may arise from its extensibility, belong to the subject of resistance to tension, which is a branch of that of strength and stiffness, the present section is confined to those motions of which a flexible cord is capable when the length, not merely of the whole cord, but of each part lying between two points fixed in the cord, is invariable, or sensibly invariable.

In order that the figure and motions of a flexible cord may be determined from cinematical considerations alone, independently of the magnitude and distribution of forces acting on the cord, its weight must be insensible compared with the tension on it, and it must everywhere be *tight;* and when that is the case, each part of the cord which is not straight is maintained in a curved figure by passing over a *convex* surface. The line in which a tight cord lies on a convex surface is the *shortest line* which it is possible to draw on that surface between each pair of points in the course of the cord. (It is a well known principle of the geometry of curved surfaces, that the *osculating plane* or *tangential plane* at each point of such a line is perpendicular to the curved surface.)

Hence it appears, that the motions of a tight flexible cord of invariable length and insensible weight are regulated by the following principles:—

I. *The length between each pair of points in the cord is constant.*

II. *That length is the shortest line which can be drawn between its extremities over the surfaces by which the cord is guided.*

112. **Motions Classed.**—The motions of a cord are of two kinds—

I. Travelling of a cord along a track of invariable form; in which case the velocities of all points of the cord are equal.

II. Alteration of the figure of the track by the motion of the guiding surfaces.

Those two kinds of motion may be combined.

The most usual problems in practice respecting the motions of cords are those in which cords are the means of transmitting motion between two pieces in a train of mechanism. Such problems will be considered in Part II. of this treatise.

Next in point of frequency in practice are the problems to be considered in the ensuing Article.

113. Cord Guided by Surfaces of Revolution.—Let a cord in some portions of its course be straight, and in others guided by the surfaces of circular drums or pulleys, over each of which its track is a circular arc in a plane perpendicular to the axis of the guiding surface. Let r be the radius of any one of the guiding surfaces, i the angle of inclination which the two straight portions of the cord contiguous to that surface make with each other, expressed in length of arc to radius unity. Then the length of the portion of the cord which lies on that surface is $r\,i$; and if s be the length of any straight portion of the cord, the total length between two given points fixed in the cord may be expressed thus:—

$$L = \Sigma\cdot s+\Sigma\cdot ri \qquad (1.)$$

Let c be the distance between the centres of a given adjacent pair of guiding surfaces, s the length of the straight portion of cord which lies between them, and r, r', their respective radii; then evidently

$$s = \sqrt{c^2-(r\pm r')^2} \qquad (2.)$$

the $\left\{\begin{matrix}\text{sum}\\\text{difference}\end{matrix}\right\}$ of the radii being employed, according as the cord $\left\{\begin{matrix}\text{crosses}\\\text{does not cross}\end{matrix}\right\}$ the line of centres c.

The case most common in practice is that in which the *plies*, or straight parts of the cord, are all parallel to each other; so that $i = 180°$ in each case, while a certain number, n, of the guiding bodies or pulleys all move simultaneously in a direction parallel to the plies of the cord with the same velocity, u; where u represents the velocity of translation of the guiding surfaces, and v the longitudinal velocity of any point in the cord

$$v = 2nu \qquad (3.)$$

SECTION 2.—MOTIONS OF FLUIDS OF CONSTANT DENSITY.

114. Velocity and Flow.—The density of a moving fluid mass may be either exactly invariable, from the constancy or the adjustment of its temperature and pressure, or sensibly invariable, from the smallness of the alterations of volume which the actual alterations of pressure and temperature are capable of producing. The latter is the case in most problems of practical mechanics affecting liquids.

Conceive an ideal surface of any figure, and of the area A, to be situated within a fluid mass, the parts of which have motion relatively to that surface; and let u denote, as the case may be, the *uniform* velocity, or the *mean* value of the varying velocity, resolved in a direction perpendicular to A, with which the particles of the fluid pass A. Then

$$Q = uA \text{ .... } (1.)$$

is the volume of fluid which passes from one side to the other of the surface A in an unit of time, and is called the *flow*, or *rate of flow*, through A.

When the particles of fluid move obliquely to A, let θ denote the angle which the direction of motion of any particle passing A makes with a normal to A, and v the velocity of that particle; then

$$u = v \cdot \cos \theta \text{ .... } (2.)$$

115. **Principle of Continuity.** Axiom. *When the motion of a fluid of constant density is considered relatively to an enclosed space of invariable volume which is always filled with the fluid, the flow into the space and the flow out of it, in any one given interval of time, must be equal*—a principle expressed symbolically by

$$\Sigma \cdot Q = 0 \text{ .... } (3.)$$

The preceding self evident principle regulates all the motions of fluids of constant density, when considered in a purely cinematical manner. The ensuing articles of this section contain its most usual applications.

116. **Flow in a Stream.**—A stream is a moving fluid mass, indefinitely extended in length, and limited transversely, and having a continuous longitudinal motion. At any given instant, let A, A′, be the areas of any two of its transverse sections, considered as fixed; u, u', the mean normal velocities through them; Q, Q′, the rates of flow through them; then in order that the principle of continuity may be fulfilled, those rates of flow must be equal; that is,

$uA = u'A' = Q = Q' =$ constant for all cross sections of the channel at the given instant; .... (1.)

consequently,

$$\frac{u'}{u} = \frac{A}{A'}; \text{ .... } (2.)$$

or, *the normal velocities at a given instant at two fixed cross sections are inversely as the areas of these sections.*

117. **Pipes, Channels, Currents, and Jets.**—When a stream of

fluid completely fills a *pipe* or *tube*, the area of each cross section is given by the figure and dimensions of the pipe, and for similar forms of section varies as the square of the diameter. Hence the mean normal velocities of a stream flowing in a full pipe, at different cross sections of the pipe, are inversely as the squares of the diameters of those sections.

A *channel* partially encloses the stream flowing in it, leaving the upper surface free; and this description applies not only to channels commonly so called, but to pipes partially filled. In this case the area of a cross section of the stream depends not only on the figure and dimensions of the channel, but on the figure and elevation of the free upper surface of the stream.

A *current* is a stream bounded by other portions of fluid whose motions are different.

A *jet* is a stream whose surface is either free all round, or is touched by a solid body in a small portion of its extent only.

118. **Steady Motion** of a fluid relatively to a given space considered as fixed is that in which the velocity and direction of the motion of the fluid at each *fixed point* is uniform at every instant of the time under consideration; so that although the velocity and direction of the motion of a given particle of the fluid may vary while it is transferred from one point to another, that particle assumes, at each fixed point at which it arrives, a certain definite velocity and direction depending on the position of that point alone; which velocity and direction are successively assumed by each particle which successively arrives at the same fixed point.

The steady motion of a stream is expressed by the two conditions, that the area of each fixed cross section is constant, and that the flow through each cross section is constant; that is to say,

$$\frac{dA}{dt} = 0; \quad \frac{dQ}{dt} = 0 \quad \text{.......................} \quad (1.)$$

If u represents the normal velocity of a fluid moving steadily, *at a given fixed point*, then the differential coefficient of a constant being equal to 0 (see Article 26, page 11),

$$\frac{du}{dt} = 0; \quad \text{...........................} \quad (2.)$$

expresses the condition of steady motion.

119. **Motion of Pistons.**—Let a mass of fluid of invariable volume be enclosed in a vessel, two portions of the boundary of which (called *pistons*) are movable inwards and outwards, the rest of the boundary being fixed. Then, if motion be transmitted between the pistons by moving one inwards and the other outwards, it follows, from the invariability of the volume of the enclosed fluid, that the velocities of the two pistons at each instant

will be to each other in the inverse ratio of the areas of the respective projections of the pistons on planes normal to their directions of motion. This is the principle of the transmission of motion in the *hydraulic press* and *hydraulic crane.*

The *flow* produced by a piston whose velocity is u, and the area of whose projection on a plane perpendicular to the direction of its motion is A, is given, as in other cases, by the equation

$$Q = u\,A \quad \ldots\ldots\ldots\ldots\ldots\ldots\ldots\ldots (1.)$$

Section 3.—Motions of Fluids of Varying Density.

120. **Flow of Volume and Flow of Mass.** In the case of a fluid of varying density, the *volume*, which in an unit of time flows through a given area A, with a normal velocity u, is still represented, as for a fluid of constant density, by

$$Q = A\,u; \quad \ldots\ldots\ldots\ldots\ldots\ldots\ldots\ldots (1.)$$

but the *absolute quantity*, or *mass* of fluid which so flows, bears no longer a constant proportion to that volume, but is proportional to the volume multiplied by the density. The density may be expressed, either in units of weight per unit of volume, or in arbitrary units suited to the particular case. Let ρ be the density; then the *flow of mass* may be thus expressed:—

$$\rho\,Q = \rho\,A\,u \quad \ldots\ldots\ldots\ldots\ldots\ldots\ldots\ldots (2.)$$

121. **The Principle of Continuity,** as applied to fluids of varying density, takes the following form: *the flow into or out of any fixed space of constant volume is that due to the variation of density alone.*

To express this symbolically, let there be a fixed space of the constant volume V, and in a given interval of time let the density of the fluid in it, which in the first place may be supposed uniform at each instant, change from ρ_1 to ρ_2. Then the mass of fluid which at the beginning of the interval occupied the volume V, occupies at the end of the interval the volume $\frac{V\rho_1}{\rho_2}$; and the difference of those volumes is the volume which flows through the surface bounding the space, *outward* if ρ_2 is less than ρ_1, *inward* if ρ_2 is greater than ρ_1. Let $t_2 - t_1$ be the length of the interval of time; then the rate of flow of volume is expressed as follows:—

$$Q = \frac{V\left(\frac{\rho_1}{\rho_2} - 1\right)}{t_2 - t_1} \quad \ldots\ldots\ldots\ldots\ldots\ldots\ldots\ldots (1.)$$

PART II.

THEORY OF MECHANISM.

CHAPTER I.

DEFINITIONS AND GENERAL PRINCIPLES.

122. **Theory of Pure Mechanism Defined.**—*Machines* are bodies, or assemblages of bodies, which transmit and modify motion and force. The word "machine," in its widest sense, may be applied to every material substance and system, and to the material universe itself; but it is usually restricted to works of human art, and in that restricted sense it is employed in this treatise. A machine transmits and modifies motion when it is the means of making one motion cause another; as when the mechanism of a clock is the means of making the descent of the weight cause the rotation of the hands. A machine transmits and modifies force when it is the means of making a given kind of physical energy perform a given kind of work; as when the furnace, boiler, water, and mechanism of a marine steam engine are the means of making the energy of the chemical combination of fuel with oxygen perform the work of overcoming the resistance of water to the motion of a ship. The acts of transmitting and modifying motion, and of transmitting and modifying force, take place together, and are connected by a certain law; and until lately, they were always considered together in treatises on mechanics; but recently great advantage in point of clearness has been gained by first considering separately the act of transmitting and modifying motion. The principles which regulate this function of machines constitute a branch of Cinematics, called the *theory of pure mechanism*. The principles of the theory of pure mechanism having been first established and understood, those of the *theory of the work of machines*, which will form the subject of Part VI. of this work, which regulate the act of transmitting and modifying force, are much more readily demonstrated and apprehended than when the two departments of the theory of machines are mingled. The establishment of the theory of pure mechanism as an independent subject has been mainly ac-

complished by the labours of Professor Willis, whose nomenclature and methods are, to a great extent, followed in this treatise.

123. **The General Problem** of the theory of pure mechanism may be stated as follows:—*Given the mode of connection of two or more movable points or bodies with each other, and with certain fixed bodies; required the comparative motions of the movable points or bodies:* and conversely, *when the comparative motions of two or more movable points are given, to find their proper mode of connection.*

The term "comparative motion" is to be understood as in Articles 70, 81, 93, and 107. In those Articles, the comparative motions of points belonging to one body have already been considered. In order to constitute *mechanism*, two or more bodies must be so connected that their motions depend on each other through cinematical principles alone.

124. **Frame; Moving Pieces; Connectors; Bearings.**—The *frame* of a machine is a structure which supports the *moving pieces*, and regulates the path or kind of motion of most of them directly. In considering the movements of machines mathematically, the frame is considered as fixed, and the motions of the moving pieces are referred to it. The frame itself may have (as in the case of a ship or of a locomotive engine) a motion relatively to the earth, and in that case the motions of the moving pieces relatively to the earth are the resultants of their motions relatively to the frame, and of the motion of the frame relatively to the earth; but in all problems of pure mechanism, and in many problems of the work of machines, the motion of the frame relatively to the earth does not require to be considered.

The *moving pieces* may be distinguished into *primary* and *secondary;* the former being those which are directly carried by the frame, and the latter those which are carried by other moving pieces. The motion of a secondary moving piece relatively to the frame is the resultant of its motion relatively to the primary piece which carries it, and of the motion of that primary piece relatively to the frame.

Connectors are those secondary moving pieces, such as links, belts, cords, and chains, which transmit motion from one moving piece to another, when that transmission is not effected by immediate contact.

Bearings are the surfaces of contact of primary moving pieces with the frame, and of secondary moving pieces with the pieces which carry them. Bearings guide the motions of the pieces which they support, and their figures depend on the nature of those motions. The bearings of a piece which has a motion of translation in a straight line, must have plane or cylindrical surfaces, *exactly straight* in the direction of motion. The bearings of rotating pieces must have surfaces accurately turned to *figures of revolution*, such as cylinders, spheres, conoids, and flat discs. The bearing of a piece whose motion is helical, must be an *exact screw*, of a pitch equal to that of the helical motion (Article 96). The parts of moving pieces which touch the bearings, should have surfaces accurately fitting those of the bearings. They may be distinguished into *slides*, for pieces which move in straight lines, *gudgeons, journals, bushes*, and *pivots*, for those which rotate, and *screws* for those which move helically.

125. **The Motions of Primary Moving Pieces** are limited by the fact, that in order that different portions of a pair of bearing surfaces may accurately fit each other during their relative motion, those surfaces must be either straight, circular, or helical; from which it follows, that the motions in question can be of three kinds only, viz.:

I. *Straight translation*, or *shifting*, which is necessarily of limited extent, and which, if the motion of the machine is of indefinite duration, must be *reciprocating*; that is to say, must take place alternately in opposite directions. See Part I., Chapter II., Section 1.

II. *Simple rotation*, or *turning* about a fixed axis, which motion may be either continuous or reciprocating, being called in the latter case *oscillation*. (See Part I., Chapter II., Section 2.

III. *Helical* or *screw like motion*, to which the same remarks apply as to straight translation. See Part I., Chapter II., Section 3, Article 96.

126. **The Motions of Secondary Moving Pieces** relatively to the pieces which carry them, are limited by the same principles which apply to the motions of primary pieces relatively to the frame. But the motions of secondary moving pieces relatively to the frame may be any motions which can be compounded of straight translations and simple rotations according to the principles already explained in Part I., Chapter II., Section 3.

127. **An Elementary Combination** in mechanism consists of a pair of *primary moving pieces*, so connected that one transmits motion to the other.

The piece whose motion is the cause is called the *driver*; that whose motion is the effect, the *follower*. The *connection* between the driver and the follower may be

I. By *rolling contact* of their surfaces, as in *toothless wheels*.

II. By *sliding contact* of their surfaces, as in *toothed wheels, screws, wedges, cams*, and *escapements*.

III. By *bands* or *wrapping connectors*, such as *belts, cords*, and *gearing-chains*.

IV. By *link-work*, such as *connecting rods, universal joints*, and *clicks*.

V. By *reduplication of cords*, as in the case of ropes and pulleys.

VI. By an *intervening fluid*, transmitting motion between two pistons.

The various cases of the transmission of motion from a driver to a follower are further classified, according as the relation between their *directions* of *motion* is constant or changeable, and according as the ratio of their velocities is constant or variable. This latter principle of classification was employed by Professor Willis, in the first edition of his *Principles of Mechanism*, as the foundation of a primary division of the subject of elementary combinations in mechanism into classes, which are subdivided according to the mode of connection of the pieces. In the present treatise, elementary combinations will be classed primarily according to the mode of connection; which is the classification employed by Professor Willis in the Edition of 1870.

128. **Line of Connection.** In every class of elementary combinations, except those in which the connection is made by reduplication of cords, or by an intervening fluid, there is at each instant a certain straight line, called the *line of connection*, or *line of mutual action* of the driver and follower. In the case of rolling contact, this is any straight line whatsoever traversing the point of contact of the surfaces of the pieces; in the case of sliding contact, it is a line perpendicular to those surfaces at their point of contact; in the case of wrapping connectors, it is the centre line of that part of the connector by whose tension the motion is transmitted; in the case of link work, it is the straight line passing through the points of attachment of the link to the driver and follower.

129. **Principle of Connection.** The line of connection of the driver and follower at any instant being known, their comparative velocities are determined by the following principle:—*The respective linear velocities of a point in the driver, and a point in the follower, each situated anywhere in the line of connection, are to each other inversely as the cosines of the respective angles made by the paths of the points with the line of connection.* This principle might be otherwise stated as follows:—*The components, along the line of connection, of the velocities of any two points situated in that line, are equal.*

130. **Adjustments of Speed.**—The velocity-ratio of a driver and its follower is sometimes made capable of being changed at will, by means of apparatus for varying the position of their line of connection, as when a pair of rotating cones are embraced by a belt which can be shifted so as to connect portions of their surfaces of different diameters.

131. **A Train of Mechanism** consists of a series of moving pieces, each of which is follower to that which drives it, and driver to that which follows it.

132. **Agregate Combinations** in mechanism are those by which compound motions are given to secondary pieces.

CHAPTER II.

ON ELEMENTARY COMBINATIONS AND TRAINS OF MECHANISM.

SECTION 1. ROLLING CONTACT.

133. **Pitch Surfaces** are th se surfaces of a pair of moving pieces, which touch each other when motion is communicated by rolling c ntact. The LINE OF CONTACT is that line which at each instant traverses all the pairs of points of the pair of pitch surfaces which are in contact.

134. **Smooth Wheels, Rollers, Smooth Racks.**—Of a pair of primary moving piec s in rolling contact, b th may rotate, or one may rotate and the other have a motion of sliding, or straight translation. A rotating piece, in rolling c ntact, is called a *smo th wheel*, and sometimes a *roller;* a sliding piece may be called a *smooth rack*.

135. **General Conditions of Rolling Contact.**—The whole of the principles which regulate the motions of a pair of pieces in rolling contact f llow from the single principle, *that each pair of poi its in the pitch surfaces, which are in contact at a given instant, must at that instant be moving in the same directi n with the same velocity;* that this must be the case is evident from the rigidity of the bodies, for did the pair of points vary in vel city, it would follow that there was motion am ng the particles, or in a particle at least, of the body, which is c ntrary to the hypothesis of rigi lity.

The direction of motion of a point in a rotating b dy being per- ndicular to a plane passing through its axis, the condition, that each pair of points in c ntact with each other must move in the same direction leads to the following consequences:

I. That when both pieces rotate, their axes, and all their points of c ntact, lie in the same plane.

II. That when one piece rotates and the other slides, the axis of the rotating piece, and all the points of contact, lie in a plane perpendicular to the direction of motion of the sliding piece.

The conditi n, that the velocities of each pair of points of contact must be equal, leads to the following consequences:—

III. That the angular vel cities of a pair of wheels, in rolling contact, must be inversely as the perpendicular distances of any pair of points of contact from the respective axes.

IV. That the linear velocity of a smo th rack in rolling contact with a wheel, is equal to the product of the angular velocity of the wheel by the perpendicular distance from its axis to a pair of points of contact.

Respecting the line of contact, the above principles III. and IV. lead to the following conclusions:—

V. That for a pair of wheels with parallel axes, and for a wheel and rack, the line of contact is straight, and parallel to the axes or axis; and hence that the pitch surfaces are either plane or cylindrical (the term "cylindrical" including all surfaces generated by the motion of a straight line parallel to itself).

VI. That for a pair of wheels, with intersecting axes, the line of contact is also straight, and traverses the point of intersection of the axes; and hence that the rolling surfaces are conical, with a common apex (the term "conical" including all surfaces generated by the motion of a straight line which traverses a fixed point).

136. **Circular Cylindrical Wheels** are employed when an uniform velocity ratio is to be communicated between parallel axes. Figs. 38, 39, and 40, of Article 102, may be taken to represent pairs of such wheels; C and O, in each figure, being the parallel axes of the wheels, and T a point in their line of contact. In fig. 38, both pitch surfaces are convex, the wheels are said to be in *outside gearing*, and their directions of rotation are contrary. In figs. 39 and 40, the pitch surface of the larger wheel is concave, and that of the smaller convex; they are said to be in *inside gearing*, and their directions of rotation are the same.

To represent the comparative motions of such pairs of wheels symbolically, let

$$\mathrm{OT} = r_1,\ \mathrm{CT} = r_2,$$

be their radii: let OC = c be the *line of centres*, or perpendicular distance between the axes, so that for

$$\left.\begin{matrix}\text{outside}\\ \text{inside}\end{matrix}\right\}\text{gearing, } c = r_1 \pm r_2 \ldots\ldots\ldots\ldots\ldots (1.)$$

Let a_1, a_2, be the angular velocities of the wheels, and v the common linear velocity of their pitch surfaces; then

$$\left.\begin{matrix} v = a_1 r_1 = a_2 r_2; \\ c : r_1 : r_2 :: a_2 \pm a_1 : a_2 : a_1; \end{matrix}\right\} \ldots\ldots\ldots\ldots (2.)$$

the sign $\pm$ applying to $\left\{\begin{matrix}\text{outside}\\ \text{inside}\end{matrix}\right\}$ gearing.

137. **A Straight Rack and Circular Wheel**, which are used when an uniform velocity-ratio is to be communicated between a sliding piece and a turning piece, may be represented by fig. 36 of Article 99, C being the axis of the wheel, P T P the plane surface of the rack, and T a point in their line of contact. Let r be the radius

of the wheel, a its angular velocity, and v the linear velocity of the rack; then

$$v \quad r a.$$

138. **Bevel Wheels**, whose pitch surfaces are frustra of regular cones are used to transmit an uniform angular velo ity-ratio between a pair of axes which intersect each other. Fig. 45 of Article 105 will serve to illustrate this case; O A and O C being the pair of axes, intersecting each other in O, O T the line of c n-tact, an l the cones described by the revolution of O T about O A and O C respectively being the pit h surfaces, of which narrow zones or frustra are used in practice.

Let a_1, a , be the angular velocities ab ut the two axes respectively; and let $i_1 \quad \angle$A O T, $i \quad \angle$C O T, be the angles ma le by those axes respectively with the line of conta t; then fr m the principle III. of Article 135 it follows, that the angul r velocity-rati is

$$\frac{a_2}{a_1} \quad \frac{\sin i_1}{\sin i};\ldots\ldots\ldots\ldots\ldots\ldots\ldots \quad (1.)$$

Which equation serves to find the angular velocity-ratio when the axes and the line of contact are given.

Conversely, let the angle between the axes,

$$\angle \mathrm{A\,O\,C} \quad i_1 \quad i_{\_} \quad j,$$

be given, and also the ratio $\frac{a_n}{a_1}$; then the position of the line of c ntact is given by either of the two following equations: –

$$\left.\begin{array}{l} \sin i_1 \quad \dfrac{a_n \sin j}{\surd\, a_1 + a_{\_} + 2\, a_1\, a_{\_} \cos j)}; \\ \sin i_2 \quad \dfrac{a_1 \sin j}{\surd(a_1^{-} + a_2 + 2\, a_1\, a_{\_} \cos j}; \end{array}\right\}\ldots\ldots\ldots\ldots(2.)$$

which are formed from equation (1 by substituting for i_1 its value $- \; j \quad i$, and for $i_{\_}$ its value $\quad j - i_1$.

As this is the first instance of the use of Trig nometrical analysis, the method of formation of these equations will be explained:

From Equation (1) it follows that—

$$\begin{aligned} \sin i_1 \cdot a_1 \quad & \sin i_2 \cdot a_2 \\ & \sin \; j \quad i_1 \;\cdot a_2 \\ & \sin j \cdot \cos i_1 \cdot a_2 - \cos j \cdot \sin i_1 \cdot a_2 \\ & \sin j \cdot \surd(1 - \sin^2 i_1) \cdot a_2 \quad \cos j \cdot \sin i_1 \cdot a_2. \end{aligned}$$

(See *Trig nometrical Rules*, Sections 19 and 21.)

Squaring both sides, and transp sing

$$\sin^2 j \cdot \sin^2 i_1 \cdot a_2^2 = (\sin i_1 \cdot a_1 + \cos j \cdot \sin i_1 \cdot a_2)^2 = \sin^2 j \cdot a_2^2$$
$$\sin^2 i_1 \cdot a_2^2 - \cos^2 j \cdot \sin^2 i_1 \cdot a_2^2 + \sin^2 i_1 \cdot a_1^2 = \cos^2 j \cdot \sin^2 i_1 \cdot a_2^2$$
$$+ 2 \sin i_1^2 \cdot a_1 \cdot \cos j \cdot a_2 = \sin^2 j \cdot a_2^2$$
$$\sin^2 i_1 \cdot a_2^2 + \sin^2 i_1 \cdot a_1^2 + 2 \sin i_1^2 \cdot a_1 \cdot \cos j \cdot a_2 = \sin^2 j \cdot a_2^2$$
$$\sin^2 i_1 = \frac{\sin^2 j \cdot a_2^2}{a_1^2 + a_2^2 + 2\, a_1 \cdot a_2 \cdot \cos j}$$
$$\therefore \sin i_1 = \frac{a_2 \cdot \sin j}{\sqrt{a_1^2 + a_2^2 + 2\, a_1 \cdot a_2 \cdot \cos j}}.$$

Graphically, the same problem is solved as follows:—On the two axes respectively, take lengths to represent the angular velocities of their respective wheels. Complete the parallelogram of which those lengths are the sides, and its diagonal will be the line of contact. As in the case of the rolling cones of Article 106, one of a pair of bevel wheels may be a flat disc, or a concave cone.

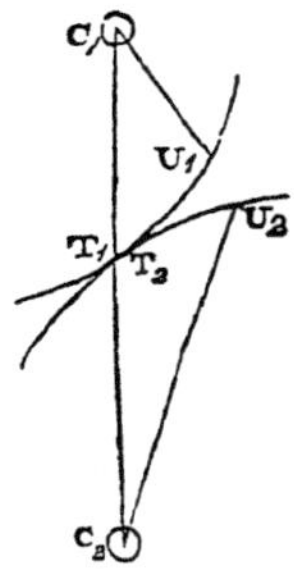

Fig. 46.

139. **Non-Circular Wheels** are used to transmit a variable velocity ratio between a pair of parallel axes. In fig. 46, let C_1, C_2, represent the axes of such a pair of wheels; T_1, T_2, a pair of points which at a given instant touch each other in the line of contact which line is parallel to the axes and in the same plane with them; and U_1, U_2, another pair of points, which touch each other at another instant of the motion; and let the four points, T_1, T_2, U_1, U_2, be in one plane perpendicular to the two axes, and to the line of contact. Then for every such set of four points, the two following equations must be fulfilled:

$$\overline{C_1 U_1} + \overline{C_2 U_2} = \overline{C_1 T_1} + \overline{C_2 T_2} = \overline{C_1 C_2}$$
$$\text{arc } T_1 U_1 = \text{arc } T_2 U_2$$

and those equations shew the geometrical relations which must exist between a pair of rotating surfaces in order that they may move in rolling contact round fixed axes.

SECTION 2.—SLIDING CONTACT.

140. **Skew-Bevel Wheels** are employed to transmit an uniform

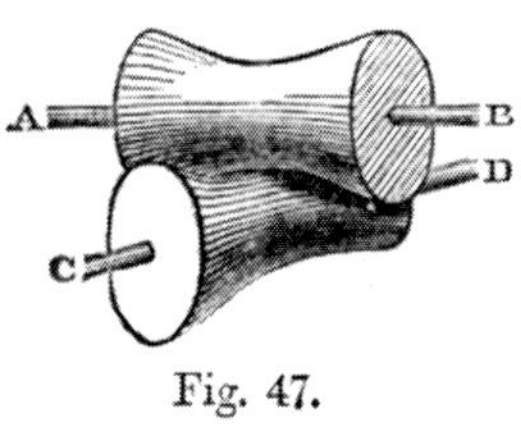

Fig. 47.

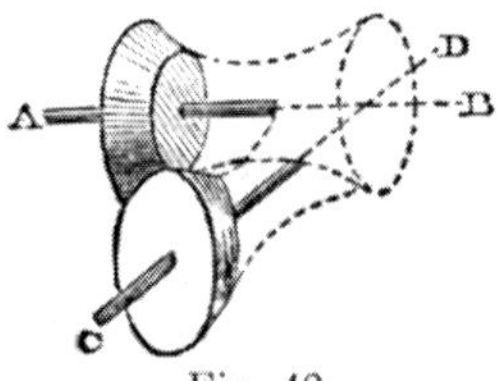

Fig. 48.

velocity-ratio between two axes which are neither parallel nor intersecting. The pitch surface of a skew bevel wheel is a frustrum or zone of a *hyperboloid of revolution*. In fig. 47, a pair of large portions of such hyperboloids are shewn, rotating about axes A B, C D. In fig. 48 are shewn a pair of narrow zones of the same figures, such as are employed in practice.

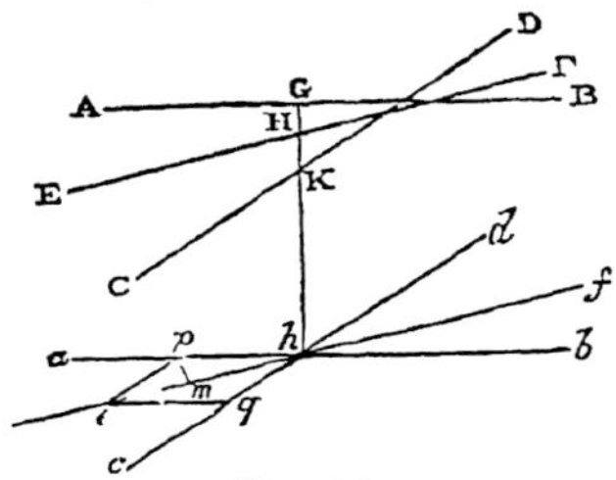

Fig. 49

A hyperboloid of revolution is a surface resembling a sheaf or a dice box, generated by the rotation of a straight line round an axis from which it is at a constant distance, and to which it is inclined at a constant angle. If two such hyperboloids, equal or unequal, be placed in the closest possible contact, as in fig. 49, they will touch each other along one of the generating straight lines of each, which will form their line of contact, and will be inclined to the axes A B, C D, in opposite directions. The axes will neither be parallel, nor will they intersect each other.

The motion of two such hyperboloids, rotating in contact with each other, has sometimes been classed amongst cases of rolling contact; but that classification is not strictly correct; for although the component velocities of a pair of points of contact in a direction at right angles to the line of contact are equal, still, as the axes are neither parallel to each other nor to the line of contact, the velocities of a pair of points of contact have components along the line of contact, which are unequal, and their difference constitutes a *lateral sliding*.

The directions and positions of the axes being given, and the required angular velocity ratio, $\frac{a_2}{a_1}$, it is required to find the *obliquities* of the generating line to the two axes, and its *radii vectores*, or least perpendicular distances from these axes.

In fig. 49, let A B, C D, be the two axes, and G K their common perpendicular.

On any plane normal to the common perpendicular G K *h*, draw *a b* A B, *c d* C D, in which take lengths in the following proportions:—

$$a_1 : a_2 : : \overline{hp} : hq;$$

complete the parallelogram *h p e q*, and draw its diagonal *e h f*; the line of contact E H F will be parallel to that diagonal.

From *p* let fall *p m* perpendicular to *h e*. Then divide the common perpendicular G K in the ratio given by the proportional equation

$$he : em : mh :: GK : GH : KH;$$

then the two segments thus found will be the least distances of the line of contact from the axes.

The first pitch surface is generated by the rotation of the line E H F about the axis A B with the radius vector G H $= r_1$; the second, by the rotation of the same line about the axis C D with the radius vector H K $= r_2$.

To draw the hyperbola which is the longitudinal section of a skew bevel wheel whose generating line has a given radius vector and obliquity, let A G B, fig. 50, represent the axis, G H ⊥ A G B, the radius vector of the generating line, and let the straight line E G F make with the axis an angle equal to the obliquity of the generating line. H will be the vertex, and E G F one of the asymptotes,† of the required hyperbola. To find any number of points in that hyperbola, proceed as follows: Draw X W Y parallel to G H, cutting G E in W, and make $XY = \sqrt{GH^2 + XW^2}$. Then will Y be a point in the hyperbola.

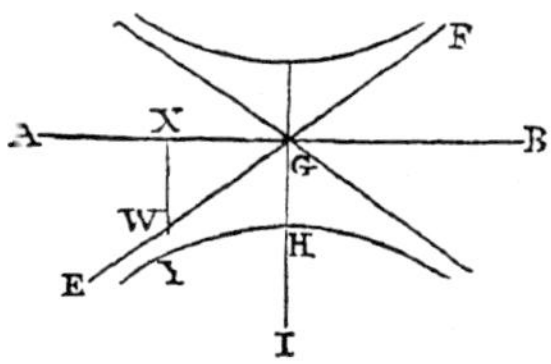

Fig. 50.

141. **Principle of Sliding Contact.** The *line of action*, or *of connection*, in the case of sliding contact of two moving pieces, is the common perpendicular to their surfaces at the point where they touch; and the principle of their comparative motion is, that *the components, along that perpendicular, of the velocities of any two points traversed by it, are equal.*

Case 1. *Two shifting pieces*, in sliding contact, have linear velocities proportional to the secants of the angles which their directions of motion make with their line of action.

Case 2. *Two rotating pieces*, in sliding contact, have angular velocities inversely proportional to the perpendicular distances from their axes of rotation to their line of action, each multiplied by the sine of the angle which the line of action makes with the particular axis on which the perpendicular is let fall.

In fig. 51, let C_1, C_2, represent the axes of rotation of the two pieces; A_1, A_2, two portions of their respective surfaces; and T_1, T_2, a pair of points in those surfaces, which, at the instant under consideration, are in contact with each other. Let $P_1 P_2$ be the common perpendicular of the surfaces at the pair of points T_1, T_2;

* The *Hyperbola* is the curve traced out by a point which moves in such a manner that its distance from a given fixed point I, continually bears the same ratio *greater than unity* to its distance from a given fixed line A B.

† An *Asymptote* is a straight line whose distance from a curve diminishes as the curve extends away from the origin.

that is, the *line of action*, and let $C_1 P_1$, $C_2 P_2$ be the common perpendiculars of the line of action and of the two axes respectively. Then at the given instant, the components along the line $P_1 P_2$ of the velocities of the points P_1, P_2, are equal. Let i_1 i_2, be the angles which that line makes with the directions of the axes respectively. Let a_1, a_2, be the respective angular velocities of the moving pieces, then

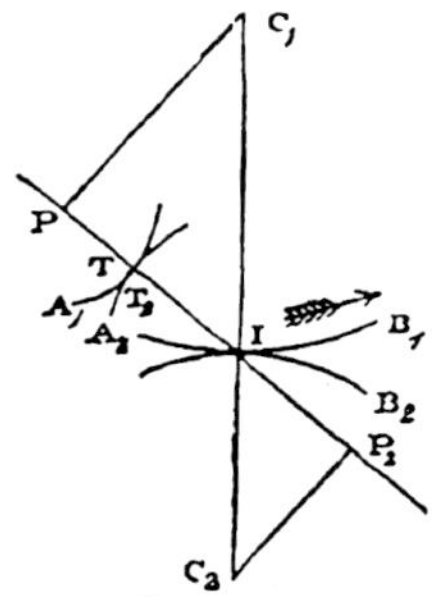

Fig. 31.

$$a_1 \cdot C_1 P_1 \cdot \sin i_1 = a_2 \cdot C_2 P_2 \cdot \sin i_2;$$

consequently,

$$\frac{a_2}{a_1} = \frac{C_1 P_1 \sin i_1}{C_2 P_2 \sin i_2}; \quad \ldots \ldots \quad (1.)$$

which is the principle stated above.

When the line of action is perpendicular in direction to both axes, then $\sin i_1 = \sin i_2 = 1$; and Equation 1 becomes

$$\frac{a_2}{a_1} = \frac{C_1 P_1}{C_2 P_2} \quad \ldots \ldots \quad (1\text{A.})$$

When the axes are parallel, $i_1 = i_2$. Let I be the point where the line of action cuts the plane of the two axes; then the triangles $P_1 C_1 I$, $P_2 C_2 I$, are similar; so that Equation 1 A is equivalent to the following

$$\frac{a_2}{a_1} = \frac{\overline{IC_1}}{\overline{IC_2}} \quad \ldots \ldots \ldots \quad (1\text{B.})$$

CASE 3. *A rotating piece and a shifting piece*, in sliding contact, have their comparative motion regulated by the following principle:—Let C P denote the perpendicular distance from the axis of the rotating piece to the line of action, i the angle which the direction of the line of action makes with that axis; a the angular velocity of the rotating piece; v the linear velocity of the sliding piece; j the angle which its direction of motion makes with the line of action, then

$$v = a \cdot \text{CP} \cdot \sin i \cdot \sec j. \quad \ldots \ldots \quad (2.)$$

When the line of action is perpendicular in direction to the axis of the rotating piece, $\sin i = 1$, and

$$v = a \cdot \text{CP} \cdot \sec j = a \cdot \overline{\text{IC}}; \quad \ldots \ldots \quad (2\text{A.})$$

where $\overline{\text{IC}}$ denotes the distance from the axis of the rotating piece to the point where the line of action cuts a perpendicular from that axis on the direction of motion of the shifting piece.

142. **Teeth of Wheels.** The most usual method of communicating motion between a pair of wheels, or a wheel and a rack, and the only method which, by preventing the possibility of the rotation of one wheel unless accompanied by the other, insures the preservation of a given velocity-ratio exactly, is by means of the projections called *teeth.*

The *pitch surface* of a wheel is an ideal smooth surface, intermediate between the crests of the teeth and the bottoms of the spaces between them, which, by rolling contact with the pitch surface of another wheel, would communicate the same velocity-ratio that the teeth communicate by their sliding contact. In designing wheels, the forms of the ideal pitch surfaces are first determined, and from them are deduced the forms of the teeth.

Wheels with cylindrical pitch surfaces are called *spur wheels;* those with conical pitch surfaces, *bevel wheels;* and those with hyperboloidal pitch surfaces, *skew bevel wheels.*

The *pitch line* of a wheel, or, in circular wheels, the *pitch circle,* is a transverse section of the pitch surface made by a surface perpendicular to it and to the axis; that is, in spur wheels, by a plane perpendicular to the axis; in bevel wheels, by a sphere described about the apex of the conical pitch surface; and in skew bevel wheels, by any oblate spheroid generated by the rotation of an ellipse whose foci are the same with those of the hyperbola that generates the pitch surface.

The *pitch point* of a pair of wheels is the point of contact of their pitch lines; that is, the transverse section of the line of contact of the pitch surfaces.

Similar terms are applied to racks.

That part of the acting surface of a tooth which projects beyond the pitch surface is called the *face,* that which lies within the pitch surface, the *flank.*

The radius of the pitch circle of a circular wheel is called the *geometrical radius;* that of a circle touching the crests of the teeth is called the *real radius;* and the difference between those radii, the *addendum.*

143. **Pitch and Number of Teeth.** The distance, measured along the pitch line, from the face of one tooth to the face of the next, is called the PITCH.

The pitch, and the number of teeth in circular wheels, are regulated by the following principles :—

I. In wheels which rotate continuously for one revolution or more, it is obviously necessary *that the pitch should be an aliquot part of the circumference.*

In wheels which reciprocate without performing a complete revolution, this condition is not necessary. Such wheels are called *sectors.*

II. In order that a pair of wheels, or a wheel and a rack, may work correctly together, it is in all cases essential *that the pitch should be the same in each.*

III. Hence, in any pair of circular wheels which work together, the numbers of teeth in a complete circumference are directly as the radii, and inversely as the angular velocities.

IV. Hence also, in any pair of circular wheels which rotate continuously for one revolution or more, the ratio of the numbers of teeth, and its reciprocal, the angular velocity ratio, must be expressible in whole numbers.

V. Let n, N, be the respective numbers of teeth in a pair of wheels, N being the greater. Let t, T, be a pair of teeth in the smaller and larger wheel respectively, which at a particular instant work together. It is required to find, first, how many pairs of teeth must pass the line of contact of the pitch surfaces before t and T work together again (let this number be called a); secondly, with how many different teeth of the larger wheel the tooth t will work at different times (let this number be called b); and thirdly, with how many different teeth of the smaller wheel the tooth T will work at different times (let this be called c).

Case 1. If n is a divisor of N,

$$a = \mathrm{N};\ b = \frac{\mathrm{N}}{n};\ c = 1 \ldots\ldots\ldots\ldots\ldots\ldots (1.)$$

Case 2. If the greatest common divisor of N and n be d a number less than n, so that $n = md$, $\mathrm{N} = \mathrm{M}d$, then

$$a = m\mathrm{N} = \mathrm{M}n = \mathrm{M}md;\ b = \mathrm{M};\ c = m \ldots\ldots\ldots (2.)$$

Case 3. If N and n be prime to each other,

$$a = \mathrm{N}n;\ b = \mathrm{N};\ c = n \ldots\ldots\ldots\ldots\ldots\ldots (3.)$$

It is considered desirable by millwrights, with a view to the preservation of the uniformity of shape of the teeth of a pair of wheels, that each given tooth in one wheel should work with as many different teeth in the other wheel as possible. They, therefore, study to make the numbers of teeth in each pair of wheels which work together such as to be either prime to each other, or to have their greatest common divisor as small as is possible consistently with the purposes of the machine.

VI. The *smallest* number of teeth which it is practicable to give to a pinion (that is, a small wheel), is regulated by the principle, that in order that the communication of motion from one wheel to another may be continuous, at least *one* pair of teeth should always be in action; and that in order to provide for the contingency of a tooth breaking, a *second* pair, at least, should be in action also. For reasons which will appear when the forms of teeth are considered, this principle gives the following as the least numbers of teeth which can be *usually* employed in pinions having teeth of the three classes of figures named below, whose properties will be explained in the sequel:—

I. Involute teeth,............................25.
II. Epicycloidal teeth,........... .... 12.
III. Cylindrical teeth, or *staves*,....... .... 6.

144. **Hunting Cog.** When the ratio of the angular velocities of two wheels, being reduced to its least terms, is expressed by small numbers, less than those which can be given to wheels in practice, and it becomes necessary to employ multiples of those numbers by a common multiplier, which becomes a common divisor of the numbers of teeth in the wheels, millwrights and engine-makers avoid the evil of frequent contact between the same pairs of teeth, by giving one additional tooth, called a *hunting cog*, to the larger of the two wheels. This expedient causes the velocity-ratio to be not exactly but only approximately equal to that which was at first contemplated; and therefore it cannot be used where the exactness of certain velocity ratios amongst the wheels is of importance as in clockwork.

145. **A Train of Wheelwork** consists of a series of axes, each having upon it two wheels, one of which is *driven* by a wheel on the preceding axis, while the other *drives* a wheel on the following axis. If the wheels are all in outside gearing, the direction of rotation of each axis is contrary to that of the adjoining axes. In some cases, a single wheel upon one axis answers the purpose both of receiving motion from a wheel on the preceding axis and giving motion to a wheel on the following axis. Such a wheel is called an *idle wheel:* it affects the direction of rotation only, and not the velocity ratio.

Let the series of axes be distinguished by numbers 1, 2, 3, &c. . . . . m; let the numbers of teeth in the *driving wheels* be denoted by N's, each with the number of its axis affixed; thus, N_1, N_2, &c. . . . . N_m ; and let the numbers of teeth in the *driven* or *following* wheels be denoted by n's, each with the number of its axis affixed; thus, n_2, n_3, &c. . . . . n_m. Then the ratio of the angular velocity a_m of the mth axis to the angular velocity a_1 of the first axis is the product of the $m-1$ velocity ratios of the successive elementary combinations, viz.:—

$$\frac{a_m}{a_1}=\frac{N_1 \cdot N_2 \cdot \&c. \ldots . N_{m-1}}{n_2 \cdot n_3 \cdot \&c. \ldots . n_m};\ldots\ldots\ldots\ldots\ldots \text{(1.)}$$

that is to say, the velocity-ratio of the last and first axes is the ratio of the product of the numbers of teeth in the drivers to the product of the numbers of teeth in the followers: and it is obvious that so long as the same drivers and followers constitute the train, the *order* in which they succeed each other does not affect the resultant velocity-ratio.

Supposing all the wheels to be in outside gearing, then as each elementary combination reverses the direction of rotation, and as the number of elementary combinations, $m-1$, is one less than the number of axes, m, it is evident that if m is odd, the direction of rotation is preserved, and if even, reversed.

It is often a question of importance to determine the numbers of teeth in a train of wheels best suited for giving a determinate velocity-ratio to two axes. It was shewn by Young, that to do this with the *least total number of teeth*, the velocity-ratio of each elementary combination should approximate as nearly as possible 3·59. This would in many cases give too many axes; and as a useful practical rule it may be laid down, that from 3 to 6 ought to be the limit of the velocity ratio of an elementary combination in wheelwork.

Let $\frac{B}{C}$ be the velocity-ratio required, reduced to its least terms, and let B be greater than C.

If $\frac{B}{C}$ is not greater than 6, and C lies between the prescribed minimum number of teeth (which may be called t), and its double $2t$, then one pair of wheels will answer the purpose, and B and C will themselves be the numbers required. Should B and C be inconveniently large, they are if possible to be resolved into factors, and those factors, or if they are too small multiples of them, used for the numbers of teeth. Should B or C, or both, be at once inconveniently large, and prime, then instead of the exact ratio $\frac{B}{C}$, some ratio approximating to that ratio, and capable of resolution into convenient factors, is to be found by the method of continued fractions. See MATHEMATICAL INTRODUCTION, page 2, Article 4.

Should $\frac{B}{C}$ be greater than 6, the best number of elementary combinations is found by dividing by 6 again and again till a quotient is obtained less than unity, when the number of divisions will be the required number of combinations, $m-1$.

Then, if possible, B and C themselves are to be resolved each into $m-1$ factors, which factors, or multiples of them, shall be not less than t, nor greater than $6t$; or if B and C contain inconveniently large prime factors, an approximate velocity-ratio, found by the method of continued fractions, is to be substituted for $\frac{B}{C}$, as before. When the prime factors of either B or C are fewer in number than $m - 1$, the required number of factors is to be made up by inserting 1 as often as may be necessary. In multiplying factors that are too small to serve for numbers of teeth, prime numbers differing from those already amongst the factors are to be preferred as multipliers; and in general, where two or more factors require to be multiplied, different prime numbers should be used for the different factors.

So far as the resultant velocity-ratio is concerned, the *order* of the drivers N, and of the followers n, is immaterial; but to secure equable wear of the teeth, as explained in Article 143, Principle V., the wheels ought to be so arranged that for each elementary combination the greatest common divisor of N and n shall be either 1, or as small as possible; and if the preceding rules have been observed in the choice of multipliers, this will be insured by so placing each driving wheel that it shall work with a following wheel whose number of teeth does not contain any of the same multipliers; for the original numbers B and C contain no common factor except 1.

The following is an example of a case requiring the use of additional multipliers: Let the required velocity-ratio, in its least terms, be

$$\frac{B}{C} = \frac{360}{7}.$$

To get a quotient less than 1, this ratio must be divided by 6 three times, therefore $m - 1 = 3$. The prime factors of 360 are $2 \cdot 2 \cdot 2 \cdot 3 \cdot 3 \cdot 5$; these may be combined so as to make three factors in various different ways; and the preference is to be given to that which makes these factors least unequal, viz., $5 \cdot 8 \cdot 9$. Hence, resolving numerator and denominator into three factors each, we have

$$\frac{B}{C} = \frac{5 \cdot 8 \cdot 9}{1 \cdot 1 \cdot 7}.$$

It is next necessary to multiply the factors of the numerator and denominator by a set of three multipliers. Suppose that the wheels to be used are of such a class that the smallest pinion has 12 teeth, then those multipliers must be such that none of their products by the existing factors shall be less than 12; and for reasons already given, it is advisable that they should be different prime numbers. Take the prime numbers, 2, 13, 17 (2 being taken to multiply 7); then the numbers of teeth in the followers will be

$$13 \times 1 = 13;\ 17 \times 1 = 17;\ 2 \times 7 = 14.$$

In distributing the multipliers amongst the factors of the numerator, let the smallest multiplier be combined with the largest factor, and so on; then we have

$$17 \times 5 = 85;\ 13 \times 8 = 104;\ 2 \times 9 = 18.$$

Finally, in combining the drivers with the followers, those numbers are to be combined which have no common factor; the result being the following train of wheels:—

$$\frac{85}{14} \cdot \frac{18}{13} \cdot \frac{104}{17} = \frac{360}{7}.$$

146. **Teeth of Spur Wheels and Racks. General Principle.** The figures of the teeth of wheels are regulated by the principle, *that the teeth of a pair of wheels shall give the same velocity ratio by their sliding contact, which the ideal smooth pitch surfaces would give by their rolling contact.* Let B_1, B_2, in fig. 51, be parts of the pitch lines (that is, of cross sections of the pitch surfaces) of a pair of wheels with parallel axes, and I the pitch point (that is, a section of the line of contact). Then the angular velocities which would be given to the wheels by the rolling contact of those pitch lines are inversely as the segments $I C_1$, $I C_2$, of the line of centres; and this also is the proportion of the angular velocities given by a pair of surfaces in sliding contact whose line of action traverses the point I (Article 141, Case 2, Equation 1 B). Hence the condition of correct working for the teeth of wheels with parallel axes is, *that the line of action of the teeth shall at every instant traverse the line of contact of the pitch surfaces;* and the same condition obviously applies to a rack sliding in a direction perpendicular to that of the axis of the wheel with which it works.

147. **Teeth Described by Rolling Curves.**—From the principle of the preceding Article it follows, that at every instant, the position of the point of contact T_1 in the cross section of the acting surface of a tooth (such as the line $A_1 T_1$ in fig. 51), and the corresponding position of the pitch point I in the pitch line $I B_1$ of the wheel to which that tooth belongs, are so related, that the line $I T_1$ which joins them is normal to the outline of the tooth $A_1 T_1$ at the point T_1. Now, this is the relation which exists between the *tracing-point* T_1, *and the instantaneous axis or line of contact* I, in a rolling curve of such a figure, that being rolled upon the pitch surface B_1, its tracing-point T_1 traces the outline of the tooth. (As to rolling curves, see Articles 100, 101, 103, and 106).

In order that a pair of teeth may work correctly together, it is necessary and sufficient that the *instantaneous radii vectores* from the pitch point to the points of contact of the two teeth should coincide at each instant, as expressed by the equation

$$I T_1 = I T_2; \qquad (1.)$$

and this condition is fulfilled *if the outlines of the two teeth be traced by the motion of the same tracing-point, in rolling the same rolling curve on the same side of the pitch surfaces of the respective wheels.*

The *flank* of a tooth is traced while the rolling curve rolls *inside* of the pitch line; the *face*, while it rolls *outside*. Hence it is evident that the *flanks* of the teeth of the driving wheel drive the *faces* of the teeth of the driven wheel; and that the *faces* of the teeth of the driving wheel drive the *flanks* of the teeth of the driven wheel. The former takes place while the point of contact of the teeth is *approaching* the pitch point, as in fig. 51, supposing the motion to be from P_1 towards P_2; the latter, after the point of contact has passed, and while it is *receding from*, the pitch point. The pitch point divides the path of the point of contact of the teeth into two parts, called the *path of approach* and the *path of recess;* and the lengths of these paths must be so adjusted, that two pairs of teeth at least shall be in action at each instant.

It is evidently necessary that the surfaces of contact of a pair of teeth should either be both convex, or that if one is convex and the other concave, the concave surface should have the flatter curvature.

148. **The Sliding of a Pair of Teeth on each Other,** that is, their relative motion in a direction perpendicular to their line of action, is found by supposing one of the wheels, such as 1, to be fixed, the line of centres $C_1 C_2$ to rotate backwards round C_1 with the angular velocity a_1, and the wheel 2 to rotate round C_2 as before with the angular velocity a_2 relatively to the line of centres $C_1 C_2$, so as to have the same motion as if its pitch surface rolled on the pitch surface of the first wheel. Thus the *relative* motion of the wheels is unchanged; but 1 is considered as fixed, and 2 has the resultant motion given by the principles of Article 102; that is, a rotation about the instantaneous axis I with the angular velocity $a_1 + a_2$. Hence the *velocity of sliding* is that due to this rotation about I, with the radius $\mathrm{IT} = r$; that is to say, its value is

$$r(a_1 + a_2); \qquad (1.)$$

so that it is greater, the farther the point of contact is from the line of centres; and at the instant when that point, passing the line of centres, coincides with the *pitch point*, the velocity of sliding is null, and the action of the teeth is, for the instant, that of rolling contact.

The roots of the teeth slide towards each other during the approach, and from each other during the recess. To find the *amount* or *total distance* through which the sliding takes place, let t_1 be the time occupied by the approach, and t_2 that occupied by the recess; then the distance of sliding is

$$s = \int_0^{t_1} r(a_1 + a_2)\, dt + \int_0^{t_2} r(a_1 + a_2)\, dt; \qquad (2.)$$

or in another form, if di denote an element of the change of angular position of one wheel relatively to the other, i_1 the amount of that change during the appr ach, and i_2 during the recess, then

$$a_1 + a_2) \, d\,t \quad d\,i; \text{ and}$$

$$s \quad \int_0^1 r\,d\,i + \int_0^2 r\,l\,i \ldots \ldots \quad 3.$$

149. **The Arc of Contact on the Pitch Lines** is the length of that portion of the pitch lines which passes the pitch point during the action of one pair of teeth; and in order that two pairs of teeth at least may be in action at each instant, its length should be at least double of the pitch. It is divided into two parts, the arc of approach and the arc of recess. In order that the teeth may be of length sufficient to give the required duration of contact, the distance moved over by the point I upon the pitch line during the rolling of a rolling curve to describe the face and flank of a tooth, must be in all equal to the length of the required arc of contact. It is usual to make the arcs of approach and recess equal.

150. **The Length of a Tooth** may be divided into two parts, that of the face and that of the flank. For teeth in the driving wheel, the length of the flank depends on the arc of approach, that of the face, on the arc of recess; for those in the following wheel, the length of the flank depends on the arc of recess,—that of the face, on the arc of approach.

151. **Involute Teeth for Circular Wheels.**—In fig. 52, let C_1, C_2, be the centres of two circular wheels, whose pitch circles are B_1, B_2. Through the pitch point I draw the intended *line of action* P_1, P_2, making the angle $C\,I\,P$ θ with the line of centres. From C_1, C_2, draw

$$\left.\begin{aligned} C_1 P_1 \quad & I\,C_1 \cdot \sin\theta, \\ \overline{C_2 P_2} \quad & I\,C_2 \cdot \sin\theta, \end{aligned}\right\} \ldots \ldots \ldots \quad 1.$$

perpendicular to $P_1 P_2$, with which two perpendiculars as radii, describe circles (called *base circles* D_1, D_2.

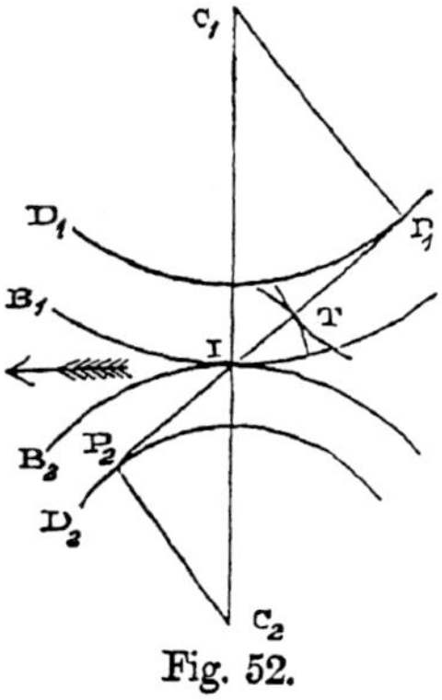

Fig. 52.

Suppose the base circles to be a pair of circular pulleys, connected by means of a cord whose course from pulley to pulley is $P_1 I P_2$. As the line of connection of these pulleys is the same with that of the proposed teeth, they will rotate with the required velocity ratio. Now suppose a tracing point T be fixed to the cord, so as to be carried along the path of contact $P_1 I P_2$. That point will trace, on a plane rotating along with the wheel 1, part of the involute of the base circle D_1, and on a plane rotating along with the wheel 2, part of the involute of the base circle D_2, and the two curves so traced will always touch each other in the required point of contact T, and will therefore fulfil the condition required by Article 146.

All involute teeth of the same pitch work smoothly together.

To find the length of the path of contact on either side of the pitch point I, it is to be observed that the distance between the fronts of two successive teeth, as measured along P_1 I P_2, is less than the pitch in the ratio sin θ : 1, for the former is proportional to $r \cdot \sin\theta$, and the latter to $r \cdot \theta$, and consequently that if distances not less than the pitch × sin θ be marked off either way from I towards P_1 and P_2 respectively, as the extremities of the path of contact, and if the addendum circles be described through the points so found, there will always be at least two pairs of teeth in action at once. In practice, it is usual to make the path of contact somewhat longer, viz., about $2\frac{1}{4}$ times the pitch; and with this length of path and the value of θ which is usual in practice, viz., $75\frac{1}{2}°$, the addendum is about $\frac{3}{10}$ of the pitch.

The teeth of a *rack*, to work correctly with wheels having involute teeth, should have plane surfaces, perpendicular to the line of connection, and consequently making, with the direction of motion of the rack, angles equal to the before mentioned angle θ.

152. **The Smallest Pinion with Involute Teeth** of a given pitch p, has its size fixed by the consideration that the path of contact of the flanks of the teeth, which must not be less than $p \cdot \sin\theta$, cannot be greater than the distance along the line of action from the pitch point to the base circle, I P $= r \cdot \cos\theta$. Then $r = \frac{\text{I P}}{\cos\theta}$; and substituting for I P its least possible value $p \cdot \sin\theta$, hence the *least radius* is

$$r = p \tan\theta; \ldots\ldots\ldots\ldots\ldots\ldots \quad (1)$$

which, for $\theta = 75\frac{1}{2}°$, gives for the radius $r = 3{\cdot}867\,p$, and for the circumference of the pitch circle, $p \times 3{\cdot}867 \times 2\pi = 24{\cdot}3\,p$; to which the next greater integer multiple of p is $25\,p$, and therefore *twenty-five*, as formerly stated, in Article 143, is the least number of *involute teeth* to be employed in a pinion.

153. **Epicycloidal Teeth.**—For tracing the figures of teeth, the most convenient rolling curve is the circle. The path of contact which a point in its circumference traces is identical with the circle itself; the flanks of the teeth are internal, and their faces external epicycloids, for wheels; and both flanks and faces are cycloids for a rack.

Wheels of the same pitch, with epicycloidal teeth traced by the same rolling circle, all work correctly with each other, whatsoever may be the numbers of their teeth; and they are said to belong to *the same set*.

For a pitch circle of twice the radius of the rolling or *describing* circle (as it is called), the internal epicycloid is a straight line, being in fact a diameter of the pitch circle; so that the flanks of the teeth for such a pitch circle are planes radiating from the axis. For a smaller pitch circle, the flanks would be convex, and *in-curved* or *under cut*, which would be inconvenient; therefore the smallest wheel of a set should have its pitch circle of twice the radius of the describing circle, so that the flanks may be either straight or concave.

In fig. 53, let B be part of the pitch circle of a wheel, C C the line of centres, I the pitch point, R the internal, and R the equal external describing circles, so placed as to touch the pitch circle and each other at I; let D I D be the path of contact, consisting of the path of approach D I, and the path of recess I D. In order that there may always be at least two pairs of teeth in action, each of those arcs should be equal to the pitch.

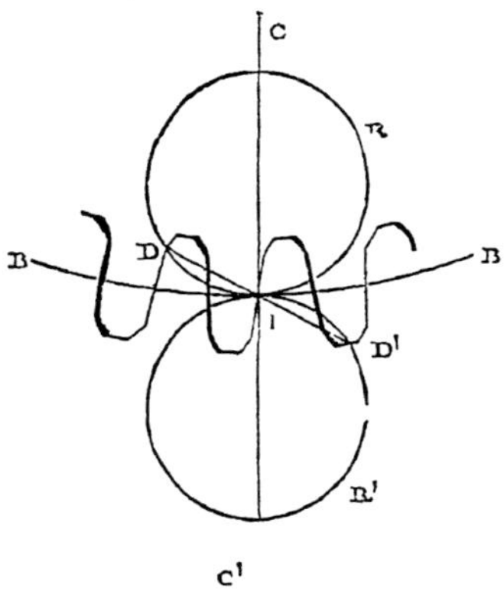

Fig. 53.

The angle θ, on passing the line of centres, is 90°; the least value of that angle is $\theta = \angle$ C I D $- \angle$ C I D.

It appears from experience that the least value of θ should be about 60°; therefore the arcs D I, I D should each be one-sixth of a circumference; therefore the circumference of the describing circle should be *six times the pitch*.

It follows that the smallest pinion of a set, in which pinion the flanks are straight, *should have twelve teeth*, as has been already stated in Article 143.

154. **Teeth of Wheel and Trundle.**—A *trundle*, as in fig. 54, has cylindrical pins called *staves* for teeth. The face of the teeth

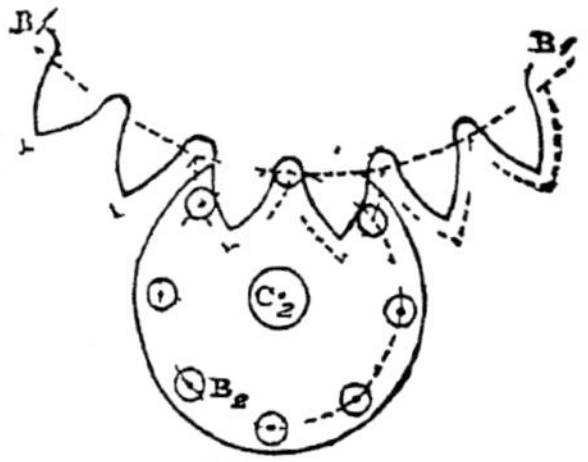

Fig. 54.

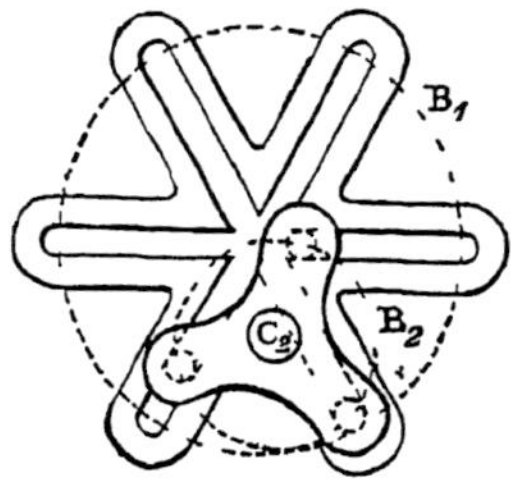

Fig. 55.

of a wheel suitable for driving it, in outside gearing, are described by first tracing external epicycloids by rolling the pitch circle B_2 of the trundle on the pitch circle B of the driving wheel, with the centre of a stave for a tracing point, as shewn by the dotted lines, and then drawing curves parallel to and within the epicycloids, at a distance from them equal to the radius of a stave. Trundles having only six staves will work with large wheels.

To drive a trundle in *inside gearing*, the outlines of the teeth of the wheel should be curves parallel to internal epicycloids. A peculiar case of this is represented in fig. 55, where the radius of the pitch circle of the trundle is exactly one-half of that of the pitch circle of the wheel; the trundle has three equi distant staves; and the internal epicycloids described by their centres while the pitch circle of the trundle is rolling within that of the wheel, are three straight lines, diameters of the wheel, making angles of 60° with each other. Hence the surfaces of the teeth of the wheel form three straight grooves intersecting each other at the centre, each being of a breadth equal to the diameter of a stave of the trundle.

155. **Dimensions of Teeth.**—Toothed wheels being in general intended to rotate either way, the *backs* of the teeth are made similar to the fronts. The *space* between two teeth, measured on the pitch circle, is made about one fifth part wider than the thickness of the tooth on the pitch circle: that is to say,

$$\text{thickness of tooth} = \frac{5}{11}\text{ pitch},$$

$$\text{width of space} = \frac{6}{11}\text{ pitch}.$$

The difference of $\frac{1}{11}$ of the pitch is called the *back lash*.

The clearance allowed between the points of teeth and the bottoms of the spaces between the teeth of the other wheel, is about one tenth of the pitch.

The *thickness* of a tooth is fixed according to the principles of strength; and the *breadth* is so adjusted, that when multiplied by pitch, the product shall contain *one square inch* for each 160 lbs. of force transmitted by the teeth.

156. **The Teeth of a Bevel Wheel** have acting surfaces of the conical kind, generated by the motion of a line traversing the apex of the conical pitch surface, while a point in it is carried round the outlines of the cross section of the teeth made by a sphere described about that apex.

The operations of describing the exact figures of the teeth of bevel-wheels, whether by involutes or by rolling curves, are in every respect analogous to those for describing the figures of the teeth of spur-wheels, except that in the case of bevel-wheels, all those operations are to be performed on the surface of a sphere described about the apex, instead of on a plane, substituting *poles* for *centres*, and *great circles* for *straight lines*.

In consideration of the practical difficulty, especially in the case of large wheels, of obtaining an accurate spherical surface, and of drawing upon it when obtained, the following *approximate* method, proposed originally by Tredgold, is generally used: Let O, fig. 56, be the apex, and O C the axis of the pitch cone of a bevel-wheel; and let the largest pitch circle be that whose radius is C B. Perpendicular to O B draw B A cutting the axis produced in A, let the outer rim of the pattern and of the wheel be made a portion of the surface of the cone whose apex is A and side A B. The narrow zone of that cone thus employed will approach sufficiently near to a zone of the sphere described about O with the radius O B, to be used in its stead. On a plane surface, with the radius A B, draw a circular arc B D; a sector of that circle will represent a portion of the surface of the cone A B C *developed*, or *spread out flat*. Describe the figures of teeth of the required pitch, suited to the pitch circle B D, as if it were that of a spur-wheel of the radius A B; those figures will be the required cross sections of the teeth of the bevel wheel, made by the conical zone whose apex is A.

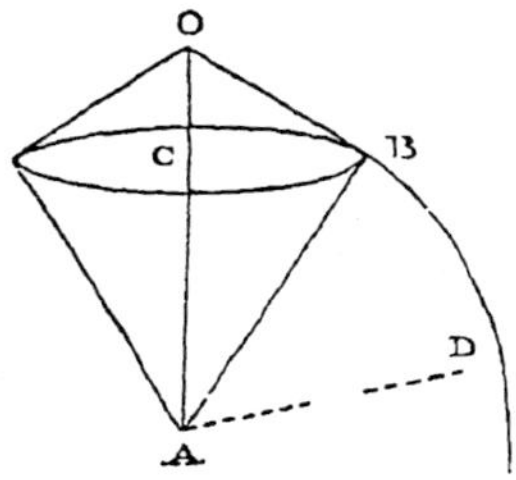

Fig. 56.

157. **The Teeth of Non-Circular Wheels** are described by rolling circles or other curves on the pitch surfaces, like the teeth of circular wheels; and when they are small compared with the wheels to which they belong, each tooth is nearly similar to the tooth of a circular wheel, having the same radius of curvature with the pitch surface of the actual wheel at the point where the tooth is situated.

158. **A Cam or Wiper** is a single tooth, either rotating continuously or oscillating, and driving a sliding or turning-piece, either constantly or at intervals. All the principles which have been stated in Article 141, as being applicable to sliding contact, are applicable to cams; but in designing cams, it is not usual to determine or take into consideration the form of the ideal pitch surface which would give the same comparative motion by rolling contact that the cam gives by sliding contact.

159. **Screws. Pitch.**—The figure of a screw is that of a convex or concave cylinder with one or more helical projections called *threads* winding round it. Convex and concave screws are distinguished technically by the respective names of *male* and *female*, or *external* and *internal;* a short internal screw is called a *nut;* and when a *screw* is not otherwise specified, *external* is understood.

The relation between the *advance* and the *rotation*, which compose the motion of a screw working in contact with a fixed nut or helical guide, has already been demonstrated in Article 96, Equation 1; and the same relation exists between the rotation of a screw about an axis fixed longitudinally relatively to the framework, and the advance of a nut in which that screw rotates, the nut being free to shift longitudinally, but not to turn. The advance of the nut in the latter case is in the direction opposite to that of the advance of the screw in the former case.

A screw is called *right handed* or *left handed*, according as its advance in a fixed nut is accompanied by right handed or left-handed rotation, when viewed by an observer *from* whom the advance takes place. Fig. 57 represents a right-handed screw, and fig. 58 a left-handed screw.

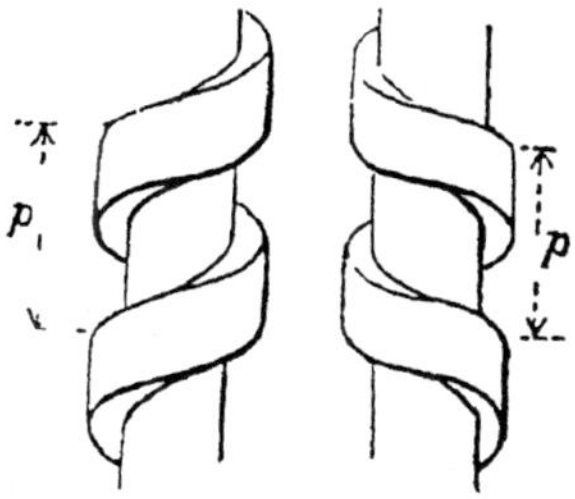

Fig. 57. Fig 58.

The *pitch* of a screw of one thread, and the *total pitch* of a screw of any number of threads, is the pitch of the helical motion of that screw, as explained in Article 96, and is the distance marked p in figs 57 and 58 measured parallel to the axis of the screw, between the corresponding points in two consecutive turns *of the same thread*.

In a screw of two or more threads, the distance measured parallel to the axis, between the corresponding points in *two adjacent threads*, may be called the *divided pitch*.

160. **Normal and Circular Pitch.** When the pitch of a screw is not otherwise specified, it is always understood to be measured parallel to the axis. But it is sometimes convenient for particular purposes to measure it in other directions; and for that purpose a *cylindrical pitch surface* is to be conceived as described about the axis of the screw, intermediate between the crests of the threads and the bottoms of the grooves between them.

If a helix be now described upon the pitch cylinder, so as to cross each turn of each thread at right angles, the distance between two corresponding points on two successive turns of the same thread, measured along this *normal helix*, may be called the *normal pitch;* and when the screw has more than one thread, the normal pitch from thread to thread may be called the *normal divided pitch*.

The distance from thread to thread measured on a circle described on the pitch cylinder, and called the *pitch circle*, may be called the *circular pitch;* for a screw of one thread it is one circumference; for a screw of n threads

$$\frac{\text{one circumference}}{n}.$$

The following set of formulæ shew the relations amongst the different modes of measuring the pitch of a screw. The *pitch*, properly speaking, as originally defined, is distinguished as the *axial pitch*, and is the same for all parts of the same screw: the normal and circular pitch depend on the radius of the pitch cylinder.

Let r denote the radius of the pitch cylinder;

n, the number of threads;

i, the obliquity of the threads to the pitch circles, and of the normal helix to the axis;

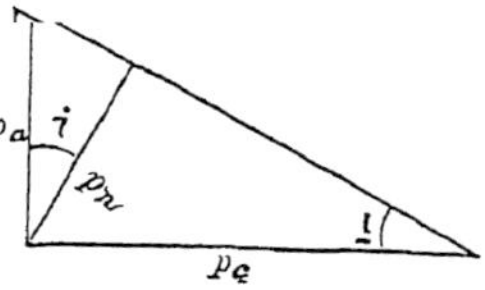

Fig. 59

$\left.\begin{matrix} P_a \\ \frac{P_a}{n} = p_a \end{matrix}\right\}$ the axial $\left\{\begin{matrix}\text{pitch;} \\ \text{divided pitch;}\end{matrix}\right.$

$\left.\begin{matrix} P_n \\ \frac{P_n}{n} = p_n \end{matrix}\right\}$ the normal $\left\{\begin{matrix}\text{pitch;} \\ \text{divided pitch;}\end{matrix}\right.$

p_c, the circular pitch;

Then

$$p_c = p_a \cdot \cot i = p_n \cdot \operatorname{cosec} i = \frac{2\pi r}{n};$$

$$p_a = p_n \cdot \sec i = p_c \cdot \tan i = \frac{2\pi r \cdot \tan i}{n};$$

$$p_n = p_c \cdot \sin i = p_a \cdot \cos i = \frac{2\pi r \cdot \sin i}{n}.$$

Fig. 59 will make these formulæ clear, in which the several lines are lettered to represent the pitches: the hypotenuse of the larger triangle is the linear development on the plane of the paper of one coil of the screw which, it will be remarked, $= \sqrt{p_a^2 + p_c^2}$; p_n the normal pitch is normal to this: it is also evident from the figure that with a constant axial pitch, the normal and radial or circumferential pitch, as well as the angle of obliquity of the threads to the pitch cylinders, vary with the radii of these cylinders.

161. **Screw Gearing.**—A pair of convex screws, each rotating about its axis, are used as an elementary combination, to transmit motion by the sliding contact of their threads. Such screws are commonly called *endless screws*. At the point of contact of the screws, their threads must be parallel; and their line of connection is the common perpendicular to the acting surfaces of the threads at their point of contact. Hence the following principles:—

I. If the screws are both right handed or both left-handed, the angle between the directions of their axes is the sum of their obliquities:—if one is right handed and the other left-handed, that angle is the difference of their obliquities.

II. The normal pitch, for a screw of one thread, and the normal divided pitch, for a screw of more than one thread, must be the same in each screw.

III. The angular velocities of the screws are inversely as their number of threads.

162. **The Wheel and Screw** is an elementary combination of two screws, whose axes are at right angles to each other, both being right handed or both left handed. As the usual object of this combination is to produce a change of angular velocity in a ratio greater than can be obtained by any single pair of ordinary wheels, one of the screws is commonly wheel like, being of large diameter and many threaded, while the other is short and of few threads; and the angular velocities are inversely as the number of threads.

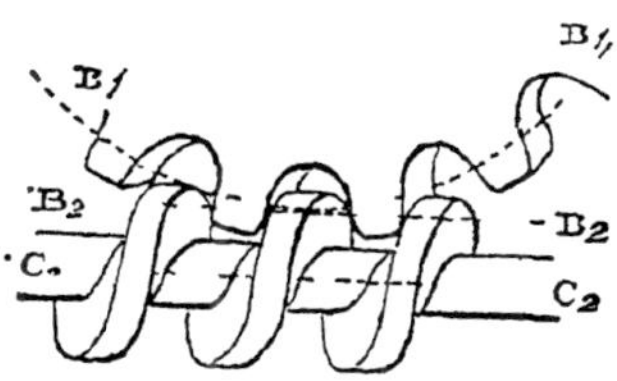

Fig. 60.

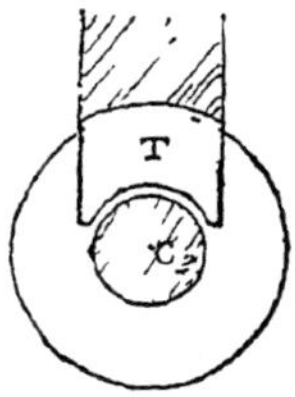

Fig. 61.

Fig. 60, represents a side view of this combination, and fig. 61 a cross section at right angles to the axis of the smaller screw. It has been shewn by Prof. Willis, that if each section of both screws be made by a plane perpendicular to the axis of the large screw or wheel, the outlines of the threads of the larger and smaller screw should be those of the teeth of a wheel and rack respectively · B_1 B_1, in fig. 60 for example, being the pitch circle of the wheel, and B_2 B_2 the pitch line of the rack.

The periphery and teeth of the wheel are usually hollowed to fit the screw, as shewn at T, fig. 61.

To make the teeth or threads of a pair of screws fit correctly and work smoothly, a hardened steel screw is made of the figure of the smaller screw, with its thread or threads notched so as to form a cutting tool; the larger screw, or wheel, is cast approximately of the required figure; the larger screw and the steel screw are fitted up in their proper relative position, and made to rotate in contact with each other by turning the steel screw, which cuts the threads of the larger screw to their true figure.

163. **The Relative Sliding of a Pair of Screws** at their point of contact is found thus:—Let r_1, r_2, be the radii of their pitch cylinders, and i_1, i_2, the obliquities of their threads to their pitch circles, one of which is to be considered as negative if the screws are contrary-handed. Let u be the common component of the velocities of a pair of points of contact along a line touching the pitch sur-

faces and perpendicular to the threads at the pitch point, and v the velocity of sliding of the threads over each other, where v may be considered to be made up of the algebraical sum of two quantities, v_1 and v_2, which act perpendicularly to u, and whose values are $v_1 = a_1 r_1 \cos i_1$, and $v_2 = a_2 r_2 \cos i_2$; the sum or difference being taken as the screws are similar or contrary handed. Then

$$u = a_1 r_1 \cdot \sin i_1 = a_2 r_2 \cdot \sin i_2;$$

so that

$$a_1 = \frac{u}{r_1 \cdot \sin i_1}; \quad a_2 = \frac{u}{r_2 \cdot \sin i_2}; \qquad \ldots\ldots\ldots\ldots (1.)$$

and

$$v = a_1 r_1 \cdot \cos i_1 + a_2 r_2 \cdot \cos i_2 = u (\text{cotan } i_1 + \text{cotan } i_2) \; . \; . \; .(2.)$$

When the screws are contrary-handed, the difference instead of the sum of the terms in Equation 2 is to be taken.

164. **Oldham's Coupling.**—A *coupling* is a mode of connecting a pair of shafts so that they shall rotate in the same direction, with the same mean angular velocity. If the axes of the shafts are in the same straight line, the coupling consists in so connecting their contiguous ends that they shall rotate as one piece; but if the axes are not in the same straight line, combinations of mechanism are required. A coupling for parallel shafts which acts by *sliding contact* was invented by Oldham, and is represented in fig. 62.

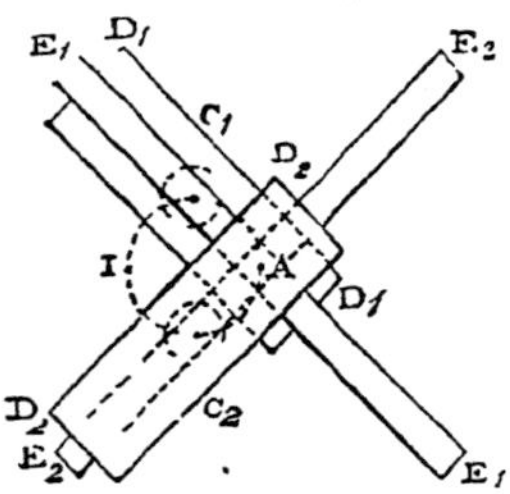

Fig. 62.

C_1, C_2, are the axes of the two parallel shafts; D_1, D_2, two cross-heads, facing each other, fixed on the ends of the two shafts respectively; E_1, E_1, a bar, sliding in a diametral groove in the face of D_1; E_2, E_2, a bar, sliding in a diametral groove in the face of D_2; those bars are fixed together at A, so as to form a rigid cross. The angular velocities of the two shafts and of the cross are all equal at every instant. The middle point of the cross, at A, revolves in the dotted circle described upon the line of centres C_1, C_2, as a diameter, twice for each turn of the shafts and cross; the instantaneous axis of rotation of the cross, at any instant, is at I, the point in the circle C_1 C_2, diametrically opposite to A.

Oldham's coupling may be used with advantage where the axes of the shafts are intended to be as nearly in the same straight line as is possible, but where there is some doubt as to the practicability or permanency of their exact continuity.

Section 3.—Connection by Bands.

165. **Bands Classed.**—Bands, or wrapping connectors, for communicating motion between pulleys or drums rotating about fixed axes, or between rotating pulleys and drums and shifting pieces, may be thus classed :—

I. *Belts,* which are made of leather or of gutta percha, are flat and thin, and require nearly cylindrical pulleys. A belt tends to move towards that part of a pulley whose radius is greatest; pulleys for belts therefore, are slightly swelled in the middle, in order that the belt may remain on the pulley unless forcibly shifted. A belt when in motion is shifted off a pulley, or from one pulley on to another of equal size alongside of it, by pressing against that part of the belt which is moving *towards* the pulley.

II. *Cords,* made of catgut, hempen or other fibres, or wire. are nearly cylindrical in section, and require either drums with ledges, or grooved pulleys.

III. *Chains,* which are composed of links or bars jointed together, require pulleys or drums, grooved, notched, and toothed, so as to fit the links of the chains.

Bands for communicating continuous motion are *endless.*

Bands for communicating reciprocating motion have usually their ends made fast to the pulleys or drums which they connect, and which in this case may be sectors.

166. **Principle of Connection by Bands.**—The *line of connection* of a pair of pulleys or drums connected by means of a band, is the central line or axis of that part of the band whose tension transmits the motion. The principle of Article 129 being applied to this case, leads to the following consequences:—

I. *For a pair of rotating pieces,* let r_1, r_2, be the perpendiculars let fall from their axes on the centre line of the band, a_1, a_2, their angular velocities, and i_1, i_2, the angles which the centre line of the band makes with the two axes respectively. Then the longitudinal velocity of the band, that is, its component velocity in the direction of its own centre line, is

$$u = r_1 a_1 \sin i_1 = r_2 a_2 \sin i_2 ;\dots\dots\dots\dots\dots\dots\dots(1.)$$

whence the angular velocity-ratio is

$$\frac{a_2}{a_1} = \frac{r_1 \sin i_1}{r_2 \sin i_2}\dots\dots\dots\dots\dots\dots\dots(2.)$$

When the axes are parallel (which is almost always the case), $i_1 = i_2$, and

$$\frac{a_2}{a_1} = \frac{r_1}{r_2}\dots\dots\dots\dots\dots\dots\dots\dots\dots(3.)$$

The same equation holds when both axes, whether parallel or n t, are perpendicular in direction to that part of the band which transmits the motion; for then $\sin i_1 \quad \sin i_2 \quad 1$.

II. *For a rotating piece and a slidinj p'ece*, let r be the perpendicular from the axis of the rotating piece on the centre line of the band, a the angular velocity, i the angle between the directi ns of the band and axis, u the longitudinal velocity of the band, j the angle between the direction of the centre line of the band and that of the motion of the sli ling piece, and v the velocity of the sliding piece; then

$$u \quad r a \sin i \quad v \cos j; \ldots\ldots (4.)$$

for $r \sin i$ is the projection on the plane of motion of r, and u the longitudinal velocity of the band must necessarily be equal to $v \cos j$, the longitudinal velocity of the sliding piece owing to the rigidity of the band; and

$$v \quad \frac{r a \, \mathrm{s\,n}\, i}{\mathrm{c\;s}\, j} \ldots\ldots 5.)$$

When the centre line of the band is parallel to the direction of motion of the sliding piece, and perpendicular to the direction of the axis of the rotating piece, $\sin i$ (90 $\quad \cos j$ (0) – 1, and

$$v = u \quad r a \ldots\ldots (6.$$

167. **The Pitch Surface of a Pulley or Drum** is a surface to which the line of connection is always a tangent; that is to say, it is a surface parallel to the acting surface of the pulley or drum, and distant from it by half the thickness of the band.

168. **Circular Pulleys and Drums** are used to commuicate a

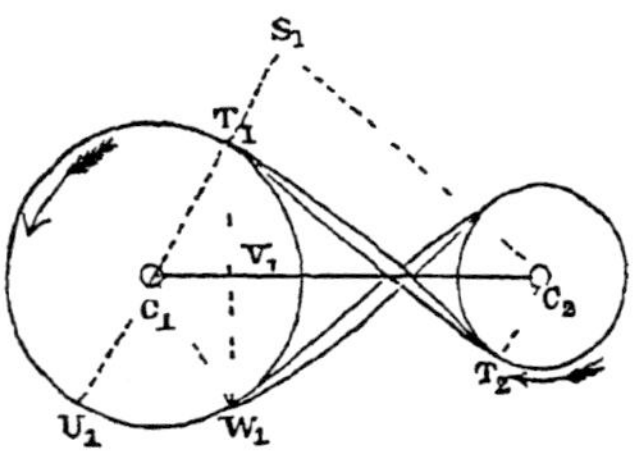

Fig. 63.

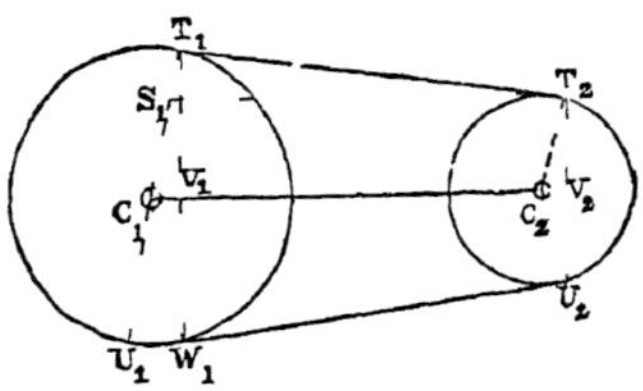

Fig. 64.

constant velocity-ratio. In each of them, the length denoted by r in the equations of Article 166 is constant, and is called the *effective radius*, being equal to the real radius of the pulley or drum added to half the thickness of the band.

A *crossed belt* connecting a pair of circular pulleys, as in fig. 63, reverses the direction of r tation; an *open belt*, as in fig. 64, preserves that direction.

169. **The length of an Endless Belt**, connecting a pair of pulleys whose effective radii are $C_1 T_1 = r_1$, $C_2 T_2 = r_2$, with parallel axes whose distance apart is $C_1 C_2 = c$, is given by formulæ founded on Equation 1 of Article 113, viz., $L = \Sigma \cdot s + \Sigma \cdot r\, i$. Each of the two equal straight parts of the belt is evidently of the length

$$\left.\begin{array}{l} s = T_1 T_2 = \sqrt{c^2 - (r_1 + r_2)^2} \text{ for a crossed belt;} \\ s = T_1 T_2 = \sqrt{c^2 - (r_1 - r_2)^2} \text{ for an open belt;} \end{array}\right\} \ldots\ldots(1.)$$

r_1 being the greater radius, and r_2 the less. Let i_1 be the arc to radius unity of the greater pulley, and i_2 that of the less pulley, with which the belt is in contact; then for a crossed belt

$$i_1 = i_2 = \left(\pi + 2 \operatorname{arc} \cdot \sin \frac{r_1 + r_2}{c}\right);$$

for the angle $V_1 C_1 W_1$ at the centre is double of the angle at the circumference $C_1 T_1 W_1$, and this is equal to the angle $S_1 C_2 C_1$ as they both differ from a right angle by the same angle $T_1 C_1 V_1$; and for an open belt, (2.)

$$i_1 = \left(\pi + 2 \operatorname{arc} \cdot \sin \frac{r_1 - r_2}{c}\right); \; i_2 = \left(\pi - 2 \operatorname{arc} \cdot \sin \frac{r_1 - r_2}{c}\right);$$

and the introduction of those values into Equation 1 of Article 113 gives the following results:—

For a crossed belt

$$L = 2\sqrt{c^2 - (r_1 + r_2)^2} + (r_1 + r_2) \cdot \left(\pi + 2 \operatorname{arc} \cdot \sin \cdot \frac{r_1 + r_2}{c}\right);$$

and if similar reasoning be applied, it may be shewn that for an open belt, (3.)

$$L = 2\sqrt{c^2 - (r_1 - r_2)^2} + \pi (r_1 + r_2) + 2(r_1 - r_2) \cdot \operatorname{arc} \cdot \sin \cdot \frac{r_1 - r_2}{c}.$$

As the last of these equations would be troublesome to employ in a practical application to be mentioned in the next Article, an approximation to it, sufficiently close for practical purposes, is obtained by considering, that if $r_1 - r_2$ is small compared with c, $\sqrt{c^2 - (r_1 - r_2)^2} = c - \frac{(r_1 - r_2)^2}{2c}$ nearly, and $\operatorname{arc} \cdot \sin \cdot \frac{r_1 - r_2}{c} = \frac{r_1 - r_2}{c}$ nearly; whence, for an open belt,

$$L \text{ nearly} = 2c + \pi (r_1 + r_2) + \frac{(r_1 - r_2)^2}{c} \ldots\ldots\ldots (3\text{A}.)$$

170. **Speed-Cones** (figs. 65, 66, 67, 68) are a contrivance for

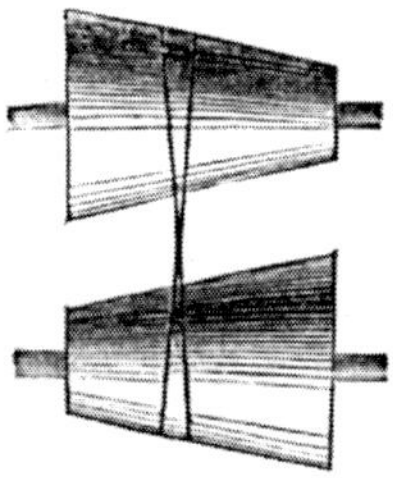

Fig. 65.

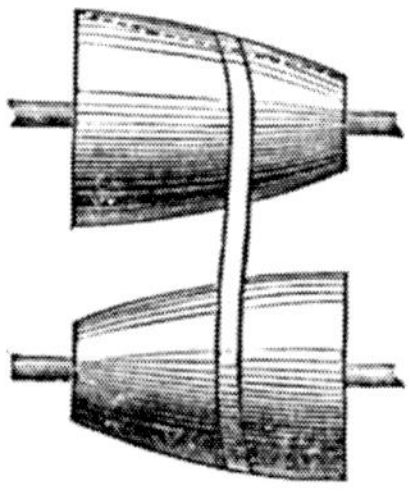

Fig. 66.

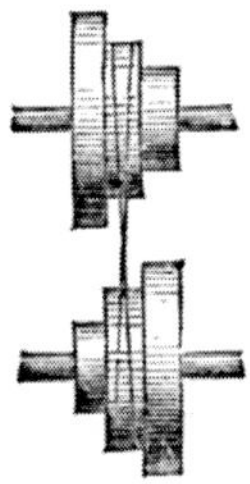

Fig. 67.

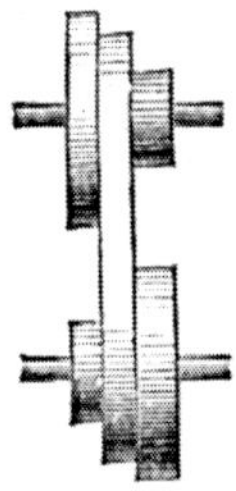

Fig. 68.

varying and adjusting the velocity ratio communicated between a pair of parallel shafts by means of a belt, and may be either continuous cones or conoids, as in figs. 65, 66, whose velocity-ratio can be varied gradually while they are in motion by shifting the belt; or sets of pulleys whose radii vary by steps, as in figs. 67, 68, in which case the velocity-ratio can be changed by shifting the belt from one pair of pulleys to another.

In order that the belt may be equally tight in every possible position on a pair of speed-cones, the quantity L in the equations of Article 169 must be constant.

For a *crossed* belt, as in figs. 65 and 66, L depends solely on c and on $r_1 + r_2$. Now c is constant, because the axes are parallel, therefore the *sum of the radii* of the pitch circles connected in every position of the belt is to be constant. That condition is fulfilled by a pair of continuous cones generated by the revolution of two straight lines inclined opposite ways to their respective axes at equal angles, and by a set of pairs of pulleys in which the sum of the radii is the same for each pair.

For an *open* belt, the following practical rule is deduced from the approximate Equation 3A of Article 169:

Let the speed-cones be equal and similar conoids, as in fig. 66, but with their large and small ends turned opposite ways. Let r_1 be the radius of the large end of each, r_2 that of the small end, r that of the middle; and let y be the *sagitta*, measured perpendicular to the axis, of the arc by whose revolution each of the conoids is generated, or, in other words, the *bulging* of the conoids in the middle of their length; then

$$y = r - \frac{r_1 + r_2}{2} = \frac{(r_1 - r_2)^2}{2\pi c} \quad \ldots\ldots \ldots\ldots \ldots\ldots (1.)$$

where the second value is obtained from the first by considering that in Equation 3A, $2\pi r_0 = \pi(r_1 + r_2) + \frac{(r_1 - r_2)^2}{c}$; $2\pi = 6{\cdot}2832$; but 6 may be used in most practical cases without sensible error.

The radii at the middle and ends being thus determined, make the generating curve an arc either of a circle or of a parabola.

For a pair of stepped cones, as in fig. 68, let a series of *differences* of the radii, or values of $r_1 - r_2$, be assumed; then for each pair of pulleys, the sum of the radii is to be computed from the difference by the formula

$$r_1 + r_2 - 2\,r_0 - \frac{(r_1 - r_2)^2}{\pi c}; \qquad \text{(2.)}$$

$2\,r$ being that sum when the radii are equal.

SECTION 4.—LINKWORK.

171. *Definitions.*—The pieces which are connected by linkwork, if they rotate or oscillate, are usually called *cranks*, *beams*, and *levers.* The *link* by which they are connected is a rigid bar, which may be straight or of any other figure; the straight figure being the most favourable to strength, is used when there is no special reason to the contrary. The link is known by various names under various circumstances, such as *coupling rod*, *connecting rod*, *crank rod*, *eccentric rod*, &c. It is attached to the pieces which it connects by two pins, about which it is free to turn. The effect of the link is to maintain the distance between the centres of those pins invariable; hence the line joining the centres of the pins is the *line of connection;* and those centres may be called the *connected points.* In a turning piece, the perpendicular let fall from its connected point upon its axis of rotation is the *arm* or *crank arm.*

172. **Principles of Connection.**—The whole of the equations already given in Article 166 for bands, are applicable to linkwork. The axes of rotation of a pair of turning pieces connected by a link are almost always parallel, and perpendicular to the line of connection; in which case the angular velocity-ratio at any instant is the reciprocal of the ratio of the common perpendiculars let fall from the line of connection upon the respective axes of rotation (Article 166, Equation 3).

173. **Dead Points.**—If at any instant the direction of one of the crank arms coincides with the line of connection, the common perpendicular of the line of connection and the axis of that crank arm vanishes, and the directional relation of the motions becomes indeterminate. The position of the connected point of the crank arm in question at such an instant is called a *dead point.* The velocity of the other connected point at such an instant is null, unless it also reaches a dead point at the same instant, so that the line of connection is in the plane of the two axes of rotation, in which case the velocity-ratio is indeterminate.

174. **Coupling of Parallel Axes.**—The only case in which an uni-

form angular velocity ratio being that of equality is communicated by linkwork, is that in which two or more parallel shafts such as those of the driving wheels of a locomotive engine) are made to rotate with constantly equal angular velocities, by having equal cranks, which are maintained parallel by a coupling rod of such a length that the line of connection is equal to the distance between the axes. The cranks pass their dead points simultaneously. To obviate the unsteadiness of motion which this tends to cause, the shafts are provided with a second set of cranks at right angles to the first, connected by means of a similar coupling rod, so that one set of cranks pass their dead points at the instant when the other set are farthest from theirs.

175. **The Comparative Motion of the Connected Points** in a piece of linkwork at a given instant is capable of determination by the method explained in Article 98; that is, by finding the instantaneous axis of the link; for the two connected points move in the same manner with two points in the link, considered as a rigid body.

If a connected point belongs to a turning piece, the direction of its motion at a given instant is perpendicular to the plane containing the axis and crank arm of the piece. If a connected point belongs to a shifting piece, the direction of its motion at any instant is given, and a plane can be drawn perpendicular to that direction.

The line of intersection of the planes perpendicular to the paths of the two connected points at a given instant, is the *instantaneous axis of the link* at that instant; and the *velocities of the connected points are directly as their distances from that axis.*

In drawing on a plane surface, the two planes perpendicular to the paths of the connected points are represented by two lines (being their sections by a plane normal to them), and the instantaneous axis by a point; and should the length of the two lines render it impracticable to produce them until they actually intersect, the velocity-ratio of the connected points may be found by the principle, that it is equal to the ratio of the segments which a line parallel to the line of connection cuts off from any two lines drawn from a given point, perpendicular respectively to the paths of the connected points.

Example I. *Two Rotating Pieces with Parallel Axes* (fig. 69)—Let C_1, C_2, be the parallel axes of the pieces; T_1, T_2, their connected points; $C_1 T_1$, $C_2 T_2$, their crank arms; $T_1 T_2$, the link. At a given instant, let v_1 be the velocity of T_1; v_2 that of T_2.

To find the ratio of those velocities, produce $C_1 T_1$, $C_2 T_2$, till they intersect in K; K is the instantaneous axis of the link or connecting rod, and the velocity ratio is

$$v_1 : v_2 :: KT_1 : KT_2 \text{ .... (1.)}$$

Should K be inconveniently far off, draw any triangle with its sides respectively parallel to $C_1 T_1$, $C_2 T_2$, and $T_1 T_2$, the ratio of the two sides first mentioned will be the velocity ratio required. For example, draw $C_2 A$ parallel to $C_1 T_1$, cutting $T_1 T_2$ in A, then

$$v_1 : v_2 :: \overline{C_2 A} : \overline{C_2 T_2} \ldots\ldots\ldots\ \ldots\ldots\ldots (2.)$$

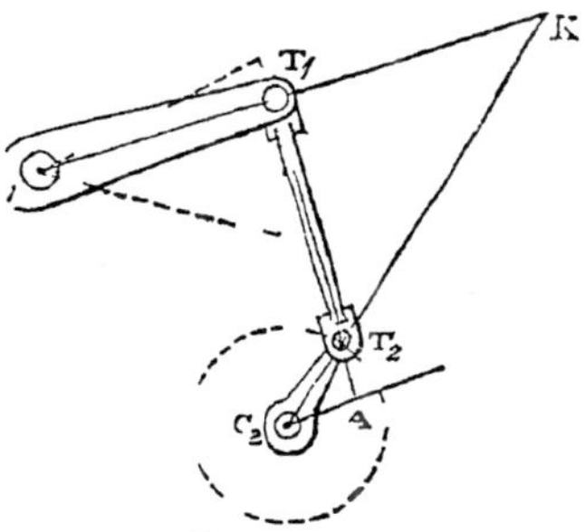

Fig. 69.

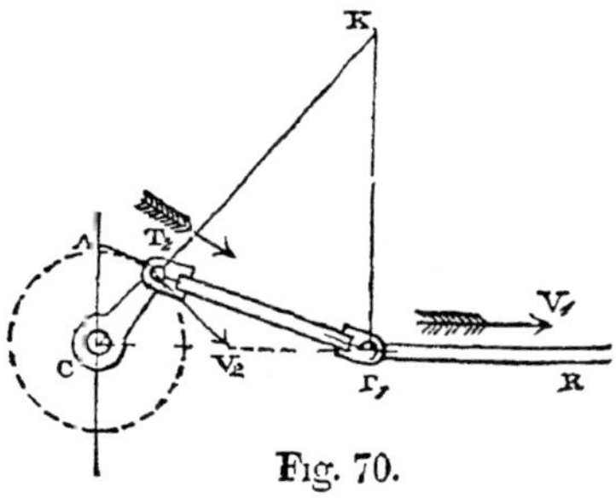

Fig. 70.

Example II. *Rotating piece and sliding piece* (fig. 70). Let C_2 be the axis of a rotating piece, and $T_1 R$ the straight line along which a sliding piece moves. Let T_1, T_2, be the connected points, $C_2 T_2$ the crank arm of the rotating piece, and $T_1 T_2$ the link or connecting rod. The point T_1, T_2, and the line $T_1 R$, are supposed to be in one plane, perpendicular to the axis C. Draw $T_1 K$ perpendicular to $T_1 R$, intersecting $C_2 T_2$ in K; K is the instantaneous axis of the link; and

$$v_1 : v_2 :: K T_1 : K T_2$$

Or otherwise draw from a point C_2, $C_2 A$ perpendicular to $T_1 R$ the direction of motion of the sliding piece, $C_2 T_2$ perpendicular to the direction of motion of the rotating piece, then the line $T_1 T_2$, or a line parallel thereto cuts off the segments $C_2 A$, $C_2 T_2$, or segments proportional thereto, and the velocity ratio of the rotating piece to the sliding piece is as $C_2 T_2$ to $C_2 A$.

176. An **Eccentric** fig. 71 being a circular disc keyed on a shaft, with whose axis its centre does not coincide, and used to give a reciprocating motion to a rod, is equivalent to a crank whose connected point is T, the centre of the eccentric disc, and whose crank arm is C T, the distance of that point from the axis of the shaft, called the *eccentricity*.

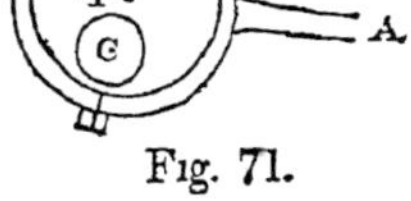

Fig. 71.

177. **The Length of Stroke** of a point in a reciprocating piece is the distance between the two ends of the path in which that point moves. When it is connected by a link with a point in a continuously rotating piece, the ends of the stroke of the reciprocating point correspond with the *dead points* of the continuously revolving piece (Article 173).

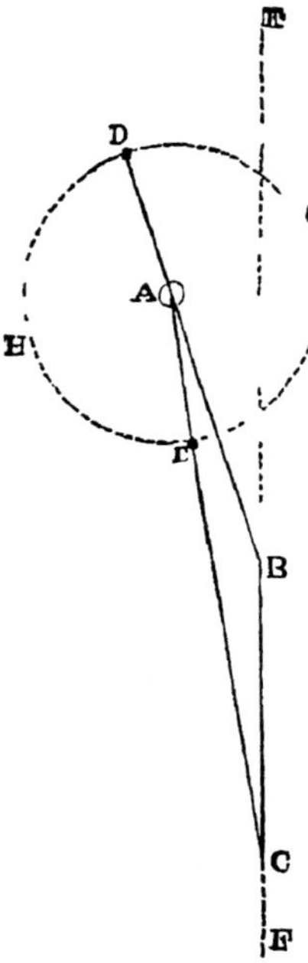

Fig. 72.

Let S B C be the length of stroke of the reciprocating piece, L E C = D B the length of the line f connection, and R A E A D the crank arm of the continuously turning piece. Then if the two ends of the stroke be in one straight line with the axis of the crank,

$$S \quad 2 R; \qquad \ldots\ldots\ldots\ldots\ldots\ldots \; 1.)$$

and if their ends be not in one straight line with that axis, then S, L − R, and L + R, are the three sides of a triangle, having the angle opposite S at that axis; so that if θ be the supplement of the arc between the dead points, D and E,

$$\left. \begin{array}{l} S^2 - (L-R)^2 + (L+R)^2 - 2\,(L-R) \\ \quad (L+R) \cos \theta; \\ S^2 = 2\,(L^2 + R^2) - 2\,(L^2 - R^2) \cos \theta; \\ \cos \theta \quad \dfrac{2\,L^2 + 2\,R^2 \quad S^2}{2(L^2 - R^2)} \end{array} \right\} \ldots \; 2.)$$

178. **Hooke's Universal Joint** fig. 73) is a contrivance for coupling shafts whose axes intersect each other in a point.

Let O be the point of intersection of the axes $O\,C_1$, $O\,C_2$, and i their angle of inclination to each other. The pair of shafts C , C_2, terminate in a pair of forks, F_1, F_2, in bearings at the extremities of which turn the gudgeons at the ends of the arms of a rectangular cross, having its centre at O. This cross is the link; the connected points are the centres of the bearings F_1, F_2. At each instant each of those points moves at right angles to the central plane of its shaft and fork, therefore the line of intersection of the central planes of the two forks, at any instant, is the instantaneous axis of the cross, and the *velocity-ratio* of the

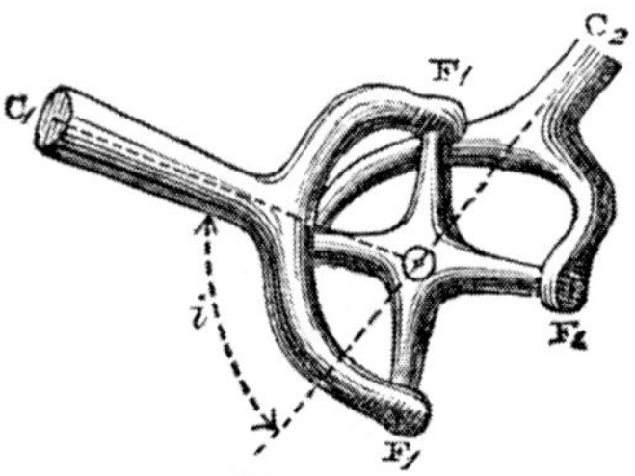

Fig. 73.

points F_1, F_2 (which, as the forks are equal, is also the *angular velocity ratio* of the shafts), is equal to the ratio of the distances of those points from that instantaneous axis. The *mean* value of that velocity-ratio is that of equality; for each successive *quarter turn* is made by both shafts in the same time; but its actual value fluctuates between the limits,

$$\left.\begin{array}{l} \frac{a_2}{a_1} = \frac{r}{r \cdot \sin 90 - i)} = \frac{1}{\cos i} \text{ when } F_1 \text{ is in the plane of the axis;} \\ \frac{a_2}{a_1} = \cos i \text{ when } F_2 \text{ is in that plane.} \end{array}\right\} \ldots\ldots\ldots(1.)$$

179. The **Double Hooke's Joint** (fig. 74) is used to obviate the vibratory and unsteady motion caused by the fluctuation of the velocity-ratio indicated in the equation of Article 178. Between the two shafts to be connected, C_1, C_3, there is introduced a short intermediate shaft C_2, making equal angles with C_1 an l C , connected with each of them by a Hooke's joint, and having both its own forks in the same plane.

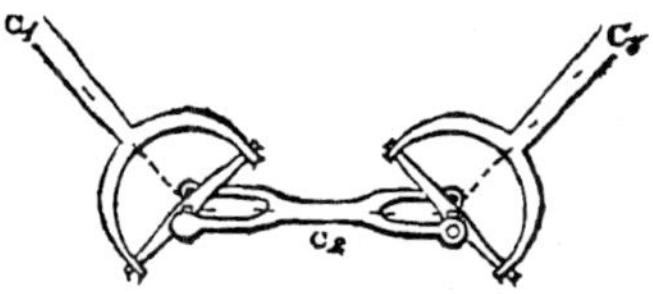

Fig. 74.

By this arrangement the angular velocities of the first and third shafts are equal to each other at every instant.

180. **A Click**, being a reciprocating bar, acting upon a rachet wheel or rack, which it pushes or pulls through a certain arc at each forward stroke, and leaves at rest at each backward stroke, is an example of intermittent linkwork. During the forward stroke, the action of the click is governed by the principles of linkwork; during the backward stroke, that action ceases. A *catch* or *pall*, turning on a fixed axis, prevents the ratchet wheel or rack from reversing its motion.

Section 5.—Reduplication of Cords.

181. *Definitions*—The combination of pieces connected by the several plies of a cord or rope consists of a pair of cases or frames called *blocks*, each containing one or more pulleys called *sheaves*. One of the blocks called the *fall-block*, B_1, is fixed; the other, or *running-block*, B_2, is movable to or from the fall-block, with which

it is connected by means of a rope of which one end is attached either to the fall block or to the running block, while the other end, T_1, called the *fall*, or *tackle fall*, is free; while the intermediate portion of the rope passes alternately round the pulleys in the fall block and running-block. The whole combination is called a *tackle* or *purchase*.

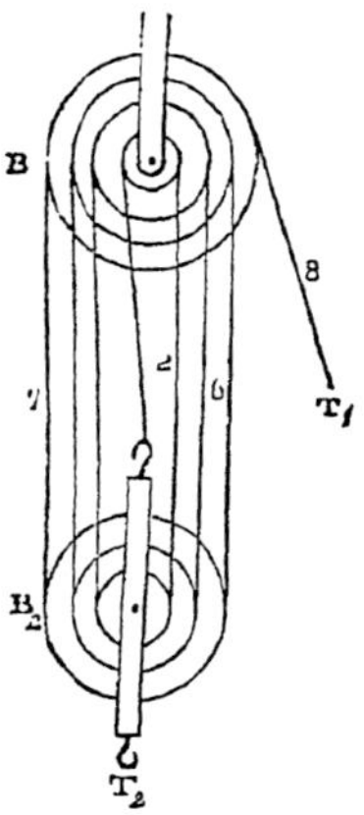

Fig. 75.

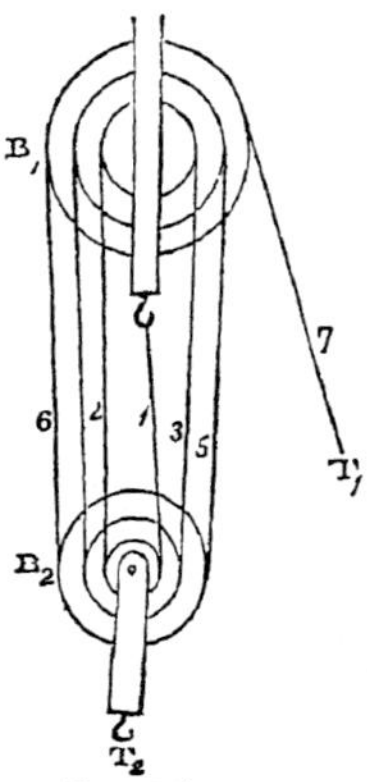

Fig. 75A.

182. The **Velocity-Ratio** chiefly considered in a tackle is that between the velocities of the running block, u, and of the tackle-fall, v. That ratio is given by Equation 3 of Article 113 (which see), viz.:—

$$v = n u; \quad\ldots\ldots\ldots\ldots (1.)$$

where n is the *number of plies* of rope by which the running block is connected with the fall block. Thus, in fig. 75 $n = 7$; and in fig. 75A $n = 6$.

182A. The **Velocity of any Ply** of the rope is found in the following manner:—

I. For a ply on the side of the fall-block next the tackle-fall, such as 2, 4, 6, fig. 75, and 3, 5, fig. 75A, it is to be considered what would be the velocity of that ply if it were itself the tackle-fall. Let that velocity be denoted by v', and let n' be the number of plies *between* the ply in question and the point of attachment by which the first ply (marked 1 in the figures) is fixed to one or other block. Then

$$v' = n' u \quad\ldots\ldots\ldots\ldots (1.)$$

II. For a ply on the side of the fall-block farthest from the tackle fall, the velocity is equal and contrary to that of the next succeeding ply, with which it is directly connected over one of the sheaves of the fall block.

III. If the first ply, as in fig. 75A, is attached to the fall-block, its velocity is nothing; if to the running block, its velocity is equal to that of the block.

183. **White's Tackle.** The sheaves in a block are usually made all of the same diameter, and turn on a fixed pin; and they have, consequently, different angular velocities. But by making the diameter of each sheaf proportional to the velocity, *relatively to the*

block, of the ply of rope which it is to carry, the angular velocities of the sheaves in one block may be rendered equal, so that the sheaves may be made all in one piece, and may have journals turning in fixed bearings. This is called *White's Tackle*, from the inventor, and is represented in figs. 75 and 75A.

SECTION 6.—COMPARATIVE MOTION IN THE "MECHANICAL POWERS."

184. **Classification of the Mechanical Powers.**—"Mechanical Powers" is a name given to certain simple or elementary machines, all of which, with the single exception of the pulley, are more simple than even an elementary combination of a driver and follower; for, with that exception, a mechanical power consists essentially of only one primary moving piece; and the comparative motion taken into consideration is simply the velocity-ratio either of a pair of points in that piece, or of two components of the velocity of one point. There are two established classifications of the mechanical powers; an older classification, which enumerates six; and a newer classification, which ranges the six mechanical powers of the older system under three heads. The following table shews both these classifications:

NEWER CLASSIFICATION.	OLDER CLASSIFICATION.
THE LEVER,	The Lever, The Wheel and Axle.
THE INCLINED PLANE,	The Inclined Plane. The Wedge. The Screw.
THE PULLEY,	The Pulley.

In the present section the comparative motions in the mechanical powers are considered alone. The relations amongst the forces which act in those machines will be treated of in the kinetic division of this Treatise.

In the lever and the wheel and axle of the older classification, which are both comprehended under the lever of the newer classification, the primary moving piece turns about a fixed axis; and the comparative motion taken into consideration is the velocity-ratio of two points in that piece, which may be called respectively the *driving point* and the *following point*. The principle upon which that velocity-ratio depends has already been stated in Article 93, page 50—viz., that the velocity of each point is proportional to the radius of the circular path which it describes; that is, to its perpendicular distance from the axis of motion.

The distinction between the lever and the wheel and axle is

this: that in the *lever*, the driving point, D, and the following point, F, are a pair of determinate points in the moving piece, as in figs. 76A to 76D; whereas in the *wheel* and *axle* they may be any pair of points which are situated respectively in a pair of cylindrical pitch-surfaces, D and F, described about the axis A, fig. 76.

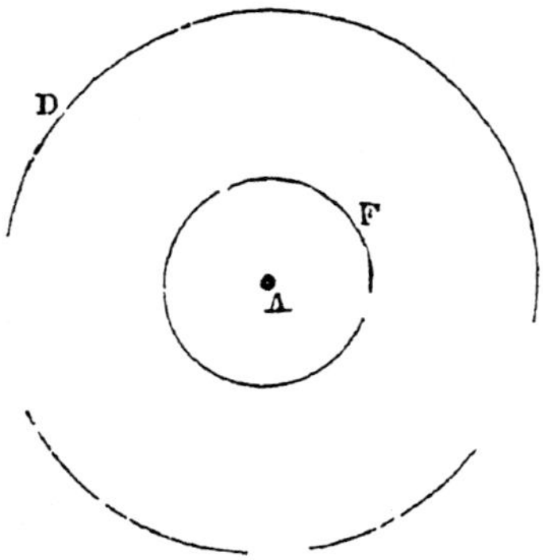

Fig. 76.

In each of these figures the plane of projection is normal to the axis, and A is the trace of the axis. In fig. 76, D and F are the traces of two cylindrical pitch-surfaces. In figs. 76A to 76D, D and F are the projections of the driving and following points respectively.

The axis of a lever is often called the *fulcrum*.

A lever is said to be *straight*, when the driving point, D, and following point, F, are in one plane traversing the axis A, as in figs. 76A, 76B, and 76C. In other cases the lever is said to be *bent*, as in fig. 76D.

Fig. 76A.

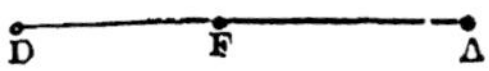

Fig. 76B.

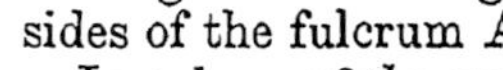

Fig. 76C.

The straight lever is said to be of one or other of three kinds, according to the following classification:—

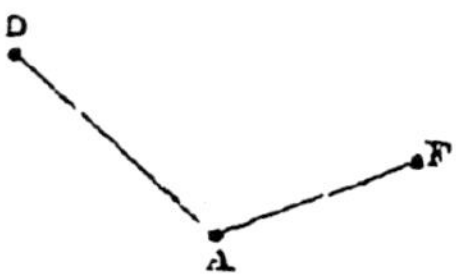

Fig. 76D.

In a *lever of the first kind*, fig. 76A, the driving and following points are at opposite sides of the fulcrum A.

In a *lever of the second kind*, fig. 76B, the driving and following points are at the same side of the fulcrum, and the driving point is the further from the fulcrum.

In a *lever of the third kind*, fig. 76C, the driving and following points are at the same side of the fulcrum, and the following point is the further from the fulcrum.

In the inclined plane, and in the wedge, the comparative motion considered is the velocity-ratio of the entire motion of a straight-sliding primary piece and one of the components of that motion; the principles of which velocity-ratio have been stated in Article 70, pages 38, 39.

In the inclined plane, fig. 76E, A A is the trace of a fixed plane; B, a block sliding on that plane in the direction B C; the plane of projection being perpendicular to the plane A A, and parallel to the direction of motion of B. B D is some direction oblique to B C. From any convenient point, C, in B C, let fall C D perpendicular to B D; then B D ÷ B C is the ratio of the component velocity in the direction B D to the entire velocity of B.

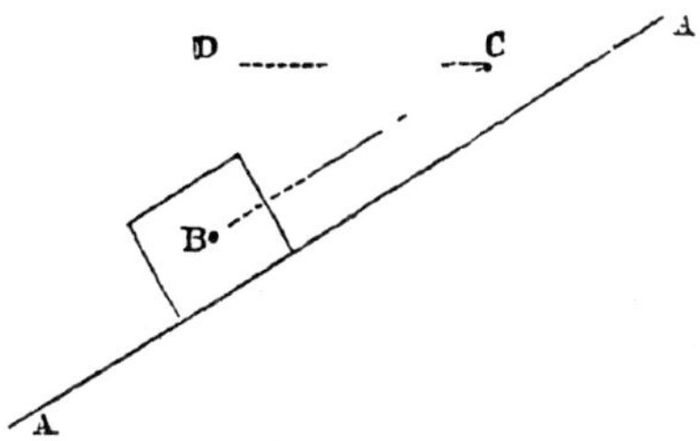

Fig. 76E.

In fig. 76F, A A is the trace of a fixed plane; B C D, the trace of a wedge which slides on that plane. While the wedge advances through the distance C c, its oblique face advances from the position C D to the position $c\ d$; and if C e be drawn normal to the plane C D, the ratio borne by the component velocity of the wedge

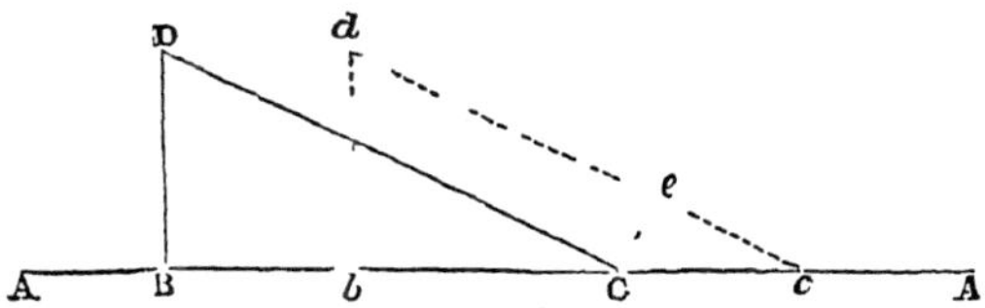

Fig, 76F.

in a direction normal to its oblique face to its entire velocity will be expressed by C e : C c.

In the screw the comparative motion considered is the ratio borne by the entire velocity of some point in, or rigidly connected with, the screw, to the velocity of advance of the screw.

The helical path of motion of a point in, or rigidly attached to, a screw may be developed as has been already explained in Article 160, page 94) into a straight line: being the hypotenuse of a right angled triangle whose height is equal to the pitch of the screw, and its base to the circumference of a circle whose radius is the distance of the given point from the axis of the screw. Then if B D in fig. 76E be taken to represent the pitch of the screw, and D C, perpendicular to B D, the circumference of the circle described by the point in question about the axis, B C will be the development of one turn of the screw-line described by that point as it revolves and advances along with the screw; and B C ÷ B D will be the ratio of its entire velocity to the velocity of advance; just as in the case of a body sliding on an inclined plane, A A, parallel to B C. This shews why the screw is comprehended under the general head of the inclined plane, in the newer classification of the mechanical powers.

The term *pulley*, in treating of the mechanical powers, means any purchase or tackle of the class already described in Section 5 of this Chapter, pages 105 to 107.

Section 7. Hydraulic Connection.

185. **The General Principle** of the communication of motion between two pistons by means of an intervening fluid of constant density has already been stated in Article 110, viz., that the velocities of the pistons are inversely as their areas, measured on planes normal to their directions of motion.

Should the density of the fluid vary, the problem is no longer one of pure mechanism; because in that case, besides the communication of motion from one piston to the other, there is an additional motion of one or other, or both pistons, due to the change of volume of the fluid.

186. **Valves** are used to regulate the communication of motion through a fluid, by opening and shutting passages through which the fluid flows; for example, a cylinder may be provided with valves which shall cause the fluid to flow in through one passage, and out through another. Of this use of valves, two cases may be distinguished.

I. *When the piston moves the fluid*, the valves may be what is called *self-acting*; that is, moved by the fluid. If there be two passages into the cylinder, one provided with a valve opening inwards, and the other with a valve opening outwards; then during the outward stroke of the piston the former valve is opened and the latter shut by the inward pressure of the fluid, which flows in through the former passage; and during the inward stroke of the piston, the former valve is shut and the latter opened by the outward pressure of the fluid, which flows out through the latter passage. This combination of cylinder, piston, and valves, constitutes a *pump*.

II. *When the fluid moves the piston*, the valves must be opened and shut by mechanism, or by hand. In this case the cylinder is a *working cylinder*.

187. **In the Hydraulic Press**, the rapid motion of a small piston in a pump causes the slow motion of a large piston in a working cylinder. The pump draws water from a reservoir, and forces it into the working cylinder: during the outward stroke of the pump piston, the piston of the working cylinder stands still; during the inward stroke of the pump piston, the piston of the working cylinder moves outward with a velocity as much less than that of the pump piston as its area is greater. When the piston of the

working cylinder has finished its outward stroke, which may be of any length, it is permitted to be moved inwards again by opening a valve by hand and allowing the water to escape.

188. **In the Hydraulic Hoist**, the slow inward motion of a large piston drives water from a large cylinder into a smaller cylinder, and causes a more rapid outward motion of the piston of the smaller cylinder. When the latter piston is to be moved inward, a valve between the two cylinders is closed, and the valve of an outlet from the smaller cylinder opened, by hand, so as to allow the water to escape from the smaller cylinder. The larger cylinder is filled and its piston moved outward, when required, by means of a pump, in a manner resembling the action of a hydraulic press.

SECTION 8.—TRAINS OF MECHANISM.

189. **Trains of Elementary Combinations** have been defined in Article 131, and illustrated in the case of wheelwork, in Article 145, and in the case of a double Hooke's joint, in Article 179. The general principle of their action is that the comparative motion of the first driver and last follower is expressed by a ratio, which is found by multiplying together the several velocity ratios of the series of elementary combinations of which the train consists, each with the sign denoting the directional relation.

Two or more trains of mechanism may *converge* into one; as when the two pistons of a pair of steam engines, each through its own connecting rod, act upon one crank shaft. One train of mechanism may *diverge* into two or more; as when a single shaft, driven by a prime mover, carries several pulleys, each of which drives a different machine. The principles of comparative motion in such converging and diverging trains are the same as in simple trains.

CHAPTER III.

ON AGGREGATE COMBINATIONS.

190. **The General Principles** of aggregate combinations have already been given in Part I., Chapter II., Section 3. The problems to which those principles are to be applied may be divided into two classes.

I. Where a secondary moving piece is connected at three, or at two points, as the case may be, with three or with two other pieces whose motions are given; so that the problem is, *from the motions of three or of two points in the secondary piece, to find its motion as a whole, and the motion of any point in it.* The solution of this problem is given in Articles 97 and 98.

II. Where a secondary piece, C, is carried by another piece, B; and denoting the frame of the machine by A, there are given two out of the three motions of A, B, and C, relatively to each other, and the third is required. The motion of C relatively to A is the resultant of the motion of C relatively to B, and of B relatively to A; and the problem is solved by the methods already explained in Articles 99 to 107, inclusive.

Professor Willis distinguishes the effects of aggregate combinations into *aggregate velocities*, whether linear or angular, produced in secondary pieces by the combined action of different drivers, and *aggregate paths*, being the curves, such as cycloids and trochoids, epicycloids and epitrochoids, described by given points in such secondary pieces.

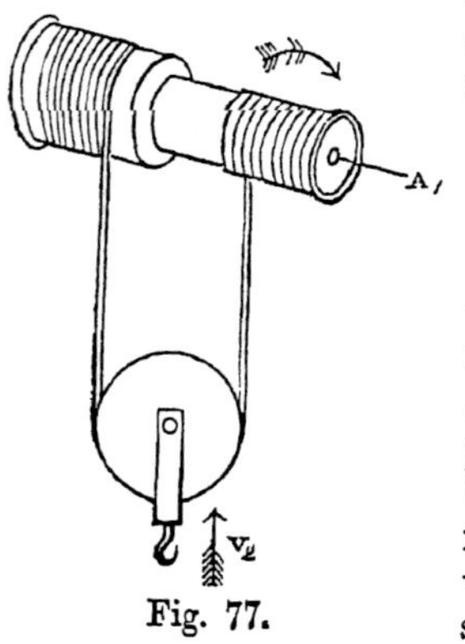

Fig. 77.

The following Articles give examples of two simple aggregate combinations.

191. **Differential Windlass.**—In fig. 77, the axis A_1 carries two barrels of different radii, r_1 being the greater, and r_2 the less. A running block containing a single pulley is hung by a rope which passes below the pulley, and has one end wound round the larger barrel, and the other wound the contrary way round the smaller barrel. When the two barrels rotate together with the common angular velocity a, the division of the rope which hangs from the larger barrel moves with the

velocity $a r_1$, and the division which hangs from the smaller barrel moves in the contrary direction with the velocity $-a r_2$ (whose direction is denoted by the negative sign). These are also the velocities of the two points at opposite extremities of a diameter of the pulley, where it is touched by the two vertical divisions of the rope. The velocity of the centre of the pulley is a mean between those two velocities; that is, their half-difference, because their signs are opposite; or denoting it by v,

$$v = \frac{a(r_1 - r_2)}{2} \text{.........................}(1.)$$

The *instantaneous axis* of the pulley may be found by the method of Article 98, as follows: In fig. 35c, let A and B be the two ends of the horizontal diameter of the pulley, and let $\mathrm{A\,V}_a = a r_1$, and $\mathrm{B\,V}_b = a r_2$ represent their velocities; join $\mathrm{V}_a \mathrm{V}_b$ cutting A B in O; this is the instantaneous axis. Now

$$\mathrm{AO - OB = AC + CO - OB = BC + CO - OB = 2\,OC},$$
$$\mathrm{AO + OB : AO - OB :: AV}_a + \mathrm{BV}_b : \mathrm{AV}_a - \mathrm{BV}_b,$$
$$\mathrm{AB} : \quad \mathrm{2\,OC} :: \quad a(r_1 + r_2) : \quad a(r_1 - r_2);$$

and hence the distance of the instantaneous axis from the centre or moving axis of the pulley is obviously

$$\mathrm{AB} \cdot \frac{r_1 - r_2}{2(r_1 + r_2)} \text{.........................}(2.)$$

The motion of the centre of the pulley is the same with that of a point in a rope wound on a barrel of the radius $\frac{r_1 - r_2}{2}$. The use of the contrivance is to obtain a slow motion of the pulley without using a small, and therefore a weak, barrel.

192. **Compound Screws.**—(Fig. 78). On the same axis let there be two screws $S_1 S_1$, and $S_2 S_2$, of the respective pitches

Fig. 78.

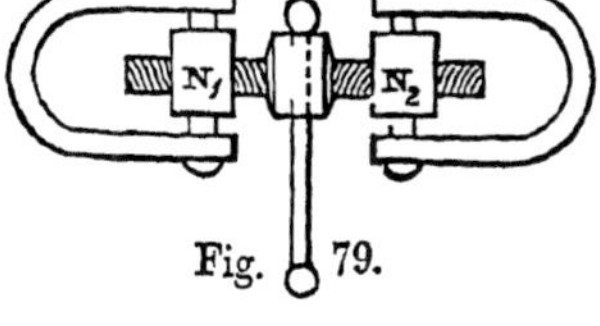

Fig. 79.

p_1 and p_2, p_1 being the greater, and let the screws in the first instance be both right handed or both left-handed. Let N_1 and N_2 be two nuts, fitted on the two screws respectively. When the compound screw rotates with the angular velocity a, the nuts approach towards or recede from each other with the relative velocity

$$v = \frac{a(p_1 - p)}{2\pi}; \qquad (1.)$$

being that due to a screw whose pitch is the *difference* of the two pitches of the compound screw. (See Article 96, Equation 1.) The object of this contrivance is to obtain the slow advance due to a fine pitch, together with the strength of large threads.

Fig. 79 represents a compound screw in which the two screws are contrary-handed, and the relative velocity of the nuts N_1 N is that due to the *sum* of the two pitches; or as they are usually equal, to *double* the pitch of each screw. This combination is used in coupling railway-carriages.

PART III

PRINCIPLES OF STATICS.

CHAPTER I.

SUMMARY OF GENERAL PRINCIPLES.

Nature and Division of the Subject.

The present Chapter contains a summary of the Principles of Statics.

193. **Forces—Action and Re-action.**—Every force is an action exerted between a pair of bodies, tending to alter their condition as to relative rest and motion; it is exerted equally, and in contrary directions, upon each body of the pair. That is to say, if A and B be a pair of bodies acting mechanically on each other, the force exerted by A upon B is equal in magnitude and contrary in direction to the force exerted by B upon A. This principle is sometimes called *the equality of action and re-action*. It is analogous to that of relative motion, explained in Article 61, page 34.

194. **Forces, how Determined and Expressed.**—A force, as respects one of the two bodies between which it acts, is determined, or made known, when the following three things are known respecting it:—*first*, the *place*, or part of the body to which it is applied; *secondly*, the *direction* of its action; *thirdly*, its *magnitude*.

The Place of the application of a force to a body may be the whole of its volume, as in the case of gravity; or the surface at which two bodies touch each other, or the bounding surface between two parts of the same body, as in the case of pressure, tension, shearing stress, and friction.

Thus every force has its action distributed over a certain space, either a volume or a surface; and a force concentrated at a single point has no real existence. Nevertheless, in investigations respecting the action of a distributed force upon the position and movements, as a whole, of a rigid body, or of a body which without error may be treated as rigid, like the solid parts of a machine, fixed or moving, that force may be treated as if it were concentrated at a point or points, determined by suitable processes; and such is the use of those numerous propositions in statics which relate to forces concentrated at points; or *single forces*, as they are called.

The DIRECTION of a force is that of the motion which it tends to produce. A straight line drawn through the points of application of a single force, and along its direction, is the LINE OF ACTION of that force.

The MAGNITUDES of two forces are equal when, being applied to the same body in opposite directions along the same line of action, they balance each other.

The magnitude of a force is expressed arithmetically by stating in numbers its ratio to a certain *unit* or *standard* of force, which, for practical purposes, is usually the *weight* or attraction towards the earth , at a certain latitude, and at a certain level, of a known mass of a certain material. Thus the British unit of force is the *standard pound avoirdupois;* which is the weight, in the latitude of London, of a certain piece of platinum kept in a public office.

For the sake of convenience, or of compliance with custom, other units of weight are occasionally employed in Britain, bearing certain ratios to the standard pound; such as—

The grain $\frac{1}{7000}$ of a pound avoirdupois.
The troy pound 5,760 grains = $0{\cdot}82\dot{2}85714\dot{4}$ pound avoirdupois.
The hundredweight = 112 pounds avoirdupois.
The ton = 2,240 pounds avoirdupois.

The French standard of weight is the *kilogramme*, which is the weight, in the latitude of Paris, of a certain piece of platinum kept in a public office. It was originally intended to be the weight of a cubic decimètre of pure water, measured at the temperature at which the density of water is greatest—viz., 4°·1 Cent., or 39°·4 Fahr., and under the pressure which supports a barometric column of 760 millimètres of mercury; but it is in reality a little heavier.

—— A kilogramme is 2·20462125 lbs. avoirdupois.
A pound avoirdupois is 0·4535926525 of a kilogramme.

For scientific purposes, forces are sometimes expressed in *Absolute Units*. The "Absolute Unit of Force" is a term used to denote the force which, acting on an unit of mass for an unit of time, produces an unit of velocity.

The unit of time employed is always a second.

The unity of velocity is in Britain one foot per second; in France one mètre per second.

The unit of mass is the mass of so much matter as weighs one unit of weight near the level of the sea, and in some definite latitude.

In Britain the latitude chosen is that of London; in France, that of Paris.

In Britain the unit of weight chosen is sometimes a grain, sometimes a pound avoirdupois; and it is equal to 32·187 of the corresponding absolute units of force. In France the unit of weight chosen is either a gramme or a kilogramme, and it is equal to 9·8087 of the corresponding absolute units of force. Each of those coefficients is denoted by the letter g.

195. **Measures of Force and Mass.**—If by the unit of force is understood the weight of a certain standard, such as the avoirdupois pound, then the mass of that standard is $1 \div g$; and the unit of mass is g times the mass of the standard; and this is the most convenient system for calculations connected with mechanical engineering, and is therefore followed in the present work.

But if we take for the unit of mass, the mass of the standard itself, then the unit of force is the *absolute unit;* and the weight of the standard in such units is expressed by g; for g is the velocity which a body's own weight, acting unbalanced, impresses on it in a second. This will be specially treated of in Part V. This is the system employed in many scientific writings, and in particular, in Thomson and Tait's *Natural Philosophy*. It has great advantages in a scientific point of view; but its use in calculations for practical purposes would be inconvenient, because of the prevailing custom of expressing forces in terms of the standard of weight.

196. **Representation of Forces by Lines.**—A single force may be represented in a drawing by a straight line; an extremity of the line indicating the point of application of the force,—the direction of the line, the direction of the force,—and the length of the line, the magnitude of the force, according to an arbitrary scale.

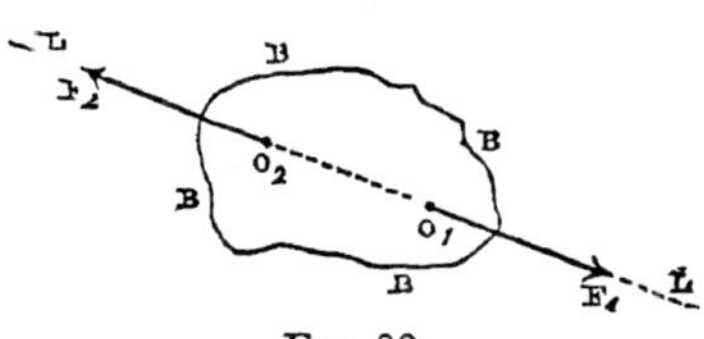

Fig. 80.

For example, in fig. 80, the fact that the body B B B B is acted upon at the point O_1 by a given force, may be expressed by drawing from O_1 a straight line $\overline{O_1 F_1}$ in the direction of the force, and of a length representing the magnitude of the force.

If the force represented by $\overline{O_1 F_1}$ is balanced by a force applied either at the same point, or at another point O_2 (which must be in the line of action L L of the force to be balanced), then the second force will be represented by a straight line $O_2 F_2$, opposite in direction, and equal in length to $\overline{O_1 F_1}$, and lying in the same line of action L L.

If the body B B B B (fig. 81), be balanced by several forces acting in the same straight line LL, applied at points O_1 O_2, &c., and represented by lines $\overline{O_1 F_1}$, $O_2 F_2$, &c.; then either direction in the line L L (such as the direction towards + L) is to be considered as positive, and the opposite direction (such as the direction towards − L) as negative; and if the sum of all the lines representing forces which point positively be equal to the sum of all those which point negatively, the algebraical sum of all the forces is nothing, and the body is balanced.

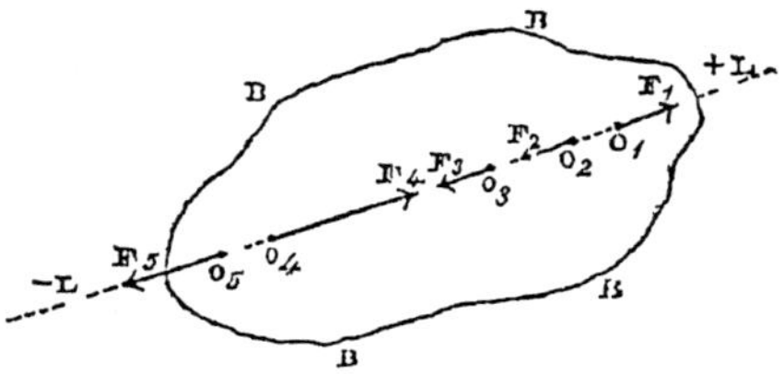

Fig. 81.

197. **Resultant and Component Forces—Their Magnitude.**—The RESULTANT of any combination of forces applied to one body is a single force capable of balancing that single force which balances the combined forces; that is to say, the resultant of the combined forces is equal and directly opposed to the force which balances the combined forces, and is *equivalent* to the combined forces so far as the balance of the body is concerned. The combined forces are called *components* of their resultant.

The resultant of a set of mutually balanced forces is nothing.

The *magnitudes* and *directions* of a resultant force and of its components are related to each other exactly in the same manner with the velocities and directions of resultant and component motions.

As to the *position* of the resultant, if the components act through one point, the resultant acts through that point also; but if the components do not act through one point, the position of the resultant is to be found by methods which will be stated further on.

198. **Equilibrium or Balance** is the condition of two or more forces which are so opposed that their combined action on a body produces no change in its rest or motion, and that each force merely *tends to* cause such change, without actually causing it.

In treatises on statics, the word *pressure* is often used to denote any balanced force; although in the popular sense that word is used to denote a force, of the nature of a thrust or push, distributed over a surface.

199. **Parallel Forces** are forces whose directions of motion are parallel, excepting couples and directly opposed forces.

200. **Couples.**—Two forces of equal magnitude applied to the same body in parallel and opposite directions, but not in the same

line of action (such as F, F, in fig. 82), constitute what is called a "*couple.*"

The *arm* or *leverage* of a couple (L, fig. 82) is the perpendicular distance between the lines of action of the two equal forces.

The tendency of a couple is to turn the body to which it is applied in the *plane* of the couple—that is, the plane which contains the lines of action of the two forces. (The plane in which a body turns is any plane parallel to those planes in the body whose position is not altered by the turning). The turning of a body is said to be *right-handed* when it appears to a spectator to take place in the same direction with that of the hands of a watch, and *left handed* when in the opposite direction; and couples are designated as right handed or left-handed according to the direction of the turning which they tend to produce. The couple represented in fig. 82 appears right-handed to the reader.

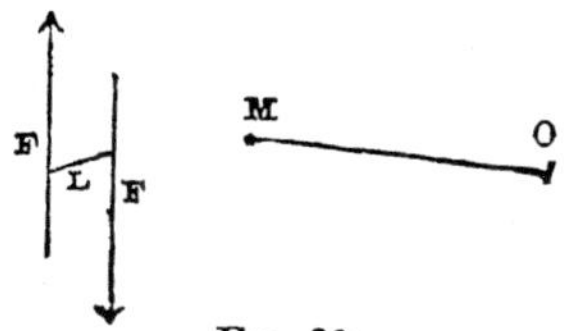

Fig. 82.

The *Moment* of a couple means the product of the magnitude of its force by the length of its arm (F L); and may be represented by the area of a rectangle whose sides are F and L. If the force be a certain number of pounds, and the arm a certain number of feet, the product of those two numbers is called the moment in *foot-pounds*, and similarly for other measures. The moment of a couple may also be represented by a single line on paper, by setting off upon its *axis* (that is, upon any line perpendicular to the plane of the couple) a length proportional to that moment (O M, fig. 82) in such a direction, that to an observer looking from O towards M the couple shall seem right-handed.

201. **The Centre of Parallel Forces** is the single point referred to in the following principle. The forces to which that principle is applied are in general either weights or pressures; and the point in question is then called the *Centre of Gravity* or the *Centre of Pressure*, as the case may be.

If there be given a system of points, and the mutual ratios of a system of parallel forces applied to those points, which forces have a single resultant, then there is one point, and one only, which is traversed by the line of action of the resultant of every system of parallel forces having the given mutual ratios and applied to the given system of points, whatsoever may be the absolute magnitudes of those forces and the angular position of their lines of action.

202. **Distributed Forces in General.**—In Article 194, page 115, it has already been explained, that the action of every real force is distributed throughout some volume, or over some surface. It is always possible, however, to find either a *single resultant*, or a

resultant couple, or a *combination of a single force with a couple*, to which a given distributed force is equivalent, so far as it affects the equilibrium of the body, or part of a body, to which it is applied.

In the application of Mechanics to Structures, the only force distributed throughout the volume of a body which it is necessary to consider, is its *weight*, or attraction towards the earth; and the bodies considered are in every instance so small as compared with the earth, that this attraction may, without appreciable error, be held to act in parallel directions at each point in each body. Moreover, the forces distributed over surfaces are either parallel at each point of their surfaces of application, or capable of being resolved into sets of parallel forces; hence, *parallel distributed forces* have alone to be considered; and every such force is statically equivalent either to a single resultant, or to a resultant couple.

The *intensity of a distributed force* is the ratio which the magnitude of that force, expressed in units of weight, bears to the space over which it is distributed, expressed in units of volume, or in units of surface, as the case may be. An *unit of intensity* is an unit of force distributed over an unit of volume or of surface, as the case may be; so that there are two kinds of units of intensity. For example, *one pound per cubic foot* is an unit of intensity for a force distributed throughout a volume, such as weight; and *one pound per square foot* is an unit of intensity for a force distributed over a surface, such as pressure or friction.

203. **Specific Gravity—Heaviness—Density—Bulkiness.**—I. *Specific Gravity* is the ratio of the weight of a given bulk of a given substance to the weight of the same bulk of pure water at a standard temperature. In Britain the standard temperature is 62° Fahr. 16°·67 Cent. In France it is the temperature of the maximum density of water—3°·94 Cent. 39°·1 Fahr.

In rising from 39°·1 Fahr. to 62° Fahr., pure water expands in the ratio of 1·001118 to 1; but that difference is of no consequence in calculations of specific gravity for engineering purposes.

II. The *heaviness* of any substance is the weight of an unit of volume of it in units of weight. In British measures heaviness is most conveniently expressed in *lbs. avoirdupois to the cubic foot*; in French measures, in *kilogrammes to the cubic decimètre* (or to the litre). The values of the heaviness of water at 39°·1 Fahr., and at 62° Fahr., are respectively 62·425 and 62·355 lbs. to the cubic foot.

III. The *density* of a substance is either the number of units of *mass* in an unit of volume, in which case it is equal to the heaviness,—or the ratio of the mass of a given volume of the substance to the mass of an equal volume of water, in which case it is equal to the specific gravity. In its application to *gases*, the term

"Density" is often used to denote the ratio of the heaviness of a given gas to that of air, at the same temperature and pressure.

IV. The *bulkiness* of a substance is the number of units of volume which an unit of weight fills; and is the *reciprocal of the heaviness.* In British measures bulkiness is most c nveniently expressed in *cubic feet to the lb. avoirdupois;* in French measures, in *cubic decimètres* (or in litres) *to the kilogramme.* Rise of temperature produces (with certain exceptions increase of bulkiness. The linear expansion of a solid body is one-third of its expansion in bulk.

204. **The Centre of Gravity** of a body or of a system of bodies, is the point always traversed by the resultant of the weight of the body or system of bodies,—in other words, the *centre of parallel forces* f r the weight of the body or system of bodies.

To *support* a b dy, that is, to balance its weight, the resultant of the supporting force must act through the centre of gravity.

When the centre of gravity of a *geometrical figure* is spoken of, it is to be understood to mean the point where the centre of gravity would be, if the figure were formed of a substance of uniform heaviness.

205. **The Centre of Pressure** in a plane surface is the point travers d by the resultant of a pressure that is exerted at that surface. When the intensity is unif rm, the centre of pressure is at the *centre of magn tude* of the pressed surface.

206. **The Centre of Buoyancy** of a solid wholly or partly immersed in a liquid is the centre of gravity of the mass of liquid displaced. The resultant pressure of the liquid on the solid is equal to the weight of liquid displaced, and is exerted vertically upwards through the centre of buoyancy.

207. **The Intensity of Pressure** is expressed in units of weight on the unit of area; as pounds on the square inch, or kilogrammes on the square mètre; or by the height of a column of some fluid; or in *atmospheres,* the unit in this case being the average pressure of the atmosphere at the level of the sea.

CHAPTER II.

COMPOSITION, RESOLUTION, AND BALANCE OF FORCES.

SECTION 1.—FORCES ACTING THROUGH ONE POINT.

208. **Resultant of Forces Acting in One Straight Line.**—The resultant of any number of forces acting on one body in the same straight line of action, acts along that line, and is equal in magnitude to the sum of the component forces, it being understood, that when some of the component forces are opposed to the others, the word "*sum*" is to be taken in the algebraical sense; that is to say, that forces acting in the same direction are to be added to, and forces acting in opposite directions subtracted from each other.

When a system of forces acting along one straight line are balanced, the sum of the forces acting in one direction is equal to the sum of the forces acting in the opposite direction.

209. **Resultant and Balance of Inclined Forces—Parallelogram of Forces.**—The smallest number of inclined forces which can balance each other is three. Those three forces must act through one point, and in one plane. Their relation to each other depends on the following theorem, called the "PARALLELOGRAM OF FORCES," from which the whole science of statics may be deduced.

If two forces whose lines of action traverse one point be represented in direction and magnitude by the sides of a parallelogram, their resultant is represented by the diagonal.

For example, through the point O (fig. 83) let two forces act, represented in direction and magnitude by $\overline{OA}$ and $\overline{OB}$. The resultant or equivalent single force of those two forces is represented in direction and magnitude by the diagonal O C of the parallelogram O A C B. Its magnitude is given algebraically by the equation.

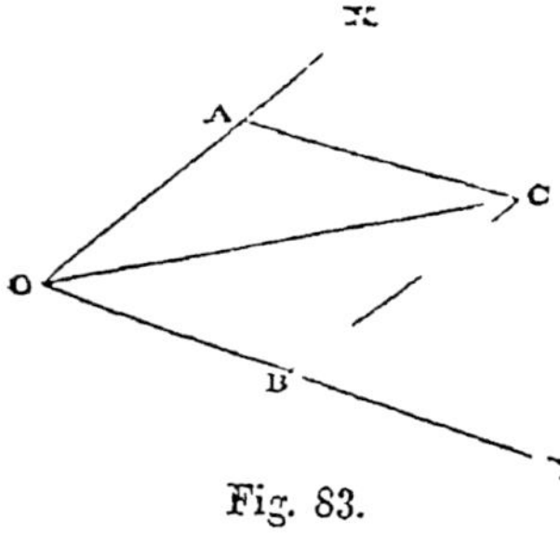

Fig. 83.

$$OC = \sqrt{\left\{ OA^2 + OB^2 + 2 OA \cdot OB \cos AOB \right\}} \qquad (1.)$$

210. **Triangle of Forces.**—To balance the forces $\overline{OA}$ and $\overline{OB}$, a force is required equal and directly opposed to their resultant O C. This may be expressed by saying, that *if the directions and*

magnitudes of three forces be represented by the three sides of a triangle, taken in the same order such as O A, A C, C O, *then those three forces, acting through one point, balance each other*, or in other words, that three forces in the same plane balance each other at one point, when each is proportional to the sine of the angle between the other two.

211. **Polygon of Forces.**—*If a number of forces acting through the same point be represented by lines equal and parallel to the sides of a closed polygon, taken in the same order, those forces balance each other.* To fix the ideas, let there be five forces acting through the point O (fig. 84), and represented in direction and magnitude by the lines F_1, F_2, F_3, F_4, F_5, which are equal and parallel to the sides of the closed polygon O A B C D O; viz :

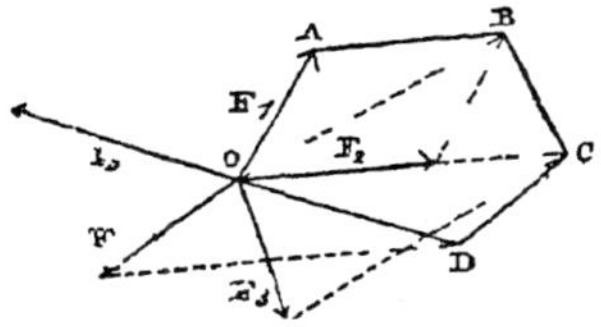

Fig. 84.

F_1 and O A; F_2 and A B; F_3 and B C;
F_4 and C D; F_5 and D O.

Then, by the principle of the parallelogram of forces, the resultant of F_1 and F_2 is O B; the resultant of F_1, F_2, and F_3 is O C; the resultant of F_1, F_2, F_3, and F_4 is O D, equal and opposite to F_5, so that the final resultant is nothing.

The closed polygon may be either plane or "gauche"—that is, not in one plane.

212. **Principles of the Parallelopiped of Forces.** The simplest gauche polygon is one of four sides. Let A O B C E F G H (fig. 85), be a parallelopiped whose diagonal is O H. Then any three successive edges so placed as to begin at O and end at H, form, together with the diagonal H O, a closed quadrilateral; consequently, if three forces F_1, F_2, F_3, acting through O, be represented by the three edges A O, O B, O C, of a parallelopiped, the diagonal O H represents their resultant, and a fourth force F_4 equal and opposite to $\overline{OH}$ balances them.

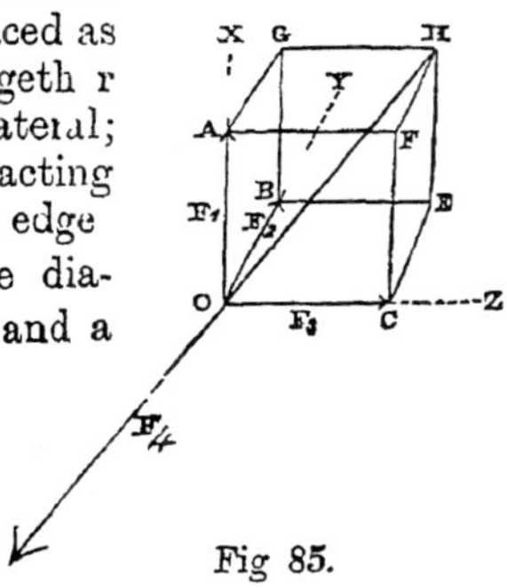

Fig 85.

213. **Resolution of a Force into two Components.**—In order that a given single force may be resolvable into two components acting in given lines inclined to each other, it is necessary, *first*, that the lines of action of those components should intersect the line of action of the given force in one point; and *secondly*, that those three lines of action should be in one plane.

Returning then to fig. 83, let $\overline{OC}$ represent the given force, which it is required to resolve into two component forces, acting in the lines O X, O Y, which lie in one plane with O C, and intersect it in one point O.

Though C draw C A O Y, cutting O X in A, and C B O X, cutting O Y in B. Then will O A and O B represent the component forces required.

Two forces respectively equal to and directly opposed to O A and O B will balance O C.

The magnitudes of the forces are in the following proportions:—

$$\text{O C : O A : O B}$$
$$\text{:: sin A O B : sin B O C : sin A O C} \quad \ldots\ldots\ldots\ldots (1.$$

214. **Resolution of a Force into three Components.**—In order that a given single force may be resolvable into three components acting in given lines inclined to each other, it is necessary that the lines of action of the components should intersect the line of action of the given force in one point.

Returning to fig. 85, let $\overline{OH}$ represent the given force which it is required to resolve into three component forces, acting in the lines O X, O Y, O Z, which intersect O H in one point O.

Through H draw three planes parallel respectively to the planes Y O Z, Z O Y, X O Y, and cutting respectively O X in A, O Y in B, O Z in C. Then will O A, O B, O C, represent the component forces required.

Three forces respectively equal to, and directly opposed to $\overline{OA}$, O B, and O C, will balance O H.

215. **Resolution of a Force. Rectangular Components.**—The rectangular components of a force are those into which it is resolved when the directions of their lines of action are at right angles to each other.

For example, in fig. 85, suppose O X, O Y, O Z, to be three axes of co-ordinates at right angles to each other. Then O H is resolved into three rectangular components, A O, O B, O C, simply by letting fall from H perpendiculars on O X, O Y, O Z, cutting them at A, B, C, respectively.

Let the three rectangular components be denoted respectively by X, Y, Z, the resultant by R, and the angles which it makes with the components by α, β, γ, respectively; then the relations between the three rectangular components and their resultant are expressed by the following equations:—

$$X = R \cos \alpha; \; Y = R \cos \beta; \; Z = R \cos \gamma; \quad \ldots\ldots\ldots (2.)$$

$$R^2 = X^2 + Y^2 + Z^2 \quad \ldots\ldots\ldots\ldots\ldots\ldots\ldots (3.)$$

When the resultant is in the same plane with two of its components (as X and Y), the third component is null, and the Equations 2 and 3 take the following form:—

$$X = R \cos \alpha = R \sin \beta; \; Y = R \cos \beta = R \sin \alpha; \; Z = 0; \ldots (4.)$$

$$R^2 = X^2 + Y^2 \ldots\ldots\ldots\ldots\ldots\ldots (5.)$$

In using Equations 2, 3, 4, and 5, it is to be remembered that cosines of obtuse angles are negative.

216. **Resultant and Balance of any number of inclined Forces acting through one Point.**—To find this resultant by calculation, assume any three directions at right angles to each other as axes; resolve each force into three components (X, Y, Z) along those axes, and consider the components along a given axis which act in one direction as positive, and those which act in the opposite direction as negative; take the algebraical sums of the components along the three axes respectively ($\Sigma \cdot X$, $\Sigma \cdot Y$, $\Sigma \cdot Z$); these will be the *rectangular components of the resultant of all the forces;* and its magnitude and direction will be given by the following equations:—

$$R^2 = (\Sigma \cdot X)^2 + (\Sigma \cdot Y)^2 + (\Sigma \cdot Z)^2; \ldots\ldots\ldots\ldots (1.)$$

$$\cos \alpha = \frac{\Sigma \cdot X}{R}; \; \cos \beta = \frac{\Sigma \cdot Y}{R}; \; \cos \gamma = \frac{\Sigma \cdot Z}{R} \ldots\ldots\ldots (2.)$$

If the forces all act in one plane, two rectangular axes in that plane are sufficient, and the terms containing Z disappear from the equations.

If the forces balance each other, the components parallel to each axis balance each other independently; that is to say, the three following conditions are fulfilled:—

$$\Sigma \cdot X = 0; \; \Sigma \cdot Y = 0; \; \Sigma \cdot Z = 0. \ldots\ldots\ldots\ldots (3.)$$

If the forces all act in one plane, these *conditions of equilibrium* are reduced to two.

SECTION 2.—RESULTANT AND BALANCE OF COUPLES.

217. **Equivalent Couples.**—*If the moments of two couples acting in the same direction and in the same or parallel planes are equal, those couples are equivalent:* that is, their tendencies to turn the body to which they are applied are the same.

The following propositions are the chief consequences of the principle just stated:—

218. **Resultant of Couples.**—The resultant of any number of couples acting in the same or parallel planes is equivalent to a couple whose moment is the algebraical sum of the moments of the combined couples.

219. **Equilibrium of Couples with same Axis.**—Two opposite couples of equal moment in the same or parallel planes balance each other. Any number of couples in the same or parallel plane balance each other when the moments of the right handed couples are together equal to the moments of the left handed couples; in other words, when the resultant moment is nothing—a condition expressed algebraically by

$$\Sigma \cdot FL = 0 \ldots\ldots\ldots\ldots\ldots\ldots\ldots\ldots (1)$$

220. **Parallelogram of Couples.**—If the two sides of a parallelogram represent the axes and moments of two couples acting on the same body in planes inclined to each other, the diagonal of the parallelogram will represent the axis and moment of the resultant couple, which is equivalent to these two.

In other words, three couples represented by the three sides of a triangle, taken in the same order, balance each other.

221. **Polygon of Couples.**—If any number of couples acting on the same body be represented by a series of lines joined end to end, and taken in the same order so as to form sides of a polygon, and if the polygon is closed, those couples balance each other.

These propositions are analogous to corresponding propositions relating to single forces; and couples, like single forces, can be resolved into components acting about two or three given axes.

222. **Resultant of a Couple and Single Force in Parallel Planes.**—Let M denote the moment of a couple applied to a body (fig. 86); and at a point O let a single force F be applied, in a plane parallel to that of the couple. For the given couple substitute an equivalent couple, consisting of a force $-F$ equal and directly opposed to F at O, and a force F acting through the point A, the arm A O perpendicular to F being $=\frac{M}{F}$, and parallel to the plane of the couple M. Then the forces at O balance each other, and F acting through A is the resultant of the single force F applied at O, and the couple M; that is to say, that if with a single force F there be combined a couple M whose plane is parallel to the force, the effect of that combination is to shift the line of action of the force parallel to itself through a distance $OA = \frac{M}{F}$;—to the left if M is right-handed—to the right if M is left-handed.

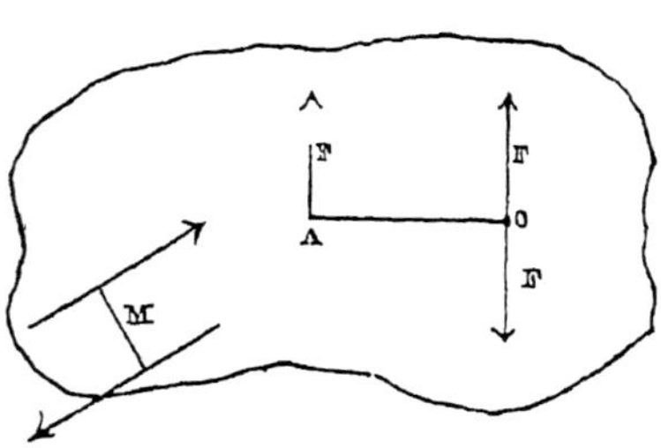

Fig. 86.

223. Moment of Force with respect to an Axis.—In fig. 87, let the straight line F represent a force. Let O X be any straight line perpendicular in direction to the line of action of the force, and not intersecting it, and let A B be the common perpendicular of those two lines. At B conceive a pair of equal and directly opposed forces to be applied in a line of action parallel to F, viz.:— $F' = F$, and $-F' = -F$. The supposed application of such a pair of balanced forces does not alter the statical condition of the body. Then the original single force F, applied in a line traversing A, is equivalent to the force F' applied in a line traversing B, the point in O X which is nearest to A, combined with the couple composed of F and −F', whose moment is F · A B. This is called the *moment of the force* F *relatively to the axis* O X, and sometimes also, the *moment of the force* F *relatively to the plane* traversing O X, parallel to the line of action of the force.

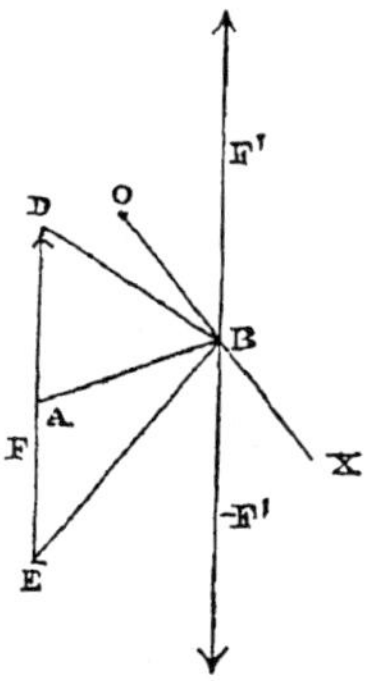

Fig. 87.

If from the point B there be drawn two straight lines B D and B E, to the extremities of the line F representing the force, the area of the triangle B D E being $\frac{1}{2}$ F · A B, represents one-half of the moment of F relatively to O X.

Section 3.—Resultant and Balance of Parallel Forces.

224. Magnitude of Resultant of Parallel Forces.—A balanced system of parallel forces consists either of pairs of directly opposed equal forces, or of couples of equal forces, or of combinations of such pairs and couples.

Hence the following propositions as to the relations amongst the *magnitudes* of systems of parallel forces.

I. In a balanced system of parallel forces the sums of the forces acting in opposite directions are equal; in other words, the algebraical sum of the magnitudes of all the forces taken with their proper signs is nothing.

II. The magnitude of the resultant of any combination of parallel forces is the algebraical sum of the magnitudes of the forces.

The relations amongst the *positions* of the lines of action of balanced parallel forces remain to be shewn; and in this inquiry all pairs of directly opposed equal forces may be left out of consideration; for each such pair is independently balanced whatsoever its position may be; so that the question in each case is to be solved by means of the theory of couples.

The following is the simplest case:—

225. Direction of Resultant of Parallel Forces—Principle of the Lever.—*If three parallel forces applied to one body balance each other, they must be in one plane; the two extreme forces must act in the same direction; the middle force must act in the opposite direction; and the magnitude of each force must be proportional to the distance between the lines of action of the other two.* Let a body fig. 88 be maintained in equilibrio by two opposite couples acting in the same plane, and of equal moments,

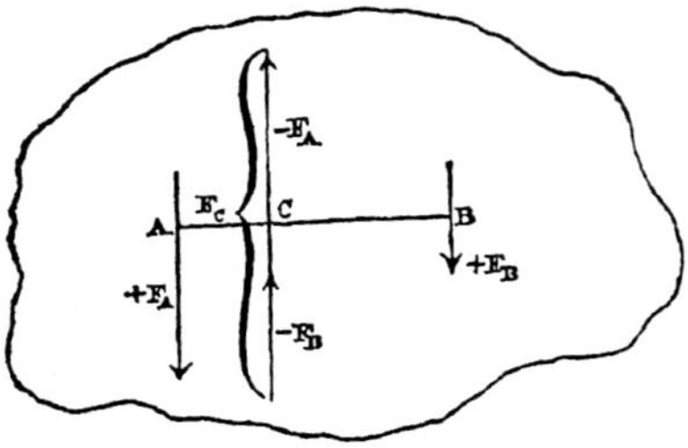

Fig. 88.

$$F_A L_A = F_B L_B,$$

and let those couples be so applied to the body that the lines of action of two of those forces, $-F_A$, $-F_B$, which act in the same direction, shall coincide. Then those two forces are equivalent to the single middle force $F_C = -(F_A + F_B)$, equal and opposite to the sum of the extreme forces $+F_A$, $+F_B$, and in the same plane with them; and if the straight line A C B be drawn perpendicular to the lines of action of the forces, then

$$\overline{AC} = L_A;\ CB = L_B;\ \overline{AB} = L_A + L_B;$$

and consequently

$$F_A : F_B : F_C :: CB : AC : \overline{AB}; \ldots\ldots\ldots\ldots (1)$$

This proposition holds also when the straight line A C B crosses the lines of action of the three forces obliquely.

226. To find the Resultant of Two Parallel Forces.—The resultant is in the same plane with, and parallel to, the components. It is their sum or difference, according as they act in the same or contrary directions; and in the latter case its direction is that of the greater component. To find its line of action by construction, proceed as follows:—Fig. 89 representing the case in which the components act in the same direction, fig. 90 that in which they act in contrary directions. Let A D and B E be the components. Join A E and B D, cutting each other in F. In B D (produced in fig. 90) take B G = D F. Through G draw a line parallel to the components; this will be the line of action of the resultant. To find its magnitude by construction: parallel to A E, draw B C and D H, cutting the line of action of the resultant in C and H; C H will represent the resultant required; and a force equal and opposite to C H will balance A D and B E.

To find the line of action of the resultant by calculation; make either

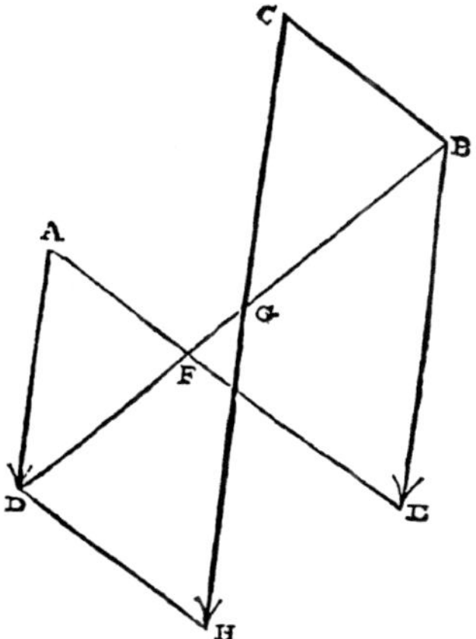

Fig. 89.

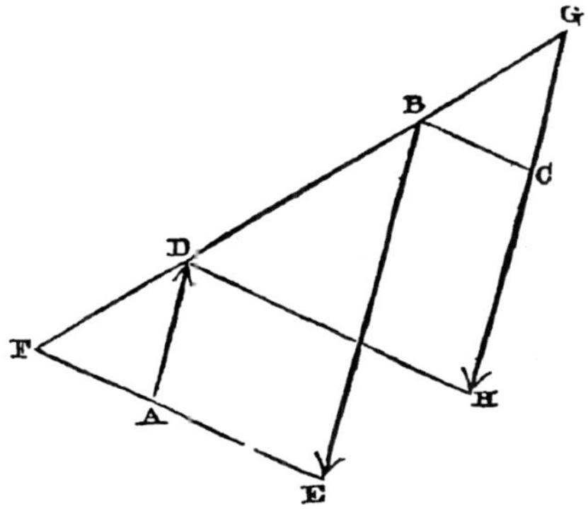

Fig. 90.

$$BG \; \frac{AD \cdot DB}{CH}; \text{ or } DG \; \frac{BE \cdot DB}{CH}.$$

When the two given parallel forces are opposite and equal, they form a couple, and have no single resultant.

227. **To find the Relative Proportions of Three Parallel Forces which Balance each other, Acting in One Plane: their Lines of Action being given.**—Across the three lines of action, in any convenient position, draw a straight line A C B, fig. 91, and measure the distances between the points where it cuts the lines of action. Then each force will be proportional to the distance between the lines of action of the other two. The direction of the middle force, C, is contrary to that of the other two forces, A and B.

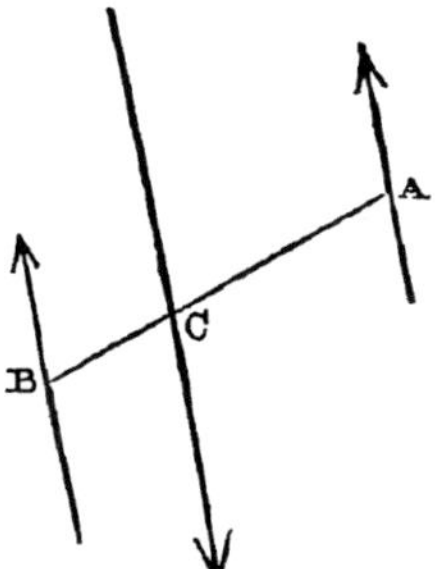

Fig. 91.

In symbols, let A, B, and C be the forces; then,

$$A + B + C - 0; \; AB : BC : CA :: C : A : B.$$

Each of the three forces is equal and opposite to the resultant of the other two; and each pair of forces are equal and opposite to the components of the third. Hence this rule serves to resolve a given force into two parallel components acting in given lines in the same plane.

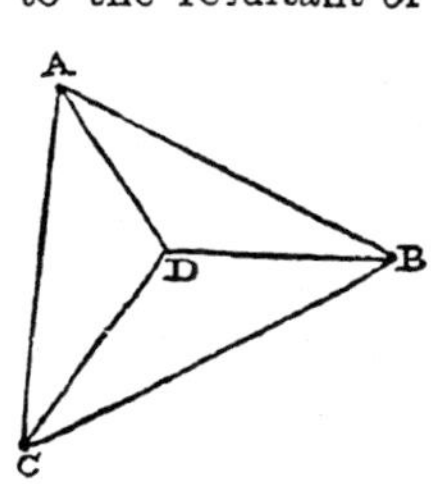

Fig. 92.

228. **To find the Relative Proportions of Four Parallel Forces which Balance each other, not Acting in One Plane: their Lines of Action being given.**—Conceive a plane to cross the lines of action in any convenient position; and in fig. 92 or fig. 93, let A, B, C, D repre-

sent the points where the four lines of action cut the plane. Draw the six straight lines joining those four points by pairs. Then the force which acts through each point will be proportional to the area of the triangle formed by the other three points.

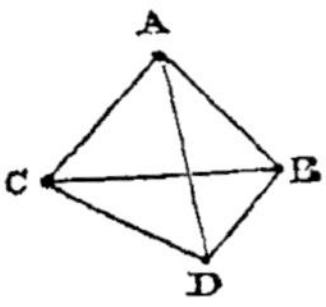

Fig. 93.

In fig. 92 the directions of the forces at A, B, and C are the same, and are contrary to that of the force at D. In fig. 93 the forces at A and D act in one direction, and those at B and C in the contrary direction.

In symbols, $A - B + C + D = 0$;

$$BCD : CDA : DAB : ABC$$
$$:: \; A \; : \; B \; : \; C \; : \; D.$$

Each of the four forces is equal and opposite to the resultant of the other three; and each set of three forces are equal and opposite to the components of the fourth. Hence the rule serves to resolve a force into three parallel components not acting in one plane.

229. **Moments of a Force with respect to a Pair of Rectangular Axes.**—In fig. 94, let F be any single force; O an arbitrarily-assumed point, called the "origin of co-ordinates;" $-$ X O $+$ X, $-$ Y O $+$ Y, a pair of axes traversing O, at right angles to each other and to the line of action of F. Let A B $= y$, be the common perpendicular of F and O X; let A C $= x$, be the common perpendicular of F and O Y. x and y are the "rectangular co-ordinates" of the line of action of F relatively to the axes $-$ X O + X, $-$ Y O + Y, respectively. According to the arrangement of the axes in the figure, x is to be considered as positive to the right, and negative to the left, of $-$ Y O + Y; and y is to be considered as positive to the left, and negative to the right, of $-$ X O + X; right and left referring to the spectator's right and left hand. In the particular case represented, x and y are both positive. Forces, in the figure, are considered as positive upwards, and negative downwards; and in the particular case represented, F is positive.

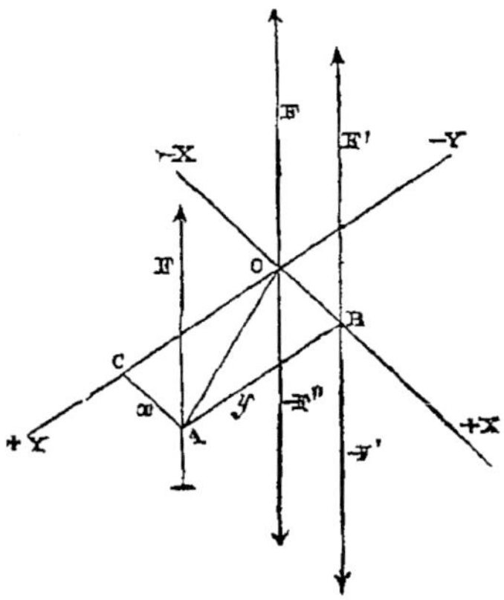

Fig. 94.

At B conceive a pair of equal and opposite forces, F′ and − F′, to be applied; F′ being equal and parallel to F, and in the same direction. Then, as in Article 223, F is equivalent to the single force

F = F applied at B, combined with the couple constituted by F and $-$ F with the arm y, whose moment is y F; being positive in the case represented, because the couple is right-handed. Next, at the origin O, conceive a pair of equal and opposite forces, F′ and $-$ F′, to be applied, F′ being equal and parallel to F and F′, and in the same direction. Then the single force F′ is equivalent to the single force F′ = F′ = F applied at O, combined with the couple constituted by F′ and $-$ F′ with the arm OB = x, whose moment is $-x$ F; being negative in the case represented, because the couple is left handed.

Hence, it appears finally, that a force F acting in a line whose co-ordinates with respect to a pair of rectangular axes perpendicular to that line are x and y, is equivalent to an equal and parallel force acting through the origin, combined with two couples whose moments are,

y F relatively to the axis O X, and $-x$ F relatively to the axis O Y; right handed couples being considered positive; and O Y lying to the left of O X, as viewed by a spectator looking from + X towards O, with his head in the direction of positive forces.

230. **Balance of any System of Parallel Forces in one Plane.**—In order that any system of parallel forces whose lines of action are in one plane may balance each other, it is necessary and sufficient that the following conditions should be fulfilled:

First,—As already stated, that the algebraical sum of the forces shall be nothing.

Secondly,—That the algebraical sum of the moments of the forces relatively to any axis perpendicular to the plane in which they act shall be nothing,

two conditions which are expressed symbolically as follows:—Let F denote any one of the forces, considered as positive or negative, according to the direction in which it acts; let y be the perpendicular distance of the line of action of this force from an arbitrarily assumed axis O X, y also being considered as positive or negative, according to its direction; then,

$$\Sigma \cdot F = 0; \ \Sigma \cdot y F = 0.$$

In summing moments, right-handed couples are usually considered as positive, and left handed couples as negative.

231. Let R denote the **Resultant of any System of Parallel Forces in one Plane**, and y_r, the distance of the line of action of that resultant from the assumed axis O X to which the positions of forces are referred; then,

$$R = \Sigma \cdot F;$$

$$y_r = \frac{\Sigma \cdot y F}{\Sigma \cdot F}.$$

In some cases the forces may have no single resultant, $\Sigma \cdot \mathrm{F}$ being 0; and then, unless the forces balance each other completely, their resultant is a couple of the moment $\Sigma \cdot y\,\mathrm{F}$.

232. **Balance of any System of Parallel Forces.**—In order that any system of parallel forces, whether in one plane or not, may balance each other, it is necessary and sufficient that the three following conditions shall be fulfilled:—

First—(As already stated) that the algebraical sum of the forces shall be nothing.

Secondl.j and Thirdl.j That the algebraical sums of the moments of the forces, relatively to a pair of axes at right angles to each other, and to the lines of action of the forces, shall each be n thing, two conditions which are expressed symbolically as follows: Let O X and O Y den te the pair of axes; let F be the magnitude of any one of the forces; y its perpendicular distance from O X, and x its perpendicular distance from O Y; then,

$$\Sigma \cdot \mathrm{F} \quad 0; \; \Sigma \cdot y\,\mathrm{F} \quad 0; \; \Sigma \cdot x\,\mathrm{F} \quad 0;$$

233. Let R d.note the **Resultant of any System of Parallel Forces**, and x_r and y_r the distances of its line of action from two rectangular axes; then,

$$\mathrm{R} \quad \Sigma \cdot \mathrm{F}; \; x_r \quad \frac{\Sigma \cdot x\,\mathrm{F}}{\Sigma \cdot \mathrm{F}}; \; y_r - \frac{\Sigma \cdot y\,\mathrm{F}}{\Sigma \cdot \mathrm{F}}.$$

In some cases the forces may have no single resultant, $\Sigma \cdot \mathrm{F}$ being 0; and then, unless the forces balance each other completely, their resultant is a cou.le, whose axis, direction, and moment, are found as follows:—

$$\text{Let} \qquad \mathrm{M}_x \quad \Sigma \cdot y\,\mathrm{F}; \; \mathrm{M}_y \quad -\Sigma \cdot x\,\mathrm{F};$$

be the moments of the pair of partial resultant couples about the axes O X and O Y respectively. From O, along those axes, set off two lines representing respectively M_x and M ; that is to say, pro-.ortional to those moments in length, and pointing in the direction from which those couples must respectively be viewed in order that they may appear right handed. Complete the rectangle whose sides are those lines; its diagonal will repres nt the axis, direction, and moment of the final resultant couple. Let M be the moment of this couple; then

$$\mathrm{M}_r \quad \sqrt{\left\{ \mathrm{M}_x^2 + \mathrm{M}_y^2 \right\}}$$

and if θ be the angle which its axis makes with O X,

$$\cos\theta \quad \frac{\mathrm{M}_x}{\mathrm{M}_r}.$$

234. To find the Centre of Parallel Forces.—Let O in fig. 95 be any convenient point, taken as the origin of co-ordinates, and O X, O Y, O Z, three axes of co-ordinates at right angles to each other.

Let A be any one of the points to which the system of parallel forces in question is applied. From A draw x parallel to O X, and perpendicular to the plane Y Z, y parallel to O Y, and perpendicular to the plane Z X, and z parallel to O Z, and perpendicular to the plane X Y. x, y, and z are the rectangular co-ordinates of A, which, being known, the position of A is determined. Let F denote either the magnitude of the force applied at A, or any magnitude proportional to that magnitude. x, y, z, and F are supposed to be known for every point of the given system of points.

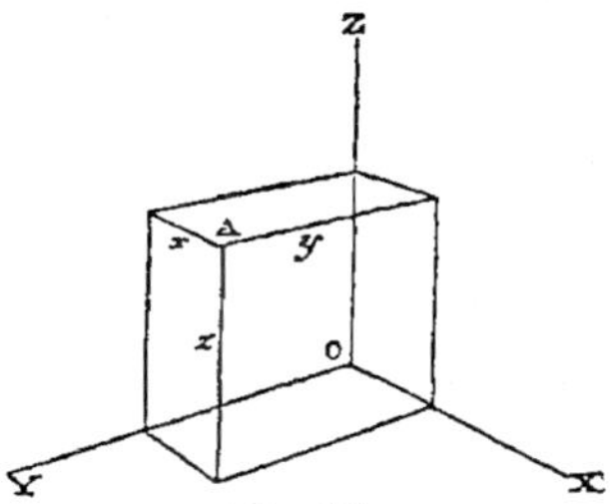

Fig. 95.

The position of the centre of parallel forces depends solely on the *proportionate* magnitudes of the parallel forces, not on their absolute magnitudes, nor on the *angular positions of their lines of actions;* so that for any system of parallel forces another may be substituted in any angular position : this is the statement of the principle of the centre of parallel forces given at Article 201, page 119. This is evident since, in considering the relations of parallel forces, they are not considered with reference to any particular plane, and hence these relations must hold for any plane.

First, conceive all the parallel forces to act in lines parallel to the plane Y Z. Then the distance of their resultant, and of the centre of parallel forces from that plane is

$$x_r = \frac{\Sigma \cdot x F}{\Sigma \cdot F} \quad \ldots\ldots\ldots\ldots\ldots\ldots\ldots\ldots (1.)$$

Secondly, conceive all the parallel forces to act in lines parallel to the plane Z X. Then the distance of their resultant, and of the centre of parallel forces from that plane is

$$y_r = \frac{\Sigma \cdot y F}{\Sigma \cdot F} \quad \ldots\ldots\ldots\ldots\ldots\ldots\ldots\ldots (2.)$$

Thirdly, conceive all the parallel forces to act in lines parallel to the plane X Y. Then the distance of their resultant, and of the centre of parallel forces from that plane is

$$z_r = \frac{\Sigma \cdot z F}{\Sigma \cdot F} \quad \ldots\ldots\ldots\ldots\ldots\ldots\ldots\ldots (3.)$$

If the forces have no single resultant, so that $\Sigma \cdot F = 0$, there is no centre of parallel forces. This may be the case with pressures, but not with weights.

If the parallel forces applied to a system of points are all equal and in the same direction, it is obvious that the distance of the centre of parallel forces from any given plane is simply the mean of the distances of the points of the system from that plane.

SECTION 4. OF ANY SYSTEM OF FORCES.

235. **Resultant and Balance of any System of Forces in One Plane.** Let the plane be that of the axes O X and O Y in fig. 95; and in looking from Z towards O, let Y lie to the right of X, so that rotation from X towards Y shall be right handed. Let x and y be the co-ordinates of the point of application of one of the forces, or of any point in its line of action, relatively to the assumed origin and axes. Resolve each force into two rectangular components X and Y, as in Article 215, page 125; then the rectangular components of the resultant are $\Sigma \cdot X$ and $\Sigma \cdot Y$; its magnitude is given by the equation

$$R^2 = (\Sigma \cdot X)^2 + (\Sigma \cdot Y)^2, \dots\dots (1.)$$

and the angle α_r which it makes with O X is found by the equations

$$\cos \alpha_r = \frac{\Sigma \cdot X}{R}; \ \sin \alpha_r = \frac{\Sigma \cdot Y}{R} \dots\dots (2.)$$

This angle is acute or obtuse according as $\Sigma \cdot X$ is positive or negative; and it lies to the right or left of O X according as $\Sigma \cdot Y$ is positive or negative.

The resultant moment of the system of forces about the axis O Z is*

$$M = \Sigma (x Y - y X), \dots\dots (3.)$$

and is right or left-handed according as M is positive or negative.

The perpendicular distance of the resultant force R from O is

$$L = \frac{M}{R} \dots\dots (4.)$$

Let x_r and y_r be the co-ordinates of any point in the line of action of that resultant; then the equation of that line is*

$$x_r \Sigma \cdot Y - y_r \Sigma \cdot X = M \dots\dots (5.)$$

* The method of obtaining this result by Co-ordinate Geometry is the

If M 0 the resultant acts through the origin O; if M has magnitude, an l R 0 (in which case $\Sigma \cdot X$ 0, $\Sigma \cdot Y$ 0 the r sultant is a couple. The conditi ns of equilibrium of the system of forces are

$$\Sigma \cdot X \quad 0; \; \Sigma \cdot Y \quad 0; \; M \quad 0. \text{ ....(6.}$$

236. **Resultant and Balance of any System of Forces.**—To find the resultant and the conditi ns of equilibrium of any system of forces acting through any system of p ints, the forces an l points are to be referred to three rectangular axes of co-ordinates.

As b f re, let O in fig. 95, p. 133, denote the origin of co or-dir at s, and O X, O Y, O Z, the three rectangular axes: and let them b arranged so that in looking from

$$\left.\begin{matrix} X \\ Y \\ Z \end{matrix}\right\} \text{towards O, rotati n fr m} \left\{\begin{matrix} Y \text{ towards } Z \\ Z \text{ towards } X \\ X \text{ towards } Y \end{matrix}\right\}$$

shall a] pear right handed.

L t X, Y, Z, den te the rectangular components of any one of the f rces; x, y, z, the c -ordinates of a point in its line of acti n.

Taking the algebraical sums of all the forces which act along the same axes, and of all the couples which act round the same axes,

f ll wing: L t O C–L, A B R, $\angle$ X A Y α_r; and let E G–x_r and O G–y_r be the co ordinates of the] int E Th n by Trig nometry sin α_r sin O A C cos C O A sin D O G c s D G O=sin E G F and –cos α_r c s O A C sin C O A c s D O G.

L D C+O D F E+OD
–E G · sin E G F+
O G · cos D O G
$x_r \cdot \sin \alpha_r \quad y_r \cdot \cos \alpha_r$

multiplying by R

L · R M $x_r \cdot R \cdot \sin \alpha_r$
$y_r \cdot R \cdot \cos \alpha_r$
$x_r \cdot \Sigma Y - y_r \cdot \Sigma X$

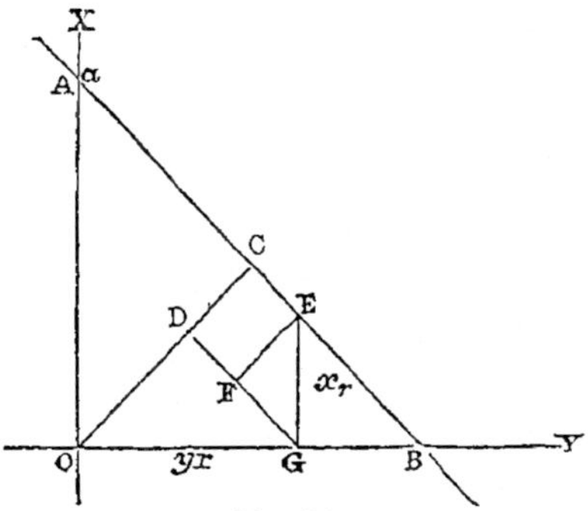

Fig. 96.

by substituting the values in Equation 2 *supra*. Equation 3 was thus obtained.

the six following quantities are found, which c mpose the resultant of the given system of forces:—

Forces.

$$\Sigma \cdot X; \; \Sigma \cdot Y; \; \Sigma \cdot Z; \; \ldots\ldots\ldots\ldots\ldots\ldots (1.)$$

Couples.

$$\left.\begin{array}{lll} \text{about } OX; & M_1 & \Sigma(yZ - zY); \\ \quad ,, \quad OY; & M_2 & \Sigma(zX - xZ); \\ \quad ,, \quad OZ; & M_3 & \Sigma(xY - yX); \end{array}\right\} \ldots\ldots\ldots\ldots (2.)$$

found as already explained in the footnote to Article 235.

The three forces are equivalent to a single f ıce

$$R \sqrt{\left\{ (\Sigma \cdot X)^2 \; (\Sigma \cdot Y)^2 + (\Sigma \cdot Z)^2 \right\}} \ldots\ldots\ldots (3.)$$

acting through O in a line which makes with the axes the angles given by the equations

$$\cos\alpha \; \frac{\Sigma \cdot X}{R}; \; \cos\beta - \frac{\Sigma \cdot Y}{R}; \; \cos\gamma - \frac{\Sigma \cdot Z}{R} \ldots\ldots\ldots (4.)$$

The three couples, M_1, M_2, M_3, are equivalent to one cou le, whose magnitude is given by the equation

$$M \; \sqrt{(M_1^2 + M_2^2 \; M_3^2)}, \ldots\ldots\ldots\ldots\ldots\ldots (5.)$$

and whose axis makes with the axes of co-ordinates the angles given by the equations

$$\cos\lambda \; \frac{M_1}{M}; \; \cos\mu \; \frac{M_2}{M}; \; \cos\nu \; \frac{M_3}{M}, \ldots\ldots\ldots\ldots\ldots\ldots (6.)$$

in which $\left\{\begin{matrix}\lambda\\ \mu\\ \nu\end{matrix}\right\}$ denote respectively the angles made by the axis of M with $\left\{\begin{matrix}OX\\ OY\\ OZ\end{matrix}\right\}$

The *onditions of equilibrium* of the system of forces may be expressed in either of the two following foıms:—

$$\Sigma \cdot X = 0; \; \Sigma \cdot Y \; 0; \; \Sigma \cdot Z = 0; \; M_1 = 0; \; M_2 = 0; \; M_3 \; 0; \; (7.)$$

or

$$R = 0; \; M = 0 \ldots\ldots\ldots\ldots\ldots\ldots\ldots\ldots (8.)$$

When the system is n t balanced, its resultant may fall under one or other of the following cases:—

CASE I.—*When M* 0, the ıesultant is the single force R acting through O.

CASE II. *When the axis of* M *is at right angles to the direction of* R,—a case expressed by the following equation:—

$$\cos \alpha \cos \lambda + \cos \beta \cos \mu + \cos \gamma \cos \nu = 0; \ldots\ldots (9.)$$

(an equation of Co-ordinate Geometry)

the resultant of M and R is a single force equal and parallel to R, acting in a plane perpendicular to the axis of M, and at a perpendicular distance from O given by the equation

$$L = \frac{M}{R} \ldots\ldots\ldots\ldots\ldots\ldots\ldots\ldots (10.)$$

CASE III. *When* R = 0, there is no single resultant; and the only resultant is the couple M.

CASE IV. *When the axis of* M *is parallel to the line of action of* R, that is, when either

$$\lambda = \alpha;\ \mu = \beta;\ \nu = \gamma, \ldots\ldots\ldots\ldots\ldots\ldots (11.)$$

or

$$\lambda = -\alpha;\ \mu = -\beta;\ \nu = -\gamma; \ldots\ldots\ldots\ldots (12.)$$

there is no single resultant; and the system of forces is equivalent to the force R and the couple M, being incapable of being farther simplified.

CASE V.—*When the axis of* M *is oblique to the direction of* R, making with it the angle given by the equation

$$\cos \theta = \cos \lambda \cos \alpha + \cos \mu \cos \beta + \cos \nu \cos \gamma, \ldots. (13.)$$

the couple M is to be resolved into two rectangular components, viz:—

M sin θ round an axis perpendicular to R, and in the plane containing the direction of R and of the axis of M; M cos θ round an axis parallel to R.	(14.)

The force R and the couple M sin θ are equivalent, as in Case II., to a single force equal and parallel to R, whose line of action is in a plane perpendicular to that containing R and axis of M, and whose perpendicular distance from O is

$$L = \frac{M \sin \theta}{R} \ldots\ldots\ldots\ldots\ldots\ldots\ldots\ldots (15.)$$

The couple M cos θ, whose axis is parallel to the line of action of R, is incapable of further combination.

Hence it appears finally, that every system of forces which is not self-balanced, is equivalent either, (A); to a single force, as in

Cases I. and II. (B); to a couple, as in Case III. (C); to a force, combined with a couple whose axis is parallel to the line of action of the force, as in Cases IV. and V. This can occur with inclined forces only; for the resultant of any number of parallel forces is either a single force or a couple.

237. **Parallel Projections or Transformations in Statics.**—If two figures be so related, that for each point in one there is a corresponding point in the other, and that to each pair of equal and parallel lines in the one, there corresponds a pair of equal and parallel lines in the other, those figures are said to be PARALLEL PROJECTIONS of each other.

The relations between such a pair of figures is expressed algebraically as follows: Let any figure be referred to axes of co-ordinates, whether rectangular or oblique; let x, y, z, denote the co-ordinates of any point in it, which may be denoted by A: let a second figure be constructed from a second set of axes of co-ordinates, either agreeing with, or differing from, the first set as to rectangularity or obliquity; let x', y', z', be the co-ordinates in the second figure, of the point A′ which corresponds to any point A in the first figure, and let those co-ordinates be so related to the co-ordinates of A, that for each pair of corresponding points, A, A′, in the two figures, the three pairs of corresponding co-ordinates shall bear to each other three constant ratios, such as

$$\frac{x'}{x} = a; \quad \frac{y'}{y} = b; \quad \frac{z'}{z} = c;$$

then are those two figures parallel projections of each other.

For example, all circles and ellipses are parallel projections of each other; so are all spheres, spheroids, and ellipsoids; so are all triangles; so are all triangular pyramids; so are all cylinders; so are all cones.

The following are the geometrical properties of parallel projections which are of most importance in statics:—

I. A parallel projection of a system of three points, lying in one straight line and dividing it in a given proportion, is also a system of three points, lying in one straight line and dividing it in the same proportion.

II. A parallel projection of a system of parallel lines, whose lengths bear given ratios to each other, is also a system of parallel lines whose lengths bear the same ratios to each other.

III. A parallel projection of a closed polygon is a closed polygon.

IV. A parallel projection of a parallelogram is a parallelogram.

V. A parallel projection of a parallelopiped is a parallelopiped.

VI. A parallel projection of a pair of parallel plane surfaces, whose areas are in a given ratio, is also a pair of parallel plane surfaces, whose areas are in the same ratio.

VII. A parallel projection of a pair of volumes having a given ratio, is a pair of volumes having the same ratio.

The following are the mechanical properties of parallel projections in connection with the principles set forth in this section:—

VIII. If two systems of points be parallel projections of each other; and if to each of those systems there be applied a system of parallel forces bearing to each other the same system of ratios, then the *centres of parallel forces* for those two systems of points will be parallel projections of each other, mutually related in the same manner with the other pairs of corresponding points in the two systems.

IX. If a *balanced system of forces* acting through any system of points be represented by a system of lines, then will any parallel projection of that system of lines represent a balanced system of forces; and if any two systems of forces be represented by lines which are parallel projections of each other, the lines, or sets of lines, representing their *resultants*, are corresponding parallel projections of each other, it being observed that *couples* are to be represented by pairs of lines, as pairs of opposite forces, or by areas, and not by single lines along their axes.

CHAPTER III.

DISTRIBUTED FORCES.

SECTION 1.—CENTRES OF GRAVITY.

238. **Centre of Gravity of a Symmetrical Homogeneous Body.** — If a body is *homogeneous*, or of equal specific gravity throughout, and so far *symmetrical* as to have a *centre of figure*; that is, a point within the body, which bisects every diameter of the body drawn through it, that point is also the centre of gravity of the body.

Amongst the bodies which answer this description, are the sphere, the ellipsoid, the circular cylinder, the elliptic cylinder, prisms whose bases have centres of figure, and parallelopipeds, whether right or oblique.

239. **The Common Centre of Gravity of a Set of Bodies** whose several centres of gravity are known, is the *centre of parallel forces* for the weights of the several bodies, each considered as acting through its centre of gravity. (See Article 234, p. 133.)

240. **Planes of Symmetry—Axes of Symmetry.** —If a homogeneous body be of a figure which is *symmetrical* on either side of a given plane, the centre of gravity is in that plane. If two or more such *planes of symmetry* intersect in one line, or *axis of symmetry*, the centre of gravity is in that axis. If three or more planes of symmetry intersect each other in a point, that point is the centre of gravity.

241. To find the **Centre of Gravity of a Homogeneous Body of any Figure,** assume three rectangular co-ordinate planes in any convenient position, as in fig. 95, p. 133.

To find the distance of the centre of gravity of the body from one of those planes for example, that of Y Z, conceive the body to be divided into indefinitely thin plane layers parallel to that plane. Let s denote the area of any one of those layers, and $d\,x$ its thickness, so that $s\,d\,x$ is the volume of the layer, and

$$V = \int s\,d\,x,$$

the volume of the whole body, being the sum of the volumes of the layers. Let x be the perpendicular distance of the centre of the layer $s\,d\,x$ from the plane of Y Z. Then the perpendicular distance x_0 of the centre of gravity of the body from that plane is given by the equation

$$x_0 = \frac{\int x\,s\,d\,x}{V} \text{ ....(1.)}$$

Find, by a similar process, the distances y_0, z_0, of the centre of gravity from the other two co-ordinate planes, and its position will be completely determined.

If the centre of gravity is previously known to be in a particular plane, it is sufficient to find by the above process its distances from *two* planes perpendicular to that plane and to each other.

If the centre of gravity is previously known to be in a particular line, it is sufficient to find its distance from *one* plane, perpendicular to that line.

242. **If the Specific Gravity of the Body Varies,** let w be the mean heaviness of the layer $s\,d\,x$, so that

$$W = \int w\,s\,d\,x,$$

is the weight of the body. Then

$$x_0 = \frac{\int x\,w\,s\,d\,x}{W} \text{ ....(2.)}$$

243. **Centre of Gravity found by Addition.**—When the figure of a body consists of parts, whose respective centres of gravity are known, the centre of gravity of the whole is to be found as in Article 239.

244. **Centre of Gravity found by Subtraction.**—When the figure of a homogeneous body, whose centre of gravity is sought, can be made by taking away a figure whose centre of gravity is known from a larger figure whose centre of gravity is known also, the following method may be used:—

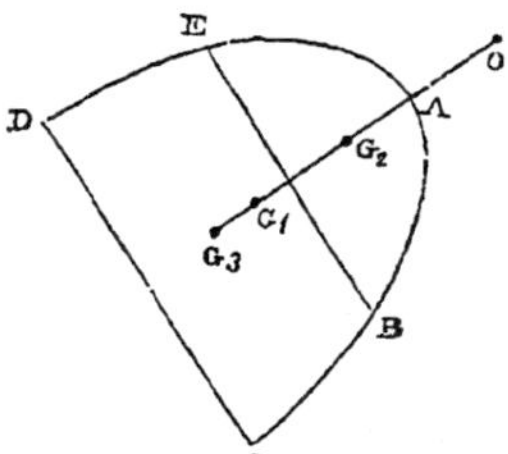

Fig. 97.

Let A C D be the larger figure, G_1 its known centre of gravity, W_1 its weight. Let A B E be the smaller figure, whose centre of gravity G_2 is known, W_2 its weight. Let E B C D be the figure whose centre of gravity G_3 is sought, made by taking away A B E from A C D, so that its weight is

$$W_3 = W_1 - W_2.$$

Join $G_1 G_2$; G_3 will be in the prolongation of that straight line beyond G_1. In the same straight line produced, take any point O as origin of co-ordinates. Make $O G_1 = x_1$; $O G_2 = x_2$, $O G_3$ (the unknown quantity) $= x_3$.

Then

$$x_3 = \frac{x_1 W_1 - x_2 W_2}{W_1 - W_2} \text{ .... (3.)}$$

245. **Centre of Gravity Altered by Transposition.**—In fig. 98, let A B C D be a body of the weight W, whose centre of gravity G is known. Let the figure of this body be altered, by transposing a part whose weight is W_1, from the position E C F to the position F D H, so that the new figure of the body is A B H E. Let G_1 be the original, and G_2 the new position of the centre of gravity of the transposed part. Then the centre of gravity of the whole body will be shifted to G_3, in a direction $G G_3$ parallel to $G_2 G_1$, and through a distance given by the formula.

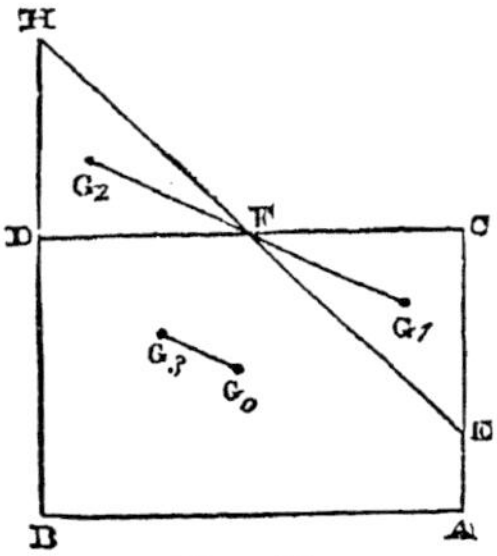

Fig. 98.

$$\overline{G G_3} = \overline{G_1 G_2} \cdot \frac{W_1}{W} \text{ .... (4.)}$$

246. **Centre of Gravity found by Projection or Transformation.**—If the figures of two homogeneous bodies are parallel projections of each other, the centres of gravity of those two bodies are corresponding points in those parallel projections.

To express this symbolically,—as in Article 237, let x, y, z, be the co-ordinates, rectangular or oblique, of any point in the figure of the first body; x', y', z', those of the corresponding point in the second body; x_0, y_0, z_0, the co-ordinates of the centre of gravity of the first body; x'_0, y'_0, z'_0, those of the centre of gravity of the second body, then

$$\frac{x'_0}{x_0} = \frac{x'}{x}; \quad \frac{y'_0}{y_0} = \frac{y'}{y}; \quad \frac{z'_0}{z_0} = \frac{z'}{z} \text{ .... (5.)}$$

This theorem facilitates much the finding of the centres of gravity of figures which are parallel projections of more simple or more symmetrical figures.

For example, let it be supposed that the centre of gravity of a sector of a circular disc has been found Case IX. Article 44), and let it be required to find the centre f gravity of a sector of an elliptic disc. In fig. 99, let A B A B be the ellipse, A O A 2 a, an l B′ O B 2 b, its axes, and C′ O D′ the sector whose centre of gravity is required. Ab ut the centre of the ellipse, O, describe the circle, A B A B, whose ra lius is the semi-axis m. j r a. Through C′ and D′ respectively draw E C C and F D D, parallel to O B, and cutting the circle in C and D respectively; the circular sector C O D is the parallel projection of the elliptic se t r C′ O D. Let G be the centre of gravity f the sector of the circular disc, its co-ordinates being

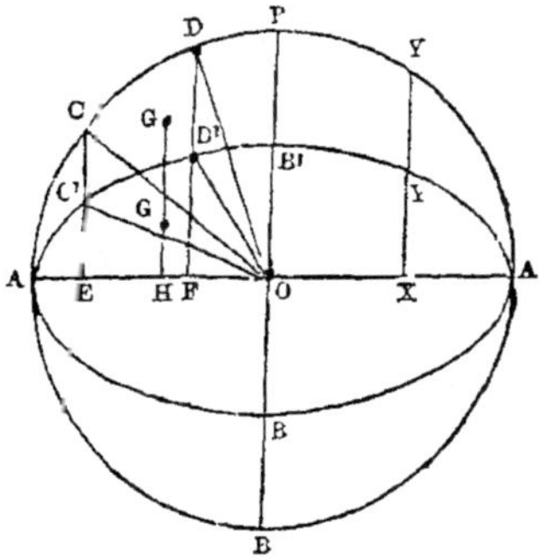

Fig. 99.

$$\overline{\mathrm{O\,H}} \quad x_0 ; \ \mathrm{H\,G} \quad y_0 .$$

Then the co ordinates of the centre of gravity G′ of the s ct r of the elliptic disc are

$$\left.\begin{array}{l} \mathrm{O\,H} \quad x' \quad x ; \\ \mathrm{H\,G'} \quad y' \quad \dfrac{b\,y}{a} ; \end{array}\right\} \cdots\cdots\cdots\cdots\cdots\cdots (6.)$$

247. **Centre of Gravity found Experimentally.**—The centre of gravity of a body f moderate size may be found appr ximately by exp riment, by hanging it up successively by a single c r l in two diff rent positi ns, and finding the single point in the body which in both positions is intersected by the axes of the cord.

SECTION 2.—OF STRESS.

248. **Stress—its Intensity.** The word STRESS has been adopted as a general term to comprehend various forces which are exerted between contiguous bodies, or parts of bodies, and which are dis tributed over the surface of contact of the masses between which they act.

The INTENSITY of a stress is its amount in units of weight, divided by the extent of the surface over which it acts, in units of area.

The following table gives a comparison of various units in which the intensity of stress is expressed :—

	Pounds on the square foot.	Pounds on the square inch
One pound on the square inch,....	144	1
One pound on the square foot,.....	1	$\frac{1}{144}$
One inch of mercury (that is, weight of a column of mercury at 32° Fahr., one inch high),...	70·73	0·4912
One foot of water (at 39°·1 Fahr.),	62·425	0 4335
One inch of water (at 39°·1 Fahr.),	5·2021	0·036125
One foot of water (at 62° Fahr.),...	62 355	0·43302
One inch of water (at 62° Fahr.), ..	5·19625	0 036085
One atmosphere, of 29·922 inches of mercury, or 760 millimètres,	2116·4	14·7
One foot of air, at 32° Fahr., and under the pressure of one atmosphere, ...	0·080728	0·0005606
One kilogramme on the square mètre,..............................	0·20481	0·00142228
One kilogramme on the square millimètre,..........................	204810	1422·28
One millimètre of mercury,........	2·7847	0·01934

249. **Classes of Stress.**—The various kinds of stress may be thus classed :—

I. *Thrust,* or *Pressure,* is the force which acts between two contiguous bodies, or parts of a body, when each pushes the other from itself.

II. *Pull,* or *Tension,* is the force which acts between two contiguous bodies, or parts of a body, when each draws the other towards itself.

Pressure and tension may be either *normal* or *oblique,* relatively to the surface at which they act.

III. *Shear,* or *Tangential Stress,* is the force which acts between two contiguous bodies, or parts of a body, when each draws the other sideways, in a direction parallel to their surface of contact.

In expressing a Thrust and a Pull in parallel directions algebraically, if one is treated as positive, the other must be treated as negative. The choice of the positive or negative sign for either is a matter of convenience.

The word "*Pressure,*" although, strictly speaking, equivalent to "*thrust,*" is sometimes applied to *stress* in general; and when this is the case, it is to be understood that thrust is treated as positive.

The following are the processes for finding the *magnitude of the resultant* of a stress distributed over a plane surface, and the *centre of stress;* that is, the point where the line of action of that resultant cuts the plane surface:

250. **In Stress of Uniform Intensity,** the magnitude of the resultant is the product of that intensity and the area of the surface; and the centre of stress is at the centre of magnitude of the surface. Or in symbols, let S be the area of the surface, p the intensity of the stress, P its resultant, then—

$$P \quad p S.$$

251. **In Stress of Varying Intensity, but of One Sign,** there is all tension, or all pressure, or all shear in one direction.

In fig. 100, let A A be the given plane surface at which the stress acts; O X, O Y, two rectangular axes of co ordinates in its plane; O Z, a third axis perpendicular to that plane. Conceive a solid to exist, bounded at one end by the given plane surface A A, laterally by a cylindrical or prismatic surface generated by the motion of a straight line parallel to O Z round the outline of A A and at the other end by a surface B B, of such a figure, that its ordinate z at any point shall be proportional to the intensity of the stress at the point a of the surface A A from which that ordinate proceeds, as shewn by the equation

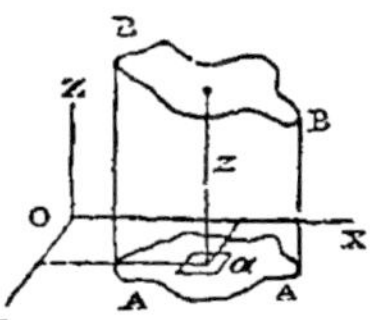

Fig. 100.

$$z \quad \frac{p}{w} \text{.............................} \quad 1.)$$

where p represents the intensity of the stress and w the heaviness, or weight per unit.

Conceive the surface A A to be divided into an indefinite number of small rectangular areas, each denoted by $dx\,dy$, and so small that the stress on each is sensibly uniform, the entire area being

$$S \quad \iint lx\,dy.$$

The volume of the ideal solid will be

$$V = \iint z \cdot dx\,dy \text{.....................} \quad 2.)$$

So that if it be conceived to consist of a material whose heaviness is $w \quad \frac{p}{z}$, the amount of the stress will be equal to the weight of the solid; that is to say,

$$P \quad \iint p\,dx\,dy - wV \text{....................} (3.)$$

The *centre of stress* is the point on the surface A A perpendicularly opposite the centre of gravity of the ideal solid.

The simplest, and at the same time the commonest, case of this kind is where the stress is *uniformly varying;* that is, where its intensity at a given point is simply proportional to the perpendicular distance of that point from a given straight line in the plane of the surface A A. To express this symbollically, take the straight line in question for the axis O Y; conceive the substance to be divided into bands by lines parallel to O Y; let y denote the length of one of these bands, and $d\,x$ its breadth, so that $y\,d\,x$ is its area, and $S = \int y\,d\,x$ the area of the whole surface. Let x be the perpendicular distance of the centre of a band from the *line of no stress* O Y, and let the intensity of the stress there be

$$p = a\,x; \quad \ldots\ldots\ldots\ldots\ldots\ldots\ldots\ldots\ldots\ldots (4.)$$

a being a constant coefficient; then the amount or resultant of the stress is

$$P = \int p\,y\,d\,x = a \int x\,y\,d\,x; \quad \ldots\ldots\ldots\ldots\ldots\ldots (5.)$$

and the perpendicular distance of the centre of stress from O Y is

$$x' = \frac{\int p\,x\,y\,d\,x}{\int p\,y\,d\,x} = a\,\frac{\int x^2\,y\,d\,x}{P}; \quad \ldots\ldots\ldots\ldots\ldots\ldots (6.)$$

252. **In Stress of Contrary Signs,** for example, pressure at one part of the surface and tension at another, the resultants and centres of stress of the pressure and tension are to be found separately. Those partial resultants are then to be treated as a pair of parallel forces acting through the two respective centres of stress; their final resultant will be equal to their difference, if any, acting through a point found as in Article 226, page 128.

If the total pressure and total tension are equal to each other, they have no single resultant and no single centre of stress: their resultant being a couple, whose moment is equal to the total stress of either kind multiplied by the perpendicular distance between the resultant of the pressure and the resultant of the tension.

SECTION 3.—PRINCIPLES OF HYDROSTATICS AND INTERNAL STRESS OF SOLIDS.

253. **Pressure and Balance of Fluids: Principles of Hydrostatics.** *Fluid* is a term opposed to *solid*, and comprehending the liquid and gaseous conditions of bodies. The property common to the liquid and the gaseous conditions is that of *not tending to preserve a definite shape*, and the possession of this property by a body in perfection throughout all its parts, constitutes that body *a perfect fluid.*

A necessary consequence of that property is the following principle, which is the foundation of the whole science of hydrostatics:

I. *In a perfect fluid, when still, the pressure exerted at a given point is normal to the surface on which it acts, and of equal intensity for all positions of that surface.*

The following are some of the most useful consequences of that principle:—

II. *A surface of equal pressure in a still fluid mass is everywhere perpendicular to the direction of gravity*; that is, horizontal throughout. In other words, the pressure at all points at the same level is of equal intensity.

III. *The intensity of the pressure at the lower of two points in a still fluid mass is greater than the intensity at the higher point, by an amount equal to the weight of a vertical column of the fluid whose height is the difference of elevation of the points, and base an unit of area.*

To express this symbolically, let p denote the intensity of the pressure at the higher of two points in a fluid mass, and p_1 the intensity at a point whose vertical depth below the former point is x. Let w be the *mean heaviness* of the layer of fluid between those two points; then

$$p_1 = p + wx \text{.............................(1.)}$$

In a gas, such as air, w varies, being nearly proportional to p; but in a liquid, such as water, the variations of w are too small to be considered in practical cases.

For example, let the upper of the two points be the surface of a mass of water where it is exposed to the air; then p_0 is the atmospheric pressure; let the depth x of the second point below the surface be given in feet, and let the temperature be 39°·1; then

$$p_1 \text{ in lbs. on the square foot} = p_0 + 62{\cdot}425\,x \text{.........(2.)}$$

In many questions relating to engineering, the pressure of the atmosphere may be left out of consideration, as it acts with sensibly equal intensity on all sides of the bodies exposed to it, and so balances its own action. The pressure calculated, in such cases, is the *excess* of the pressure of the water abo\e the atm spheric pressure, which may be thus expressed,—

$$p' \quad p_1 - p \quad 62{\cdot}425\, x \text{ nearly} \ldots\ldots\ldots\ldots\ldots\ldots (3.)$$

IV. The pressure of a liquid on a *fl ating* or *i mersed b d*, is equal to the weight of the volume of fluid displaced by that body; and the resultant of that pressure acts vertically upwaıds through the centre of gravity of that volume; which centre of gravity is called the "*centre of buoyanc.*"

V. The pressure of a liquid against *a plane surface i mersed i it* is perpendicular to that surface in direction: its magnitude is equal to the weight of a volume of the liquid, found by multi lying the area of the surface by the depth to which its centre of gravity is immersed.

VI. The *centre of p essu e* on such a surface, if the surface is horiz ntal, c incides with its centre of gravity; if the surface is vertical or sloping, the centre of pressure is always b low the centre of gravity of the surface, and is found by considering that the pressure is an *uniformly-var ing* stress, whose intensity at a given point varies as the distance of that oint from the line where the given plane surface (pr duced if necessary) intersects the upper surface of the liquid.

To express the last two principles by symbols in the case in which the pressed surface is vertical or sl ping, let the line where the plane of that surface cuts the upper surface of the liquid be taken as the axis O Y. Let θ denote the angle of inclination of the pressed surface to the horizon. Conceive that surface to be divided by parallel horiz ntal lines into an in lefinite number of narrow bands. Let y be the length of any one of th se bands, $d x$ its breadth, x the distan e of its centre from O Y; then $y\, d x$ is its area, $x \sin \theta$ the depth at which it is immersed; and if w be the weight of unity of volume of the fluid, the intensity of the pressure on that band is

$$p \quad w\, x \sin \theta \ldots\ldots\ldots\ldots\ldots\ldots (4.)$$

The whole area of the pressed surface, being the sum of the areas of all the bands, is S $\int y\, d x$; the whole pressure u on it is

$$\text{P} \quad \int p\, y\, d x \quad w \sin \theta \int x\, y\, d x; \ldots\ldots\ldots\ldots (5.)$$

the mean intensity of the pressure is

$$\frac{\text{P}}{\text{S}} = \frac{\int p\, y\, d x}{\int y\, d x} = w \sin \theta \frac{\int x\, y\, d x}{\int y\, d x}; \ldots\ldots\ldots\ldots (6.)$$

and the distance of the centre of pressure from O Y is

$$x_0 = \frac{\int x p y \, d x}{P} = \frac{\int x^2 y \, d x}{\int x y \, d x} \quad \text{.................(7.)}$$

For example, let the sloping pressed surface be rectangular, like a sluice, or the back of a reservoir-wall; and in the first instance, let it extend from the surface of a mass of water down to a distance x_1, measured along the slope, so that its lower edge is immersed to the depth $x_1 \sin \theta$. Then its centre of gravity is immersed to the depth $x_1 \sin \theta \div 2$, and the mean intensity of the pressure in lbs. on the square foot, is

$$\frac{P}{S} = \frac{62\cdot4 \, x_1 \sin \theta}{2} \quad \text{........................(8.)}$$

The breadth y is constant; so that the area of the surface is $S = x_1 y$; and the total pressure is

$$P = \frac{62\cdot4 \, x_1^2 y \sin \theta}{2} \quad \text{.......................(9.)}$$

The distance of the centre of pressure from the upper edge is

$$x_0 = \frac{2}{3} x_1 \quad \text{........................(10.)}$$

Next, let the upper edge, instead of being at the surface of the water, be at the distance x_2 from it, so as to be immersed to the depth $x_2 \sin \theta$. Then the centre of gravity of the pressed surface is immersed to the depth $(x_1 + x_2) \sin \theta \div 2$, and the mean intensity of the pressure upon it, in lbs. on the square foot, is

$$\frac{P}{S} = \frac{62\cdot4 \, (x_1 + x_2) \sin \theta}{2}; \quad \text{.......... .....(11.)}$$

the area of the surface is $(x_1 - x_2) y$, and the total pressure on it

$$P = \frac{62\cdot4 \, (x_1^2 - x_2^2) y \sin \theta}{2} \quad \text{.................(12.)}$$

The distance of the centre of pressure from the line O Y is

$$x_0 = \frac{2}{3} \cdot \frac{x_1^3 - x_2^3}{x_1^2 - x_2^2} \quad \text{......................(13.)}$$

254. **Compound Internal Stress of Solids.**—If a body be conceived to be divided into two parts by an ideal plane traversing it in any direction, the force exerted between those two parts at the plane of division is an *internal stress.*

According to the principles stated in the preceding article, the internal stress at a given point in a fluid is normal and of equal intensity for all positions of the ideal plane of division. In a solid body, on the other hand, the stress may be either normal, oblique, or shearing; and it may vary in direction and intensity, as the position of the ideal plane of division varies.

255. **Conjugate Stresses Principal Stresses.** If two planes traverse a point in a body, and the direction of the stress on the first plane is parallel to the second plane, then the direction of the stress on the second plane is parallel to the first plane. Such a pair of stresses are said to be *conjugate;* and if they are both normal to their planes of application (and consequently perpendicular to each other) they are called *principal stresses.* Three conjugate stresses, or three principal stresses, may act through one point; but in the present treatise it is sufficient to consider two.

Fig. 101 represents a pair of conjugate oblique tensions acting in the direction XX and YY through a prismatic particle A B C D.

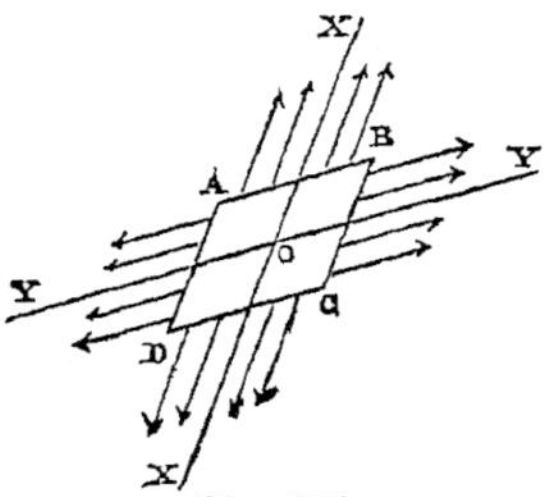

Fig. 101.

The rectangular directions in which principal stresses—that is, direct pulls and thrusts—act, through a given point in a solid, are called *axes of stress.*

In a fluid, the stress at a given point being of equal intensity in all directions, every direction has the property of an axis of stress. A solid may be in the same condition with a fluid as to stress; but it may also have the principal stresses at a given point of different intensities. In a mass of loose grains, the ratio of those intensities has a limit depending on friction:—in a firm continuous solid, the principal stresses at a point may bear any ratio to each other, and may be either of the same or of opposite kinds.

256. **The Shearing Stress,** on two planes traversing a point in a solid at right angles to each other, is of equal intensity.

257. **A Pair of Equal and Opposite Principal Stresses**; that is, a pull and a thrust of equal intensity acting through a particle of a solid in directions at right angles to each other, are equivalent to a pair of shearing stresses of the same intensity on a pair of planes at right angles to each other, and making angles of 45° with the first pair of planes.

258. **Combination of any Two Principal Stresses.**

Problem.—A pair of principal stresses of any intensities, and of the same or opposite kinds, being given, it is required to find the direction and intensity of the stress on a plane in any position at right angles to the plane parallel to which the two principal stresses act.

Let O X and O Y figs. 102 and 103 be the directions of the two principal stresses; O X being the direction of the greater stress.

Let p_1 be the intensity of the greater stress;
and p_2 that of the less.

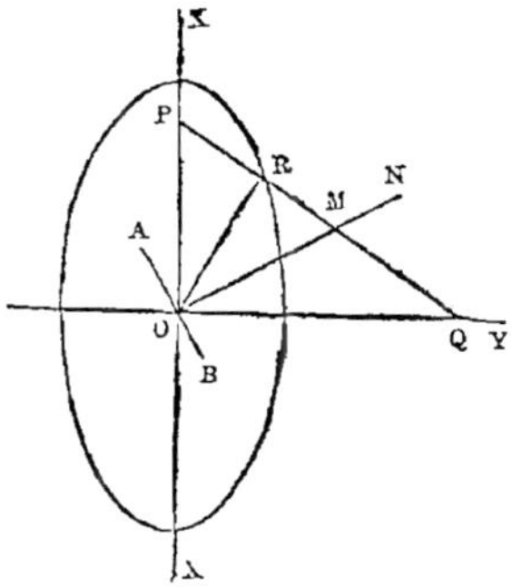

Fig. 102.

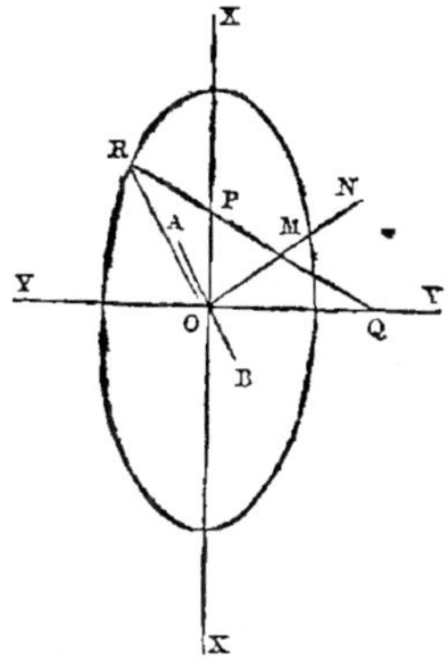

Fig. 103.

The kind of stress to which each of these belongs, pull or thrust, is to be distinguished by means of the algebraical signs. If a pull is considered as positive, a thrust is to be considered as negative, and *vice versâ*. It is in general convenient to consider that kind of stress as positive to which the greater principal stress belongs. Fig. 102 represents the case in which p_1 and p_2 are of the same kind; fig. 103 the case in which they are of opposite kinds. In all the following equations, the sign of p_2 is held to be *implied* in that symbol; that is to say, when p_2 is of the contrary kind to p_1, the sign applied to its arithmetical value, in computing by means of the equations, is to be reversed.

Let A B be the plane on which it is required to ascertain the direction and intensity of the stress, and O N a normal to that plane, making with the axis of greatest stress the angle

$$\angle \mathrm{X\,O\,N} = \widehat{x\,n}.$$

On O N take O M $\frac{p_1 + p_2}{2}$; this will represent a normal stress on A B of the same kind with the greater principal stress, and of an intensity which is a mean between the intensities of the two principal stresses.

Through M draw P M Q, making with the axes of stress the same angles which O N makes, but in the opposite direction; that is to say, take M P − M Q M O. On the line thus found set off from M tow, take axis of greatest stress, $\overline{\mathrm{MR}} = \frac{p_1 - p_2}{2}$.

Join $\overline{\mathrm{OR}}$. Then will that line represent the direction and intensity of the stress on A B.

In fig. 102, p_1 and p_2 are represented as being of the same kind; and M R is consequently less than O M, so that $\overline{\mathrm{OR}}$ falls on the same side of O X with O N; that is to say, $\widehat{nr} < \widehat{xn}$. In fig. 103, p_1 and p_2 are of opposite kinds, M R is greater than $\overline{\mathrm{OM}}$, and O R falls on the opposite side of O X to O M; that is to say, $\widehat{nr} > \widehat{xn}$.

The locus of the point M is a circle of the radius $\frac{p_1 + p_2}{2}$, and that of the point R, an ellipse whose semi axes are p_1 and p_2, and which may be called the Ellipse of Stress, because its semi-diameter in any direction represents the intensity of the stress in that direction.

259. **Deviation of Principal Stresses by a Shearing Stress.**—Problem. Let p_x and p_y denote the original intensities of a pair of principal stresses acting at right angles to each other through one particle of a solid. Suppose that with these there is combined a shearing stress of the intensity q, acting in the same plane with the original pulls or thrusts; it is required to find the new intensities and new directions of the principal stresses.

To assist the conception of this problem, the original stresses referred to are represented in fig. 104, as acting through a particle of the form of a square prism. The principal stresses, both original and new, are represented as tensions, although any or all of them might be pressures. In the formulæ annexed, tensions are considered positive, pressures negative; angles lying to the right of A A are considered as positive, to the left as negative; and a shearing stress is considered as positive or negative according as it tends to make the upper right-hand and lower left hand corner of the square particle acute or obtuse.

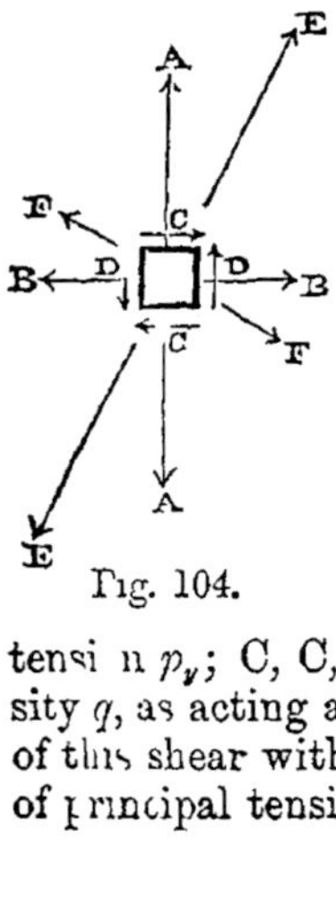

Fig. 104.

The arrows A A represent the greater original tension p_x; the arrows B B, the less original tension p_y; C, C, D, D, represent the positive shear of the intensity q, as acting at the four faces of the particle. The combination of this shear with the original tensions is equivalent to a new pair of principal tensions, oblique to the original pair. The greater new principal tension, p_1, is represented by the arrows E, E; it deviates to the right of p_x through an angle which will be denoted by θ. The less new principal tension p_2 is represented by the arrows F, F; it deviates through the same angle to the right of p_y.

Then the intensities of the new principal stresses are given by the equations,

$$\left.\begin{aligned} p_1 &= \frac{p_x+p_y}{2}+\sqrt{\left\{\frac{(p_x-p_y)^2}{4}+q^2\right\}};\\ p_2 &= \frac{p_x+p_y}{2}-\sqrt{\left\{\frac{(p_x-p_y)^2}{4}+q^2\right\}}; \end{aligned}\right\}\ldots\ldots\ldots\ldots(3.)$$

and the double of the angle of deviation by either of the following,

$$\tan 2\theta = \frac{2q}{p_x-p_y};\ \text{or}\ \cot\tan 2\theta = \frac{p_x-p_y}{2q}\ldots\ldots\ldots\ldots(4.)$$

The greatest value of θ is 45°, when $p_x = p_y$.

The new principal stresses are to be conceived as acting normally on the faces of a new square prism.

260. **Parallel Projection of Distributed Forces.**—In applying the principles of parallel projection to distributed forces, it is to be borne in mind that these principles, as stated in Article 237, are applicable to lines representing the *amounts* or *resultants* of distributed forces, and *not their intensities*. The relations amongst the intensities of a system of distributed forces, whose resultants have been obtained by the method of projection, are to be arrived at by a subsequent process of dividing each projected resultant by the projected space over which it is distributed.

261. **Friction** is that force which acts between two bodies at their surface of contact so as to resist their sliding on each other, and which depends on the force with which the bodies are pressed together. It is a kind of shearing stress. The following law respecting the friction of solid bodies has been ascertained by experiment:—

The friction which a given pair of solid bodies, with their surfaces in a given condition, are capable of exerting, is simply proportional to the force with which they are pressed together.

If a body be acted upon by a force tending to make it slide on another, then so long as that force does not exceed the amount fixed by this law, the friction will be equal and opposite to it, and will balance it.

There is a limit to the exactness of the above law, when the pressure becomes so intense as to crush or indent the parts of the bodies at and near their surface of contact. At and beyond that limit the friction increases more rapidly than the pressure; but that limit ought never to be attained in any structure. For some substances, especially those whose surfaces are sensibly indented by a moderate pressure, such as timber, the friction between a pair of surfaces which have remained for some time at rest relatively to each other, is somewhat greater than that between the same pair of surfaces when sliding on each other. That excess, however, of the *friction of rest* over the *friction of motion,* is instantly destroyed by a slight vibration; so that the *friction of motion* is alone to be taken into account, as contributing to the stability of a structure.

The friction between a pair of surfaces is calculated by multiplying the force with which they are directly pressed together, by a factor called the *coefficient of friction*, which has a special value depending on the nature of the materials and the state of the surfaces. Let F denote the friction between a pair of surfaces; N, the force, in a direction perpendicular to the surfaces, with which they are pressed together; and f the coefficient of friction; then

$$F = fN \quad \ldots\ldots\ldots\ldots\ldots\ldots \quad (1.)$$

The coefficient of friction of a given pair of surfaces is the *tangent* of an angle called the *angle of repose,* being the greatest angle which an oblique pressure between the surfaces can make with a perpendicular to them, without making them slide on aceh other.

Let P denote the amount of an oblique pressure between two plane surfaces, inclined to their common normal at the angle of repose φ; then

$$F = fN = N \tan \varphi = P \sin \varphi = \frac{fP}{\sqrt{1 + f^2}} \quad \ldots\ldots \quad (2.)$$

The angle of repose is the steepest inclination of a plane to the horizon, at which a block of a given substance will remain balanced on it without sliding down.

The *intensity* of the friction between two surfaces bears the same proportion to the intensity of the pressure that the whole friction bears to the whole pressure.

The following is a table of the angle of repose φ, the coefficient of friction $f = \tan \varphi$, and its reciprocal $1 : f$, for various materials—condensed from the tables of General Morin, and other sources, and arranged in a few comprehensive classes. The values of those constants which are given in the table have reference to the *friction of motion.*

SURFACES.	φ	f	$\frac{1}{f}$
Dry masonry and brickwork, ...	31° to 35°	0·6 to 0·7	1·67 to 1·43
Masonry and brickwork with wet mortar, ...	25½°	0·47	2·1
Masonry and brickwork, with slightly damp mortar, ...	36½°	0·74	1·35
Wood on stone, ...	22°	about 0·4	2·5
Iron on stone, ...	35° to 16⅔°	0·7 to 0·3	1·43 to 3·33
Masonry on dry clay, ...	27°	0·51	1·96
,, on moist clay, ...	18¼°	0·33	3
Earth on earth, ...	14° to 45°	0·25 to 1·0	4 to 1
,, ,, dry sand, clay, and mixed earth, ...	21° to 37°	0·38 to 0·75	2·63 to 1·33
Earth on earth, damp clay, ...	45°	1·0	1
,, ,, wet clay, ...	17°	0·31	3·23
,, ,, shingle and gravel,	35° to 48°	0·7 to 1·11	1·43 to 0·9
Wood on wood, dry, ...	14° to 26½°	·25 to ·5	4 to 2
,, ,, soaped, ...	11½° to 2°	·2 to ·04	5 to 25
Metals on oak, dry, ...	26½° to 31°	·5 to ·6	2 to 1·67
,, ,, wet, ...	13½° to 14½°	·24 to ·26	4·17 to 3·85
,, ,, soapy, ...	11½°	·2	5
Metals on elm, dry, ...	11½° to 14	.2 to ·25	5 to 4
Bronze on lignum vitæ, constantly wet, ...	3°?	·05?	20?
Hemp on oak, dry, ...	28°	·53	1·89
,, ,, wet, ...	18½°	·33	3
Leather on oak, ...	15° to 19½°	·27 to ·38	3·7 to 2·86
Leather on metals, dry, ...	29½°	·56	1·79
,, ,, wet, ...	20°	·36	2·78
,, ,, greasy, ...	13°	·23	4·35
,, ,, oily, ...	8½°	·15	6·67
Metals on metals, dry, ...	8½° to 11½°	·15 to ·2	6·67 to 5
,, ,, wet and clean, ..	16½°	·3	3·33
,, ,, damp and slimy,	8°	·14	7·14
Smooth surfaces, occasionally greased	4° to 4½°	·07 to ·08	14·3 to 12·5
Smooth surfaces, continually greased, ...	3°	·05	20
Smoothest and best greased surfaces,	1½° to 2°	·03 to ·036	33·3 to 27·6

PART IV.

THEORY OF STRUCTURES.

CHAPTER I.

SUMMARY OF PRINCIPLES OF STABILITY AND STRENGTH.

SECTION 1.—OF STRUCTURES IN GENERAL.

262. **A Structure** consists of portions of solid materials, put together so as to preserve a definite form and arrangement of parts, and to withstand external forces tending to disturb such form and arrangement. As the parts of a structure are intended to remain at rest relatively to each other, the forces which act on the whole structure, and on each of its parts, should be *balanced*, so that the mechanical principles on which the permanence and efficiency of structures depend for the most part belong to STATICS, or the science of balanced forces.

The *materials* of a structure may be more or less stiff, like stone, timber, and metals, or loose, like earth.

In the present chapter are given a summary of mechanical principles applicable to structures.

263. **Pieces — Joints — Supports — Foundations.**—A structure consists of two or more solid bodies, called its *pieces*, which touch each other and are connected at portions of their surfaces, called *joints*. This statement may appear to be applicable to structures of stiff materials only; but, nevertheless, it comprehends masses of earth also, if they are considered as consisting of a very great number of very small pieces, touching each other at innumerable joints.

Although the pieces of a structure are fixed relatively to each other, the structure as a whole may be either fixed or movable relatively to the earth.

A fixed structure is supported on a part of the solid material of the earth, called the *foundation* of the structure; the pressures by which the structure is supported, being the resistances of the various parts of the foundation, may be more or less oblique.

A movable structure may be supported, as a ship, by floating in water, or as a carriage, by resting on the solid ground through wheels. When such a structure is actually in motion, it partakes to a certain extent of the properties of a machine; and the determination of the forces by which it is supported requires the consideration of kinetic as well as of statical principles; but when it is not in actual motion, though capable of being moved, the pressures which support it are determined by the principles of statics; and it is obvi us that they have their resultant equal and directly opp sed to the weight of the struct ıre.

264. **The Conditions of Equilibrium of a Structure** are the three following:

I. *That the forces exerted on the wl o e structure by external bodies shall balance each other.* The forces to be c nsidered under th's head are— 1. the *Attracti n of the Earth* that is, the *weight* of the structure; 2.) the *External Load*, arising from the pressures exerted against the structure by bodies not f rming part of it nor of its foundation; these two kinds of forces constitute the *gross* or *total load*); (3.) the *Supp rting Pressures*, or resistance of the f undation. Those three classes of forces will be spoken of together as the *External Forc s.*

II. *That the forces exerted on each piece of the structur shall balance each other.*—These c nsist of 1.) the *Weight* of the piece, and 2. the *External L ad* on it, making together the *Gr ss Load;* and 3. the *Resistances,* or forces exerted at the joints, between the piece under consideration and the pieces in contıct with it.

III. *That the forces exerted on each of the parts int wl ich each piece of the structure can be conceived to be divided shall balance each other.*—Suppose an ideal surface to divide any part of any one of the pieces of the structure from the remainder of the piece; the forces which act on the part so considered are—(1.) its weight, and (2. if it is at the external surface of the piece the external force applied to it, if any, making t gether its *g oss load;* 3) the *stress,* or force, exerted at the ideal surface of division, between the part in question and the other parts of the ı iece.

265. **Stability, Strength, and Stiffness**—It is necessary to the permanence of a structure, that the three foregoing conditions of equilibrium should be fulfilled, not only under one amount and one mode of distribution of load, but under all the variations of the load as to amount and mode of distribution which can occur in the use of the structure.

Stability consists in the fulfilment of the *first* and *second* conditions of equilibrium of a structure under all variations of the load within given limits. A structure which is deficient in stability gives way by the displacement of its pieces from their proper positions.

When a structure, or one of its parts, is *flexible*, like the chain of a suspension bridge, or in any other way free to move, its stability consists in a tendency to recover its original figure and position after having been disturbed.

Strength consists in the fulfilment of the *third* condition of equilibrium of a structure for all loads not exceeding prescribed limits; that is to say, the greatest internal stress produced in any part of any piece of the structure, by the prescribed greatest load, must be such as the material can bear, not merely without immediate breaking, but without such injury to its texture as might endanger its breaking in the course of time.

A piece of a structure may be rendered unfit for its purpose, not merely by being broken, but by being stretched, compressed, bent, twisted, or otherwise strained out of its proper shape. It is necessary, therefore, that each piece of a structure should be of such dimensions that its alteration of figure under the greatest load applied to it shall not exceed given limits. This property is called *stiffness*, and is so connected with strength that it is necessary to consider them together.

SECTION 2.—BALANCE AND STABILITY OF FRAMES, CHAINS, AND BLOCKS.

266. **A Frame** is a structure composed of bars, rods, links, or cords, attached together or supported by *joints*, such as occur in carpentry, in frames of metal bars, and in structures of ropes and chains, fixing the ends of two or more pieces together, but offering little or no resistance to change in the relative angular positions of those pieces. In a joint of this class, the *centre of resistance*, or point through which the resultant of the resistance to displacement of the pieces connected at the joint acts, is at or near the middle of the joint, and does not admit of any variation of position consistently with security.

The *line of resistance* of a frame is a line traversing the *centres of resistance* of the joints, and is in general a polygon, having its angles at these centres.

267. **A Single Bar** in a frame may act as a TIE, a STRUT, or a BEAM.

I. A *tie* has equal and directly opposite forces applied to its two ends, acting outwards, or *from* each other. The bar is in a state of *tension*, and the stress exerted between any two divisions of it is a *pull*, equal and opposite to the applied forces. A *rope* or *chain* will answer the purpose of a tie.

The equilibrium of a movable tie is stable; for if its angular position be deviated, the forces applied to its ends, which originally were directly opposed, now constitute a *couple* tending to restore the tie to its original position.

II. A *strut* has equal and directly opposite forces applied to its two ends, acting inwards, or towards each other. The bar is in a state of compression, and the stress exerted between any two divisions of it is a *thrust* equal and opposite to the applied forces. It is obvious that a flexible body will not answer the purpose of a strut.

The equilibrium of a movable strut is unstable; for if its angular position be deviated, the forces applied to its ends, which originally were directly opposed, now constitute a couple tending to make it deviate still farther from its original position.

In order that a strut may have stability, its ends must be prevented from deviating laterally. Pieces connected with the ends of a strut for this purpose are called *stays.*

III. A *beam* is a bar supported at two points, and loaded in a direction perpendicular or oblique to its length.

CASE I.—Let the supporting pressures be parallel to each other and to the direction of the load; and let the load act *between* the points of support, as in fig. 105; where P represents the resultant of the gross load, including the weight of the beam itself; L, the point where the line of action of that resultant intersects the axis of the beam; R_1, R_2, the two supporting pressures or resistances of the props parallel to, and in the same plane with P, and acting through the points S_1, S_2, in the axis of the beam.

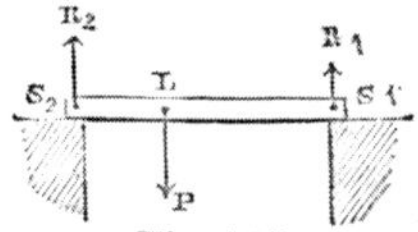

Fig. 105.

Then, according to the principle of the lever, Article 225, page 128, each of these three forces is proportional to the distance between the lines of action of the other two; and the load is equal to the sum of the two supporting pressures; that is to say,

$$P : R_1 : R_2 :: S_1 S_2 : \overline{L S_2} : L S_1; \ldots\ldots\ldots\ldots(1.)$$

$$\text{and } P = R_1 + R_2 \ldots\ldots\ldots\ldots\ldots\ldots\ldots\ldots(2.)$$

CASE II.—Let the load act *beyond* the points of support, as in fig. 106, which represents a cantilever or projecting beam, held up by a wall or other prop at S_1, held down by a notch in a mass of masonry or otherwise at S_2, and loaded so that P is the resultant of the load, including the weight of the beam. Then the proportional Equation 1. remains exactly as before; but the load is equal to the difference of the supporting pressures, that is to say,

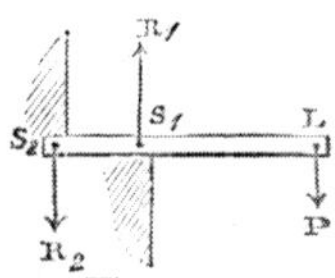

Fig. 106.

$$P = R_1 - R_2.$$

In these examples the beam is represented as horizontal; but the same principles would hold if it were inclined.

CASE III.—Let the directions of the supporting forces R_1, R_2, be now inclined to that of the resultant of the load, P, as in fig 107. This case is that of the equilibrium of three forces treated of in Article 2 9, page 122, and consequently the following principles apply to it:—

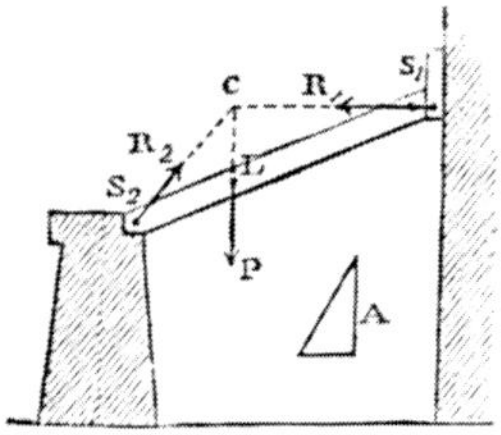

Fig. 107.

The lines of action of the supporting forces and of the resultant of the load must be in one plane.

They must intersect in one point C, fig 107).

Those three forces must be proportional to the three sides of a triangle A, respectively parallel to their directions.

PROBLEM.—Given, the resultant of the load in magnitude and position, P, the line of action of one of the supporting forces, R_1, and the centre of resistance of the other, S ; required, the line of action of the second supporting force, and the magnitudes of both.

Produce the line of action of R_1, till it cuts the line of action of P at the point C; join CS_2; this will be the line of action of R_2; construct a triangle A with its sides respectively parallel to those three lines of action; the ratios of the sides of that triangle will give the ratios of the forces.

To express this algebraically, let i_1, i_2, be the angles made by the lines of action of the supporting forces with that of the resultant of the load; then

$$P : R_1 : R_2 : : \sin (i_1 + i_2) : \sin i_2 : \sin i_1 \text{ ....} \quad (4.)$$

The same piece in a frame may act at once as a beam and tie, or as a beam and strut; or it may act alternately as a strut and as a tie, as the action of the load varies.

The load tends to break a tie by tearing it asunder, a strut by crushing it, and a beam by breaking it across. The power of materials to resist those tendencies will be considered in a later section.

268. **Distributed Loads.**—Before applying the principles of the present section to frames in which the load, whether external or arising from the weight of the bars, is distributed over their length, it is necessary to reduce that distributed load to an equivalent load, or series of loads, applied at the centres of resistance. The steps in this process are as follows:—

I. Find the resultant load on each single bar.

II. Resolve that load, as in Article 267, Equation 1, page 159, into two parallel components acting through the centres of resistance at the two ends of the bar.

III. At each centre of resistance where two bars meet, combine the component loads due to the loads on the two bars into one resultant, which is to be considered as the total load acting through that centre of resistance.

IV. When a centre of resistance is also a point of support, the component load acting through it, as found by step II. of the process, is to be left out of consideration until the supporting force required by the system of loads at the other joints has been determined; with this supporting force is to be compounded a force equal and opposite to the component load acting directly through the point of support, and the resultant will be the total supporting force.

In the following Articles of this section, all the frames will be supposed to be loaded only at those centres of resistance which are *not* points of support; and, therefore, in those cases in which components of the load act directly through the points of support also, forces equal and opposite to such components must be combined with the supporting forces as determined in the following Articles, in order to complete the solution.

269. **Frames of Two Bars.**—Figures 108, 109, and 110, represent cases in which a frame of two bars, jointed to each at the point L, is loaded at that point with a given force, P, and is sup-

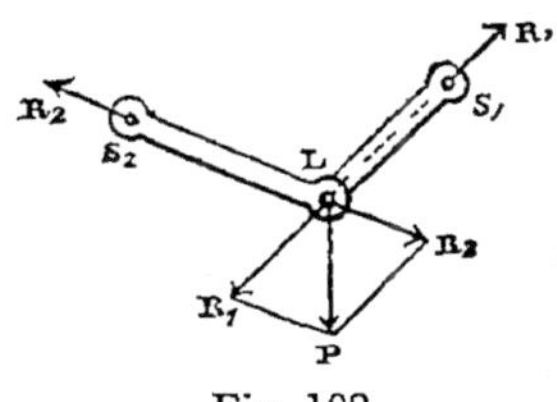

Fig. 108.

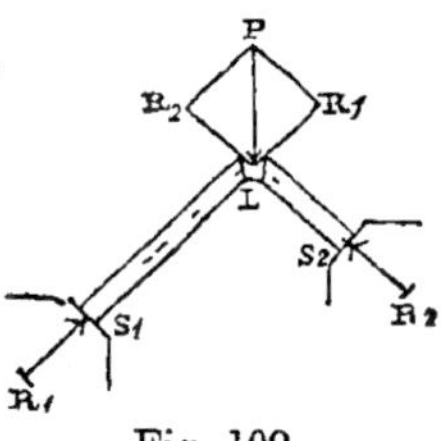

Fig. 109.

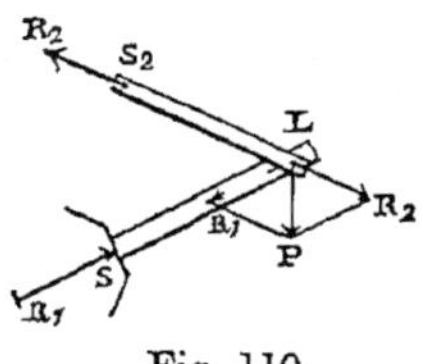

Fig. 110.

ported by the connection of the bars at their farther extremities, S_1, S_2, with fixed bodies. It is required to find the stress on each bar, and the supporting forces at S_1 and S_2.

Resolve the load P as in Article 213, page 123 into two components, R_1, R_2, acting along the respective lines of resistance of the two bars. Those components are the loads borne by the two bars respectively; to which loads the supporting forces at S_1, S_2, are equal and directly opposed.

The symbolical expression of this solution is as follows: Let i_1, i_2, be the respective angles made by the lines of resistance of the bars with the line of action of the load; then

$$P : R_1 : R_2 : : \sin (i_1 + i_2) : \sin i_2 : \sin i_1.$$

The inward or outward direction of the forces acting along each bar indicates that the stress is a thrust or a pull, and the bar a strut or a tie, as the case may be. Fig. 108 represents the case of two ties; fig. 109 that of two struts (such as a pair of rafters abutting against two walls); fig. 110 of a strut, $L S_1$, and a tie, $L S_2$ (such as the jib and the tie rod of a crane).

A frame of two bars is *stable* as regards deviations in the plane of its lines of resistance.

With respect to *lateral* deviations of angular position, in a direction perpendicular to that plane, a frame of two ties is stable; so also is a frame consisting of a strut and a tie, when the direction of the load inclines *from* the line $S_1 S_2$, joining the points of support.

A frame consisting of a strut and a tie, when the direction of the load inclines *towards* the line $S_1 S_2$, and a frame of two struts in all cases, are unstable laterally, unless provided with lateral stays.

These principles are true of *any pair of adjacent bars whose farther centres of resistance are fixed;* whether forming a frame by themselves, or a part of a more complex frame.

270. **Triangular Frames.**—Let fig. 111 represent a frame, consisting of three bars, A, B, C, connected at the three joints 1, 2, 3,—viz., C and A at 1, A and B at 2, B and C at 3. Let a load P_1 be applied at the joint 1 in any given direction; let supporting forces, P_2, P_3, be applied at the joints 2, 3; the lines of action of those two forces must be in the same plane with that of P_1, and must either be parallel to it or intersect it in one point. The latter case is taken first, because its solution comprehends that of the former.

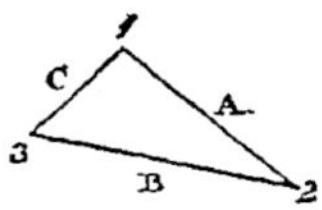

Fig. 111.

The three external forces balance each other, and are therefore proportional to the three sides of a triangle respectively parallel to their directions. In fig. 112, let A B C be such a triangle, in which

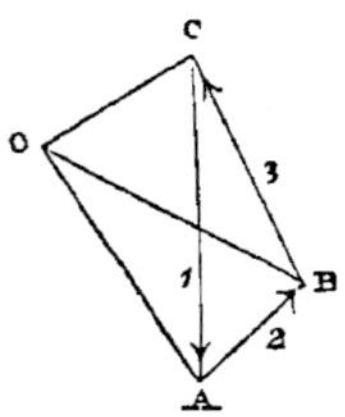

Fig. 112.

$\overline{CA}$ represents P_1,
$\overline{AB}$ „ P_2,
B C „ P_3,

Draw C O parallel to the bar C, and A O parallel to the bar A, meeting in the point O, and join B O, which will be parallel to B.

The lengths of the three lines radiating from O will represent the stresses on the bars to which they are respectively parallel.

When the three external forces are parallel to each other, the triangle of forces A B C of fig. 112, becomes a straight line C A, as in fig. 113, divided into two segments by the point B. Let straight lines radiate from O to A, B, C, respectively parallel to the bars of the frame; then if the load C A be applied at 1 (fig. 111), A B applied at 2, and B C applied at 3, are the supporting forces required to balance it; and the radiating lines O A, O B, O C, represent the stresses on the bars A, B, C, respectively, as before.

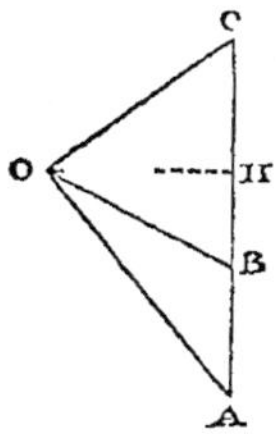

Fig. 113.

From O let fall O H perpendicular to C A, the common direction of the external forces. Then that line will represent a component of the stress, which is of equal amount in each bar. When C A, as is usually the case, is vertical, O H is horizontal; and the force represented by it is called the "*horizontal thrust*" of the frame. *Horizontal Stress or Resistance* would be a more precise term; because the force in question is a pull in some parts of the frame, and a thrust in others.

In fig. 111, A and C are *struts*, and B a *tie*. If the frame were exactly inverted, all the forces would bear the same proportions to each other; but A and C would be *ties*, and B a strut.

The trigonometrical expression of the relations amongst the forces acting in a triangular frame, under parallel forces, is as follows:—

Let a, b, c, denote the respective angles of inclination of the bars A, B, C, to the line O H (that is, in general, to a horizontal line); viz., the angles A O H, B O H, C O H of fig. 113, then

$$\text{Horizontal Stress O H} = \frac{\text{load C A}}{\tan c \pm \tan a} \text{.........} (1.)$$

$$\begin{array}{c}\text{Supporting}\\ \text{Forces}\end{array}\left\{\begin{array}{l}\text{A B} = \text{O H}\cdot(\tan a \mp \tan b);\\ \text{B C} = \text{O H}\cdot(\tan b \pm \tan c);\end{array}\right\} \text{.....} (2)$$

The sign $\left\{\begin{array}{c}+\\ -\end{array}\right\}$ is to be used when the two inclinations are in $\left\{\begin{array}{l}\text{opposite directions,}\\ \text{the same direction.}\end{array}\right.$

$$\text{Stresses}\left\{\begin{array}{l}\text{O A} = \text{O H}\cdot\sec a\\ \text{O B} = \text{O H}\cdot\sec b\\ \text{O C} = \text{O H}\cdot\sec c\end{array}\right\} \text{...............} (3.)$$

271. **Polygonal Frame.**—In fig. 114, let A, B, C, D, E, be the lines of resistance of the bars of a frame connected together at the joints, whose centres of resistance are, 1 between A and B, 2 between B and C, 3 between C and D, 4 between D and E, and 5 between E and A. In the

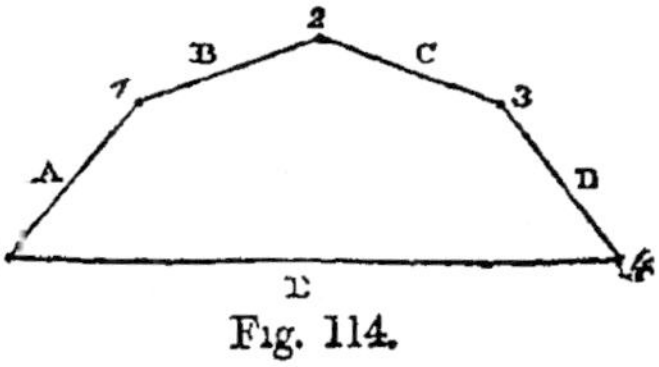

Fig. 114.

figure, the frame consists of five bars; but the principle is applicable to any number. From a point O, in fig. 115, (which may be called the *Diagram of Forces*), draw radiating lines O A, O B, O C, O D, O E, parallel respectively to the lines of resistance of the bars; and on those radiating lines take any lengths whatsoever, to represent the stresses on the several bars, which may have any magnitudes within the limits of strength of the material. Join the points thus found by straight lines, so as to form a closed polygon A B C D E A; then the sides of the polygon will represent a system of forces, which, being applied to the joints of the frame, will balance each other; each such force being applied to the joint between the bars whose lines of resistance are parallel to the pair of radiating lines that enclose the side of the polygon of forces representing the force in question.

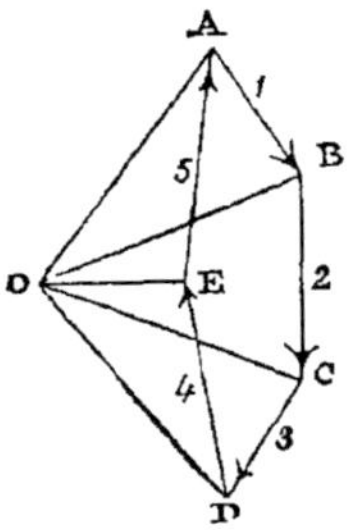

Fig. 115.

When the external forces are parallel to each other, the polygon of forces of fig. 115 becomes a straight line A D, as in fig. 116, divided into segments by the radiating lines; and each segment represents the external force which acts at the joint of the bars whose lines of resistance are parallel to the radiating lines that bound the segment. Moreover, the segment of the line A D which is intercepted between the radiating lines parallel to the lines of resistance of *any two bars whether contiguous or not*, represents the resultant of the external forces which act at points *between the bars*.

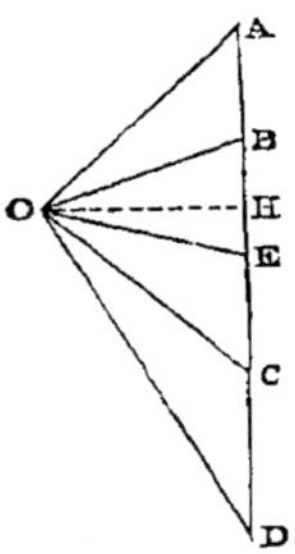

Fig, 116.

Thus, A D represents the total load, consisting of the three portions A B, B C, C D, applied at 1, 2, 3, respectively. D A represents the total supporting force, equal and opposite to the load, consisting of the two portions D E, E A, applied at 4 and 5 respectively. A C represents the resultant of the load applied between the bars A and C; and similarly for any other pair of bars.

From O draw O H perpendicular to A D; then that line represents a component of the stress, whose amount is the same in each bar of the frame. When the load, as is usually the case, is vertical, that component is called the "*horizontal thrust*" of the frame, and, as in Article 270, might more correctly be called *horizontal stress* or *resistance*, seeing that it is a pull in some of the bars and a thrust in others.

The trigonometrical expression of those principles is as follows: —

Let the force O H be denoted simply by H.

Let i, i', denote the inclinations to O H of the lines of resistance of *any two bars*, contiguous or not.

Let R, R′, be the respective stresses which act along those bars.

Let P be the resultant of the external forces acting through the joint or joints between those two bars.

Then

$$P = H (\tan i \pm \tan i');\ldots\ldots\ldots\ldots\ldots\ldots(1.)$$

$$R = H \cdot \sec i;\ R' = H \cdot \sec i' \ldots\ldots\ldots\ldots\ldots\ldots(2.)$$

The $\left\{\begin{array}{c}\text{sum}\\\text{difference}\end{array}\right\}$ of the tangents of the inclinations is to be used, according as the inclinations are $\left\{\begin{array}{c}\text{opposite}\\\text{similar.}\end{array}\right\}$

272. **Open Polygonal Frame.**—When the frame, instead of being closed, as in fig. 114, is converted into an open frame, by the omission of one bar, such as E, the corresponding modification is made in the diagram of inclined forces, fig. 115, by omitting the lines O E, D E, E A, and in the diagram of parallel forces, fig. 116, by omitting the line O E. Then, in both diagrams, D O and O A represent the *supporting forces* respectively, equal and directly opposed to the stresses along the extreme bars of the frame, D and A, which must be exerted by the supports (called in this case *abutments*), at the points 4 and 5, against the ends of those bars, in order to maintain the equilibrium.

In the case of parallel loads, the following formulæ give the horizontal stress and supporting pressures.

Let i_d and i_a denote the angles of inclination of the bars D and A respectively.

Let $R_d = O\,D$ and $R_a = O\,A$ be the stresses along them.

Let $\Sigma \cdot P = A\,D$ denote the total load on the frame; then,

$$H = \frac{\Sigma \cdot P}{\tan i_d + \tan i_a};\ldots\ldots\ldots\ldots\ldots\ldots\ldots\ldots(1.)$$

$$R_d = H \cdot \sec i_d;\ R_a = H \cdot \sec i_a \ldots\ldots\ldots\ldots\ldots\ldots(2.)$$

273. **Polygonal Frame—Stability.**—The stability or instability of a polygonal frame depends on the principles stated in Article 267, page 159, viz., that if a bar be free to change its angular position, then if it is a tie it is stable, and if a strut, unstable; and that a strut may be rendered stable by fixing its ends.

For example, in the frame of fig. 114, E is a tie, and stable; A, B, C, and D, are struts, free to change their angular position, and therefore unstable.

But these struts may be rendered stable in the plane of the frame by means of stays; for example, let two stay-bars connect the joints 1 with 4, and 3 with 5; then the points 1, 2, and 3, are all fixed, so that none of the struts can change their angular positions. The same effect might be produced by two stay-bars connecting the joint 2 with 5 and 4.

The frame, as a whole, is unstable, as being liable to overturn laterally, unless provided with lateral stays, connecting its joints with fixed points.

Now, suppose the frame to be exactly inverted, the loads at 1, 2, and 3, and the supporting forces at 4 and 5, being the same as before. Then E becomes a strut; but it is stable, because its ends are fixed in position; and A, B, C, and D becomes ties, and are stable without being stayed.

An open polygon consisting of ties, such as is formed by A, B, C, and D, when inverted, is called by mathematicians, a *funicular polygon*, because it may be made of ropes.

It is to be observed, that the stability of an *unstayed* polygon of ties is of the kind which admits of *oscillation* to and fro about the position of equilibrium. That oscillation may be injurious in practice, and stays may be required to prevent it.

274. **Bracing of Frames.**—A *brace* is a stay-bar on which there is a permanent stress. If the distribution of the loads on the joints of a polygonal frame, though consistent with its equilibrium as a whole, be not consistent with the equilibrium of each bar, then, in the diagram of forces, when converging lines respectively parallel to the lines of resistance are drawn from the angles of the polygon of external forces, those converging lines, instead of meeting in one point, will be found to have gaps between them. The lines necessary to fill up those gaps will indicate the forces to be supplied by means of the resistance of braces.*

The resistance of a brace introduces a pair of equal and opposite forces, acting along the line of resistance of the brace, upon the pair of joints which it connects. It therefore does not alter the *resultant* of the forces applied to that pair of joints in amount nor in position, but only the *distribution* of the components of that resultant on the pair of joints.

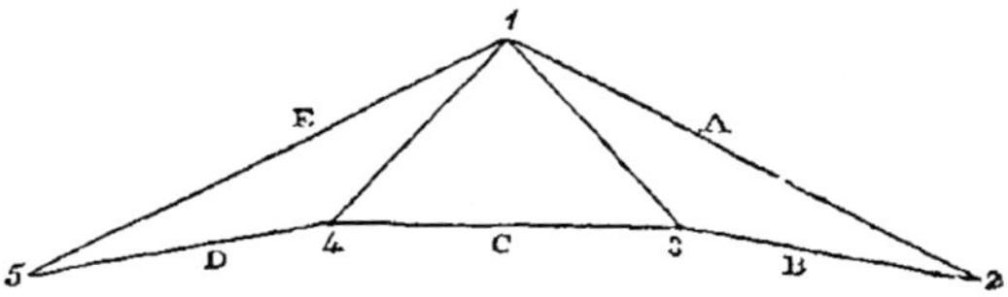

Fig. 117.

To exemplify the use of braces, and the mode of determining the stresses on them, let fig. 117 represent a frame such as frequently

* This method of treating *braced* frames contains an improvement suggested by Prof. Clerk Maxwell in 1867.

occurs in iron roofs, consisting of two struts or rafters, A and E, and three tie-bars, B, C, and D, forming a polygon of five sides, jointed at 1, 2, 3, 4, 5, loaded vertically at 1, and supported by the vertical resistance of a pair of walls at 2 and 5. The joints 3 and 4 having no loads applied to them, are connected with 1 by the braces 1 4 and 1 3.

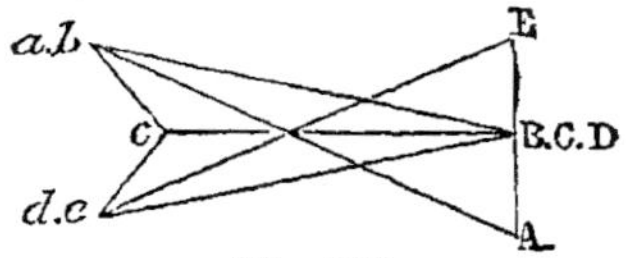

Fig. 118.

To make the diagram of forces (fig. 118), draw the vertical line E, A, as in Article 271, to represent the direction of the load and of the supporting forces.

The two segments of that line, A B and D E, are to be taken to represent the supporting forces at 2 and 5; and the whole line E A will represent the load at 1. From the ends, and from the point of division of the *scale of external forces*, E A, draw straight lines parallel respectively to the lines of resistance of the frame, each line being drawn from the point in E A that is marked with the corresponding letter. Then A *a* and B *b*, meeting at *a*, *b*, will represent the stresses along A and B respectively; and E *e* and D *d*, meeting in D *e*, will represent the stresses along D and E respectively; but those four lines, instead of meeting each other and C *c* parallel to C in one point, leave *gaps*, which are to be filled up by drawing straight lines parallel to the *braces:* that is to say, from *a*, *b*, to *c*, parallel to 1 3; and from *d*, *e*, to *c* parallel to 4 1. Then those straight lines will represent the stresses along the braces to which they are respectively parallel; and C *c* will represent the tension along C. To each joint in the frame, fig. 117, there corresponds, in fig. 118, a triangle, or other closed polygon, having its sides respectively parallel, and therefore proportional, to the forces that act at that joint. For example,

Joints,	1,	2,	3,	4,	5,
Polygons,	E A *a c e* E;	A B *b* A;	B *c b* B;	D *d c* D;	D E *e*D.

The order of the letters indicates the directions in which the forces act relatively to the joints.

Another method of treating simple cases of bracing is illustrated by fig. 119. A and B are two struts, forming the two halves of

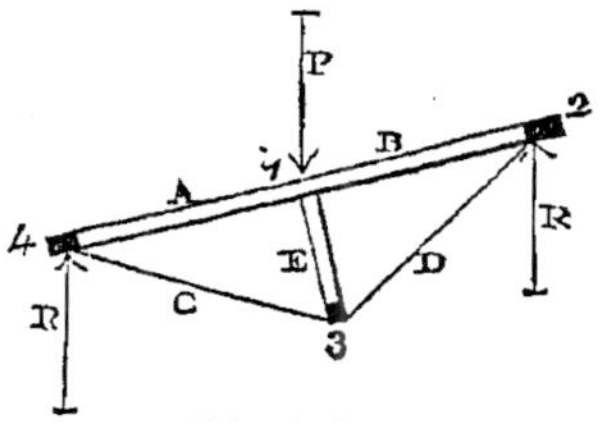

Fig. 119.

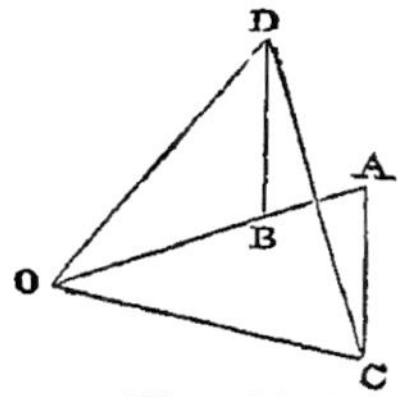

Fig. 120.

one straight bar; C and D are two equal tie-rods; E, a strut brace. A vertical rod P is applied at the joint 1, between A and B; two vertical supporting pressures, each denoted by $R = P \div 2$, act at the joints 4 and 2. The joint 3 has no external load.

Fig. 120 is the diagram of forces, constructed as follows:—Through a point O draw O B A parallel to A and B, O C parallel to C, and O D parallel to D. Make O D = O C; join C D; this line will be parallel to the brace E, and perpendicular to O A.

Through D and C draw vertical lines D B, C A; these, being equal to each other, are to be taken to represent the two supporting pressures R; and their sum D B + A C will represent the load P. The equal tensions on C and D will be represented by O C and O D, and the thrusts along A, B, and E, by O A, O B, and C D.

The polygon of external forces in this case is the crossed quadrilateral A C D B, in which C A and B D represent (as already stated) the supporting pressures, and D C and A B the components of the load P respectively parallel and perpendicular to the brace E. When A and B are horizontal, and E vertical, A B in fig. 120 vanishes, and B D and C A coincide with the two halves of C D.

275. **Rigidity of a Truss.**—The word *truss* is applied in carpentry to a triangular frame, and to a polygonal frame to which rigidity is given by staying and bracing, so that its figure shall be incapable of alteration by turning of the bars about their joints. If each joint were like a hinge, incapable of offering any resistance to alteration of the relative angular position of the bars connected by it, it would be necessary, in order to fulfil the condition of rigidity, that every polygonal frame should be divided by the lines of resistance of stays and braces into triangles and other polygons, so arranged that every polygon of four or more sides should be surrounded by triangles on all but two sides and the included angle at farthest: for every unstayed polygon of four sides or more, with flexible joints, is flexible, unless all the angles except one be fixed by being connected with triangles.

Sometimes, however, a certain amount of stiffness in the joints of a frame, and sometimes the resistance of its bars to bending, is relied upon to give rigidity to the frame, when the load upon it is subject to small variations only in its mode of distribution. For example, in the truss of fig. 121, the tie-beam A A is made in one piece, or in two or more pieces so connected together as to act like one piece; and part of its weight is suspended from the joints C, C, by the rods C B, C B. These rods also serve to make the resistance of the tie-beam A A to being bent act so as to prevent the

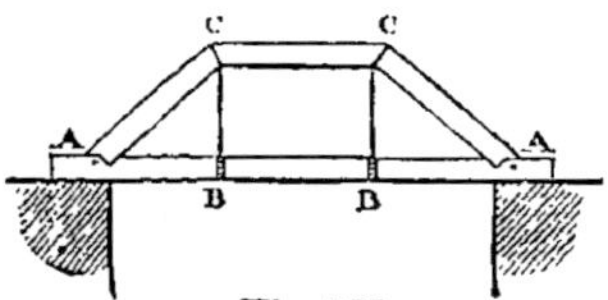

Fig. 121.

struts A C, C C, C A, from deviating from their proper angular positions, by turning on the joints A, C, C, A. If A B, B B, and B A, were three distinct pieces, with flexible joints at B B, it is evident that the frame might be disfigured by distortion of the quadrangle B C C B.

The object of stiffening a truss by braces is to enable it to sustain loads variously distributed; for were the load always distributed in one way, a frame might be designed of a figure exactly suited to that load, so that there should be no need of bracing.

The variations of load produce variations of stress on all the pieces of the frame, but especially on the braces; and each piece must be suited to withstand the greatest stress to which it is liable.

Some pieces, and especially braces, may have to act sometimes as struts and sometimes as ties, according to the mode of distribution of the load.

276. **Secondary and Compound Trussing.**—A *secondary truss* is a truss which is supported by another truss.

When a load is distributed over a great number of centres of resistance, it may be advantageous, instead of connecting all those centres by one polygonal frame, to sustain them by means of several small trusses, which are supported by larger trusses, and so on, the whole structure of secondary trusses resting finally on one large truss, which may be called the *primary truss*. In such a combination the same piece may often form part of different trusses; and then the stress upon it is to be determined according to the following principle:—

When the same bar forms at the same time part of two or more different frames, the stress on it is the resultant of the several stresses to which it is subject by reason of its position in the several frames.

In a *Compound Truss*, several frames, without being distinguishable into primary and secondary, are combined and connected in such a manner that certain pieces are common to two or more of them, and require to have their stresses determined by the principle above stated.

Example.—Fig. 122, represents a kind of secondary trussing common in the framework of iron roofs.

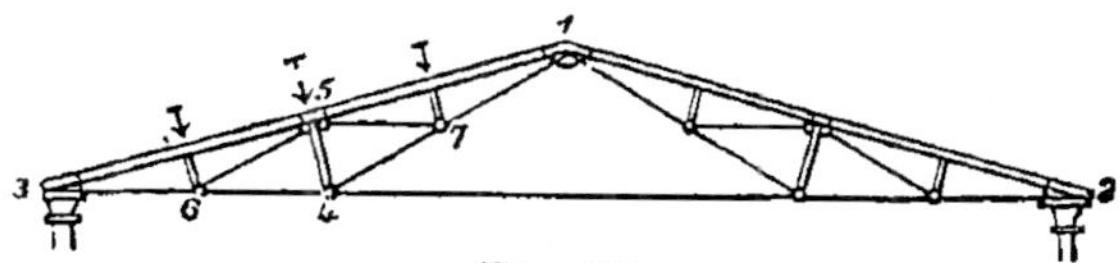

Fig. 122.

The entire frame is supported by pillars at 2 and 3, each of which sustains in all, half the weight.

1 2 3 is the *primary truss*, consisting of two rafters 1 3, 1 2, and a tie-rod 2 3.

The weight of a division of the roof is distributed over the rafters.

The middle point of each rafter is supported by a *secondary truss;* one of those is marked 1 4 3; it consists of a strut, 1 3 (the rafter itself), two ties 4 1, 4 3, and a strut-brace, 5 4, for transmitting the load, applied at 5, to the point where the ties meet.

Each of the two larger secondary trusses just described supports two *smaller secondary trusses* of similar form and construction to itself; two of those are marked 1 7 5, 5 6 3; and the subdivision of the load might be carried still farther.

In determining the stresses on the pieces of this structure, it is indifferent, so far as mathematical accuracy is concerned, whether we commence with the primary truss or with the secondary trusses; but by commencing with the primary truss, the process is rendered more simple.

(1.) *Primary Truss* 1 2 3. Let W denote the weight of the roof; then $\frac{1}{2}$ W is distributed over each rafter, the resultants acting through the middle points of the rafters. Divide each of those resultants into two equal and parallel components, each equal to $\frac{1}{4}$ W, acting through the ends of the rafter; then $\frac{1}{4}$ W is to be considered as directly supported at 3, $\frac{1}{4}$ W at 2, and $\frac{1}{4}$ W + $\frac{1}{4}$ W = $\frac{1}{2}$ W at 1; therefore the load at the joint 1 is

$$P = \tfrac{1}{2} W.$$

Let i be the inclination of the rafters to the horizon; then by the equations of Article 270.

$$H = \frac{\frac{1}{2} W}{2 \tan i} = \frac{W}{4 \tan i}; \ldots\ldots\ldots\ldots\ldots\ldots (1.)$$

This is the pull upon the horizontal tie-rod of the primary truss, 2 3; and the thrust on each of the rafters 1 3, 1 2, is given by the equation

$$R = H \sec i = \frac{W \operatorname{cosec} i}{4} \ldots\ldots\ldots\ldots\ldots\ldots (2.)$$

(2.) *Secondary Truss* 1 4 3 5. The rafter 1 3 has the load $\frac{1}{2}$ W dsstributed over it; and reasoning as before, we are to leave two quarters of this out of the calculation, as being directly supported at 1 and 3, and to consider one-half, or $\frac{1}{4}$ W, as being the vertical load at the point 5. The truss is to be considered as consisting of a polygon of four pieces, 5 1, 1 4, 4 3, 3 5, two of which happen to be in the same straight line, and of the strut-brace, 5 4, which exerts obliquely upwards against 5, and obliquely downwards against 4, a thrust equal to the component perpendicular to the rafter of the load $\frac{1}{4}$ W; which thrust is given by the equation

$$R_{54} = \frac{1}{4} W \cos i \ldots\ldots\ldots\ldots\ldots\ldots\ldots\ldots (3.)$$

Then we easily obtain the following values of the stresses on the rafter and ties, in which each stress is distinguished by having affixed to the letter R the numbers denoting the two joints between which it acts.

$$\left.\begin{array}{ll} \text{Pulls on ties} & \left\{ R_{43} = R_{41} = \frac{R_{54}}{2 \sin i} = \frac{1}{8} W \operatorname{cotan} i; \right. \\ \text{Thrusts on rafter} & \left\{ \begin{array}{l} R_{35} = \frac{R_{54}}{2 \tan i} + \frac{1}{8} W \sin i = \frac{1}{8} W \operatorname{cosec} i, \\ R_{51} = \frac{R_{54}}{2 \tan i} - \frac{1}{8} W \sin i = \frac{1}{8} W (\operatorname{cosec} i - 2 \sin i); \end{array} \right. \end{array} \right\} (4.)$$

The difference between the thrusts on the two divisions of the rafter,

$$R_{35} - R_{51} = \frac{1}{4} W \sin i,$$

is the component *along the rafter* of the load at the point 5.

(3.) *Smaller Secondary Trusses*, 1 7 5, 5 6 3.—These trusses are similar in every respect to the larger secondary trusses, except that the load on each point is one-half, and consequently each of the stresses is reduced to one-half of the corresponding stress in the Equations 3 and 4.

(4.) *Resultant Stresses.* The pull on the middle division of the great tie-rod 2 3 is simply that due to the primary truss, 1 2 3. The pull on the tie 4 7 is simply that due to the secondary truss 1 4 3. The pulls on the ties 5 7, 5 6, are simply those due to the smaller secondary trusses, 1 5 7, 5 6 3. But agreeably to the Theorem stated at the commencement of this article, the pull on the tie 1 7 is the sum of those due to the larger secondary truss 1 4 3, and the smaller secondary truss 1 7 5. The pull on 6 4 is the sum of those due to the primary truss 1 2 3, and to the larger secondary truss 1 4 3. The pull on 6 3 is the sum of those due to the primary truss 1 2 3, to the larger secondary truss 1 4 3, and to the smaller secondary truss 5 6 3. The thrust on each of the four divisions of the rafter 1 3, is the sum of three thrusts, due respectively to the primary truss, the larger secondary truss, and one or other of the smaller secondary trusses.

277. **Resistance of a Frame at a Section.**—The labour of calculating the stress on the bars of a frame may sometimes be abridged by the application of the following principle:—

If a frame be acted upon by any system of external forces, and if that frame be conceived to be completely divided into two parts by an ideal surface, the stresses along the bars which are intersected by that surface, balance the external forces which act on each of the two parts of the frame.

In most cases which occur in practice, the lines of resistance of the bars, and the lines of action of the external forces, are all in one vertical plane, and the external forces are vertical. In such cases the most convenient position for an assumed plane of section is vertical, and perpendicular to the plane of the frame. Take the vertical line of intersection of these two planes for an axis of co-ordinates,—say for the axis of y, and any convenient point in it for the origin O; let the axis of x be horizontal, and in the plane of the frame, and the axis of z horizontal, and in the plane of section.

The external forces applied to the part of the frame at one side of the plane of section (either may be chosen), being combined, as in Article 235, page 134, give three data—viz., the total force along $x = \Sigma \cdot \mathrm{X}$; the total force along $y = \Sigma \cdot \mathrm{Y}$; and the moment of the couple acting round $z = \mathrm{M}$; and the bars which are cut by the plane of section must exert resistances capable of balancing those two forces and that couple. If not more than three bars are cut by the plane of section, there are not more than three unknown quantities, and three relations between them and given quantities, so that the problem is determinate; if more than three bars are cut by the plane of section, the problem is or may be indeterminate.

The formulæ to which this reasoning leads are as follows:—Let x be positive in a direction from the plane of section towards the part of the structure which is considered in determining $\Sigma \cdot \mathrm{X}$, $\Sigma \cdot \mathrm{Y}$, and M; let $+y$ be measured upwards; let angles measured from Ox towards $+y$, that is, upwards, be positive; and let the lines of resistance of the three bars cut by the plane of section make the angles i_1, i_2, i_3, with x. Let n_1, n_2, n_3, be the perpendicular distances of those three lines of resistance from O, distances lying

$$\left\{\begin{matrix}\text{upwards}\\ \text{downwards}\end{matrix}\right\} \text{ from O}x \text{ being considered as } \left\{\begin{matrix}\text{positive}\\ \text{negative.}\end{matrix}\right\}$$

Let R_1, R_2, R_3, be the resistances, or total stresses, along the three bars, pulls being positive, and thrusts negative. Then we have the following three equations:—

$$\left.\begin{aligned}\Sigma \cdot \mathrm{X} &= \mathrm{R}_1 \cos i_1 + \mathrm{R}_2 \cos i_2 + \mathrm{R}_3 \cos i_3;\\ \Sigma \cdot \mathrm{Y} &= \mathrm{R}_1 \sin i_1 + \mathrm{R}_2 \sin i_2 + \mathrm{R}_3 \sin i_3;\\ -\mathrm{M} &= \mathrm{R}_1 n_1 + \mathrm{R}_2 n_2 + \mathrm{R}_3 n_3;\end{aligned}\right\} \ldots\ldots\ldots(1.)$$

from which the three quantities sought, R_1, R_2, R_3 can be found.

Speaking with reference to the given plane of section, $\Sigma \cdot \mathrm{X}$ may be called the *normal stress*, $\Sigma \cdot \mathrm{Y}$, the *shearing stress*, and M, the *moment of flexure*, or *bending stress;* for it tends to bend the frame at the section under consideration. M is to be considered as $\left\{\begin{matrix}\text{positive}\\ \text{negative}\end{matrix}\right\}$ according as it tends to make the frame become concave $\left\{\begin{matrix}\text{upwards}\\ \text{downwards.}\end{matrix}\right\}$

The following is one of the simplest examples of the solution of a problem by the *method of polygons*, and the *method of sections.* Fig. 121 represents a truss of a form very common in carpentry (already referred to in Article 275), and consisting of three struts, A C, C C, C A, a tie-beam A A, and two suspension-rods, C B, C B, which serve to suspend part of the weight of the tie-beam from the joints C C, and also to stiffen the truss in the manner mentioned in Article 275.

Let i denote the equal and opposite inclinations of the rafters A C, C A, to the horizontal tie-beam A A; and leaving out of consideration the portions of the load directly supported at A, A, let P, P, denote equal vertical loads applied at C, C, and - P, $-$P, equal upward vertical supporting forces applied at A, A, by the resistance of the props. Let H denote the pull on the tie-beam, R the thrust on each of the sloping rafters, and T the thrust on the horizontal strut C C.

Proceeding by the *method of polygons*, as in Article 271, we find at once,

$$\left.\begin{aligned} \mathrm{H} = -\mathrm{T} = \mathrm{P} \operatorname{cotan} i; \\ \mathrm{R} = -\mathrm{P} \operatorname{cosec} i. \end{aligned}\right\} \quad \text{.....................(2.)}$$

(Thrusts being considered as negative.)

To solve the same question by the *method of sections*, suppose a vertical section to be made by a plane traversing the centre of the right hand joint C; take that centre for the origin of co-ordinates; let x be positive towards the right, and y positive downwards; let x_1, y_1, be the co-ordinates of the centre of resistance at the right hand point of support A. When the plane of section traverses the centre of resistance of a joint, we are at liberty to suppose either of the two bars which meet at that joint on opposite sides of the plane of section to be cut by it at an insensible distance from the joint.

First, consider the plane of section as cutting C A. The forces and couple acting on the part of the frame to the right of the section are

$$\mathrm{F}_x = 0;\ \mathrm{F}_y = -\mathrm{P}$$
$$\mathrm{M} = -\mathrm{P}x_1.$$

Then, observing that for the strut A C, $n = 0$, and that for the tie A A, $n = y_1$, we have, by the equations 1 of this Article

$$R \cos i + H = F_x = 0 ;$$
$$R \sin i = -P ;$$
$$H y_1 = -M = +P x_1 ;$$

whence we obtain, from the last equation,

$$H = \frac{P x_1}{y_1} = P \text{ cotan } i$$

from the first, or from the second

$$R = -\frac{H}{\cos i} = -P \text{ cosec } i \quad \text{..............(3.)}$$

Next, conceive the section to cut C C at an insensible distance to the left of C. Then the equal and opposite applied forces + P at C, and - P at A, have to be taken into account; so that

$$F_x = 0 ; \ F_y = 0 ; \ M = -P x_1 ;$$

from the first of which equations we obtain

$$H + T = F_x = 0, \text{ and}$$
$$T = -H = -P \text{ cotan } i \text{....................(4.)}$$

In the example just given, the method of sections is tedious and complex as compared with the method of polygons, and is introduced for the sake of illustration only.

278. **Balance of a Chain or Cord.**—A loaded chain may be looked upon as a polygonal frame whose pieces and joints are so numerous that its figure may without sensible error be treated as a continuous curve. The following are the principles respecting the equilibrium of loaded chains and cords which are of most importance in practice.

I. *Balance of a Chain in general.*—Let D A C, in fig. 123, represent a flexible cord or chain supported at the points C and D, and loaded by forces in any direction, constant or varying, distributed over its whole length with constant or varying intensity.

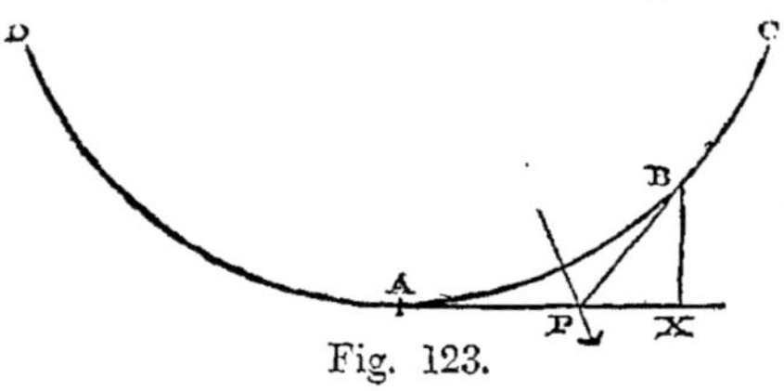

Fig. 123.

Let A and B be any two points in this chain; from those points draw tangents to the chain, A P and B P, meeting in P. The load acting on the chain between the points A and B is balanced by the pulls along the chain at those two points respectively; those pulls must respectively act along the tangents A P, B P; hence the resultant of the load between A and B acts through the point of intersection of the tangents at A and B; and that load, and the tensions on the chain at A and B, are respectively proportional to the sides of a triangle parallel to their directions.

II. *Chain under Vertical Load.—Curve of Equilibrium.*—If the direction of the load be everywhere parallel and vertical, draw a vertical straight line, C D, fig. 124, to represent the total load, and from its ends draw C O and D O, parallel to two tangents at the points of support of the chain, and meeting in O; those lines will represent the tensions on the chain at its points of support.

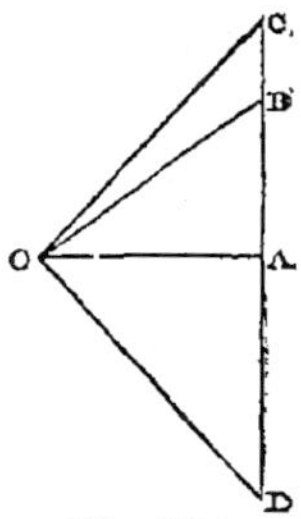

Fig. 124.

Let A, in fig. 123, be the lowest point of the chain. In fig. 124, draw the horizontal line O A; this will represent the horizontal component of the tension of the chain at every point, and if O B be parallel to a tangent to the chain at B (fig. 123), A B will represent the portion of the load supported between A and B, and O B the tension at B.

To express this algebraically, let

H = O A = *horizontal tension* along the chain at A;

R = O B = pull along the chain at B;

P = A B = load on the chain between A and B;

$i = \angle$ X P B (fig. 123) = $\angle$ A O B fig. 124) = inclination of chain at B;

then,

$$P = H \tan i;\ R = \sqrt{(P^2 + H^2)} = H \sec i \quad \text{.........(1.)}$$

To deduce from these formulæ an equation by which the form of the curve assumed by the chain can be determined when the distribution of the load is known, let that curve be referred to rectangular, horizontal, and vertical co-ordinates, measured from the lowest point A, fig. 123, the co-ordinates of B being, A X = x, X B = y, then $\tan i = \frac{dy}{dx} = \frac{P}{H}$, a differential equation, which enables the form assumed by the cord (or "curve of equilibrium") to be determined when the distribution of the load is known.

279. **Stability of Blocks.**—The conditions of stability of a single block supported upon another body at a plane joint may be thus summed up:—

In fig. 125, let A A represent the upper block, B B part of the supporting body, e E the joint, C its centre of pressure, P C the resultant of the whole pressure distributed over the joint, N C, T C, its components perpendicular and parallel to the joints respectively. Then the conditions of stability are the following:—

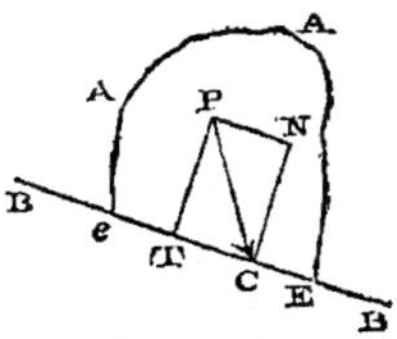

Fig. 125.

I. *In order that the block may not slide, the obliquity of the pressure must not exceed the angle of repose* (Article 261, page 154), that is to say,

$$\angle \mathrm{P\,C\,N} \leqq \phi \ldots\ldots\ldots\ldots\ldots\ldots(1.)$$

II. *In order that the block may be in no danger of overturning, the ratio which the deviation of the centre of pressure from the centre of figure of the joint bears to the length of the diameter of the joint traversing those two centres, must not exceed a certain fraction.* The value of that fraction varies, according to circumstances, from one-eighth to three-eighths.

The first of these conditions is called that of *stability of friction,* the second, that of *stability of position.*

In a structure composed of a series of blocks, or of a series of courses so bonded that each may be considered as one block, which blocks or courses press against each other at plane joints, the two conditions of stability must be fulfilled at each joint.

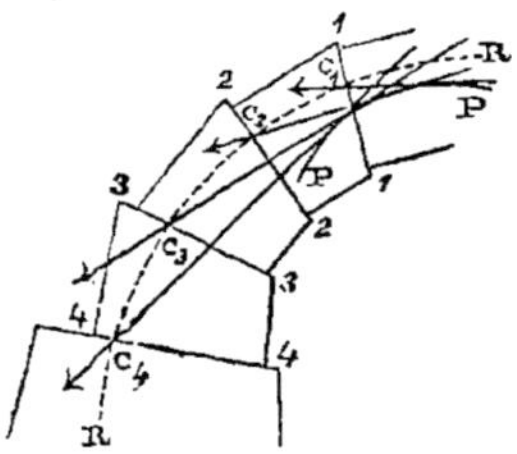

Fig. 126.

Let fig. 126 represent part of such a structure, 1, 1, 2, 2, 3, 3, 4, 4, being some of its plane joints.

Suppose the centre of pressure C_1 of the joint 1, 1, to be known, and also the amount and direction of the pressure, as indicated by the arrow traversing C_1. With that pressure combine the weight of the block 1, 2, 2, 1, together with any other external force which may act on that block; the resultant will be the total pressure to be resisted at the joint 2, 2, which will be given in magnitude, direction, and position, and will intersect that joint in the centre of pressure C_2. By continuing this process there are found the centres of pressure C_3, C_4, &c., of any number of successive joints, and the directions and magnitudes of the resultant pressures acting at those joints.

The magnitude and position of the resultant pressure at any joint whatsoever, and consequently the centre of pressure at that joint, may also be found simply by taking the resultant of all the forces which act on one of the parts into which that joint divides the structure.

The centres of pressure at the joints are sometimes called *centres of resistance.* A line traversing all those centres of resistance, such as the dotted line R R, in fig. 126, has received from Mr. Moseley the name of the "*line of resistance;*" and that author has also shewn how in many cases the equation which expresses the form of that line may be determined, and applied to the solution of useful problems.

The straight lines representing the resultant pressures may be all parallel, or may all lie in the same straight line, or may all intersect in one point. The more common case, however, is that in which those straight lines intersect each other in a series of points, so as to form a polygon. A curve, such as P P, in fig. 126 touching all the sides of that polygon, is called by Mr. Moseley the "*line of pressures.*"

The properties which the line of resistance and line of pressures must have, in order that the conditions of stability may be fulfilled, are, as already stated, the following:—

To insure stability of position, *the line of resistance must not deviate from the centre of figure of any joint by more than a certain fraction of the diameter of the joint, measured in the direction of deviation.*

To insure stability of friction, *the normal to each joint must not make an angle greater than the angle of repose with a tangent to the line of pressures drawn through the centre of resistance of that joint.*

Conceive a line to pass through all the limiting positions of the centre of resistance of the joint, so as to enclose a space beyond which that centre must not be found.

The product of the weight of the structure into the horizontal distance of a point in this line from a vertical line traversing the centre of gravity of the structure is the MOMENT OF STABILITY *of the structure, when the applied thrust acts in a vertical plane parallel to that horizontal distance, and tends to overturn the structure in the direction of the given point in the line limiting the position of the centre of resistance;* for that, according to Article 222, is the moment of the couple, which, being combined with a single force equal to the weight of the structure, transfers the line of action of that force parallel to itself through a distance equal to the given horizontal distance of the centre of resistance from the centre of gravity of the structure. The applied couple usually consists of the thrust of a frame, or an arch, or the pressure of a fluid, or of a mass of earth, against the structure, together with the equal, opposite, and parallel, but not directly opposed, resistance of the joint to that lateral force.

To express this symbolically, let t be the length of the diameter of the joint where it is cut by the vertical plane traversing the centre of gravity of the structure and parallel to the applied thrust; let j be the inclination of that diameter to the horizon; let $q\,t$ be the distance of the given limiting centre of resistance from the middle point of that diameter, and $q'\,t$ the distance from the same middle point to the point where the diameter is cut by the vertical line through the centre of gravity of the structure, and let W be the weight of the structure. Then the moment of stability is

$$W\,(q \pm q')\,t \cos j; \qquad (1.)$$

the sign $\left\{ {+ \atop -} \right\}$ being used according as the centre of resistance, and the vertical line through the centre of gravity, lie towards $\left\{\begin{array}{l}\text{opposite sides}\\ \text{the same side}\end{array}\right\}$ of the middle of the diameter.

Let h denote the height of the structure above the middle of the plane joint which is its base, b the breadth of that joint in a direction perpendicular or conjugate to the diameter t, and w the weight of an unit of volume of the material. Then we shall have

$$W = n \cdot w\, h\, b\, t \ldots\ldots\ldots\ldots\ldots\ldots\ldots\ldots (2.)$$

where n is a *numerical factor* depending on the *figure* of the structure, and on the angles which the dimensions, h, b, t, make with each other; that is, the angles of obliquity of the co-ordinates to which the figure of the structure is referred. Introducing this value of the weight of the structure into the formula 1, we find the following value for the moment of stability:—

$$n\,(q \pm q')\cos j \cdot w \cdot h\, b\, t^2 \ldots\ldots\ldots\ldots\ldots\ldots (3.)$$

This quantity is divided by points into three factors, viz. :—

(1.) $n\,(q \pm q')\cos j$, a *numerical factor*, depending on the *figure* of the structure, the *obliquities* of its co-ordinates, and the *direction* in which the applied force tends to overturn it.

(2.) w, the specific gravity of the material.

(3.) $h\, b\, t^2$, a geometrical factor, depending on the dimensions of the structure.

Now the first factor is the same in all structures having figures of the same class, with co-ordinates of equal obliquity, and exposed to similarly applied external forces; that is say, to all structures whose figures, together with the lines of action of the applied forces, are *parallel projections of each other, with co-ordinates of equal obliquity;* hence for any set of structures which fulfil that condition, the moments of stability are proportional to—

I. The specific gravity of the material;

II. The height;

III. The breadth;

IV. The *square* of the thickness; that is, of the dimension of the base which is parallel to the vertical plane of the applied force.

280. **Transformation of Blockwork Structures.**—If a structure composed of blocks have stability of position when acted on by forces represented by a given system of lines, then will a structure whose figure is a parallel projection of the original structure have stability of position when acted on by forces represented by the corresponding parallel projection of the original system of lines; also, the centres of pressure in the new structure will be the corresponding projections of the centres of pressure in the original structure.

The question, whether the new structure obtained by transformation will possess *stability of friction* is an independent problem.

CHAPTER II.

PRINCIPLES AND RULES RELATING TO STRENGTH AND STIFFNESS.

281. **The Object of this Chapter** is to give a summary of the principles, and of the general rules of calculation, which are applicable to problems of strength and stiffness, whatsoever the particular material may be.

Section I.—Of Strength and Stiffness in General.

282. **Load, Stress, Strain, Strength.**—The *load,* or combination of external forces, which is applied to any piece, moving or fixed, in a structure or machine, produces *stress* amongst the particles of that piece, being the combination of forces which they exert in resisting the tendency of the load to disfigure and break the piece, accompanied by *strain,* or alteration of the volumes and figures of the whole piece, and of each of its particles.

If the load is continually increased, it at length produces either *fracture* or (if the material is very tough and ductile) such a disfigurement as is practically equivalent to fracture, by rendering the piece useless.

The *Ultimate Strength* of a body is the load required to produce fracture in some specified way. The *Proof Strength* is the load required to produce the greatest strain of a specific kind consistent with safety; that is, with the retention of the strength of the material unimpaired. A load exceeding the proof strength of the body, although it may not produce instant fracture, produces fracture eventually by long-continued application and frequent repetition.

The *Working Load* on each piece of a machine is made less than the ultimate strength, and less than the proof strength, in certain ratios determined partly by experiment and partly by practical experience, in order to provide for unforeseen contingencies.

Each solid has as many different kinds of strength as there are different ways in which it can be strained or broken, as shewn in the following classification:—

	Strain.	Fracture.
Elementary	Extension	......Tearing.
	Compression	......Crushing.
Compound........	Distortion	......Shearing.
	Twisting	......Wrenching.
	Bending	......Breaking across

283. **Coefficients or Moduli of Strength** are quantities expressing the *intensity* of the stress under which a piece of a given material gives way when strained in a given manner; such intensity being expressed in units of weight for each unit of sectional area of the layer of particles at which the body first begins to yield. In Britain, the ordinary unit of intensity employed in expressing the strength of materials is the *pound avoirdupois on the square inch.*

Coefficients of strength are of as many different kinds as there are different ways of breaking a body. Their use will be explained in the sequel.

Coefficients of strength, when of the same kind, may still vary according to the direction in which the stress is applied to the body. Thus the tenacity, or resistance to tearing, of most kinds of wood is much greater against tension exerted along than across the grain.

284. **Factors of Safety.**—A factor of safety, in the ordinary sense, is the ratio in which the load that is just sufficient to overcome instantly the strength of a piece of material is greater than the greatest safe ordinary working load.

The proper value for the factor of safety depends on the nature of the material; it also depends upon how the load is applied. The load upon any piece in a structure or in a machine is distinguished into *dead load* and *live load.* A *dead load* is a load which is put on by imperceptible degrees, and which remains steady; such as the weight of a structure, or of the fixed framing in a machine. A *live load* is one that is or may be put on suddenly, or accompanied with vibration; like a swift train travelling over a railway bridge; or like most of the forces exerted by and upon the moving pieces in a machine.

It can be shewn that in most cases which occur in practice a live load produces, or is liable to produce, *twice,* or very nearly twice, the effect, in the shape of stress and strain, which an equal dead load would produce. The *mean* intensity of the stress produced by a suddenly applied load is no greater than that produced by the same load acting steadily; but in the case of the suddenly applied load, the stress begins by being insensible, increases to double its mean intensity, and then goes through a series of fluctuations, alternately below and above the mean, accompanied by vibration of the strained body. Hence the ordinary practice is to make the factor of safety for a live load *double* of the factor of safety for a dead load.

A distinction is to be drawn between *real* and *apparent* factors of safety. A real factor of safety is the ratio in which the ultimate or breaking stress is greater than the real working stress at the time when the straining action of the load is greatest. The apparent factor of safety has to be made greater than the real factor of safety in those cases in which the calculation of strength is based, not upon the greatest straining action of the load, but upon a mean straining action, which is exceeded by the greatest straining action in a certain proportion. In such cases the apparent factor of safety is the product obtained by multiplying the real factor of safety by the ratio in which the greatest straining action exceeds the mean.

Another class of cases in which the apparent exceeds the real factor of safety is when there are additional straining actions besides that due to the transmission of motive power, and when those additional actions, instead of being taken into account in detail, are allowed for in a rough way by means of an increase of the fact r of safety. A third class of cases is when there is a possibility of an increased load coming by accident to act upon the piece under consideration. For example, a steam engine may drive two lines of shafting, exerting half its power on each; one may suddenly break down, or be thrown out of gear, and the engine may for a short time exert its whole power on the other.

The following table shews the ordinary values of real factors of safety:—

	Real Factors of Safety. Dead L ad	Live L ad.
Perf ct materials and workmanship,.. .	2	4
Ordinary materials and workmanship—		
Metals, ...	3	6
Wood, Hempen Ropes,..................from	3 to 5	10
Masonry and Brickwork,.	4	8

The following are examples of ap| arent factors of safety:—

Real Fact r of Safety, 6	Rati in which Greatest Effort ex eeds Mean Effort, nearly.	Apparent Fact r of Safety.
Steam engines acting against a constant resistance—		
Single engine, ...	1·6	9·6
Pair of engines driving cranks at right angles,..................................	1·1	6·6
Three engines driving equiangular cranks, ...	1·05	6·3
Ordinary cases of varying effort and resistance,....	2·0	12·0
Lines of shafting in millwork; apparent factor of safety for twisting stress due to motive power, to cover allowances for bending actions, accidental extra load, &c., ...		from 18 to 36

Almost all the experiments hitherto made on the strength of materials give coefficients or moduli of *ultimate strength;* that is, coefficients expressing the intensity of the stress exerted by the most severely strained particles of the material just before it gives way. In calculations for the purpose of designing framework or machinery to bear a given working load, there are two ways of using the factor of safety,—one is, to multiply the working load by the factor of safety, so as to determine the breaking load, and use this load in the calculation, along with the modulus of ultimate strength: the other is, to divide the modulus of ultimate strength by the factor of safety, and thus to find a modulus or coefficient of *working stress,* which is to be used in the calculation, along with the *working load.* It is obvious that the two methods are mathematically equivalent, and must lead to the same result; but the latter is on the whole the more convenient in designing machines.

285. The **Proof** or **Testing** by experiment of the strength of a piece of material is conducted in two different ways, according to the object in view.

I. If the piece is to be *afterwards used,* the testing load must be so limited that there shall be no possibility of its impairing the strength of the piece; that is, it must not exceed the *proof strength,* being from one-third to one-half of the ultimate strength. About double or treble of the working load is in general sufficient. Care should be taken to avoid vibrations and shocks when the testing load approaches near to the proof strength.

II. If the piece is to be *sacrificed* for the sake of ascertaining the strength of the material, the load is to be increased by degrees until the piece breaks, care being taken, especially when the breaking point is approached, to increase the load by small quantities at a time, so as to get a sufficiently precise result.

The *proof strength* requires much more time and trouble for its determination than the ultimate strength. One mode of approximating to the proof strength of a piece is to apply a moderate load and remove it, apply the same load again and remove it, two or three times in succession, observing at each time of application of the load the *strain* or alteration of figure of the piece when loaded, by stretching, compression, bending, distortion, or twisting, as the case may be. If that alteration does *not sensibly increase* by repeated applications of the same load, the load is within the limit of proof strength. The effects of a greater and a greater load being successively tested in the same way, a load will at length be reached whose successive applications produce increasing disfigurements of the piece; and this load will be greater than the proof strength, which will lie between the last load and the last load but one in the series of experiments.

It was formerly supposed that the production of a *set*—that is, a disfigurement which continues after the removal of the load—was a test of the proof strength being exceeded; but Mr. Hodgkinson shewed that supposition to be erroneous, by proving that in most materials a set is produced by almost any load, how small soever.

The strength of bars and beams to resist breaking across, and of axles to resist twisting, can be tested by the application of known weights either directly or through a lever.

To test the tenacity of rods, chains, and ropes, and the resistance of pillars to crushing, more powerful and complex mechanism is required. The apparatus most commonly employed is the hydraulic press. In computing the stress which it produces, no reliance ought to be placed on the load on the safety valve, or on a weight hung to the pump handle, as indicating the intensity of the pressure, which should be ascertained by means of a pressure gauge. This remark applies also to the proving of boilers by water pressure. From experiments by Messrs. Hick and Luthy it appears that, in calculating the stress produced on a bar by means of a hydraulic press, the friction of the collar may be allowed for by deducting a force equivalent to the pressure of the water upon an area of a length equal to the circumference of the collar, and one-eightieth of an inch broad.

For the exact determination of general laws, although the load may be applied at one end of the piece to be tested by means of a hydraulic press, it ought to be resisted and measured at the other end by means of a combination of levers.

286. **Stiffness or Rigidity, Pliability, their Moduli or Coefficients.**—Rigidity or stiffness is the property which a solid body possesses of resisting forces tending to change its figure. It may be expressed as a quantity, called a *modulus or coefficient of stiffness*, by taking the ratio of the intensity of a given stress of a given kind to the strain, or alteration of figure, with which that stress is accompanied; that strain being expressed as a quantity by dividing the alteration of some dimension of the body by the original length of that dimension. In most materials which are used in machinery, the moduli of stiffness, though not exactly constant, are nearly constant for stresses not exceeding the proof strength.

The reciprocal of a modulus of stiffness may be called a "*modulus of pliability*;" that is to say,

$$\text{Modulus of Stiffness} = \frac{\text{Intensity of Stress}}{\text{Strain}};$$

$$\text{Modulus of Pliability} = \frac{\text{Strain}}{\text{Intensity of Stress}}.$$

287. **The Elasticity of a Solid** consists of stiffness, or resistance to change of figure, combined with the power of recovering the

original figure when the straining force is withdrawn. If that recovery is complete and immediate, the body is *perfectly elastic;* if there is a *set*, or permanent change of figure, after the removal of the straining force, the body is *imperfectly elastic.* The elasticity of no solid substance is absolutely perfect, but that of many substances is nearly perfect when the stress does not exceed the proof strength, and may be made sensibly perfect by restricting the stress within small enough limits.

Moduli or *Coefficients of Elasticity* are the values of moduli of stiffness when the stress is so limited that the value of each of those moduli is sensibly constant, and the elasticity of the body sensibly perfect.

288. **Resilience or Spring** is the quantity of *mechanical work** required to produce the proof stress on a given piece of material, and is equal to the product of the *proof strain*, or alteration of figure, into the mean load which acts during the production of that strain; that is to say, in general, very nearly one-half of the proof load.

289. **Heights or Lengths of Moduli of Stiffness and Strength.**—The term *height* or *length*, as applied to a modulus or coefficient of strength or of stiffness, means the length of an imaginary vertical column of the material to which the modulus belongs, whose weight would cause a pressure on its base equal in intensity to the stress expressed by the given modulus. Hence

Height of a modulus in feet

$$= \frac{\text{Modulus in lbs. on the square foot}}{\text{Heaviness of material in lbs. to the cubic foot}};$$

$$= \frac{\text{Modulus in lbs. on the square inch}}{\text{Weight of 12 cubic inches of the material}};$$

Height of a modulus in inches

$$= \frac{\text{Modulus in lbs. on the square inch}}{\text{Heaviness of material in lbs. to the cubic inch}};$$

Height of a modulus in mètres

$$= \frac{\text{Modulus in kilogrammes on the square mètre}}{\text{Heaviness of material in kilogrammes to the cubic mètre}}.$$

Section 2.—Of Resistance to Direct Tension.

290. **Strength, Stiffness, and Resilience of a Tie.**—The word *tie* is here used to denote any piece in framing or in mechanism, such

* *Mechanical Work*, which will be fully treated of in Part VI., may be defined as the product of a *force* into the *space* through which it acts.

as a rod, bar, band, cord, or chain, which is under the action of a pair of equal and opposite longitudinal forces tending to stretch it, and to tear it asunder. The common magnitude of those two forces is the load; and it is equal to the product of the sectional area of the piece into the intensity of the tensile stress. The values of that intensity, corresponding to the immediate breaking load, the proof load, and the working load, are called respectively the moduli or coefficients of *ultimate tenacity*, of *proof tension*, and of *working tension*.

In symbols, let P be the load, S the sectional area, and p the intensity of the tensile stress; then

$$\mathrm{P} = p\,\mathrm{S} \qquad \text{(1.)}$$

If the sectional area varies at different points, the *least* area is to be taken into account in calculations of strength.

The elongation of a tie produced by any load, P, not exceeding the proof load, is found as follows, provided the sectional area is uniform:—

Let x denote the original length of the tie, $\triangle x$ the elongation, and $\alpha = \frac{\triangle x}{x}$ the extension; that is, the *proportion* which that elongation bears to the original length of the bar, being the numerical measure of the strain.

Let E denote the *modulus of direct elasticity*, or resistance to stretching. Then

$$\alpha = \frac{p}{\mathrm{E}};\ \triangle x = \alpha x = \frac{p}{\mathrm{E}} x. \qquad \text{(2.)}$$

Let f' denote the *proof tension* of the material, so that f' S is the proof load of the tie; then the *proof extension* is $f' \div \mathrm{E}$.

The Resilience or Spring of the tie, or the work done in stretching it to the limit of proof strain, is computed as follows. The length, as before, being x, the elongation of the tie produced by the proof load is $f' x \div \mathrm{E}$. The force which acts through this space has for its least value 0, for its greatest value $\mathrm{P} = f'\,\mathrm{S}$, and for its mean value $f'\,\mathrm{S} \div 2$; so that the work done in stretching the tie to the proof strain, that is, its *resilience* or *spring*, is

$$\frac{f'\,\mathrm{S}}{2} \cdot \frac{f' x}{\mathrm{E}} = \frac{f'^2}{\mathrm{E}} \cdot \frac{\mathrm{S}\,x}{2}. \qquad \text{(3.)}$$

The coefficient $f'^2 \div \mathrm{E}$, by which one-half of the volume of the tie is multiplied in the above formula, is called the Modulus of Resilience.

A *sudden pull* of $f'\,\mathrm{S} \div 2$, or one-half of the proof load, being applied to the bar, will produce the entire proof strain of $f' \div \mathrm{E}$, which is produced by the *gradual* application of the proof load itself; for the work performed by the action of the constant force

$f' S \div 2$, through a given space, is the same with the work performed by the action, through the same space, of a force increasing at an uniform rate from 0 up to $f' S$. Hence a tie, to resist with safety the sudden application of a given pull, requires to have twice the strength that is necessary to resist the gradual application and steady action of the same pull. This is an illustration of the principle, that the factor of safety for a live load is twice that for a dead load.

291. **Thin Cylindrical and Spherical Shells.**—Let r denote the radius of a thin hollow cylinder, such as the shell of a high-pressure boiler;

t, the thickness of the shell;

f, the ultimate tenacity of the material, in pounds per square inch;

p, the intensity of the pressure, in pounds per square inch, required to burst the shell. This ought to be taken at SIX TIMES the effective working pressure—*effective pressure* meaning the excess of the pressure from within above the pressure from without, which last is usually the atmospheric pressure, of 14·7 lbs. on the square inch or thereabouts.

Then

$$p = \frac{f\,t}{r}; \qquad (1.)$$

and the proper proportion of thickness to radius is given by the formula,—

$$\frac{t}{r} = \frac{p}{f}. \qquad (2.)$$

Thin spherical shells are *twice as strong* as cylindrical shells of the same radius and thickness.

The tenacity of good wrought-iron boiler-plates is about 50,000 lbs.

SECTION 3.—OF RESISTANCE TO DISTORTION AND SHEARING.

292. **Distortion and Shearing Stress in General.**—In framework and mechanism many cases occur in which the principal pieces, such as plates, links, bars, or beams, being themselves subjected to tension, pressure, twisting, or bending, are connected with each other at their joints by rivets, bolts, pins, keys, or screws, which are under the action of a shearing force, tending to make them give way by the sliding of one part over another.

Every shearing stress is equivalent to a pair of direct stresses of the same intensity, one tensile and the other compressive, exerted in directions making angles of 45° with the shearing stress. Hence it follows that a body may give way to a shearing stress either by actual shearing, at a plane parallel to the direction of the shearing force, or by tearing, in a direction making an angle of 45° with that force. The manner of breaking depends on the structure of the material, hard and brittle materials giving way by tension, and soft and tough materials by shearing.

When a shearing force does not exceed the limit within which moduli of stiffness are sensibly constant, it produces distortion of the body on which it acts. Let q denote the intensity of shearing stress applied to the four lateral faces of an originally square prismatic particle, so as to distort it; and let ν be the *distortion*, expressed by the *tangent of the difference between each of the distorted angles of the prism and a right angle;* then

$$\frac{q}{\nu} = \mathrm{C}, \qquad (1.)$$

is the *modulus of transverse elasticity*, or *resistance to distortion.*

One mode of expressing the distortion of an originally square prism is as follows:—Let α denote the proportionate elongation of one of the diagonals of its end, and $-\alpha$ the proportionate shortening of the other; then the distortion is

$$\nu = 2\,\alpha.$$

The ratio $\frac{\mathrm{C}}{\mathrm{E}}$ of the modulus of transverse elasticity to the modulus of direct elasticity defined in Article 287, page 184, has different values for different materials, ranging from 0 to $\frac{1}{2}$. For wrought-iron and steel it is about $\frac{1}{3}$.

SECTION 4.—OF RESISTANCE TO TWISTING AND WRENCHING.

293. **Twisting or Torsion in General.**—Torsion is the condition of strain into which a cylindrical or prismatic body is put when a pair of couples of equal and opposite moment, tending to make it rotate about its axis in contrary directions, are applied to its two ends. Such is the condition of shafts which transmit motive power. The moment is called the *twisting moment*, and at each cross-section of the bar it is resisted by an equal and opposite moment of stress. Each particle of the shaft is in a state of distortion, and exerts shearing stress.

In British measures, twisting moments are expressed in *inch-lbs.*

294. **Strength of a Cylindrical Shaft.**—A cylindrical shaft, A B,

fig. 127, being subjected to the twisting moment of a pair of equal and opposite couples applied to the cross-sections, A and B, it is required to find the condition of stress and strain at any intermediate cross-section, such as S, and also the angular displacement of any cross-section relatively to any other.

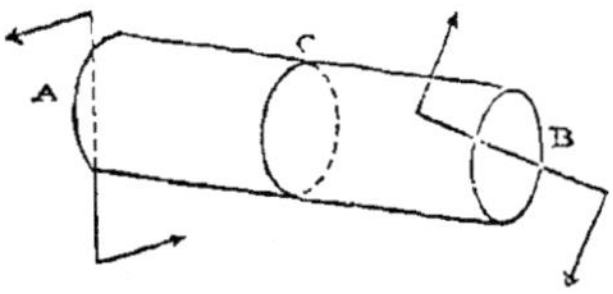

Fig. 127.

From the uniformity of the figure of the bar, and the uniformity of the twisting moment, it is evident that the condition of stress and strain of all cross-sections is the same; also, because of the circular figure of each cross-section, the condition of stress and strain of all particles at the same distance from the axis of the cylinder must be alike.

Suppose a circular layer to be included between the cross-section S, and another cross-section at the longitudinal distance $d\,x$ from it. The twisting moment causes one of those cross-sections to rotate relatively to the other, about the axis of the cylinder, through an angle which may be denoted by $d\,\theta$. Then if there be two points at the same distance, r, from the axis of the cylinder, one in the one cross-section and the other in the other, which points were originally in one straight line parallel to the axis of the cylinder, the twisting moment shifts one of those points laterally, relatively to the other, through the distance $r\,d\,\theta$. Consequently, the part of the layer which lies between those points is in a condition of *distortion*, in a plane perpendicular to the radius r; and the distortion is expressed by the ratio

$$\nu = r \cdot \frac{d\,\theta}{d\,x}; \qquad \text{(1.)}$$

which varies *proportionally to the distance from the axis.* There is therefore a *shearing stress* at each point of the cross-section, whose direction is perpendicular to the radius drawn from the axis to that point, and whose intensity is *proportional to that radius*, being represented by

$$q = \mathrm{C}\nu = \mathrm{C}\,r \cdot \frac{d\,\theta}{d\,x} \qquad \text{(2.)}$$

The STRENGTH of the shaft is determined in the following manner:—Let q_1 be the limit of the shearing stress to which the material is to be exposed, being the *ultimate* resistance to wrenching if it is to be broken, the *proof* resistance if it is to be tested, and the *working* resistance if the working moment of torsion is to be determined. Let r_1 be the external radius of the axle. Then

q_1 is the value of q at the distance r_1 from the axis; and at any other distance, r, the intensity of the shearing stress is

$$q = \frac{q_1 r}{r_1} \ldots\ldots\ldots\ldots\ldots\ldots(3.)$$

Conceive the cross-section to be divided into narrow concentric rings, each of the breadth $d\,r$. Let r be the *mean radius* of one of these rings. Then its area is $2\,\pi\,r\,d\,r$; the intensity of the shearing stress on it is that given by Equation 3, and the leverage of that stress relatively to the axis of the cylinder is r; consequently the moment of the shearing stress of the ring in question, being the product of the three quantities,

$$\frac{q_1 r}{r_1},\ r, \text{ and } 2\,\pi\,r\,d\,r \quad \text{is} \quad \frac{2\,\pi\,q_1}{r_1} \cdot r^3\,d$$

which being integrated for all the rings from the centre to the circumference of the cross-section, gives for the moment of torsion, and of resistance to torsion,

$$\mathrm{M} = \frac{\pi}{2}\,q_1\,r_1^3 = \frac{\pi}{16}\,q_1\,h_1^3;\ \ldots\ldots\ldots\ldots\ldots(4.)$$

if $h = 2\,r_1$ be the diameter of the shaft,

$$\left(\frac{\pi}{2} = 1{\cdot}5708\,;\ \frac{\pi}{16} = 0{\cdot}196 \text{ nearly}\right).$$

If the axle is *hollow*, h_0 being the diameter of the hollow, the moment of torsion becomes

$$\mathrm{M} = \frac{\pi}{16} \cdot q_1 \frac{h_1^4 - h_0}{h_1} \ldots\ldots\ldots\ldots\ldots\ldots(5.)$$

The following formulæ serve to calculate the diameters of shafts when the twisting moment and stress are given; solid shafts:—

$$h_1 = \left(\frac{5{\cdot}1\ \mathrm{M}}{q_1}\right)^{\frac{1}{3}};\ \ldots\ldots\ldots\ldots\ldots(6.)$$

hollow shafts—

$$h_1 = \left\{\frac{5{\cdot}1\ \mathrm{M}}{q_1\left(1 - \frac{h_0^4}{h_1^4}\right)}\right\}^{\frac{1}{3}} \ldots\ldots\ldots\ldots\ldots\ldots(7.)$$

SECTION 5.—OF RESISTANCE TO BENDING AND CROSS-BREAKING.

295. **Resistance to Bending in General.**—In explaining the principles of the resistance which bodies oppose to bending and cross-breaking, it is convenient to use the word *beam* as a general term

to denote the body under consideration; but those principles are applicable, not only to beams for supporting weights, but to levers, cross-heads, cross-tails, shafts, journals, cranks, and all pieces in machinery or framework to which forces are applied tending to bend them and to break them across; that is to say, forces transverse to the axis of the piece.

Conceive a beam which is acted upon by a combination of parallel transverse forces that balance each other, to be divided into two parts by an imaginary transverse section; and consider separately the conditions of equilibrium of one of those parts. The external transverse forces which act on that part, and constitute the load on it, do not necessarily balance each other. Their resultant may be found by the rule of Article 233, page 132. That resultant is called the *Shearing Load* at the cross-section under consideration, and it is balanced by the *Shearing Stress* exerted by the particles which that cross-section traverses. The resultant moment of the same set of forces, relatively to the same cross-section, may be found by the same rule; it is called the *Bending Moment* at that cross-section, and it is balanced (if the beam is strong enough) by the *Moment of Stress* exerted by the particles which the cross-section traverses, called also the *Moment of Resistance*. That moment of stress is due wholly to longitudinal stress, and it is exerted in the following way:—The bending of the beam causes the originally straight layers of particles to become curved; those near the concave side of the beam become shortened; those near the convex side, lengthened; the shortened layers exert longitudinal thrust; the lengthened layers, longitudinal tension; the resultant thrust and the resultant tension are equal and opposite, and compose a couple, whose moment is the moment of stress, equal and opposite to the bending moment.

In the solution of problems respecting the transverse strength of beams, it is necessary to determine the shearing load and bending moment produced by the transverse external forces at different cross-sections, and especially at those cross-sections at which they act most intensely, and the relations between the dimensions and figure of a cross-section of the beam, and the moment of stress which that cross-section is capable of exerting, so that each cross-section, and especially that at which the bending moment is greatest, may have sufficient strength.

296. **Calculation of Shearing Loads and Bending Moments.**—In the formulæ which follow, the shearing load at a given cross-section will be denoted by F, and the bending moment by M. In British measures it is most convenient to express the bending moment in *inch-lbs.*, because of the transverse dimensions of pieces in machines being expressed in inches.

The mathematical process for finding F and M at any given

cross-section of a beam, though always the same in principle, may be varied considerably in detail. The following is on the whole the most convenient way of conducting it:—

Fig. 128 represents a beam *supported* at both ends, and loaded between them. Fig. 129 represents a *bracket;* that is, a beam *supported* and *fixed* at one end, and loaded on a projecting portion. P, Q, represent in each case the supporting forces; in fig. 128, W_1,

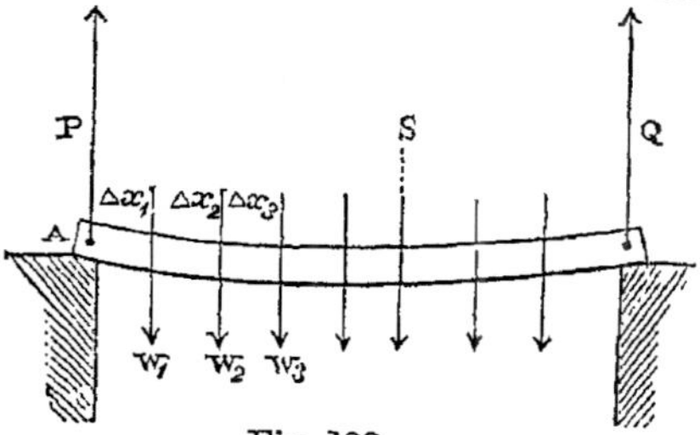

Fig. 128.

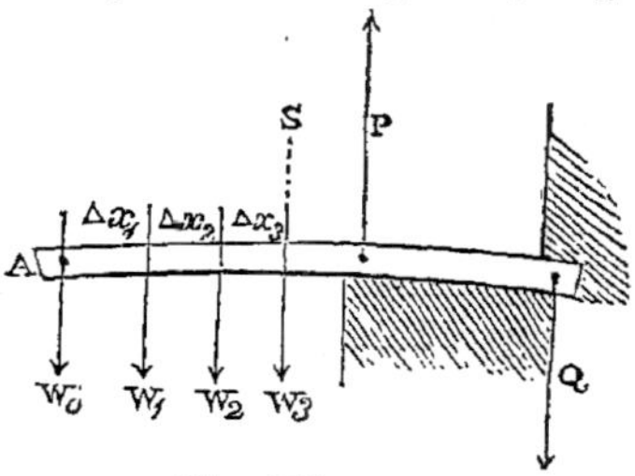

Fig. 129.

W_2, W_3, &c., represent portions of the load; in fig. 129, W_0 represents the endmost portion of the load, and W_1, W_2, W_3, other portions; in both figures, Δx_1, Δx_2, Δx_3, &c., denote the lengths of the intervals into which the lines of action of the portions of the load divide the longitudinal axis of the beam. The forces marked W may be the weights of parts of the beam itself, or of bodies carried by it; or they may be forces exerted by moving pieces in a machine on each other; or, in short, they may be any external transverse forces. If the body called the beam is a shaft, P and Q will be the bearing pressures.

The figures represent the load as applied at detached points; but when it is continuously distributed, the length of any indefinitely short portion of the beam may be denoted by $d\,x$, the intensity of the load upon it *per unit of length* by w, and the amount of the load upon it by $w\,d\,x$.

The process to be gone through will then consist of the following steps:—

STEP I. *To find the Supporting Forces or Bearing Pressures,* P *and* Q.—Assume any convenient point in the longitudinal axis as origin of co-ordinates, and find the distance x_0 of the resultant of the load from it, by the rule of Article 233, page 132; that is to say,

$$\left.\begin{aligned} x_0 &= \frac{\Sigma \cdot x\,W}{\Sigma \cdot W}; \text{ or} \\ x_0 &= \frac{\int x\,w\,d\,x}{\int w\,d\,x}. \end{aligned}\right\} \dots\dots\dots\dots\dots\dots (2.)$$

Then, by the rule of Article 227, page 129, find the two supporting forces or bearing pressures, P and Q; that is to say, let R be the resultant load, and P R and R Q its distances from the points of support; and make

$$\left.\begin{array}{l} PQ : PR : QR \\ :: R : Q : P. \end{array}\right\} \ldots\ldots\ldots\ldots\ldots\ldots (3.)$$

STEP II. *To find the shearing loads at a series of sections.*—In what position soever the origin of co-ordinates may have been during the previous step, assume it now, in a beam supported at both ends, to be at one of the points of support (as A, fig. 128), and in a bracket to be at the loaded point farthest from the fixed end (as A, fig. 129). Consider P as positive and W as negative.

Then the shearing load in any given interval of the length of the beam is the resultant of all the forces acting on the beam from the origin to that interval; so that it has the series of values,

In Fig. 128.	In Fig. 129.
$F_{01} = P$;	$-F_{01} = W_0$;
$F_{12} = P - W_1$;	$-F_{12} = W_0 + W_1$;
$F_{23} = P - W_1 - W_2$;	$-F_{23} = W_0 + W_1 + W_2$;
$F_{34} = P - W_1 - W_2 - W_3$; &c.;	$-F_{34} = W_0 + W_1 + W_2 + W_3$, &c.;
and generally,	and generally,
$F = P - \Sigma \cdot W$;......(4.)	$-F = \Sigma \cdot W$;......(5.)

so that the shearing loads which act in a series of intervals of the length of the beam can be computed by successive subtractions or successive additions, as the case may be.

For a continuously distributed load, these equations become respectively,

$$\text{In a beam supported at both ends, } F = P - \int_0^{x'} w\,dx; \ldots (6.)$$

$$\text{In a bracket, } -F = \int_0^{x'} w\,dx; \ldots\ldots\ldots\ldots\ldots (7.)$$

in which expressions, x' denotes the distance from the origin, A, to the plane of section under consideration.

The positive and negative signs distinguish the two contrary directions of the distortion which the shearing load tends to produce.

The Greatest Shearing Load acts in a beam supported at both ends, close to one or other of the points of support, and its value is either P or Q. In a bracket, the greatest shearing load on the projecting part acts close to the outer point of support, and its value is equal to the entire load.

In a beam supported at both ends the **Shearing Load vanishes**, or changes from positive to negative at some intermediate section, whose position may be found from Equation 4 or Equation 6, by making $F = 0$. At the second point of support, $F = -Q$.

STEP III. *To find the bending moments at a series of sections.*—At the origin A there is no bending moment. Multiply the length of each of the intervals Δx of the longitudinal axis of the beam by the shearing load F, which acts throughout that interval; the first of the products so obtained is the bending moment at the inner end of the first interval; and by adding to it the other products successively, there are obtained the bending moments at the inner ends of the other intervals in succession.*

That is to say,—bending moment

at the origin A; $M_0 = 0$;
at the line of action of W_1; $M_1 = F_{01} \cdot \Delta x_1$;
" " " W_2; $M_2 = F_{01} \cdot \Delta x_1 + F_{12} \Delta x_2$;
" " " W_3; $M_3 = F_{01} \cdot \Delta x_1 + F_{12} \Delta x_2 + F_{23} \cdot \Delta x_3$;
&c. &c.

and generally, $M = \Sigma \cdot F \Delta x$............(8.)

If the divisions Δx are of equal length, this becomes

$$M = \Delta x \cdot \Sigma F; \qquad (9.)$$

and for a continuously distributed load,

$$M = \int_0^x F\, d\, x \qquad (10.)$$

The three preceding Equations 8, 9, and 10, are applicable to beams whether supported at both ends or fixed at one end. By substituting for F in Equation 10 its values as given by Equations 6 and 7 respectively, we obtain the following results:—

For a beam supported at both ends,

$$M = P_1 x' - \int_0^{x'} \int_0^x w\, d\, x^2$$

$$= P_1 x' - \int_0^{x'} (x' - x)\, w\, d\, x; \qquad (11.)$$

For a beam fixed at one end,

$$-M = \int_0^{x'} \int_0^x w\, d\, x^2 = \int_0^{x'} (x' - x)\, w\, d\, x; \qquad (12.)$$

in the latter of which equations the symbols $-M$ denotes that the bending moment acts downwards.

* This process is substantially the same with that employed by Mr. Herbert Latham, in his work *On Iron Bridges*, to compute the stress in a half-lattice girder.

The Greatest Bending Moment acts, in a bracket, at the outer point of support; and in a beam supported at both ends, at the section where the shearing load vanishes; found, as already stated in Step II., from the Equation $F = 0$.

When the transverse forces applied to a beam supported at both ends are symmetrically distributed relatively to its middle section, the Greatest Bending Moment acts at that section; and it is sometimes convenient to assume a point in that section as the origin of co-ordinates.

STEP IV. *To deduce the shearing load and bending moment in one beam from those in another beam similarly supported and loaded.*—This is done by the aid of the following principle:—

When beams differing in length and in the amounts of the loads upon them are similarly supported, and have their loads similarly distributed, the shearing loads at corresponding sections in them vary as the total loads, and the bending moments as the products of the loads and lengths.

This principle may be expressed by symbols in either of the two following ways:—

First, Let l, l', denote the lengths of two beams, similarly supported; let W, W′, denote their total loads, similarly distributed; let F, F′, be the shearing forces, and M, M′, the bending moments, at sections similarly situated in the two beams; then

$$W : W' :: F : F'; \qquad (13.)$$

$$lW : l'W' :: M : M'. \qquad (14.)$$

Secondly, Let k and m be two numerical factors, depending on the way in which a beam is supported, the mode of distribution of its load, and the position of the cross-section under consideration; then

$$F = kW; \qquad (15.)$$

$$M = mWl. \qquad (16.)$$

The length between the points of support of a beam supported at the ends, as in fig. 128, is often called the *span.*

297. **Examples.**—In the following formulæ, which are examples of the application of the principles of the preceding Article to the cases which occur most frequently in practice, W denotes the total load;

w, when the load is distributed, the load per unit of length of the beam;

c, in brackets, the length of the free part of the bracket;

c, in beams either loaded or supported at both ends, the *half span,* between the extreme points of load or support and the middle;

M, the greatest bending moment.

I. Bracket fixed at one end and loaded at the other, ... $M = c\,W$......(1.)

II. Bracket fixed at one end and uniformerly loaded, ... $M = \frac{c\,W}{2} = \frac{w\,c^2}{2}$....(2.)

III. Beam supported at both ends and loaded at an intermediate point, whose distance from the middle of the span is x, ... $M = \frac{(c^2 - x^2)\,W}{2\,c}$....(3.)

IV. Beam supported at both ends and loaded in the middle, ... $M = \frac{c\,W}{2}$........(4.)

V. Beam supported at both ends and uniformly loaded, ... $M = \frac{c\,W}{4} = \frac{w\,c^2}{2}$.....(5.)

In Example III. the greatest force exerted is $\frac{c \pm x}{2\,c}W$, and the leverage with which it acts is $c \mp x$; and Examples IV. and V. follow from it by making $x = o$.

VI. If a beam has equal and opposite couples applied to its two ends; for example, if the beam in fig. 130 has the couple of equal and opposite forces P_1 applied at A and B, and the couple of equal and opposite forces P_2 at C and D, and if the opposite moments $P_1 \cdot A\,B = P_2 \cdot C\,D = M$ are equal, then each of the endmost divisions, A B and C D, is in the condition of a bracket fixed at one end and loaded at the other (Example 1.); and the middle division B C is acted upon by the *uniform bending moment* M, and by no shearing load.

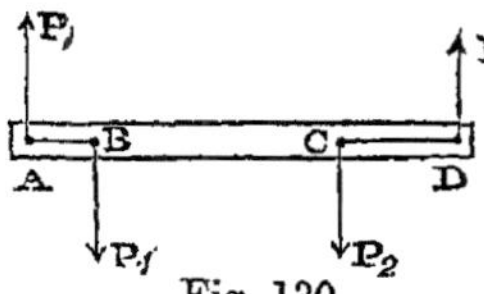

Fig. 130.

VII. Let a beam of the half span c be loaded with an uniformly distributed load of w units of weight per unit of span; and at a point whose distance from the middle of the span is a, let there be applied an additional load W. It is required to find x, the distance from the middle of the span at which the greatest bending moment is exerted, and M, that greatest moment.

Make

$$\frac{W}{2\,c\,w} = m;$$

then the solutions are as follows:—

CASE 1.—When $\frac{a}{c} =$ or $> \frac{m}{1+m}$; $x = m\,(c - a)$; and

$$M = \frac{w\,c^2}{2}\left(1 + m - \frac{m\,a}{c}\right) \text{ ....(6.)}$$

CASE 2.—When $\frac{a}{c} = \text{or} < \frac{m}{1 + m}$; $x = a$; and

$$M = \frac{w\,c^2}{2}\,(1 + 2\,m)\left(1 - \frac{a^2}{c^2}\right) \text{.....................(7.)}$$

In the following case both sets of formulæ give the same result; when $\frac{a}{c} = \frac{m}{1 + m}$; $x = a = m\,(c - a)$; and

$$M = \frac{w\,c^2}{2}\left(\frac{1 + 2\,m}{1 + m}\right)^2 \text{ ....(8.)}$$

298. **Bending Moments produced by Longitudinal and Oblique Forces.**—When a bar is acted upon at a given cross-section by any external force, whose line of action, whether transverse, oblique, or parallel to the axis of the bar, does not traverse the centre of magnitude of that cross-section, that force exerts a moment upon that cross-section equal to the product of the force into the perpendicular distance of its line of action from the centre of the cross-section, and that moment is to be balanced by the moment of longitudinal stress at the cross-section.

The external force may be resolved into a longitudinal and a transverse component. The longitudinal component is balanced by an uniform longitudinal tension or pressure, as the case may be, exerted at the cross-section, and combined with the stress which resists the bending moment; and the transverse component is resisted by shearing stress.

299. **Moment of Stress—Transverse Strength.**—The bending moment at each cross-section of a beam bends the beam so as to make any originally plane longitudinal layer of the beam perpendicular to the plane in which the load acts, become concave in the direction towards which the moment acts, and convex in the opposite direction. Thus, fig. 131 represents a side view of a short portion of a bent beam; C C′ is a layer, originally plane, which is now bent so as to become concave at one side and convex at the other.

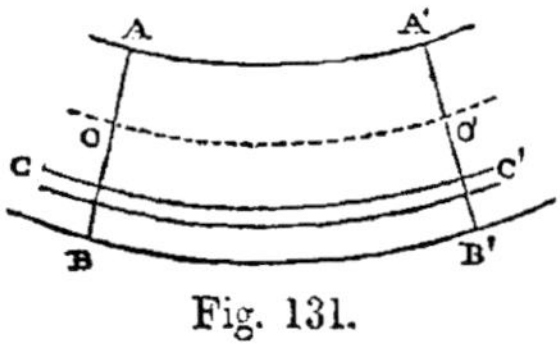

Fig. 131.

The layers at and near the concave side of the beam, A A′, are shortened, and the layers near the convex side, B B′ lengthened, by the bending action of the load. There is one intermediate surface, O O′, which is neither lengthened nor shortened; it is called the "*neutral surface.*" The particles at that surface are not necessarily, however, in a state devoid of strain; for, in common with the other particles of the beam, they are compressed and extended in a pair of diagonal directions, making angles of 45° with the neutral surface, by the shearing action of the load, when such action exists.

The condition of the particles of a beam, produced by the combined bending and shearing actions of the load, is illustrated by fig. 132, which represents a vertical longitudinal section of a rectangular beam, supported at the ends, and loaded at intermediate points. It is covered with a network consisting of two sets of curves cutting each other at right angles. The curves convex upwards are *lines of direct thrust;* those convex downwards are *lines of direct tension.* A pair of tangents to the pair of curves which traverse any particle are the *axes of stress* of that particle. The *neutral surface* is cut by both sets of curves at angles of 45°. At that vertical section of the beam where the shearing load vanishes, and the bending moment is greatest, both sets of curves become parallel to the neutral surface.

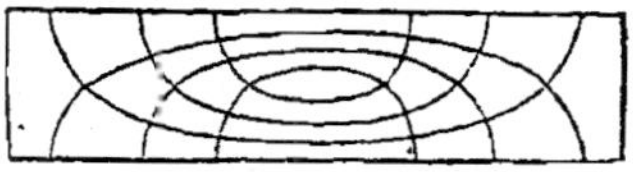

Fig. 132.

When a beam breaks under the bending action of its load, it gives way, either by the crushing of the compressed side, A A′, or by the tearing of the stretched side, B B′.

In fig. 133, A represents a beam of a granular material, like cast iron, giving way by the crushing of the compressed side, out of which a sort of wedge is forced. B represents a beam giving way by the tearing asunder of the stretched side.

Fig. 133.

The *resistance* of a beam to bending and cross-breaking at any given cross-section is the moment of a couple, consisting of the thrust along the longitudinally-compressed layers, and the equal and opposite tension along the longitudinally-stretched layers.

It has been found by experiment, that in most cases which occur in practice, the longitudinal stress of the layers of a beam may, without material error, be assumed to be *uniformly varying,* its intensity being simply proportional to the distance of the layer from the neutral surface.

Let fig. 134 represent a cross-section of a beam (such as that represented in fig. 131), A the compressed side, B the extended side, C any layer, and O O the *neutral axis* of the section, being the line in which it is cut by the neutral surface. Let p denote the intensity of the stress along the layer C, and y the distance of that layer from the neutral axis. Because the stress is uniformly varying, $p \div y$ is a constant quantity. Let that constant be denoted for the present by a.

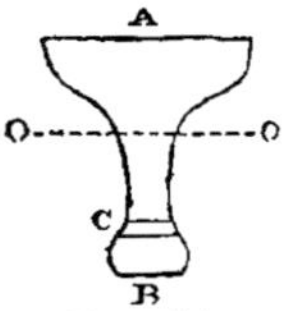

Fig. 134.

Let z be the breadth of the layer C, and dy its thickness;

Then the amount of stress along it is

$$p z d y = a y z d y;$$

the amount of the stress along all the layers at the given cross-section is

$$a \int y z d y;$$

and this amount must be nothing,—in other words, the total thrust and total tension at the cross-section must be equal,—because the forces applied to the beam are wholly transverse; from which it follows that

$$\int y z d y = 0, \dots\dots\dots\dots\dots\dots\dots\dots (1.)$$

and the *neutral axis traverses the centre of magnitude of the cross-section.* This principle enables the neutral axis to be found by the aid of the methods explained in Section 1, Chapter III., Part III.

To find the greatest value of the constant $p \div y$ consistent with the strength of the beam at the given cross-section, let y_a be the distance of the compressed side, and y_b that of the extended side from the neutral axis; f_a the greatest thrust, and f_b the greatest tension which the material can bear in the form of a beam; compute $f_a \div y_a$, and $f_b \div y_b$, and adopt the *less* of those two quantities as the value of $p \div y$, which may now be denoted by $f \div y_1$; f being f_a or f_b, and y_1 being y_a or y_b, according as the beam is liable to give way by crushing or by tearing.

For the best economy of material, the two quotients ought to be equal; that is to say,

$$\frac{f}{y_1} = \frac{f_a}{y_a} = \frac{f_b}{y_b} = \frac{f_a + f_b}{h}; \dots\dots\dots\dots (1 \text{ A.})$$

and this gives what is called a *cross-section of equal strength.*

The moment relatively to the neutral axis, of the stress exerted along any given layer of the cross-section, is

$$y p z d y = \frac{f}{y_1} y^2 z d y;$$

and the sum of all such moments, being the MOMENT OF STRESS, or MOMENT OF RESISTANCE of the given cross-section of the beam to breaking across, is given by the formula,

$$M = \int p\, y\, z\, d\, y = \frac{f}{y_1} \int y^2\, z\, d\, y; \qquad (2.)$$

or making $\int y^2\, z\, d\, y = I$,

$$M = \frac{f I}{y_1} \qquad (2\text{ A.})$$

When the *breaking* load is in question, the coefficient f is what is called the MODULUS OF RUPTURE of the material.

When the *proof load* or *working load* is in question, the coefficient f is the modulus of rupture divided by a suitable *factor of safety*, which, for the working stress in parts of machinery that are made of metal, is usually 6, and for the parts made of wood, 10. Thus, the *working modulus* f is usually 9,000 lbs. on the square inch for wrought iron, 4,500 for cast iron, and from 1,000 to 1,200 for wood.

The factor denoted by I in the preceding equation is what is called the "*geometrical moment of inertia*" of the cross-section of the beam. For sections whose figures are similar, or are parallel projections of each other, the moments of inertia are to each other as the breadths, and as the cubes of the depths of the sections; and the values of y_1 are as the depths. If, therefore, b be the breadth and h the depth of the rectangle circumscribing the cross-section of a given beam at the point where the moment of stress is greatest, we may put

$$I = n' b h^3, \qquad (3.)$$

$$y_1 = m' h, \qquad (4.)$$

n and m' being numerical factors depending on the form of section, and making $n' \div m' = n$, the moment of resistance may be thus expressed,—

$$M = n f b h^2 \qquad (5.)$$

Hence it appears that the *resistances of similar cross-sections to cross-breaking are as their breadths and as the squares of their depths.*

The relation between the load and the dimensions of a beam is found by equating the value of the greatest bending moment in terms of the load and span of the beam, as given in Article 296, Equations 10, 11, 12, 16, to the value of the moment of resistance of the beam, at the cross-section where that greatest bending moment acts, as given in Equation 5 of this Article.

The depth h is usually fixed by considerations of stiffness, and then the unknown quantity is the breadth, b. Sometimes, as when the cross-section is circular or square, we have $b = h$; and then we have h^3, instead of $b h^2$ in Equation 5, which is solved so as to give h by extraction of the cube root. The following are the formulæ for these calculations:—

$$b = \frac{M}{n f h^2}; \qquad (6.)$$

and when $h = b$,

$$h = \left(\frac{M}{n f}\right)^{\frac{1}{3}}, \qquad (6\text{ A.})$$

Examples of the Numerical Factors in Equations 3, 4, 5 and 6.

Form of Cross-Sections.	$n' = \frac{I}{b h^3}$.	$m' = \frac{y_1}{h}$	$n = \frac{M}{f b h^2}$.
I. Rectangle $b h$, (including square)	$\frac{1}{12}$	$\frac{1}{2}$	$\frac{1}{6}$
II. Ellipse— Vertical axis h, Horizontal axis b, (including circle)	$\frac{\pi}{64} = \frac{1}{20 \cdot 4} = 0{\cdot}0491$	$\frac{1}{2}$	$\frac{\pi}{32} = \frac{1}{10 \cdot 2} = 0{\cdot}0982$
III. Hollow rectangle, $b h - b' h'$; also I-formed section, where b' is the sum of the breadths of the lateral hollows,	$\frac{1}{12}\left(1 - \frac{b' h'^3}{b h^3}\right)$	$\frac{1}{2}$	$\frac{1}{6}\left(1 - \frac{b' h'^3}{b h^3}\right)$
IV. Hollow square— $h^2 - h'^2$,	$\frac{1}{12}\left(1 - \frac{h'^4}{h^4}\right)$	$\frac{1}{2}$	$\frac{1}{6}\left(1 - \frac{h'^4}{h^4}\right)$
V. Hollow ellipse,	$\frac{\pi}{64}\left(1 - \frac{b' h'^3}{b h^3}\right)$	$\frac{1}{2}$	$\frac{\pi}{32}\left(1 - \frac{b' h'^3}{b h^3}\right)$
VI. Hollow circle,	$\frac{\pi}{64}\left(1 - \frac{h'^4}{h^4}\right)$	$\frac{1}{2}$	$\frac{\pi}{32}\left(1 - \frac{h'^4}{h^4}\right)$
VII. Isosceles triangle; base b, height h; y_1 measured from summit,	$\frac{1}{36}$	$\frac{2}{3}$	$\frac{1}{24}$

300. **Allowance for Weight of Beam—Limiting Length of Beam.**—When a beam is of great span, its own weight may bear a proportion to the load which it has to carry, sufficiently great to require to be taken into account in determining the dimensions of the beam. The following is the process to be performed for that purpose, when the load is uniformly distributed, and the

beam of uniform cross-section. Let W′ be the external working load, s_1 its factor of safety, s_2 a factor of safety suited to a steady load, like the weight of the beam.

Let b' denote the breadth of any part of the beam, as computed by considering the *external breaking load alone*, s_1 W′. Compute the weight of the beam from that *provisional* breadth, and let it be denoted by B.′ Then $\frac{s_1 \mathrm{W}'}{s_1 \mathrm{W}' - s_2 \mathrm{B}'}$ is the proportion in which the *gross* breaking load exceeds the external part of that load. Consequently, if for the *provisional* breadth b' there be substituted the *exact* breadth,

$$b = \frac{b' s_1 \mathrm{W}'}{s_1 \mathrm{W}' - s_2 \mathrm{B}'}, \qquad (1.)$$

the beam will now be strong enough to bear both the proposed external load W′, and its own weight, which will now be

$$\mathrm{B} = \frac{\mathrm{B}' s_1 \mathrm{W}'}{s_1 \mathrm{W}' - s_2 \mathrm{B}'}; \qquad (2.)$$

and the true gross breaking load will be

$$\mathrm{W} = s_1 \mathrm{W}' + s_2 \mathrm{B} = \frac{s_1^2 \mathrm{W}'^2}{s_1 \mathrm{W}' - s_2 \mathrm{B}'}. \qquad (3.)$$

As the factor of safety for a steady load is in general one-half of that for a moving load, s_1 may be made $= 2 s_2$; in which case the preceding formulæ become

$$b = \frac{2 b' \mathrm{W}'}{2 \mathrm{W}' - \mathrm{B}'}; \qquad (4.)$$

$$\mathrm{B} = \frac{2 \mathrm{B}' \mathrm{W}'}{2 \mathrm{W}' - \mathrm{B}'}; \qquad (5.)$$

$$\mathrm{W} = \frac{2 s_1 \mathrm{W}'^2}{2 \mathrm{W}' - \mathrm{B}'}. \qquad (6.)$$

In all these formulæ, both the external load and the weight of the beam are treated as if uniformly distributed—a supposition which is sometimes exact, and always sufficiently near the truth for the purposes of the present Article.

The gross load of beams of similar figures and proportions, varying as the breadth and square of the depth directly, and inversely as the length, is proportional to the square of a given linear dimension. The weights of such beams are proportional to the cubes of corresponding linear dimensions. Hence the weight increases at a faster rate than the gross load; and for each particular figure of a beam of a given material and proportion of its dimensions, there must be a certain size at which the beam will bear its own weight only, without any additional load.

To reduce this to calculation, let the uniformly distributed gross breaking load of a beam of a given figure be expressed as follows:—

$$W = s_1 W' + s_2 B = \frac{M}{m\,l} = \frac{8\,n\,f\,h\,A}{l}\ ;\ \ldots\ldots\ldots\ldots (7.)$$

the value of m for an uniformly distributed load and rectangular cross-section being $\frac{1}{8}$; and $n\,f\,h\,A$ being $= n\,f\,b\,h^2$, Equation 5, Article 299; l, h and A being the length, depth, and sectional area of the beam, f the modulus of rupture, and n a factor depending on the form of cross-section. The weight of the beam will be expressed by

$$B = k\,w'\,l\,A\ ;\ \ldots\ldots\ldots\ldots (8.)$$

w' being the weight of an unit of volume of the material, and k a factor depending on the figure of the beam. Then the ratio of the weight of the beam multiplied by its proper factor of safety to the gross breaking load is

$$\frac{s_2 B}{W} = \frac{s_2\,k\,w'\,l^2}{8\,n\,f\,h}\ ;\ \ldots\ldots\ldots\ldots (9.)$$

which increases in the simple ratio of the length, if the proportion $l \times h$ is fixed. When this is the case, the length L of a beam, whose weight (treated as uniformly distributed) is its working load, is given by the condition $s_2 B = W$; that is,

$$L = \frac{8\,n\,f\,h}{s_2\,k\,w'\,l}\ ;\ \ldots\ldots\ldots\ldots (10.)$$

This *limiting length* having once been determined for a given class of beams, may be used to compute the ratios of the gross breaking load, weight of the beam, and external working load to each other, for a beam of the given class, and of any smaller length, l, according to the following proportional equation:—

$$L : \frac{l}{s_2} : \frac{L - l}{s_1} :: W : B : W'\ ;\ \ldots\ldots\ldots\ldots (11.)$$

Section 6.—Of Resistance to Thrust or Pressure.

301. **Resistance to Compression and Direct Crushing.**—Resistance to *longitudinal compression*, when the proof stress is not

exceeded, is sensibly equal to the resistance to stretching, and is expressed by the same modulus of elasticity, denoted by E. When that limit is exceeded, it becomes irregular.

The present Article has reference to direct and simple crushing only, and is limited to those cases in which the pillars, blocks, struts, or rods along which the thrust acts are not so long in proportion to their diameter as to have a sensible tendency to give way by bending sideways. Those cases comprehend—

Stone and brick pillars and blocks of ordinary proportions;

Pillars, rods, and struts of cast iron, in which the length is not more than five times the diameter, approximately;

Pillars, rods, and struts of wrought iron, in which the length is not more than ten times the diameter, approximately;

Pillars, rods, and struts of dry timber, in which the length is not more than about five times the diameter.

In such cases the rules for the strength of ties (Article 290) are approximately applicable, substituting *thrust* for *tension*, and using the proper modulus of resistance to direct crushing instead of the tenacity.

Blocks whose lengths are less than about once-and-a-half their diameters offer greater resistance to crushing than that given by the rules; but in what proportion is uncertain.

The modulus of resistance to direct crushing often differs considerably from the tenacity. The nature and amount of those differences depend mainly on the modes in which the crushing takes place. These may be classed as follows:—

I. *Crushing by splitting* (fig. 135) into a number of nearly prismatic fragments, separated by smooth surfaces whose general direction is nearly parallel to the direction of the load, is characteristic of very hard homogeneous substances, in which the resistance to direct crushing is greater than the tenacity; being in many examples about double.

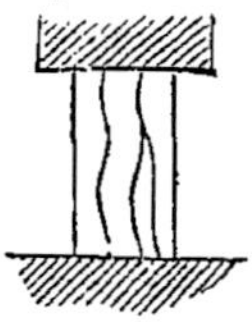

Fig. 135.

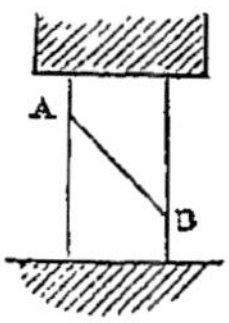

Fig. 136.

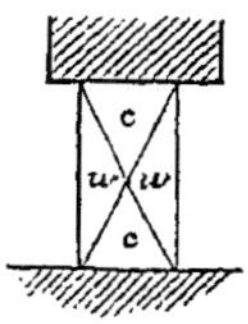

Fig. 137.

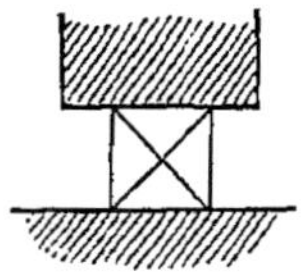

Fig. 138.

II. *Crushing by shearing or sliding* of portions of the block along oblique surfaces of separation is characteristic of substances of a granular texture, like cast iron, and most kinds of stone and brick. Sometimes the sliding takes place at a single plane surface, like A B in fig. 136; sometimes two cones or pyramids are formed, like *c*, *c* in fig, 137, which are forced towards each other, and split or drive outwards a number of wedges surrounding them, like *w*, *w*, in the same figure. Sometimes the block splits into four wedges, as in fig. 138. In substances which are crushed by shearing, the resistance to crushing is always much greater than the tenacity; for example, in cast iron it is from four times to six times.

III. *Crushing by bulging,* or lateral swelling and spreading of the block which is crushed, is characteristic of ductile and tough materials, such as wrought iron. Owing to the gradual manner in which materials of this nature give way to a crushing load, it is difficult to determine their resistance to that load exactly. That resistance is in general less, and sometimes considerably less, than the tenacity. In wrought iron, the resistance to the direct crushing of pillars or struts of moderate length, as nearly as it can be ascertained, is from $\frac{2}{3}$ to $\frac{4}{5}$ of the tenacity.

IV. *Crushing by buckling or crippling* is characteristic of fibrous subtances, such as wood, under the action of a thrust along the fibres. It consists in a lateral bending and wrinkling of the fibres, sometimes accompanied by a splitting of them asunder.

V. *Crushing by cross-breaking* is the mode of fracture of columns and struts in which the length greatly exceeds the diameter, under the breaking load they yield sideways, and are broken across like beams under a transverse load.

PART V.

PRINCIPLES OF KINETICS.

CHAPTER I.

SUMMARY OF GENERAL PRINCIPLES.

NATURE AND DIVISION OF THE SUBJECT.

THE present Chapter contains a summary of the Principles of Kinetics.

302. **Effort; Resistance; Lateral Force.**—Let F denote a force applied to a moving point, and θ the angle made by the direction of that force with the direction of the motion of the point. Then, by the principles of Article 215, the force F may be resolved into two rectangular components, one along, and the other across, the direction of motion of the point, viz:—

The *direct* force, F cos θ.
The *lateral* force, F sin θ.

A direct force is further distinguished, according as its acts *with* or *against* the motion of the point (that is, according as θ is acute or obtuse), by the name of *effort*, or of *resistance*, as the case may be. Hence, each force applied to a moving point may be thus decomposed:—

Effort, $\mathrm{P} = \mathrm{F} \cos \theta$, if θ is acute;
Resistance, $\mathrm{R} = \mathrm{F} \cos (\pi - \theta)$ if θ is obtuse;
Lateral Force, $\mathrm{Q} = \mathrm{F} \sin \theta$.

303. **The Conditions of Uniform Motion** of a pair of points are, that the forces applied to each of them shall balance each other; that is to say, *that the lateral forces applied to each point shall balance each other, and that the efforts applied to each point shall balance the resistances.*

The direction of a force being, as stated in Article 194, that of the motion which it tends to produce, it is evident that the balance of lateral forces is the condition of *uniformity of direction* of motion, that is, of motion in a straight line; and that the balance of efforts and resistances is the condition of *uniformity of velocity*.

304. **Work** consists in moving against resistance. The work is said to be *performed*, and the resistance *overcome*. Work is measured by the product of the resistance into the distance through which its point of application is moved. The *unit of work* commonly used in Britain is a resistance of one pound overcome through a distance of one foot, and is called a *foot-pound*.

305. **Energy** means *capacity for performing work*. The *energy of an effort*, or *potential energy*, is measured by the product of the effort into the distance through which its point of application is *capable* of being moved. The unit of energy is the same with the unit of work.

When the point of application of an effort *has been moved* through a given distance, energy is said to have been *exerted* to an amount expressed by the product of the effort into the distance through which its point of application has been moved.

306. **The Conservation of Energy**, in the case of uniform motion, means the fact, that *the energy exerted is equal to the work performed*.

307. **The Principle of Virtual Velocities** is the name given to the application of the principle of the conservation of energy to the determination of the conditions of equilibrium amongst the forces externally applied to any connected system of points.

308. **The Mass, or Inertia**, of a body, is a quantity proportional to the unbalanced force which is required in order to produce a given definite change in the motion of the body in a given interval of time.

It is known that the weight of a body, that is, the attraction between it and the earth, at a fixed locality on the earth's surface, acting unbalanced on the body for a fixed interval of time (*e. g.*, for a second), produces a change in the body's motion, which is the same for all bodies whatsoever. Hence it follows, that the *masses of all bodies are proportional to their weights at a given locality on the earth's surface.*

This fact has been learned by experiment; but it can also be shewn that it is necessary to the permanent existence of the universe; for if the gravity of all bodies whatsoever were not proportional to their respective masses, it would not produce similar and equal changes of motion in all bodies which arrive at similar positions with respect to other bodies, and the different parts which make up stars and systems would not accompany each other in their motions, never departing beyond certain limits, but would be dispersed and reduced to chaos. Neither an imponderable body, nor a body whose gravity, as compared with its mass, differs in the slightest conceivable degree from that of other bodies, can belong to the system of the universe.*

* See the Rev. Dr. Whewell's demonstration "that all matter gravitates."

309. **The Centre of Mass** of a body is its centre of gravity, found in the manner explained in Part III., Chapter III., Section 1.

310. **The Momentum** of a body means, the product of its mass into its velocity relatively to some point assumed as fixed. The momentum of a body, like its velocity, can be resolved into components, rectangular or otherwise, in the manner already explained for motions in Part I., Chapter I.

311. **The Resultant Momentum** of a system of bodies is the resultant of their separate momenta, compounded as if they were motions or statical couples.

312. **Variations and Deviations of Momentum** are the products of the mass of a body into the rates of variation of its velocity and deviation of its direction, found as explained in Part I., Chapter I., Section 3.

313. **Impulse** is the product of an unbalanced force into the *time* during which it acts unbalanced, and can be resolved and compounded exactly like force. If F be a force, and $d\,t$ an interval of time during which it acts unbalanced, F $d\,t$ is the impulse exerted by the force during that time. The impulse of an unbalanced force in an unit of time is the magnitude of the force itself.

314. **Impulse, Accelerating, Retarding, Deflecting.**—Corresponding to the resolution of a force applied to a moving body into effort or resistance, as the case may be, and lateral stress, as explained in Article 302, there is a resolution of impulse into accelerating or retarding impulse, which acts with or against the body's motion, and deflecting impulse, which acts across the direction of the body's motion. Thus, if θ, as before, be the angle which the unbalanced force F makes with the body's path during an indefinitely short interval, $d\,t$.

$\mathrm{P}\,d\,t = \mathrm{F} \cos\theta \cdot d\,t$ is accelerating impulse if θ is acute;
$\mathrm{R}\,d\,t = \mathrm{F} \cos(\pi - \theta) \cdot d\,t$ is retarding impulse if θ is obtuse;
$\mathrm{Q}\,d\,t = \mathrm{F} \sin\theta \cdot d\,t$ is deflecting impulse.

315. **A Deviating Force** is one which acts unbalanced in a direction perpendicular to that of a body's motion, and changes that direction without changing the velocity of the body.

316. **Centrifugal Force** is the force with which a revolving body reacts on the body that guides it, and is equal and opposite to the deviating force with which the guiding body acts on the revolving body.

In fact, as has been stated in Article 193, every force is an action between two bodies; and *deviating force* and *centrifugal force* are but two different names for the same force, applied to it according as its action on the revolving body or on the guiding body is under consideration at the time.

317. **The Actual Energy** of a moving body relatively to a fixed

point is the product of the *mass* of the body into *one-half* of the *square of its velocity*, that is to say, it is represented by

$$\frac{m v^2}{2} = \frac{W v^2}{2 g} = W h$$

The product $m v^2$, the double of the actual energy of a body, was formerly called its *vis-viva.* Actual energy, being the product of a *weight* into a *height*, is expressed, like potential energy and work, in *foot-pounds* (Articles 304, 305.)

318. **Energy Stored and Restored.**—A body alternately accelerated and retarded, so as to be brought back to its original speed, performs work by means of its retardation exactly equal in amount to the potential energy exerted in producing its acceleration; and that amount of energy may be considered as *stored* during the acceleration, and *restored* during the retardation.

319. **The Transformation of Energy** is a term applied to such processes as the expenditure of potential energy in the production of an equal amount of actual energy, and *vice versa.*

320. **Periodical Motion.**—If a body moves in such a manner that it periodically returns to its original velocity, then at the end of each period, the entire variation of its actual energy is nothing; and in each such period the whole potential energy exerted is equal to the whole work performed, exactly as in the case of a body moving uniformly (Article 306.)

321. **A Reciprocating Force** is a force which acts alternately as an effort and as an equal and opposite resistance, according to the direction of motion of the body. The work which a body performs in moving against a reciprocating force is employed in increasing its own potential energy, and is not lost by the body.

322. **Collision** is a pressure of inappreciably short duration between two bodies.

323. **The Moment of Inertia** of an indefinitely small body, or "physical point," relatively to a given axis, is the product of the mass of the body, or of some quantity proportional to the mass, such as the weight, into the square of its perpendicular distance from the axis.

324. **The Radius of Gyration** of a body about a given axis is that length whose square is the *mean of all the squares* of the distances of the indefinitely small equal particles of the body from the axis, and is found by dividing the moment of inertia by the mass.

325. **The Centre of Percussion** of a body, for a given axis, is a point so situated, that if part of the mass of the body were concentrated at that point, and the remainder at the point directly opposite in the given axis, the statical moment of the weight so distributed, and its moment of inertia about the given axis, would

be the same as those of the actual body in every position of the body.

326. The subjects to which the principles of kinetics relate will be classed in the following manner:—

I. Uniform Motion.
II. Varied Translation of Points and Rigid Bodies.
III. Rotations of Rigid Bodies.
IV. Motions of Fluids.

CHAPTER II.

ON UNIFORM MOTION UNDER BALANCED FORCES.

327. **First Law of Motion.**—*A body under the action of no force, or of balanced forces, is either at rest, or moves uniformly.* (Uniform motion has been defined in Article 66.)

Such is the first law of motion as usually stated; but in that statement is implied something more than the literal meaning of the words; for it is understood, that the *rest* or *motion of the body* to which the law refers, is its rest or motion *relatively to another body which is also under the action of no force or of balanced forces.* Unless this implied condition be fulfilled, the law is not true. Therefore the complete and explicit statement of the first law of motion is as follows:—

If a pair of bodies be each under the action of no force, or of balanced forces, the motion of each of those bodies relatively to the other is either none or uniform.

The first law of motion has been learned by experience and observation: not directly, for the circumstances supposed in it never occur; but indirectly, from the fact that its consequences, when it is taken in conjunction with other laws, are in accordance with all the phenomena of the motions of bodies.

The first law of motion may be regarded as a consequence of the definitions of *force* and of *balance* (Articles 55, 56); at the same time it is to be observed, that the framing of those definitions has been guided by experimental knowledge.

CHAPTER III.

ON THE VARIED TRANSLATION OF POINTS AND RIGID BODIES.

Section 1.—Law of Varied Translation.

328. **Second Law of Motion.**—*Change of momentum is proportional to the impulse producing it.* In this statement, as in that of the first law of motion, Article 327, it is implied that the motion of the moving body under consideration is referred to a fixed point or body whose motion is uniform. In questions of applied mechanics, the motion of any part of the earth's surface may be treated as uniform without sensible error in practice. The units of mass and of force may be so adapted to each other as to make *change of momentum equal to the impulse producing it.* (See Articles 330, 331.)

329. **General Equations of Dynamics.**—To express the second law of motion algebraically, two methods may be followed: the first method being to resolve the change of momentum into direct variation and deviation, and the impulse into direct and deflecting impulse; and the second method being to resolve both the change of momentum and the impulse into components parallel to three rectangular axes.

First method. m being the mass of the body, v its velocity, and r the radius of curvature of its path, it follows from Articles 73 and 75 that the *rate of direct variation* of its momentum is

$$m\frac{d\,v}{d\,t} = m \cdot \frac{d^2\,s}{d\,t^2};$$

and from Articles 77 and 78, that the rate of deviation of its momentum is

$$m\frac{v^2}{r}.$$

Equating these respectively to the direct and lateral impulse per unit of time, exerted by an unbalanced force F, making an angle θ with the direction of the body's motion, we find the two following equations:—

$$\text{P or } -\text{R} = \text{F} \cos\theta = m \cdot \frac{d\,v}{d\,t} = m\frac{d^2\,s}{d\,t^2}; \quad \ldots\ldots\ldots\ldots\ldots(1.)$$

$$\text{Q} = \text{F} \sin\theta = \frac{m\,v^2}{r}. \quad \ldots\ldots\ldots\ldots\ldots\ldots\ldots\ldots(2.)$$

The radius of curvature r is in the direction of the deviating force Q.

Second method. As in Article 80, let the velocity of the body be resolved into three rectangular components, $\frac{dx}{dt}, \frac{dy}{dt}, \frac{dz}{dt}$; so that the three component rates of variation of its momentum are

$$m\frac{d^2x}{dt^2}, m\frac{d^2y}{dt^2}, m\frac{d^2z}{dt^2}.$$

Also let the unbalanced force F, making the angles α, β, γ, with the axes of co-ordinates, and its impulse per unit of time, be resolved into three components, F_x, F_y, F_z. Then we obtain

$$\left.\begin{aligned} F_x &= F\cos\alpha = m\cdot\frac{d^2x}{dt^2}; \\ F_y &= F\cos\beta = m\frac{d^2y}{dt^2}; \\ F_z &= F\cos\gamma = m\frac{d^2z}{dt^2}; \end{aligned}\right\} \dots\dots\dots\dots\dots\dots(3.)$$

three equations, which are substantially identical with the Equations 1 and 2.

330. **Mass in Terms of Weight.**—A body's own weight, acting unbalanced on the body, produces velocity towards the earth, increasing at a rate per second denoted by the symbol g, whose numerical value is as follows:—Let λ denote the latitude of the place, h its elevation above the mean level of the sea,

$$g_1 = 32{\cdot}1695 \text{ feet, or } 9{\cdot}8051 \text{ mètres, per second;}$$

being the value of g for $\lambda = 45°$ and $h = 0$, and

$$R = 20900000 \text{ feet, or } 6370000 \text{ mètres, nearly,}$$

being the earth's mean radius; then

$$g = g_1\cdot(1 - 0{\cdot}00284\cos 2\lambda)\cdot\left(1 - \frac{2h}{R}\right)\dots\dots\dots\dots(1.)$$

For latitudes exceeding 45°, it is to be borne in mind that $\cos 2\lambda$ is negative, and the terms containing it as a factor have their signs reversed.

For practical purposes connected with ordinary machines, it is sufficiently accurate to assume

$$g = 32{\cdot}2 \text{ feet, or } 9{\cdot}81 \text{ mètres, per second nearly}\dots\dots(2.)$$

If, then, a body of the weight W be acted upon by an unbalanced force F, the change of velocity in the direction of F produced in a second will be

$$\frac{\mathrm{F}}{m} = \frac{\mathrm{F}\,g}{\mathrm{W}};$$

whence

$$m = \frac{\mathrm{W}}{g} \text{..........................} (3.)$$

is the expression for the *mass* of a body in terms of its weight, suited to make a change of momentum *equal* to the impulse producing it. m being absolutely constant for the same body, g and W vary in the same proportion at different elevations and in different latitudes.

331. **An Absolute Unit of Force** is the force which, acting during an unit of time on an arbitrary unit of mass, produces an unit of velocity. In Britain, the unit of time being a second (as it is elsewhere), and the unit of velocity one foot per second, the unit of mass employed is the mass whose weight in vacuo at London and at the level of the sea is a standard avoirdupois pound.

The *weight* of an unit of mass, in any given locality, has for its value, in absolute units of force, the coefficient g. When the *unit of weight* is employed as the unit of force, instead of the *absolute unit*, the corresponding unit of mass becomes g times the unit just mentioned: that is to say, in British measures, the mass of 32·2 lbs.; or in French measures, the mass of 9·81 kilogrammes.

332. **The Motion of a Falling Body,** under the unbalanced action of its own weight, a sensibly uniform force, is a case of the uniformly varied velocity described in Article 73. In the equations of that Article, for the rate of variation of velocity a, is to be substituted the coefficient g, mentioned in the last Article. Then if v_0 be the velocity of the body at the beginning of an interval of time t, its velocity at the end of that time is

$$v = v_0 + g\,t, \text{..........................} (1.)$$

the mean velocity during that time is

$$\frac{v_0 + v}{2} = v_0 + \frac{g\,t}{2}, \text{......................} (2.)$$

and the vertical height fallen through is

$$h = v_0\,t + \frac{g\,t^2}{2} \text{........................} (3.)$$

The preceding equations give the final velocity of the body, and the height fallen through, each in terms of the initial velocity and the time. To obtain the height in terms of the initial and final velocities, or *vice versa*, Equation 2 is to be multiplied by $v - v_0 = g\,t$, the acceleration, and compared with Equation 3; giving the following results :—

$$\left.\begin{aligned}\frac{v^2 - v_0^2}{2} = v_0\, g\, t + \frac{g^2\, t^2}{2} = g\, h\,; \\ h = \frac{v^2 - v_0^2}{2\, g}.\end{aligned}\right\} \text{..............(4.)}$$

When the body falls from a state of rest, v_0 is to be made $= 0$; so that the following equations are obtained :—

$$v = g\, t\,;\ h = \frac{g\, t^2}{2} = \frac{v^2}{2\, g} \text{.....................(5.)}$$

The height h in the last equation is called *the height or fall due to the velocity v;* and that velocity is called *the velocity due to the height or fall h.*

Should the body be at first projected vertically upwards, the initial velocity v_0 is to be made negative. To find the height to which it will rise before reversing its motion and beginning to fall, v is to be made $= 0$ in the last of the Equations 4; then

$$h = -\frac{v_0^2}{2\, g}, \text{..........................(6.)}$$

being a rise equal to the fall due to the initial velocity v_0.

333. **An Unresisted Projectile**, or a projectile to whose motion there is no sensible resistance, has a motion compounded of the vertical motion of a falling body, and of the horizontal motion due to the horizontal component of its velocity of projection. In fig. 139, let O represent the point from which the projectile is originally projected in the direction O A, making the angle X O A $= \theta$ with a horizontal line O X in the same vertical plane with O A. Let horizontal distances parallel to O X be denoted by x, and vertical ordinates parallel to O Z by z, positive upwards, and negative downwards. In the equations of vertical motion, the symbol h of the equations of Article 332 is to be replaced by $-z$, because of h and z being measured in opposite directions.

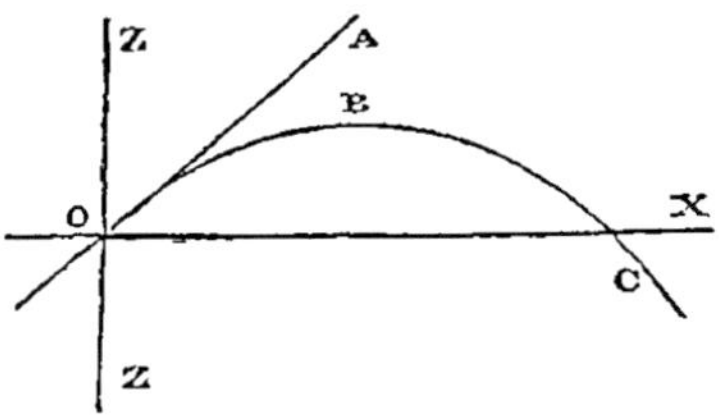

Fig. 139.

Let v_0 be the velocity of projection. Then at the instant of projection, the components of that velocity are,

$$\text{horizontal, } \frac{d\,x}{d\,t} = v_0 \cos\theta\,;\ \text{vertical, } \frac{d\,z}{d\,t} = v_0 \sin\theta\,;$$

and after the lapse of a given time t, those components have become

$$\left.\begin{aligned}\frac{dx}{dt}&=v_0\cos\theta=\text{constant};\\ \frac{dz}{dt}&=v_0\sin\theta-g\,t.\end{aligned}\right\}\ \ldots\ldots\ldots\ldots\ldots(1.)$$

Hence the co-ordinates of the body at the end of the time t are

$$\left.\begin{aligned}&\text{horizontal,}\quad x=v_0\cos\theta\cdot t;\\ &\text{vertical,}\quad z=v_0\sin\theta\cdot t-\frac{g\,t^2}{2};\end{aligned}\right\}\ \ldots\ldots\ldots\ldots\ldots(2.)$$

the Equations 2 being those of which the differential coefficients are Equations 1, and because $t=\frac{x}{v_0\cos\theta}$, those co-ordinates are thus related,

$$z=x\cdot\tan\theta-\frac{g}{2v_0^2\cos^2\theta}\cdot x^2;\ldots\ldots\ldots\ldots(3.)$$

an equation which shews the path O B C of the projectile to be a parabola with a vertical axis, touching O A in O.

The total velocity of the projectile at a given instant, being the resultant of the components given by Equation 1, has for the value of its square (remembering that $\sin^2\theta+\cos^2\theta=1$),

$$v^2=\frac{dx^2}{dt^2}+\frac{dz^2}{dt^2}=v_0^2-2v_0\sin\theta\cdot g\,t+g^2\,t^2=v_0^2-2\,g\,z;\ldots..(4.)$$

from the last form of which is obtained the equation

$$z=\frac{v_0^2-v^2}{2\,g};\ldots\ldots\ldots\ldots\ldots\ldots\ldots\ldots\ldots(5.)$$

which, being compared with Equation 4 of Article 332, shews that *the relation between the variation of vertical elevation, and the variation of the square of the resultant velocity, is the same, whether the velocity is in a vertical, inclined, or horizontal direction.*

The resistance of the air prevents any actual projectile near the earth's surface from moving exactly as an unresisted projectile. The approximation of the motion of an actual projectile to that of an unresisted projectile is the closer, the slower is the motion, and the heavier the body, because of the resistance of the air increasing with the velocity, and because of its proportion to the body's weight being dependent upon that of the body's surface to its weight.

334. **An Uniform Effort or Resistance**, unbalanced, causes the velocity of a body to vary according to the law expressed by this equation,

$$\frac{dv}{dt}=f\,g;\ldots\ldots\ldots\ldots\ldots\ldots\ldots\ldots\ldots(1.)$$

where f is the constant ratio which the unbalanced force bears to the weight of the moving body, positive or negative according to the direction of the force; so that by substituting $f\,g$ for g in the equations of Article 332, those equations are transformed into the equations of motion of the body in question, h being taken to represent the distance traversed by it in a positive direction.

In the apparatus known by the name of its inventor, Attwood, for illustrating the effect of uniform moving forces, this principle is applied in order to produce motions following the same law with those of falling bodies. Two weights, P and R, of which P is the greater, are hung to the opposite ends of a cord passing over a finely constructed pulley. Considering the masses of the cord and pulley to be insensible, the weight of the mass to be moved is P + R, and the moving force P − R, being less than the weight in the ratio,

$$f = \frac{P - R}{P + R};$$

consequently the two weights move according to the same law with a falling body, but more slowly in the ratio of f to 1.

335. **Deviating Force of a Single Body.**—It is part of the first law of motion, that if a body moves under no force, or balanced forces, it moves in a straight line.

It is one consequence of the second law of motion, that in order that a body may move in a curved path, it must be continually acted upon by an unbalanced force at right angles to the direction of its motion, the direction of the force being that towards which the path of the body is curved, and its magnitude bearing the same ratio to the weight of the body that the height due to the body's velocity bears to half the radius of curvature of its path.

This principle is expressed symbolically as follows:—

Half radius of curvature.		Height due to velocity.		Body's weight.		Deviating force.
$\frac{r}{2}$	:	$\frac{v^2}{2\,g}$	: :	W	:	$Q = \frac{W v^2}{g\,r}$ ......(1.)

or otherwise that the acceleration produced by gravity, bears the same ratio to the rate of deviation, that the weight bears to the magnitude of the deviating force, which may be symbolically expressed $g : \frac{v^2}{r} :: W : Q = \frac{W\,v^2}{g\,r}$.

In the case of projectiles, just described, and of the heavenly bodies, deviating force is supplied by that component of the mutual attraction of two masses which acts perpendicular to the direction of their relative motion. In machines, deviating force is supplied by the strength or rigidity of some body, which *guides* the deviating mass, making it move in a curve.

A pair of free bodies attracting each other have both deviated motions, the attraction of each guiding the other; and their deviations of momentum are equal in equal times; that is, their deviations of motion are inversely as their masses.

In a machine, each revolving body tends to press or draw the body which guides it away from its position, in a direction from the centre of curvature of the path of the revolving body; and that tendency is resisted by the strength and stiffness of the guiding body, and of the frame with which it is connected.

336. **A Revolving Simple Pendulum** consists of a small mass A, suspended from a point C by a rod or cord C A of insensibly small weight as compared with the mass A, and revolving in a circle about a vertical axis C B. The tension of the rod is the resultant of the weight of the mass A, acting vertically, and of its centrifugal force, acting horizontally; and therefore the rod will assume such an inclination that

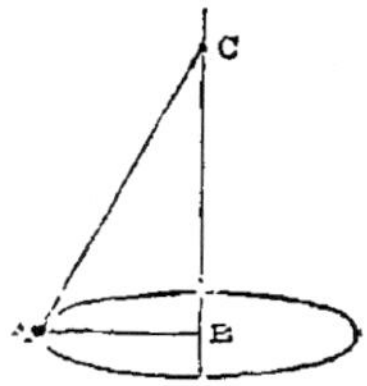

Fig. 140.

$$\frac{\text{height } \overline{\mathrm{B\,C}}}{\text{radius } \overline{\mathrm{A\,B}}} = \frac{\text{weight}}{\text{centrifugal force}} = \frac{g\,r}{v^2} \dots\dots\dots\dots (1.)$$

where $r = \mathrm{A\,B}$. Let n be the *number of turns per second* of the pendulum; then

$$v = 2\,\pi\,n\,r;$$

and therefore, making $\overline{\mathrm{B\,C}} = h$,

$$h = \frac{g\,r^2}{v^2} = \frac{g}{4\,\pi^2\,n^2}$$

$$= (\text{in the latitude of London})\ \frac{0{\cdot}8154 \text{ foot}}{n^2} = \frac{9{\cdot}7848 \text{ inches}}{n^2} \dots (2.)$$

When the speed of revolution varies, the inclination of the pendulum varies so as to adjust the height to the varying speed.

337. **Deviating Force in Terms of Angular Velocity.**—If the radius of curvature of the path of a revolving body be regarded as a sort of *arm* of constant or variable length at the end of which the body is carried, the angular velocity of that arm is given by the expression,

$$a = \frac{v}{r} \dots\dots\dots\dots\dots\dots\dots\dots\dots\dots (1.)$$

Let $a\,r$ be substituted for v in the value of deviating force of Article 335, and that value becomes

$$\mathrm{Q} = \frac{\mathrm{W}\,a^2\,r}{g} \dots\dots\dots\dots\dots\dots\dots\dots\dots (2.)$$

In the case of a body revolving with uniform velocity in a circle, like the bob A of the revolving pendulum of Article 336, $a = 2 \pi n$, where n is the number of revolutions per second, so that

$$Q = \frac{4 \pi^2 W n^2 r}{g}; \quad \text{.......................} (3.)$$

from which equation the height of a revolving pendulum might be deduced with the same result as in the last Article.

338. **A Simple Oscillating Pendulum** consists of an indefinitely small weight A, fig. 141, hung by a cord or rod of insensible weight A C from a point C, and swinging in a vertical plane to and fro on either side of a central point D vertically below C. The path of the weight or *bob* is a circular arc, A D E.

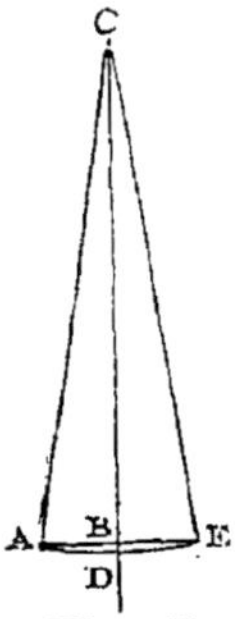
Fig. 141.

The weight W of the bob, acting vertically, may be resolved at any instant into two components, viz.:—

$$W \cdot \cos \angle DCA = W \cdot \frac{\overline{BC}}{\overline{CA}},$$

acting along C A, and balanced by the tension of the rod or cord, and

$$W \cdot \sin \angle DCA = W \cdot \frac{\overline{AB}}{\overline{CA}},$$

acting in the direction of a tangent to the arc, towards D, and unbalanced. The motion of A depends on the latter force.

When the arc A D E is small compared with the length of the pendulum A C, it very nearly coincides with the chord A B E; and the horizontal distance A B, to which the moving force is proportional, is very nearly equal to the distance of the bob from D, the central point of its oscillations. Then if the length of the pendulum, $\overline{CA}$, be denoted by l, we have approximately, for small arcs of oscillation,

$$\left. \begin{aligned} \frac{1}{n} &= 2 \pi \sqrt{\frac{l}{g}}; \text{ and} \\ l &= \frac{g}{4 \pi^2 n^2}; \end{aligned} \right\} \quad \text{.................} (1.)$$

and the following statement shews the connection between a simple oscillating and revolving pendulum, viz., *that the length of a simple oscillating pendulum, making a given number of small double oscillations in a second, is sensibly equal to the height of a revolving pendulum, making the same number of revolutions in a second.*

SECTION 2.—VARIED TRANSLATION OF A SYSTEM OF BODIES.

339. **Conservation of Momentum.**—THEOREM. *The mutual actions of a system of bodies cannot change their resultant momentum.* (Resultant momentum has been defined in Article 311.) Every force is a pair of equal and opposite actions between a pair of bodies; in any given interval of time it constitutes a pair of equal and opposite impulses on those bodies, and produces equal and opposite momenta. Therefore the momenta produced in a system of bodies by their mutual actions neutralize each other, and have no resultant, and cannot change the resultant momentum of the system.

340. **Motion of Centre of Gravity.**—COROLLARY. *The variations of the motion of the centre of gravity of a system of bodies are wholly produced by forces exerted by bodies external to the system;* for the motion of the centre of gravity is that which, being multiplied by the total mass of the system, gives the resultant momentum, and this can be varied by external forces only.

It follows that in all dynamical questions in which the mutual actions of a certain system of bodies are alone considered, the centre of gravity of that system of bodies may be correctly treated as a point whose motion is none or uniform; because its motion cannot be changed by the forces under consideration.

341. **The Angular Momentum,** relatively to a fixed point, of a body having a motion of translation, is the product of the momentum of the body into the perpendicular distance of the fixed point from the line of direction of the motion of the body's centre of gravity at the instant in question. Let m be the mass of the body, v its velocity, l the length of the before-mentioned perpendicular; then

$$m\,v\,l = \frac{W\,v\,l}{g}$$

is the angular momentum relatively to the given point.

Angular momenta are compounded and resolved like forces, each angular momentum being represented by a line whose length is proportional to the magnitude of the angular momentum, and whose direction is perpendicular to the plane of the motion of the body and of the fixed point, and such, that when the motion of the body is viewed from the extremity of the line, the radius vector of the body seems to have right-handed rotation. The direction of such a line is called the *axis* of the angular momentum which it represents. The *resultant angular momentum* of a system of bodies is the resultant of all their angular momenta relatively to their common centre of gravity; and the axis of that resultant angular momentum is called the *axis of angular momentum* of the system. The term *angular momentum* was introduced by Mr. Hayward.

342. **Angular Impulse** is the product of the moment of a couple of forces (Article 200) into the time during which it acts. Let F be the force of a couple, l its leverage, and $d\,t$ the time during which it acts, then

$$\mathrm{F}\,l\,d\,t$$

is the angular impulse. Angular impulses are compounded and resolved like the moments of couples.

343. **Relations of Angular Impulse and Angular Momentum.**—THEOREM. *The variation, in a given time, of the angular momentum of a body, is equal to the angular impulse producing that variation, and has the same axis.* This is a consequence which is deduced from the second law of motion in the following manner:—Conceive an unbalanced force F to be applied to a body m, and an equal, opposite, and parallel force, to a fixed point, during the interval $d\,t$; and let l be the perpendicular distance from the fixed point to the line of action of the first force. Then the couple in question exerts the angular impulse

$$\mathrm{F}\,l\,d\,t.$$

At the same time, the body m acquires a variation of momentum in the direction of the force applied to it, of the amount

$$m\,d\,v = \mathrm{F}\,d\,t;$$

so that relatively to the fixed point, the variation of the body's angular momentum is

$$m\,l\,d\,v = \mathrm{F}\,l\,d\,t;$$

being equal to the angular impulse, and having the same axis.—Q. E. D.

344. **Conservation of Angular Momentum.**—THEOREM. *The resultant angular momentum of a system of bodies cannot be changed in magnitude, nor in the direction of its axis, by the mutual actions of the bodies.*

Considering the common centre of gravity of the system of bodies as a fixed point, conceive that for each force with which one of the bodies of the system is urged in virtue of the combined action of all the other bodies upon it, there is an equal, opposite, and parallel force applied to the common centre of gravity, so as to form a couple. The forces with which the bodies act on each other are equal and opposite in pairs, and their resultant is nothing; therefore, the resultant of the ideal forces conceived to act at the common centre of gravity is nothing, and the supposition of these forces does not effect the equilibrium or motion of the system. Also, the resultant of all the couples thus formed is nothing; therefore, the resultant of their angular impulses is nothing; therefore, the resultant of the several variations of angular momentum produced by those angular impulses is nothing; therefore, the resultant angular momentum of the system is invariable in amount and in the direction of its axis.—Q. E. D.

345. **Collision.**—The most useful problem in cases of collision is, when two bodies whose masses are given move before the collision in one straight line with given velocities, and it is required to find their velocities after the collision. The two bodies form a system whose resultant momentum and internal energy are each unaltered by the collision; but a certain fraction of the internal energy disappears as visible motion, and appears as vibration and heat. If the bodies are equal, similar, and perfectly elastic, that fraction is nothing.

Let m_1, m_2, be the masses of the two bodies, and u_1, u_2, their velocities before the collision, whose directions should be indicated by their signs. Then the velocity of their common centre of gravity is

$$u_0 = \frac{u_1 m_1 + u_2 m_2}{m_1 + m_2}; \qquad \text{(1.)}$$

and this is not altered by the collision.

CHAPTER IV.

ROTATIONS OF RIGID BODIES.

346. **The Motion of a Rigid Body**, or of a body which sensibly preserves the same figure, has already been shewn in Part I., Chapter II., to be always capable of being resolved at each instant into a translation and a rotation; and by the aid of the principles explained in Section 3 of that chapter, the component rotation can always be conceived to take place about an axis traversing the centre of gravity of the body, and to be combined, if necessary, with a translation of the whole body in a curved or straight path along with its centre of gravity. The variations of the *momentum* of the translation, whether in amount or in direction, are due to the resultant force acting through the centre of gravity of the body, and are exactly the same with those of the momentum of the entire mass if it were concentrated at that centre; the variations of the *angular momentum* of the rotation are due to the resultant couple which is combined with that resultant force. The variations of *actual energy* are due to both causes.

When the translation of the centre of gravity of a rotating body, and its rotation about an axis traversing that centre, are known, the motion of every point in the body is determined by cinematical principles, which have been explained in Part I., Chapter II., Section 3.

Section 1.—On Moments of Inertia, Radii of Gyration, and Centres of Percussion.

347. **The Moment of Inertia** of an indefinitely small body, or "physical point," relatively to a given axis, is the product of the mass of the body, or of some quantity proportional to the mass, such as the weight, into the square of its perpendicular distance from the axis: thus in the following equation:—

$$\frac{\mathrm{I}}{g} = m\, r^2 = \frac{\mathrm{W}\, r^2}{g}, \qquad (1.)$$

r is the perpendicular distance of the mass m, whose weight is W, from a given axis; and the moment of inertia, according to the unit employed, is either I, or $\mathrm{I} \div g$; the former, when the unit is the moment of inertia of an unit of *weight* at the end of an arm whose length is unity; and the latter, when the unit is the moment of inertia of an unit of *mass* at the end of the same arm. The former is the more convenient unit, and will be employed in this treatise.

By an extension of the term "moment of inertia," it is applied to the product of any quantity, such as a volume, or an area, into the square of the distance of the point to which that quantity relates from a given axis; but in the remainder of this treatise the term will be used in its strict sense, and according to the unit of measure already specified; that is, in British measures, moment of inertia will be expressed by the product of a certain number of *pounds avoirdupois* into the square of a certain number of *feet.*

The geometrical relations amongst moments of inertia, to which the present section refers, are independent of the unit of measure.

348. **The Moment of Inertia of a System of Physical Points,** relatively to a given axis, is the sum of the moments of inertia of the several points; that is,

$$I = \Sigma \cdot W r^2 \ldots\ldots\ldots\ldots (1.)$$

349. **The Moment of Inertia of a Rigid Body** is the sum of the moments of inertia of all its parts, and is found by integration; that is, by conceiving the body to be divided into small parts of regular figure, multiplying the mass of each of those parts into the square of the distance of its centre of gravity from the axis, adding the products together, and finding the value towards which their sum converges when the size of the small parts is indefinitely diminished. For example, let the body be conceived to be built up of rectangular molecules, whose dimensions are $d\,x$, $d\,y$, and $d\,z$, the volume of each $d\,x\,d\,y\,d\,z$, and the mass of unity of volume w. Then

$$I = \int\int\int r^2 w \cdot d\,x\,d\,y\,d\,z \ldots\ldots\ldots\ldots (1.)$$

Hence follows the general principle that propositions relative to the geometrical relations amongst the moments of inertia of systems of points are made applicable to continuous bodies by substituting integration for ordinary summation; that is, for example, by putting $\int\int\int$ for Σ, and $w \cdot d\,x\,d\,y\,d\,z$ for W.

350. **The Radius of Gyration** of a body about a given axis is that length whose square is the *mean of all the squares* of the distances of the indefinitely small equal particles of the body from the axis, and is found by dividing the moment of inertia by the mass, thus,

$$\mathfrak{k}^2 = \frac{I}{\Sigma \cdot W} = \frac{\Sigma \cdot W r^2}{\Sigma W} \ldots\ldots\ldots\ldots (1.)$$

When symbols of integration are used, this becomes

$$\varrho^2 = \frac{\iiint r^2 w \cdot d x d y d z}{\iiint w \cdot d x d y d z} \dots\dots\dots\dots\dots (2.)$$

351. **Components of Moment of Inertia.**—Let the positions of the particles of a body be referred to three rectangular axes, one of which, O X, is that about which the moment of inertia is to be taken. Then the square of the radius vector of any particle is

$$r^2 = y^2 + z^2;$$

so that the moment of inertia round the axis of x is

$$I_x = \Sigma \cdot W y^2 + \Sigma \cdot W z^2; \dots\dots\dots\dots\dots (1.)$$

that is to say, *the moment of inertia of a body round a given axis may be found by adding together the sum of the products of the masses of the particles, each multiplied by the square of each of its distances from a pair of planes cutting each other at right angles in the given axis.*

In the same manner it may be shewn that the moments of inertia of the same body round the other two axes are given by the equations

$$I_y = \Sigma \cdot W z^2 + \Sigma \cdot W x^2; \; I_z = \Sigma \cdot W x^2 + \Sigma \cdot W y^2 \dots\dots (2.)$$

352. **Moments of Inertia Round Parallel Axes Compared.**—THEOREM. *The moment of inertia of a body about any given axis is equal to its moment of inertia about an axis traversing its centre of gravity parallel to the given axis, added to the moment of inertia about the given axis due to the whole mass of the body concentrated at its centre of gravity.*

This theorem may be expressed as follows:—Let I_0 be the moment of inertia of a body about an axis traversing its centre of gravity in any given direction, and I the moment of inertia of the same body about an axis parallel to the former at the perpendicular distance r_0; then

$$I = r_0^2 \cdot \Sigma W + I_0 \dots\dots\dots\dots\dots (1.)$$

COROLLARY I. The radius of gyration (ϱ) of a body about any axis is equal to the hypotenuse of a right-angled triangle, of which the two sides are respectively equal to the radius of gyration of the body about an axis traversing the centre of gravity parallel to the given axis (ϱ_0), and to the perpendicular distance between these axes (r_0). That is to say,

$$\varrho^2 = r_0^2 + \varrho_0^2 \ldots\ldots\ldots\ldots\ldots\ldots\ldots\ldots\ldots\ldots (2.)$$

COROLLARY II. The moment of inertia of a body about an axis traversing its centre of gravity in a given direction, is less than the moment of inertia of the same body about any other axis parallel to the first.

COROLLARY III. The moments of inertia of a body about all axes parallel to each other, which lie at equal distances from its centre of gravity, are equal.

353. **Combined Moments of Inertia.**—THEOREM. *The combined moment of inertia of a rigidly connected system of bodies about a given axis, is equal to the combined moment of inertia which the system would have about the given axis, if each body were concentrated at its own centre of gravity, added to the sum of the several moments of inertia of the bodies, about axes traversing their respective centres of gravity, parallel to the given axis.*

Let W now denote the mass of *one of the bodies*, I_0 its moment of inertia about an axis traversing its own centre of gravity parallel to the given common axis, and r_0 the distance of its centre of gravity from that common axis. Then the moment of inertia of that body about the common axis, according to Article 352, Equation 1, is

$$I = W\, r_0^2 + I_0.$$

Consequently, the combined moment of inertia of the system of bodies is

$$\Sigma\, I = \Sigma \cdot W\, r_0^2 + \Sigma\, I_0; \ldots\ldots\ldots\ldots\ldots\ldots (1).$$

—Q. E. D.

354. **Examples of Moments of Inertia and Radii of Gyration** of homogeneous bodies of some of the more simple and ordinary figures, are given in the following tables. In each case, the axis is supposed to traverse the *centre of gravity* of the body; for the principles of Article 352 enable any other case to be easily solved. The axes are also supposed, in each case, to be *axes of symmetry* of the figure of the body.

The column headed W gives the mass of the body; that headed I_0 gives the moment of inertia; that headed ϱ_0^2, the *square* of the radius of gyration. The mass of an unit of volume is in each case denoted by *w*.

Body.	Axis.	W	I_0	ϱ_0^2
I. Sphere of radius r,........	Diameter	$\frac{4\pi w r^3}{3}$	$\frac{8\pi w r^5}{15}$	$\frac{2r^2}{5}$
II. Spheroid of revolution—polar semi-axis a, equatorial radius r,..........	Polar axis	$\frac{4\pi w a r^2}{3}$	$\frac{8\pi w a r^4}{15}$	$\frac{2r^2}{5}$
III. Ellipsoid — semi-axes, a, b, c,......................	Axis, $2a$	$\frac{4\pi wabc}{3}$	$\frac{4\pi wabc(b^2+c^2)}{15}$	$\frac{b^2+c^2}{5}$
IV. Spherical shell—external radius r, internal r',....	Diameter	$\frac{4\pi w(r^3-r'^3)}{3}$	$\frac{8\pi w(r^5-r'^5)}{15}$	$\frac{2(r^5-r'^5)}{5(r^3-r'^3)}$
V. Spherical shell, insensibly thin — radius r, thickness dr,....................	Diameter	$4\pi w r^2 dr$	$\frac{8\pi w r^4 dr}{3}$	$\frac{2r^2}{3}$
VI. Circular cylinder—length $2a$, radius r, ...	Longitudinal axis, $2a$	$2\pi w a r^2$	$\pi w a r^4$	$\frac{r^2}{2}$
VII. Elliptic cylinder—length $2a$, transverse semi-axes b, c,......................	Longitudinal axis, $2a$	$2\pi wabc$	$\frac{\pi wabc(b^2+c^2)}{2}$	$\frac{b^2+c^2}{4}$
VIII. Hollow circular cylinder—length $2a$, external radius r, internal r', ...	Longitudinal axis, $2a$	$2\pi w a(r^2-r'^2)$	$\pi w a(r^4-r'^4)$	$\frac{r^2+r'^2}{2}$
IX. Hollow circular cylinder, insensibly thin — length $2a$, radius r, thickness dr,	Longitudinal axis, $2a$	$4\pi w a r dr$	$4\pi w a r^3 dr$	r^2
X. Circular cylinder—length $2a$, radius r, ...	Transverse diameter	$2\pi w a r^2$	$\frac{\pi w a r^2(3r^2+4a^2)}{6}$	$\frac{r^2}{4}+\frac{a^2}{3}$
XI. Elliptic cylinder—length $2a$, transverse semi-axes b, c,......................	Transverse axis, $2b$	$2\pi wabc$	$\frac{\pi wabc(3c^2+4a^2)}{6}$	$\frac{c^2}{4}+\frac{a^2}{3}$
XII. Hollow circular cylinder—length $2a$, external radius r, internal r', ...	Transverse diameter	$2\pi w a(r^2-r'^2)$	$\frac{\pi w a}{6}\left\{3(r^4-r'^4)+4a^2(r^2-r'^2)\right\}$	$\frac{r^2+r'^2}{4}+\frac{a^2}{3}$
XIII. Hollow circular cylinder, insensibly thin — radius r, thickness dr,..........	Transverse diameter	$4\pi w a r dr$	$\pi w a(2r^3+\frac{4}{3}a^2 r)dr$	$\frac{r^2}{2}+\frac{a^2}{3}$
XIV. Rectangular prism — dimensions $2a$, $2b$, $2c$,.....	Axis, $2a$	$8wabc$	$\frac{8wabc(b^2+c^2)}{3}$	$\frac{b^2+c^2}{3}$
XV. Rhombic prism — length $2a$, diagonals $2b$, $2c$,....	Axis, $2a$	$4wabc$	$\frac{2wabc(b^2+c^2)}{3}$	$\frac{b^2+c^2}{6}$
XVI. Rhombic prism, as above,	Diagonal, $2b$	$4wabc$	$\frac{2wabc(c^2+2a^2)}{3}$	$\frac{c^2}{6}+\frac{a^2}{3}$

355. **The Centre of Percussion** of a body, for a given axis, is a point so situated, that if part of the mass of the body were concentrated at that point, and the remainder at the point directly opposite in the given axis, the statical moment of the weight so distributed (Article 223), and its moment of inertia about the given axis, would be the same as those of the actual body in every position of the body.

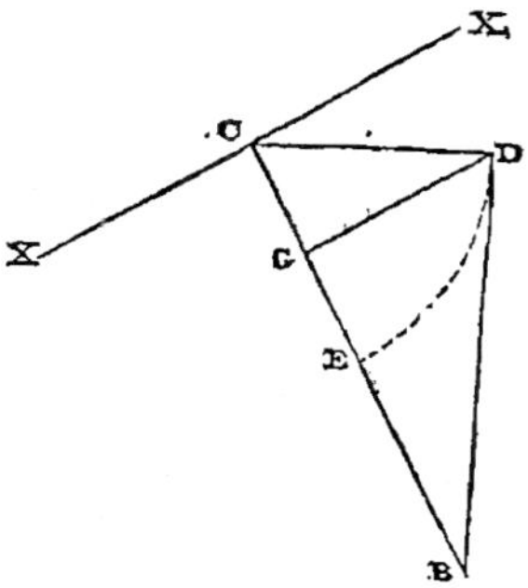

Fig. 142.

In fig. 142 let X X be the given axis, and G the centre of gravity of the body. It is evident, in the first place, that the centre of percussion must be somewhere in the perpendicular C G B let fall from the centre of gravity on the given axis. Secondly, in order that the statical moment of the whole mass, concentrated partly at C, and partly at the centre of percussion B (still unknown), may be the same with that of the actual body, the centre of gravity must be unaltered by that concentration of mass; that is to say, the masses concentrated at B and C must be inversely as the distances of those points from G. Hence denoting the weights of those masses by the letters B and C respectively, and the weight of the whole body by W, we have the proportion

$$\mathrm{W} : \mathrm{C} : \mathrm{B} : : \overline{\mathrm{BC}} : \overline{\mathrm{GB}} : \overline{\mathrm{GC}} \quad \ldots\ldots\ldots\ldots\ldots\ldots (1.)$$

Lastly, in order that the moment of inertia of the mass as supposed to be concentrated at B and C, about the axis X X, may be the same with that of the actual body, we must have

$$\mathrm{B} \cdot \overline{\mathrm{BC}}^2 = \mathrm{W}\rho^2 = \mathrm{W}\,(\rho_0^2 + r_0^2) \quad \ldots\ldots\ldots\ldots\ldots\ldots (2.)$$

where $r_0 = \overline{\mathrm{GC}}$, and ρ_0 is the radius of gyration of the body about an axis parallel to X X and traversing G; and substituting for B its value from Equation 1, viz., $\mathrm{B} = \mathrm{W} r_0 \div \overline{\mathrm{BC}}$, we find, for the distance of the centre of percussion from the axis,

$$\overline{\mathrm{BC}} = \frac{\rho^2}{r_0} = \frac{\rho_0^2}{r_0} + r_0; \quad \ldots\ldots\ldots\ldots\ldots\ldots\ldots\ldots (3.)$$

and for its distance from the centre of gravity,

$$\overline{\mathrm{GB}} = \overline{\mathrm{BC}} - r_0 = \frac{\rho_0^2}{r_0} \quad \ldots\ldots\ldots\ldots\ldots\ldots\ldots (4.)$$

The last equation may also be expressed in the form

$$\overline{GB} \cdot \overline{GC} = \varrho_0^2; \quad \ldots\ldots\ldots\ldots\ldots\ldots\ldots\ldots (5.)$$

which preserves the same value when $\overline{GB}$ and $\overline{GC}$ are interchanged; thus shewing, that if a new axis parallel to the original axis X X be made to traverse the original centre of percussion, the new centre of percussion is the point C in the original axis.

The proportion in which the mass of the body is to be considered as distributed between B and C takes the following form, when each of the last three terms of the proportion 1 is multiplied by $r_0 = \overline{GC}$:—

$$W : C : B :: \varrho_0^2 + r_0^2 : \varrho_0^2 : r_0^2 . \ldots\ldots\ldots\ldots\ldots\ldots (6.)$$

The preceding solution is represented by the following geometrical construction:—Draw $\overline{GD}$ perpendicular to $\overline{CG}$ and $= \varrho_0$; join $\overline{CD}$, perpendicular to which draw $\overline{DB}$ cutting $\overline{CG}$ produced in B; this point is the centre of percussion.

Also, $\overline{CD} = \varrho$, the radius of gyration about X X; and D B is the radius of gyration about an axis traversing B parallel to X X.

If $\overline{CE}$ be taken $= \overline{CD}$, E is sometimes called the **Centre of Gyration** of the body for the axis X X.

Section 2.—On Uniform Rotation.

356. **The Momentum** of a body rotating about its centre of gravity is nothing, according to the principle of Article 344. As every motion of a rigid body can be resolved into a translation, and a rotation about its centre of gravity, the rotation will be supposed to take place about the centre of gravity of the body throughout this section.

357. **The Angular Momentum** is found in the following manner:—Let x denote the axis of rotation, and y and z any two axes fixed in the body, perpendicular to it and to each other. Let a be the angular velocity of rotation. Then the velocity of any particle W, whose radius vector is $r = \sqrt{y^2 + z^2}$, is

$$a\, r = a \sqrt{y^2 + z^2},$$

and the angular momentum of that particle, *relatively to the axis of rotation*, is

$$\frac{W a r^2}{g} = \frac{W a}{g} (z^2 + y^2);$$

being the *product of its moment of inertia into its angular velocity*, divided by g, because of the weights of the particles having been used in computing the moment of inertia.

358. **The Actual Energy of Rotation** of a body rotating about its centre of gravity, being the sum of the masses of its particles, each multiplied into one-half of the square of its velocity, is found as follows:—a being the angular velocity of rotation, the linear velocity of any particle whose distance from the axis of rotation is r, is

$$v = a\, r;$$

and the actual energy of that particle, its weight being W, is

$$\frac{\mathrm{W}\, v^2}{2\, g} = \frac{\mathrm{W}\, a^2\, r^2}{2\, g}; \quad \ldots\ldots\ldots\ldots\ldots\ldots\ldots\ldots (1.)$$

being the *moment of inertia* of the particle multiplied by $\frac{a^2}{2\, g}$. Hence for the whole body the actual energy of rotation is

$$\mathrm{E} = \frac{a^2\, \mathrm{I}}{2\, g}; \quad \ldots\ldots\ldots\ldots\ldots\ldots\ldots\ldots (2.)$$

that is to say, *actual energy bears the same relation to angular velocity and moment of inertia that it does to linear velocity and weight.*

CHAPTER V.

MOTIONS OF FLUIDS.

359. **Division of the Subject.**—The mode of division that will be employed in this chapter in treating of the motions of fluids is founded on the distinction between motions not sensibly affected by friction, and those which are so affected. The motions of fluids not sensibly affected by friction, and therefore governed by pressure and weight only, take place according to laws which are exactly known; so that any difficulty which exists in tracing their consequences, in particular cases, arises from mathematical intricacy alone. The laws of the friction of fluids, on the other hand, are only known approximately and empirically; and the mode of operation of that force amongst the particles of a fluid is not yet thoroughly understood; so that the solution of a particular problem has often to be deduced, not from first principles representing the condensed results of all experience, but from experiments of a special class, suited to the problem under consideration.

The following is the division of the subject of this chapter:—

I. Motions of Liquids under Gravity and Pressure alone.

II. Motions of Liquids affected by Friction.

SECTION 1.—MOTIONS OF LIQUIDS WITHOUT FRICTION.

360. **Dynamic Head.**—Let p denote the intensity of the pressure of the liquid at a given point and ϱ the *weight of an unit of volume;* then the quotient $\frac{p}{\varrho}$ is what is called the *height, or head, due to the pressure;* that is, the height of a column of the liquid, of the uniform specific gravity ϱ, whose weight per unit of base would be equal to the pressure p. Now, let a vertical ordinate z be measured *positively downwards* from a datum horizontal plane, ϱz is the weight of a column of liquid per unit of base extending down from that plane to a particle under consideration; $p - \varrho z$ is the difference between the intensity of the actual pressure at that particle and the pressure due to its depth below the datum horizontal plane; and

$$\frac{p}{\varrho} - z = h \text{.............................} (1.)$$

is the *height* or *head due to that difference of intensity*, being what will be termed the *dynamic head.* When z is measured *positively upwards* from a datum horizontal plane, its sign is to be changed; so that the expression for the dynamic head in that case becomes

$$\frac{p}{\varrho} + z = h \dots\dots\dots\dots\dots\dots\dots\dots\dots\dots (2.)$$

361. **Law of Dynamic Head for Steady Motion.**—This principle may be stated thus:—*In steady motion, the sum of the height due to the velocity of a particle and of its dynamic head is constant,* or symbolically

$$\frac{V^2}{2g} + h = \text{constant}.$$

This equation applies to the particles which successively occupy the same fixed point, as well as to each individual particle.

362. **The Total Energy** of a particle of a moving liquid without friction is expressed by multiplying the expression in the previous equation by the weight of the particle W, thus:—

$$\frac{W V^2}{2g} + W h;$$

in which $\frac{W V^2}{2g}$ is the *actual energy* of the particle, and $W h$ is its *potential energy;* because, from the last Article it appears, that by the diminution of $W h$, $\frac{W V^2}{2g}$ may be increased by an equal amount, and *vice versa;* so that *the dynamic head of a particle is its potential energy per unit of weight.* In the case of *steady motion,* the total energy of each particle is constant; and the total energy of each of the equal particles which successively occupy the same position is the same.

363. **The Free Surface** of a moving liquid mass, being that which is in contact with the air only, is characterized by the pressure being uniform all over it, and equal to that of the atmosphere. Let p_1 be the atmospheric pressure, z_1 the vertical ordinate, measured *positively upwards* from a given horizontal plane, of any point in the free surface of the liquid, and h_1 the dynamic head at the same point; then it appears from Article 360, Equation 2, that for that surface,

$$h_1 - z_1 = \frac{p_1}{\varrho} = \text{constant} \dots\dots\dots\dots\dots\dots (1.)$$

364. **A Surface of Equal Pressure** is characterized by an analogous equation,

$$h - z = \frac{p}{\varrho} = \text{constant}; \dots\dots\dots\dots\dots\dots (1.)$$

and all surfaces of equal pressure fulfil the differential equation,

$$d\,h = d\,z; \quad\text{..............................(2.)}$$

for the differential coefficient of a constant being equal to 0 $d\,h - d\,z = 0$ Equation 1, and $\therefore d\,h = d\,z$ which, for *steady motion*, becomes

$$d\,z = d\,h = -\,d \cdot \frac{V^2}{2\,g}; \quad\text{......................(3.)}$$

found by differentiating the equation of Article 361, expressing that the variations of actual energy are those due to the variations of level simply.

365. **Motion in Plane Layers** is a state which is either exactly or approximately realized in many ordinary cases of liquid motion;

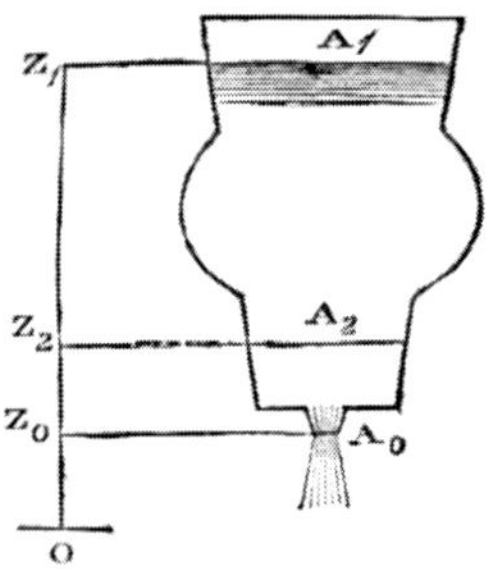

Fig. 143.

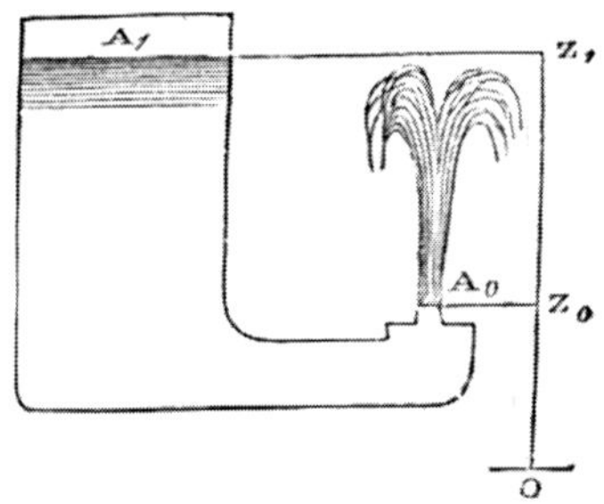

Fig. 144.

and the assumption of which is often used as a first approximation to the solution of various questions in hydraulics. It consists in the motions of all the particles in one plane being parallel to each other, perpendicular to the plane, and equal in velocity. It is illustrated by the three figures 143, 144, and 145, each of which represents a reservoir containing liquid up to the elevation $\overline{O\,Z_1} = z_1$ above a given datum, and discharging the liquid from an orifice A_0 at the smaller elevation $\overline{O\,Z_0} = z_0$. The liquid moves exactly or nearly in plane layers at the upper surface A_1 and at the orifice A_0. Let these symbols denote the areas of the upper surface and of the issuing stream respectively.

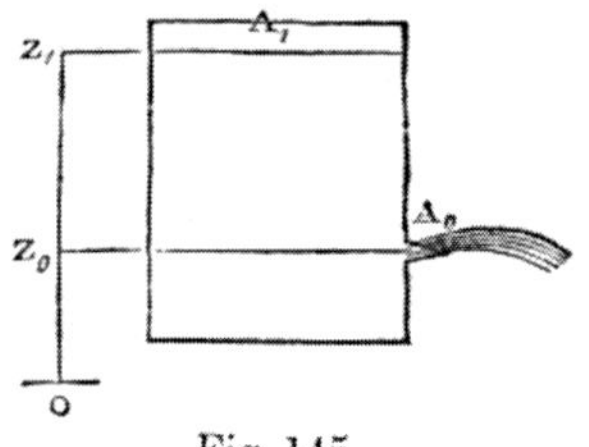

Fig. 145.

Let Q denote the rate of flow per second, v_1 the velocity of descent of the liquid at the upper surface, v_0 its velocity of outflow from the orifice; then, according to Article 116, the equation of continuity is

$$\left.\begin{aligned} v_1 A_1 = v_0 A_0 = Q; \\ \text{or } v_1 = \frac{Q}{A_1};\ v_0 = \frac{Q}{A_0}. \end{aligned}\right\} \text{......................(1.)}$$

The pressures at the upper surface and at the orifice respectively are each equal to the atmospheric pressure; hence the difference of dynamic head is simply the difference of elevation; that is to say,

$$h_1 - h_0 = z_1 - z_0;$$

therefore, according to Article 361 and Article 364, Equations 2 and 3,

$$\frac{v_0^2 - v_1^2}{2g} = \frac{v_0^2}{2g}\left(1 - \frac{A_0^2}{A_1^2}\right) = z_1 - z_0 \text{..............(2.)}$$

This gives for the velocity of outflow,

$$v_0 = \sqrt{\left\{\frac{2g(z_1 - z_0)}{1 - \frac{A_0^2}{A_1^2}}\right\}}; \text{................(3.)}$$

from which can be computed the rate of flow or discharge by means of Equation 1.

366. **The Contracted Vein** is the name given to a portion of a jet of fluid at a short distance from an orifice in a plate, which is smaller in diameter and in area than the orifice, owing to a spontaneous contraction which the jet undergoes after leaving the orifice.

The area of the narrowed part of the contracted vein is in every case to be considered as the *virtual* or *effective outlet*, and used for A_0 in the equations of the last Article.

The ratio of the area of the contracted vein, or effective orifice, to that of the actual orifice, is called the *coefficient of contraction*. For sharp edged orifices in thin plates, it has different values for different figures and proportions of the orifice, ranging from about 0·58 to 0·7, and being on an average about $\frac{5}{8}$. It diminishes somewhat for great pressures, and for dynamic heads of six feet and upwards may be taken at about 0·6. The most elaborate table of those coefficients is that of Poncelet and Lesbros.

For orifices with edges that are not sharp and thin, the discharge is modified sensibly by friction.

Section 2.—Motions of Liquids with Friction.

367. **General Laws of Fluid Friction.**—It is known by experiment, that between a fluid, and a solid surface over which it glides,

there is exerted a resistance to their relative motion which is proportional to their surface of contact, and to the density of the fluid, and is approximately proportional to the square of the velocity of the relative motion; that is, the resistance is approximately proportional *to the weight of a prism of the fluid, whose base is the surface of contact, and its height the height due to the relative velocity.*

Let S be the surface of contact, v the velocity, ϱ the weight of an unit of volume of the fluid, and f a factor called the *coefficient of friction;* then

$$R = f \varrho S \frac{v^2}{2g}, \ldots\ldots\ldots\ldots\ldots\ldots\ldots\ldots (1.)$$

is the amount of the friction at the surface S.

The coefficient f is not absolutely constant at different velocities. The mode of calculation employed in practice, where the velocity is one of the unknown quantities to be determined, is to find an approximate value of the velocity from the *mean* value of f; then to compute the value of f corresponding to that approximate velocity, and use it to compute the velocity more exactly.

The following are some of the values of the coefficients of friction, according to different authorities, for streams of WATER, gliding over various surfaces; v being the *mean velocity of the stream*, in feet per second:—

Iron pipes (Darcy). Let d = diameter of pipe in feet; then,

$$f = 0{\cdot}0043 \left(1 + \frac{1}{9\,d}\right) + \frac{0{\cdot}001}{v} \left(1 + \frac{1}{18\,d}\right);$$

or for velocities that are not very small,

$$f = 0{\cdot}005 \left(1 + \frac{1}{12\,d}\right).$$

Iron pipes, value of f for first approximation, 0·0064

Beds of rivers (Weisbach),......$f = a + \frac{b}{v}$; $a = 0{\cdot}0074$.

$b = 0{\cdot}00023$ foot.

Beds of rivers, value of f for first approximation,........................... 0·0076.

A collection of numerous formulæ for fluid friction, proposed by different authors, together with tables of the results of the best formulæ, is contained in Mr. Neville's work on hydraulics. The formulæ of many authors, though differing in appearance, are founded on the same, or nearly the same, experimental data, being chiefly those of Du Buat, with additions by subsequent inquirers; and their practical results do not materially differ. The two formulæ given above, on the authority of Darcy, for iron pipes, are based on his experiments as recorded in his treatise *du Mouvement de l'Eau dans les Tuyaux.*

368. **Internal Fluid Friction.**—Although the particles of fluids have no transverse elasticity—that is, no tendency to recover a certain figure after having been distorted—it is certain that they resist being made to slide over each other, and that there is a lateral communication of motion amongst them; that is, that there is a tendency of particles which move side by side in parallel lines to assume the same velocity. The laws of this lateral communication of motion, or internal friction of fluids, are not known exactly; but its effects are known thus far:—that the energy due to differences of velocity, which it causes to disappear, is replaced by heat in the proportion of one thermal unit of Fahrenheit's scale for 772 foot pounds of energy, and that it causes the friction of a stream against its channel to take effect, not merely in retarding the film of fluid which is immediately in contact with the sides of the channel, but in retarding the whole stream, so as to reduce its motion to one approximating to a motion in plane layers perpendicular to the axis of the channel (Article 365).

369. **Friction in an Uniform Stream.**—It is this last fact which renders possible the existence of an open stream of uniform section, velocity, and declivity. In hydraulic calculations respecting the resistance of this, or any other stream, the value given to the velocity is its mean value throughout a given cross-section of the stream A,

$$v = \frac{Q}{A} \qquad \text{(1.)}$$

The greatest velocity in each cross-section of a stream takes place at the point most distant from the rubbing surface of the channel. Its ratio to the mean velocity is given by the following empirical formula of Prony, where V is the greatest velocity in feet per second:—

$$\frac{v}{V} = \frac{7 \cdot 71 + V}{10 \cdot 25 + V} \qquad \text{(2.)}$$

In an uniform stream, the dynamic head which would otherwise have been expended in producing increase of actual energy, is wholly expended in overcoming friction. Consider a portion of the stream whose length is l, and fall z. The loss of head is equal to the fall of the surface of the stream, according to Article 363; and the expenditure of potential energy in a second is accordingly

$$z \varrho Q = z \varrho v A.$$

Equating this to the work performed in a second in overcoming friction, viz., v R, Equation 1, Article 367, we find

$$z \varrho v A = f \varrho S \frac{v^3}{2g};$$

or dividing by common factors, and by the area of section A, we find for the value of the fall in terms of the velocity

$$z = f \cdot \frac{S}{A} \cdot \frac{v^2}{2g} \text{..........................(3.)}$$

Let s be what is called the *wetted perimeter* of the cross-section of the stream; that is, the cross-section of the rubbing surface of the stream and channel; then

$$S = l s;$$

and dividing both sides of Equation 3 by l, we find for the relation between the rate of declivity and the velocity,

$$\sin i = \frac{z}{l} = f \frac{s}{A} \cdot \frac{v^2}{2g} \text{..................(4.)}$$

$\frac{A}{s}$ is what is called the "HYDRAULIC MEAN DEPTH" of the stream; and as the friction is inversely proportional to it, it is evident that the figure of cross-section of channel which gives the least friction is that whose hydraulic mean depth is greatest, viz., a semicircle. When the stability of the material limits the side-slope of the channel to a certain angle, Mr. Neville has shewn that the figure of least friction consists of a pair of straight side-slopes of the given inclination connected at the bottom by an arc of a circle whose radius is the depth of liquid in the middle of the channel; or, if a flat bottom be necessary, by a horizontal line touching that arc. For such a channel, the hydraulic mean depth is half of the depth of liquid in the middle of the channel.

370. **Varying Stream.**—In a stream whose area of cross-section varies, and in which, consequently, the mean velocity varies at different cross-sections, the loss of dynamic head is the sum of that expended in overcoming friction, and of that expended in producing increased velocity, when the velocity increases, or the difference of those two quantities when the velocity diminishes, which difference may be positive or negative, and may represent either a loss or a gain of head. The following method of representing this principle symbolically is the most convenient for practical purposes. In fig. 146, let the origin of co-ordinates be taken at a point O *completely below* the part of the stream to be considered; let horizontal abscissæ x be measured *against* the direction of flow, and vertical ordinates to the surface of the stream, z, upwards. Consider any indefinitely short portion of the stream whose

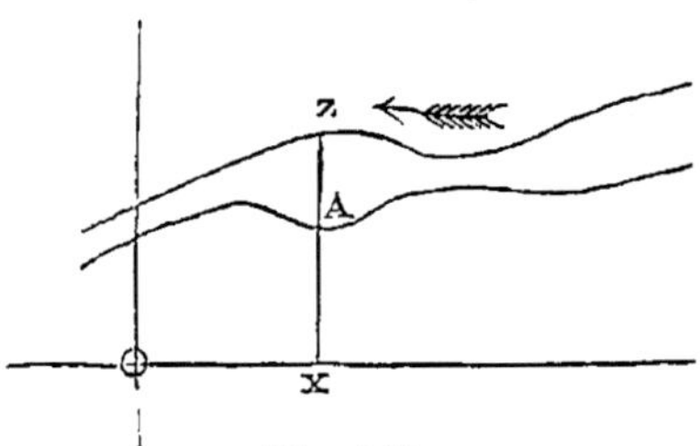

Fig. 146.

horizontal length is $d\,x$; in practice this may almost always be considered as equal to the actual length. The fall in that portion of the stream is $d\,z$, and the acceleration $-d\,v$, because of v being opposite to x. Then modifying the expression for the loss of head due to friction in Equation 3 of Article 369 to meet the present case, and adding the loss of head due to acceleration, we find

$$d\,z = f \cdot \frac{s\,d\,x}{\mathrm{A}} \cdot \frac{v^2}{2\,g} - \frac{v\,d\,v}{g} \text{..................(1.)}$$

In applying this differential equation to the solution of any particular problem, for v is to be put Q ÷ A, and for A and s are to be put their values in terms of x and z. Thus is obtained a differential equation between z and x, and the constant quantity Q, the flow per second. If Q is known, then it is sufficient to know the value of z for one particular value of x, in order to be able to determine the integral equation between z and x If Q is unknown, the values of z for two particular values of x, or of z and $\frac{d\,z}{d\,x}$ (the declivity), for one particular value of x, are required for the solution, which comprehends the determination of the value of O.

371. **The Friction in a Pipe Running Full** produces loss of dynamic head according to the same law with the friction in a channel, except that the dynamic head is now the sum of the elevation of the pipe above a given level, and of the height due to the pressure within it. The differential equation which expresses this is as follows:—Let $d\,l$ be the length of an indefinitely short portion of a pipe measured *in* the direction of flow, s its internal circumference, A its area of section, z its elevation above a given level, p the pressure within it, h the dynamic head. Then the loss of head is

$$-d\,h = -d\,z - \frac{d\,p}{\rho} = \frac{v\,d\,v}{g} + f \cdot \frac{s\,d\,l}{\mathrm{A}} \cdot \frac{v^2}{2\,g} \text{...........(1.)}$$

The ratio $\frac{d\,h}{d\,l}$ is called the *virtual* or *hydraulic declivity*, being the rate of declivity of an open channel of the same flow, area, and hydraulic mean depth. This may differ to any extent from the *actual declivity* of the pipe, $\frac{d\,z}{d\,l}$.

When the pipe is of uniform section, $d\,v = 0$, and the first term of the right-hand side of Equation 1 vanishes.

When the section of the pipe varies, s and A are given functions of l. If Q is given, $v = \mathrm{Q} \div \mathrm{A}$ is also a given function of l; and to solve the equation completely, there is only required in addition the value of h for one particular value of l. If Q is unknown, the values of h for two particular values of l, or of h and $\frac{dh}{dl}$ for one particular value of l, are required for the solution, which comprehends the determination of Q.

372. **Resistance of Mouthpieces.**—A mouthpiece is the part of a channel or pipe immediately adjoining a reservoir. The internal friction of the fluid on entering a mouthpiece causes a loss of head equal to the height due to the velocity multiplied by a constant depending on the figure of the mouthpiece, whose values for certain figures have been found empirically; that is to say, let $-\Delta h$ be the loss of head; then

$$-\Delta h = \frac{f' v^2}{2g}, \qquad \text{(1.)}$$

f' being a constant.

For the mouthpiece of a cylindrical pipe, issuing from the flat side of a reservoir, and making the angle i with a normal to the side of the reservoir, according to Weisbach,

$$f' = 0{\cdot}505 + 0{\cdot}303 \sin i + 0{\cdot}226 \sin^2 i \qquad \text{(2.)}$$

373. **The Resistance of Curves and Knees** in pipes causes a loss of head equal to the height due to the velocity multiplied by a coefficient, whose values, according to Weisbach, are given by the following formulæ:—For *curves*, let i be the arc to radius unity, r the radius of curvature of the centre line of the pipe, and d its diameter.

Then for a circular pipe,

$$f'' = \frac{i}{\pi}\left\{0{\cdot}131 + 1{\cdot}847\left(\frac{d}{2r}\right)^{\frac{7}{2}}\right\};$$

and for a rectangular pipe,

$$f'' = \frac{i}{\pi}\left\{0{\cdot}124 + 3{\cdot}104\left(\frac{d}{2r}\right)^{\frac{7}{2}}\right\}; \qquad \text{(1.)}$$

for *knees*, or sudden bends, let i be the angle made by the two portions of the pipe at either side of the knee with each other; then

$$f'' = 0{\cdot}9457 \sin^2 \frac{i}{2} + 2{\cdot}047 \sin^4 \frac{i}{2} \qquad \text{(2.)}$$

374. **A Sudden Enlargement** of the channel in which a stream of liquid flows, causes a sudden diminution of the mean velocity in the same proportion as that in which the area of section is in-

creased. Thus, let v_1 be the velocity in the narrower portion of the channel, and let m be the number expressing the ratio in which the channel is suddenly enlarged: the velocity in the enlarged part is $\frac{v_1}{m}$. Now it appears from experiment, that the actual energy due to the velocity of the narrow stream *relatively* to the wide stream, that is, to the difference $v_1\left(1-\frac{1}{m}\right)$, is expended in overcoming the internal fluid friction of eddies, and so producing heat; so that there is a *loss of total head*, represented by

$$\frac{v_1^2}{2g}\left(1-\frac{1}{m}\right)^2 \ldots\ldots\ldots\ldots (1.)$$

375. **The General Problem** of the flow of a stream with friction is thus expressed:—Let $h_1+\frac{v_1^2}{2g}$, and $h_2+\frac{v_2^2}{2g}$, be the total heads at the beginning and end of the stream respectively; then the loss of total head is represented by

$$h_1-h_2+\frac{v_1^2-v_2^2}{2g}=\Sigma\cdot \mathrm{F}\,\frac{v^2}{2g} \ldots\ldots\ldots\ldots (1.)$$

where the right-hand side of the equation represents the sum of all the losses of head due to the friction in various parts of the channel.

PART VI.

THEORY OF MACHINES.

CHAPTER I.

DEFINITIONS AND GENERAL PRINCIPLES.

376. **Nature and Division of the Subject.**—In the present Part of this work, machines are to be considered not merely as modifying motion, but also as modifying force, and transmitting energy from one body to another. The theory of machines consists chiefly in the application of the principles of dynamics to trains of mechanism; and therefore much of the present Part of this treatise will consist of references back to Parts II. and V.

There are two fundamentally different ways of considering a machine, each of which must be employed in succession, in order to obtain a complete knowledge of its working.

I. In the first place is considered the action of the machine during a certain period of time, with a view to the determination of its EFFICIENCY; that is, the ratio which the *useful* part of its work bears to the whole expenditure of energy. The motion of every ordinary machine is either uniform or periodical; and therefore the principle of the equality of energy and work is fulfilled, either constantly, or periodically at the end of each period or cycle of changes in the motion of the machine.

II. In the second place is to be considered the action of the machine during intervals of time less than its period or cycle, if its motion is periodic, in order to determine the law of the periodic changes in the motions of the pieces of which the machine consists, and of the periodic or reciprocating forces by which such changes are produced.

377. **A Prime Mover** is an engine, or combination of moving pieces, which serves to transfer energy from those bodies which naturally develop it, to those by means of which it is to be employed, and to transform energy from the various forms in which

it may occur, such as chemical affinity, heat, or electricity, into the form of mechanical energy, or energy of force and motion. The mechanism of a prime mover comprehends all those parts by means of which it regulates its own operations.

The *useful work* of a prime mover is the energy which it transmits to any machine driven by it; and its *efficiency* is the ratio of that useful work to the whole energy received by it from a natural source of energy.

The *effect* or *available power* of a prime mover is its useful work in some given unit of time, such as a second, a minute, an hour, or a day.

378. **The Regulator** of a prime mover is some piece of apparatus by which the rate at which it receives energy from the source of energy can be varied.

379. **A Governor** is a self-acting adjusting apparatus, usually consisting of a pair of rotating pendulums, whose angle of deviation from their axis depends upon the speed.

380. **Fluctuations of Speed** in a machine are caused by the alternate excess of the energy received above the work performed, and of the work performed above the energy received, which produce an alternate increase and diminution of actual energy.

381. **A Fly-Wheel** is a wheel with a heavy rim, whose great moment of inertia reduces the coefficient of fluctuation of speed to a certain fixed amount.

382. **A Brake** is employed to stop a machine in a shorter time than can be done by simply suspending the effort of the prime mover.

383. **Useful and Lost Work.**—The whole work performed by a machine is distinguished into *useful work*, being that performed in producing the effect for which the machine is designed, and *lost work* being that performed in producing other effects.

384. **Useful and Prejudicial Resistance** are overcome in performing useful work and lost work respectively.

385. **The Efficiency** of a machine is a fraction expressing the ratio of the useful work to the whole work performed, which is equal to the energy expended. The limit to the efficiency of a machine is *unity*, denoting the efficiency of a perfect machine in which no work is lost. The object of improvements in machines is to bring their efficiency as near to unity as possible.

386. **Power and Effect; Horse Power.**—The *power* of a machine is the energy exerted, and the *effect*, the useful work performed, in some interval of time of definite length.

The unit of power called conventionally a *horse power*, is 550 foot pounds per second, or 33,000 foot pounds per minute, or 1,980,000 foot pounds per hour. The effect is equal to the power multiplied by the efficiency.

387. **Driving Point; Train; Working Point.**—The driving point is that through which the resultant effort of the prime mover acts. The train is the series of pieces which transmit motion and force from the driving point to the working point, through which acts the resultant of the resistance of the useful work.

388. **Points of Resistance** are points in the train of mechanism through which the resultants of prejudicial resistances act.

389. **Efficiencies of Pieces of a Train.**—The useful work of an intermediate piece in a train of mechanism consists in driving the piece which follows it, and is less than the energy exerted upon it by the amount of the work lost in overcoming its own friction. Hence the efficiency of such an intermediate piece is the ratio of the work performed by it in driving the following piece, to the energy exerted on it by the preceding piece; and it is evident that *the efficiency of a machine is the product of the efficiencies of the series of moving pieces which transmit energy from the driving point to the working point.* The same principles apply to a train of *successive machines,* each driving that which follows it.

CHAPTER II.

OF THE PERFORMANCE OF WORK BY MACHINES.

SECTION 1.—OF WORK.

390. **The Action of a Machine** is to produce motion against Resistance. For example, if the machine is one for lifting solid bodies, such as a crane, or fluid bodies, such as a pump, its action is to produce upward motion of the lifted body against the resistance arising from gravity; that is, against its own weight: if the machine is one for propulsion, such as a locomotive engine, its action is to produce horizontal or inclined motion of a load against the resistance arising from friction, or from friction and gravity combined: if it is one for shaping materials, such as a planing machine, its action is to produce relative motion of the tool and of the piece of material shaped by it, against the resistance which that material offers to having part of its surface removed; and so of other machines.

391. **Work.**—The action of a machine is measured, or expressed as a definite quantity, by multiplying the motion which it produces into the resistance, or force directly opposed to that motion, which it overcomes; the product resulting from that multiplication being called WORK.

In Britain, the distances moved through by pieces of mechanism are usually expressed in feet; the resistances overcome, in pounds avoirdupois; and quantities of work, found by multiplying distances in feet by resistances in pounds, are said to consist of so many *foot-pounds*. Thus the work done in lifting a weight of one pound, through a height of one foot, is *one foot-pound;* the work done in lifting a weight of twenty pounds, through a height of one hundred feet, is $20 \times 100 = 2{,}000$ foot-pounds.

In France, distances are expressed in mètres, resistances overcome in kilogrammes, and quantities of work in what are called *kilogrammètres*, one kilogrammètre being the work performed in lifting a weight of one killogramme through a height of one mètre.

392. **The Rate of Work** of a machine means, the quantity of work which it performs in some given interval of time, such as a second, a minute, or an hour. It may be expressed in units of work (such as foot-pounds) per second, per minute, or per hour, as the case

may be; but there is a peculiar unit of power appropriated to its expression, called a HORSE-POWER, which is, in Britain,

550 foot-pounds per second,
or 33,000 foot-pounds per minute,
or 1,980,000 foot-pounds per hour.

In France, the term FORCE DE CHEVAL is applied to the following rate of work:—

		Foot-pounds.
75 kilogrammètres per second	=	$542\frac{1}{2}$
or 4,500 kilogrammètres per minute	=	32,549
or 270,000 kilogrammètres per hour	=	1,952,932

being about one-seventieth part less than the British horse-power.

393. **Velocity.**—If the *velocity of the motion* which a machine causes to be performed against a given resistance be given, then the product of that velocity into the resistance obviously gives the rate of work, or effective power. If the velocity is given in feet per second, and the resistance in pounds, then their product is the rate of work in foot-pounds per second, and so of minutes, or hours, or other units of time.

It is usually most convenient, for purposes of calculation, to express the velocities of the parts of machines either in feet per second or in feet per minute. For certain kinetic calculations the second is the more convenient unit of time: in stating the performance of machines for practical purposes, the minute is the unit most commonly employed.

394. **Work in Terms of Angular Motion.**—When a resisting force opposes the motion of a part of a machine which moves round a fixed axis, such as a wheel, an axis, or a crank, the product of the amount of that resistance into its *leverage* (that is, the perpendicular distance of the line along which it acts from the fixed axis) is called the *moment*, or *statical moment*, of the resistance. If the resistance is expressed in pounds, and its leverage in feet, then its moment is expressed in terms of a measure which may be called a *foot-pound*, but which, nevertheless, is a quantity of an entirely different kind from a foot-pound of work.

Suppose now that the body to whose motion the resistance is opposed turns through any number of revolutions, or parts of a revolution; and let T denote the angle through which it turns, expressed in revolutions, and parts of a revolution, also, let

$$2\pi = 6{\cdot}2832 = \frac{710}{113}$$

denote, as is customary, the ratio of the circumference of a circle to its radius. Then the distance through which the given resistance is overcome is expressed by

$$\text{the leverage} \times 2\,\pi \times \text{T};$$

that is, by the product of the circumference of a circle whose radius is the leverage, into the number of turns and fractions of a turn made by the rotating body.

The distance thus found being multiplied by the resistance overcome, gives the work performed; that is to say,

The work performed

= the resistance × the leverage × 2 π × T:

But the product of the resistance into the leverage is what is called the *moment* of the resistance, and the product 2 π T is called the *angular motion* of the rotating body; consequently,

The work performed

= the moment of the resistance × the angular motion.

The mode of computing the work indicated by this last equation is often more convenient than the direct mode already explained in Article 391.

The angular motion 2 π T of a body during some definite unit of time, as a second or a minute, is called its *angular velocity;* that is to say, *angular velocity is the product of the turns and fractions of a turn made in an unit of time into the ratio of the circumference of a circle to its radius.* Hence it appears that

The rate of work

= the moment of the resistance × the angular velocity.

395. **Work in Terms of Pressure and Volume.**—If the resistance overcome be a pressure uniformly distributed over an area, as when a piston drives a fluid before it, then the amount of that resistance is equal to the intensity of the pressure, expressed in units of force on each unit of area (for example, in pounds on the square inch, or pounds on the square foot) multiplied by the area of the surface at which the pressure acts, if that area is perpendicular to the direction of the motion; or, if not, then by the projection of that area on a plane perpendicular to the direction of motion. In practice, when the *area of a piston* is spoken of, it is always understood to mean the projection above mentioned.

Now, when a plane area is multiplied into the distance through which it moves in a direction perpendicular to itself, if its motion is straight, or into the distance through which its centre of gravity moves, if its motion is curved, the product is the *volume of the space traversed* by the piston.

Hence the work performed by a piston in driving a fluid before it, or by a fluid in driving a piston before it, may be expressed in either of the following ways:—

Resistance × *distance traversed*
= *intensity of pressure* × *area* × *distance traversed;*
= *intensity of pressure* × *volume traversed.*

In order to compute the work in foot-pounds, if the pressure is stated in pounds on the square foot, the area should be stated in square feet, and the volume in cubic feet; if the pressure is stated in pounds on the square inch, the area should be stated in square inches, and the volume in units, each of which is a prism of one foot in length, and one square inch in area; that is, of $\frac{1}{144}$ of a cubic foot in volume.

396. **Algebraical Expressions for Work.**—To express the results of the preceding articles in algebraical symbols, let

s denote the distance in feet through which a resistance is overcome in a given time;

R, the amount of the resistance overcome in pounds.

Also, supposing the resistance to be overcome by a piece which turns about an axis, let

T be the number of turns and fractions of a turn made in the given time, and $i = 2\pi T = 6{\cdot}2832\, T$ the angular motion in the given time; and let

l be the leverage of the resistance; that is, the perpendicular distance of the line along which it acts from the axis of motion; so that $s = i\,l$, and $R\,l$ is the statical moment of the resistance. Supposing the resistance to be a pressure, exerted between a piston and a fluid, let A be the area or projected area of a piston, and p the intensity of the pressure in pounds per unit of area.

Then the following expressions all give quantities of work in the given time in foot-pounds:—

$$R\,s;\ i\,R\,l;\ p\,A\,s;\ i\,p\,A\,l.$$

The last of these expressions is applicable to a piston turning on an axis, for which l denotes the distance from the axis to the centre of gravity of the area A.

397. **Work against an Oblique Force.**—The resistance directly due to a force which acts against a moving body in a direction oblique to that in which the body moves, is found by resolving that force into two components, one at right angles to the direction of motion, which may be called a *lateral force*, and which must be balanced by an equal and opposite lateral force, unless it takes effect by altering the direction of the body's motion, and the other component directly opposed to the body's motion, which is the

resistance required. That resolution is effected by means of the well-known principle of the parallelogram of forces as follows:—

In fig. 147, let A represent the point at which a resistance is overcome, A B the direction in which that point is moving, and let $\overline{AF}$ be a line whose direction and length represent the direction and magnitude of a force obliquely opposed to the motion of A.

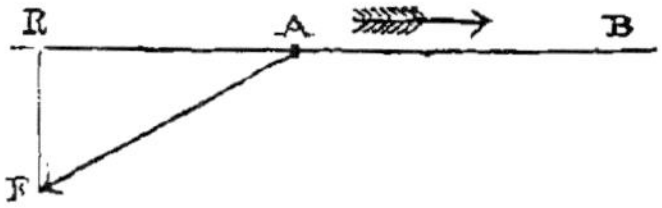

Fig. 147.

From F upon B A produced, let fall the perpendicular $\overline{FR}$; the length of that perpendicular will represent the magnitude of the lateral component of the oblique force, and the length $\overline{AR}$ will represent the direct component or resistance.

The work done against an oblique resisting force may also be calculated by resolving the motion into a direct component in the line of action of the force, and a transverse component, and multiplying the whole force by the direct component of the motion.

398. **Summation of Quantities of Work.**—In every machine, resistances are overcome during the same interval of time, by different moving pieces, and at different points in the same moving piece; and the whole work performed during the given interval is found by adding together the several products of the resistances into the respective distances through which they are simultaneously overcome. It is convenient, in algebraical symbols, to denote the result of that summation by the symbol—

$$\Sigma \cdot R s; \quad \ldots\ldots\ldots\ldots\ldots\ldots\ldots\ldots (1.)$$

in which Σ denotes the operation of taking the sum of a set of quantities of the kind denoted by the symbols to which it is prefixed.

When the resistances are overcome by pieces turning upon axes, the above sum may be expressed in the form—

$$\Sigma \; i R l; \quad \ldots\ldots\ldots\ldots\ldots\ldots\ldots\ldots (2.)$$

and so of other modes of expressing quantities of work.

The following are particular cases of the summation of quantities of work performed at different points:—

I. In a *shifting piece*, or one which has the kind of movement called *translation* only, the velocities of every point at a given instant are equal and parallel; hence, in a given interval of time, the motions of all the points are equal; and the work performed is to be found by multiplying the *sum of the resistances* into the motion as a common factor; an operation expressed algebraically thus—

$$s \Sigma R; \quad \ldots\ldots\ldots\ldots\ldots\ldots\ldots\ldots (3.)$$

II. For a *turning piece*, the angular motions of all the points during a given interval of time are equal; and the work performed is to be found by multiplying the *sum of the moments of the resistances relatively to the axis* into the angular motion as a common factor—an operation expressed algebraically thus—

$$i \Sigma \cdot \mathrm{R}\, l; \qquad (4.)$$

The sum denoted by $\Sigma \cdot \mathrm{R}\, l$ is the *total moment of resistance* of the piece in question.

III. In every *train of mechanism*, the *proportions* amongst the motions performed during a given interval of time by the several moving pieces, can be determined from the mode of connection of those pieces, independently of the absolute magnitudes of those motions, by the aid of the *Theory of Pure Mechanism*, Part II. This enables a calculation to be performed which is called *reducing the resistances to the driving point;* that is to say, determining the resistances, which, if they acted directly at the point where the motive power is applied to the machine, would require the same quantity of work to overcome them with the actual resistances.

Suppose, for example, that by the principles of pure mechanism it is found, that a certain point in a machine, where a resistance R is to be overcome, moves with a velocity bearing the ratio $n : 1$ to the velocity of the driving point. Then the work performed in overcoming that resistance will be the same as if a resistance n R were overcome directly at the driving point. If a similar calculation be made for each point in the machine where resistance is overcome, and the results added together, as the following symbol denotes:—

$$\Sigma \cdot n\, \mathrm{R}, \qquad (5.)$$

that sum is the *equivalent resistance at the driving point;* and if in a given interval of time the driving point moves through the distance s, then the work performed in that time is—

$$s\, \Sigma \cdot n\, \mathrm{R}. \qquad (6.)$$

The process above described is often applied to the steam engine, by reducing all the resistances overcome to equivalent resistances acting directly against the motion of the piston.

A similar method may be applied to the moments of resistances overcome by rotating pieces, so as to reduce them to *equivalent moments at the driving axle.* Thus, let a resistance R, with the leverage l, be overcome by a piece whose angular velocity of rotation bears the ratio $n : 1$ to that of the driving axle. Then the equivalent moment of resistance at the driving axle is $n\, \mathrm{R}\, l$; and if a similar calculation be made for each rotating piece in the machine which overcomes resistance, and the results added together, the sum—

$$\Sigma \cdot n\, \mathrm{R}\, l \qquad (7.)$$

is the total *equivalent moment of resistance* at the driving axle; and if in a given interval of time the driving axle turns through the arc i to radius unity, the work performed in that time is—

$$i \Sigma \cdot n R l \text{..........................} (8.)$$

IV. *Centre of gravity.*—The work performed in lifting a body is *the product of the weight of the body into the height through which its centre of gravity is lifted.*

If a machine lifts the centres of gravity of several bodies at once to heights either the same or different, the whole quantity of work performed in so doing is the sum of the several products of the weights and heights; but that quantity can also be computed by *multiplying the sum of all the weights into the height through which their common centre of gravity is lifted.*

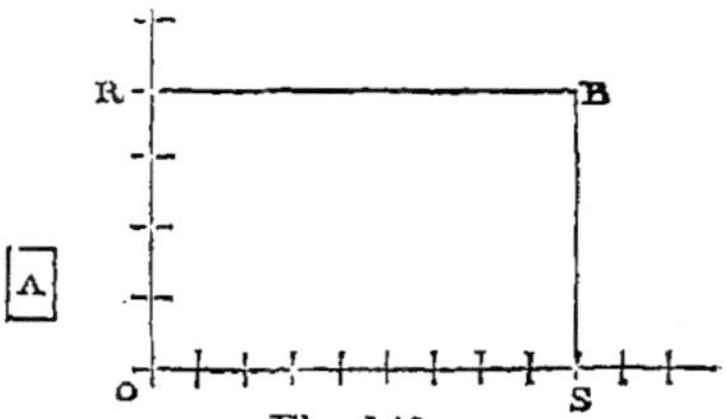

Fig. 148.

399. **Representation of Work by an Area.**—As a quantity of work is the product of two quantities, a force and a motion, it may be represented by the area of a plane figure, which is the product of two dimensions. Let the base of the rectangle A, fig. 148, represent *one foot* of motion, and its height *one pound* of resistance; then will its area represent one foot-pound of work.

In the larger rectangle, let the base $\overline{OS}$ represent a certain motion s, on the same scale with the base of the unit-area A; and let the height $\overline{OR}$ represent a certain resistance R, on the same scale with the height of the unit-area A; then will the number of times that the rectangle $\overline{OS} \cdot \overline{OR}$ contains the unit-rectangle A, express the number of foot-pounds in the quantity of work R s, which is performed in overcoming the resistance R through the distance s.

400. **Work against Varying Resistance.**—In fig. 149, let distances, as before, be represented by lengths measured along the base line O X of the figure; and let the magnitudes of the resistance overcome at each instant be represented by the lengths of ordinates drawn perpendicular to O X, and parallel to O Y:—For example, when the working body has moved through the distance represented by $\overline{OS}$, let the resistance be represented by the ordinate $\overline{SR}$.

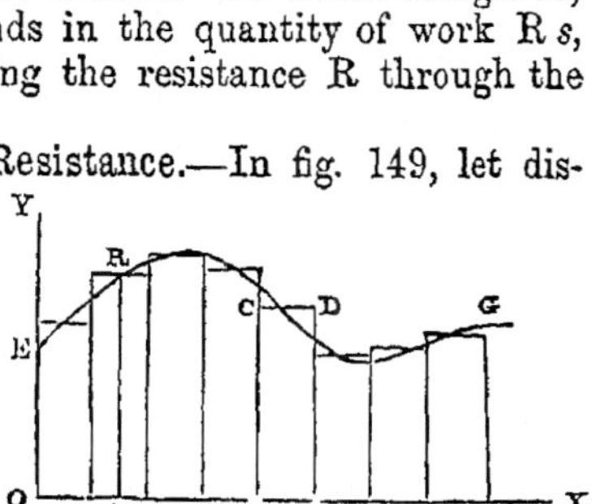

Fig. 149.

If the resistance were constant, the summits of those ordinates would lie in a straight line parallel to O X, like R B in fig. 148; but if the resistance varies continuously as the motion goes on, the summits of the ordinates will lie in a line, straight or curved, such as that marked E R G, fig. 149, which is not parallel to O X.

The values of the resistance at each instant being represented by the ordinates of a given line E R G, let it now be required to determine the work performed against that resistance during a motion represented by $\overline{\mathrm{O\,F}} = s$.

Suppose the area O E G F to be divided into bands by a series of parallel ordinates, such as A C and B D, and between the upper ends of those ordinates let a series of short lines, such as C D, be drawn parallel to O X, so as to form a stepped or serrated outline, consisting of lines parallel to O X and O Y alternately, and *approximating* to the given continuous line E G.

Now conceive the resistance, instead of varying continuously, to remain constant during each of the series of divisions into which the motion is divided by the parallel ordinates, and to change abruptly at the instants between those divisions, being represented for each division by the height of the rectangle which stands on that division: for example, during the division of the motion represented by A B, let the resistance be represented by A C, and so for other divisions.

Then the work performed during the division of the motion represented by $\overline{\mathrm{A\,B}}$, on the supposition of alternate constancy and abrupt variation of the resistance, is represented by the rectangle $\overline{\mathrm{A\,B}} \cdot \overline{\mathrm{A\,C}}$; and the whole work performed, on the same supposition during the whole motion $\overline{\mathrm{O\,F}}$, is represented by the sum of all the rectangles lying between the parallel ordinates; and inasmuch as the supposed mode of variation of the resistance represented by the stepped outline of those rectangles is an approximation to the real mode of variation represented by the continuous line E G, and is a closer approximation the closer and the more numerous the parallel ordinates are, so the sum of the rectangles is an approximation to the exact representation of the work performed against the continuously varying resistance, and is a closer approximation the closer and more numerous the ordinates are, and by making the ordinates numerous and close enough, can be made to differ from the exact representation by an amount less than any given difference.

But the sum of those rectangles is also an approximation to the area O E G F, bounded above by the continuous line E G, and is a closer approximation the closer and the more numerous the ordinates are, and by making the ordinates numerous and close enough, can be made to differ from the area O E G F by an amount less than any given difference.

Therefore the area O E G F, *bounded by the straight line* O F, *which represents the motion, by the line* E G, *whose ordinates represent the values of the resistance, and by the two ordinates* O E *and* F G, *represents exactly the work performed.* (See Article 34, page 17).

The MEAN RESISTANCE during the motion is found by dividing the area O E G F by the motion $\overline{OF}$.

401. **Useful Work and Lost Work.**—The useful work of a machine is that which is performed in effecting the purpose for which the machine is designed. The lost work is that which is performed in producing effects foreign to that purpose. The resistances overcome in performing those two kinds of work are called respectively *useful resistance* and *prejudicial resistance.*

The useful work and the lost work of a machine together make up its *total* or *gross work.*

In a pumping engine, for example, the useful work in a given time is the product of the weight of water lifted in that time into the height to which it is lifted : the lost work is that performed in overcoming the friction of the water in the pumps and pipes, the friction of the plungers, pistons, valves, and mechanism, and the resistance of the air pump and other parts of the engine.

For example, the useful work of a marine steam engine in a given time is the product of the resistance opposed by the water to the motion of the ship, into the distance through which she moves: the lost work is that performed in overcoming the resistance of the water to the motion of the propeller through it, the friction of the mechanism, and the other resistances of the engine, and in raising the temperature of the condensation water, of the gases which escape by the chimney, and of adjoining bodies.

There are some cases, such as those of muscular power and of windmills, in which the useful work of a prime mover can be determined, but not the lost work.

402. **The Work Performed against Friction** in a given time, between a pair of rubbing surfaces, is the product of that friction into the distance through which one surface slides over the other.

When the motion of one surface relatively to the other consists in rotation about an axis, the work performed may also be calculated by multiplying the relative *angular motion* of the surfaces to radius unity into the *moment of friction;* that is, the product of the friction into its leverage, which is the mean distance of the rubbing surfaces from the axis.

For a cylindrical journal, the leverage of the friction is simply the radius of the journal.

For a *flat pivot,* the leverage is two-thirds of the radius of the pivot.

For a *collar*, let r and r' be the inner and outer radii; then the leverage of the friction is

$$\frac{2}{3} \cdot \frac{r^3 - r'^3}{r^2 - r'^2} \quad \text{.......................}(1.)$$

In the *cup and ball* pivot, the end of the shaft, and the step on which it presses, present two recesses facing each other, into which are fitted two shallow cups of steel or hard bronze. Between the concave spherical surfaces of those cups is placed a steel ball, being either a complete sphere, or a lens having convex surfaces of a somewhat less radius than the concave surfaces of the cups. The moment of friction of this pivot is at first almost inappreciable, from the extreme smallness of the radius of the circles of contact of the ball and cups; but as they wear, that radius and the moment of friction increase.

By the rolling of two surfaces over each other without sliding, a resistance is caused, which is called sometimes "rolling friction," but more correctly *rolling resistance*. It is of the nature of a *couple* resisting rotation; its *moment* is found by multiplying the normal pressure between the rolling surfaces by an *arm* whose length depends on the nature of the rolling surfaces; and the work lost in an unit of time in overcoming it is the product of its moment by the *angular velocity* of the rolling surfaces relatively to each other. The following are approximate values of the arm in *decimals of a foot*:—

Oak upon oak,.........................	0·006	(Coulomb).
Lignum-vitæ on oak,..................	0·004	,,
Cast-iron on cast-iron,.................	0·002	(Tredgold).

The work lost in friction produces HEAT in the proportion of one British thermal unit, being so much heat as raises the temperature of a pound of water 1° of Fahr., for every 772 foot-pounds of lost work.

The heat produced by friction, when moderate in amount, is useful in softening and liquefying unguents; but when excessive it is prejudicial by decomposing the unguents, and sometimes even by softening the metal of the bearings, and raising their temperature so high as to set fire to neighbouring combustible matters.

Excessive heating is prevented by a constant and copious supply of a good unguent. When the velocity of rubbing is about four or five feet per second, the elevation of temperature is found to be, with good fatty and soapy unguents, 40° to 50° Fahr., with good mineral unguents, about 30°. The effect of friction upon the efficiency of machines will be considered at the end of this Part.

403. **Work of Acceleration.**—In order that the velocity of a body's motion may be changed, it must be acted upon by some other body with a force in the direction of the change of velocity, which force is proportional directly to the change of velocity, and to the mass of the body acted upon, and inversely to the time occupied in producing the change. If the change is an acceleration or increase of velocity, let the first body be called the *driven body*, and the second

the *driving body*. Then the force must act upon the driven body in the direction of its motion. Every force being a pair of equal and opposite actions between a pair of bodies, the same force which accelerates the driven body is a *resistance* as respects the driving body.

For example, during the commencement of the stroke of the piston of a steam engine, the velocity of the piston and of its rod is accelerated; and that acceleration is produced by a certain part of the pressure between the steam and the piston, being the excess of that pressure above the whole resistance which the piston has to overcome. The piston and its rod constitute the driven body; the steam is the driving body; and the same part of the pressure which accelerates the piston, acts as a *resistance* to the motion of the steam, in addition to the resistance which would have to be overcome if the velocity of the piston were uniform.

The resistance due to acceleration is computed in the following manner:—It is known by experiment, that if a body near the earth's surface is accelerated by the attraction of the earth,—that is, by its own weight, or by a force equal to its own weight, its velocity goes on continually increasing very nearly at the rate of 32·2 *feet per second of additional velocity, for each second during which the force acts.* This quantity varies in different latitudes, and at different elevations, but the value just given is near enough to the truth for purposes of mechanical engineering. For brevity's sake, it is usually denoted by the symbol g; so that, if at a given instant the velocity of a body is v_1 feet per second, and if its own weight, or an equal force, acts freely on it in the direction of its motion for t seconds, its velocity at the end of that time will have increased to

$$v_2 = v_1 + g\,t \text{..............................(1.)}$$

If the acceleration be at any different rate per second, *the force necessary to produce that acceleration, being the resistance on the driving body due to the acceleration of the driven body, bears the same proportion to the driven body's weight which the actual rate of acceleration bears to the rate of acceleration produced by gravity acting freely.* (In mètres per second, $g = 9·81$ nearly.)

To express this by symbols, let the weight of the driven body be denoted by W. Let its velocity at a given instant be v_1 feet per second; and let that velocity increase at an uniform rate, so that at an instant t seconds later, it is v_2 feet per second.

Let f denote the rate of acceleration; then

$$f = \frac{v_2 - v_1}{t}; \text{..............................(2.)}$$

and the force R necessary to produce it will be given by the proportion,

$$g : f :: W : R;$$

that is to say,

$$R = \frac{f W}{g} = \frac{W (v_2 - v_1)}{g t} \quad \text{.......................(3.)}$$

The factor $\frac{W}{g}$, in the above expression, is called the MASS of the driven body; and being the same for the same body, in what place soever it may be, is held to represent the *quantity of matter* in the body. (See Article 195, page 117.)

The product $\frac{W v}{g}$ of the mass of a body into its velocity at any instant, is called its MOMENTUM; so that the resistance due to a given acceleration is equal to *the increase of momentum divided by the time which that increase occupies.*

If the product of a force by which a body is accelerated, equal and opposite to the resistance due to acceleration, into the time during which it acts, be called IMPULSE, the same principle may be otherwise stated by saying, that the *increase of momentum is equal to the impulse by which it is caused.*

If the rate of acceleration is not constant, but variable, the force R varies along with it. In this case, the value, at a given instant of the rate of acceleration, is represented by $f = \frac{d v}{d t}$, and the corresponding value of the force is

$$R = \frac{f W}{g} = \frac{W}{g} \cdot \frac{d v}{d t} \quad \text{.......................(4.)}$$

The WORK PERFORMED in accelerating a body is the product of the resistance due to the rate of acceleration into the distance moved through by the driven body while the acceleration is going on. The resistance is equal to the mass of the body, multiplied by the increase of velocity, and divided by the time which that increase occupies. The distance moved through is the product of the mean velocity into the same time. Therefore, the work performed is equal to the mass of the body multiplied by the increase of the velocity, and by the mean velocity; that is, *to the mass of the body, multiplied by the increase of the half-square of its velocity.*

To express this by symbols, in the case of an uniform rate of acceleration, let s denote the distance moved through by the driven body during the acceleration; then

$$s = \frac{v_2 + v_1}{2} t; \quad \text{.............................(5.)}$$

which being multiplied by Equation 3, gives for the work of acceleration,

$$R s = \frac{W}{g} \cdot \frac{v_2 - v_1}{t} \cdot \frac{v_2 + v_1}{2} \cdot t = \frac{W}{g} \cdot \frac{v_2^2 - v_1^2}{2} \text{.............(6.)}$$

In the case of a variable rate of acceleration, let v denote the mean velocity, and $d\,s$ the distance moved through, in an interval of time $d\,t$ so short that the increase of velocity $d\,v$ is indefinitely small compared with the mean velocity. Then

$$d\,s = v\,d\,t; \text{...........................(7.)}$$

which being multiplied by Equation 4, gives for the work of acceleration during the interval $d\,t$,

$$R\,d\,s = \frac{W}{g} \cdot \frac{d\,v}{d\,t} \cdot v\,d\,t$$
$$= \frac{W}{g} \cdot v\,d\,v; \text{..........................(8.)}$$

and the *integration* of this expression (see Article 29) gives for the work of acceleration during a finite interval,

$$\int R\,d\,s = \frac{W}{g} \int v\,d\,v = \frac{W}{g} \cdot \frac{v_2^2 - v_1^2}{2} \text{..............(9.)}$$

being the same with the result already arrived at in Equation 6.

From Equation 9 it appears that *the work performed in producing a given acceleration depends on the initial and final velocities,* v_1 *and* v_2*, and not on the intermediate changes of velocity.*

If a body falls freely under the action of gravity from a state of rest through a height h, so that its initial velocity is 0, and its final velocity v, the work of acceleration performed by the earth on the body is simply the product $W\,h$ of the weight of the body into the height of fall. Comparing this with Equation 6, we find—

$$h = \frac{v^2}{2\,g} \text{...............................(10.)}$$

This quantity is called the *height, or fall, due to the velocity* v; and from Equations 6 and 9 it appears that *the work performed in producing a given acceleration is the same with that performed in lifting the driven body through the difference of the heights due to its initial and final velocities.*

If work of acceleration is performed by a prime mover upon bodies which neither form part of the prime mover itself, nor of the machines which it is intended to drive, that work is lost; as when a marine engine performs work of acceleration on the water that is struck by the propeller.

Work of acceleration performed on the moving pieces of the prime mover itself, or of the machinery driven by it, is not necessarily lost, as will afterwards appear. (Article 413.)

404. **Summation of Work of Acceleration.**—If several pieces of a machine have their velocities increased at the same time, the work performed in accelerating them is the sum of the several quantities of work due to the acceleration of the respective pieces; a result expressed in symbols by

$$\Sigma \left\{ \frac{W}{g} \cdot \frac{v_2^2 - v_1^2}{2} \right\} \quad \text{.......................(1.)}$$

The process of finding that sum is facilitated and abridged in certain cases by special methods.

I. *Accelerated Rotation.*—Let a denote the angular velocity of a solid body rotating about a fixed axis;—that is, as explained in Article 87, the velocity of a point in the body whose radius-vector, or distance from the axis, is unity.

Then the velocity of a particle whose distance from the axis is r is

$$v = a\,r; \quad \text{..............................(2.)}$$

and if in a given interval of time the angular velocity is accelerated from the value a_1, to the value a_2, the increase of the velocity of the particle in question is

$$v_2 - v_1 = r\,(a_2 - a_1) \quad \text{......................(3.)}$$

Let w denote the weight, and $\frac{w}{g}$ the mass of the particle in question. Then the work performed in accelerating it, being equal to the product of its mass into the increase of the half-square of its velocity, is also equal to *the product of its mass into the square of its radius-vector, and into the increase of the half-square of the angular velocity;* that is to say, in symbols,

$$\frac{w}{g} \cdot \frac{v_2^2 - v_1^2}{2} = \frac{w\,r^2}{g} \cdot \frac{a_2^2 - a_1^2}{2} \quad \text{.....................(4.)}$$

To find the work of acceleration for the whole body, it is to be conceived to be divided into small particles, whose velocities at any given instant, and also their accelerations, are proportional to their distances from the axis; then the work of acceleration is to be found for each particle, and the results added together. But in the sum so obtained, the increase of the half-square of the angular velocity is a common factor, having the same value for each particle of the body; and the rate of acceleration produced by gravity, $g = 32{\cdot}2$ is a common divisor. It is therefore sufficient *to add together the products of the weight of each particle* (w) *into the square of its radius-vector* (r^2), *and to multiply the sum so obtained* ($\Sigma \cdot w r^2$) by *the increase of the half-square of the angular velocity* $\left(\frac{1}{2}(a_2^2 - a_1^2)\right)$, *and divide by the rate of acceleration due to gravity* (g). The result, viz.:—

$$\Sigma \left\{ \frac{w}{g} \cdot \frac{v_2^2 - v_1^2}{2} \right\} = \frac{a_2^2 - a_1^2}{2 g} \cdot \Sigma w r^2 \text{.............}(5.)$$

is the work of acceleration sought. In fact, the sum $\Sigma w r^2$ is *the weight of a body, which, if concentrated at the distance unity from the axis of rotation, would require the same work to produce a given increase of angular velocity which the actual body requires.*

405. **Reduced Inertia.**—If in a certain machine, a moving piece whose weight is W has a velocity always bearing the ratio $n : 1$ to the velocity of the driving point, it is evident that when the driving point undergoes a given acceleration, the work performed in producing the corresponding acceleration in the piece in question is the same with that which would have been required if a weight n^2 W had been concentrated at the driving point, the work performed in producing the acceleration depending on the *square* of the velocity.

If a similar calculation be performed for each moving piece in the machine, and the results added together, the sum

$$\Sigma \cdot n^2 W \text{............................}(1.)$$

gives the weight which, being concentrated at the driving point, would require the same work for a given acceleration of the driving point that the actual machine requires; so that if v_1 is the initial, and v_2 the final velocity of the driving point, the work of acceleration of the whole machine is

$$\frac{v_2^2 - v_1^2}{2 g} \cdot \Sigma \cdot n^2 W \text{........................}(2.)$$

This operation may be called *the reduction of the inertia to the driving point.* Mr. Moseley, by whom it was first introduced into the theory of machines, calls the expression (1.) the "*coefficient of steadiness.*"

In finding the reduced inertia of a machine, the mass of each rotating piece is to be treated as if concentrated at a distance from its axis equal to its radius of gyration ϱ; so that if v represents the velocity of the driving point at any instant, and a the corresponding angular velocity of the rotating piece in question, we are to make

$$n^2 = \frac{a^2 \varrho^2}{v^2} \text{.................................}(3.)$$

in performing the calculation expressed by the formula (1.)

406. **Summary of Various Kinds of Work.**—In order to present at one view the symbolical expression of the various modes of performing work described in the preceding articles, let it be supposed that in a certain interval of time $d\,t$ the driving point of a machine moves through the distance $d\,s$; that during the same time its centre of gravity is elevated through the height $d\,h$; that resistances, any one of which is represented by R, are overcome at points, the respective ratios of whose velocities to that of the driving point are denoted by n; that the weight of any piece of the mechanism is W, and that n' denotes the ratio of its velocity (or if it rotates, the ratio of the velocity of the end of its radius of gyration) to the velocity of the driving point; and that the driving point, whose mean velocity is $v = \frac{d\,s}{d\,t}$, undergoes the acceleration $d\,v$. Then the *whole work performed* during the interval in question is

$$d\,h \cdot \Sigma\,\mathrm{W} + d\,s \cdot \Sigma\,n\,\mathrm{R} + \frac{v\,d\,v}{g} \cdot \Sigma\,n'^2\,\mathrm{W} \ldots\ldots\ldots\ldots (1.)$$

The *mean total resistance, reduced to the driving point,* may be computed by dividing the above expression by the motion of the driving point $d\,s = v\,d\,t$, giving the following result:—

$$\frac{d\,h}{d\,s} \cdot \Sigma\,\mathrm{W} + \Sigma\,n\,\mathrm{R} + \frac{d\,v}{g\,d\,t} \cdot \Sigma\,n'^2\,\mathrm{W} \ldots\ldots\ldots\ldots (2.)$$

Section 2.—Of Energy, Power, and Efficiency.

407. **Condition of Uniform Speed.**—According to the first law of motion, in order that a body may move uniformly, the forces applied to it, if any, must balance each other; and the same principle holds for a machine consisting of any number of bodies.

When the *direction* of a body's motion varies, but not the *velocity*, the lateral force required to produce the change of direction depends on the principles set forth in Article 335; but the condition of balance still holds for the forces which act *along* the direction of the body's motion, that is, for the *efforts* and *resistances*; so that, whether for a single body or for a machine, the condition of *uniform velocity* is, that the *efforts shall balance the resistances.*

In a machine, this condition must be fulfilled for each of the single moving pieces of which it consists.

It also follows, from the principles of statics, that in any body, system, or machine, that condition is fulfilled when *the sum of the products of the efforts into the velocities of their respective points of action is equal to the sum of the products of the resistances into the velocities of the points where they are overcome.*

Thus, let v be the velocity of a *driving point*, or point where an effort P is applied; v' the velocity of a *working point*, or point where a resistance R is overcome; the condition of uniform velocity for any body, system, or machine is

$$\Sigma \cdot \mathrm{P}\, v = \Sigma \cdot \mathrm{R}\, v' \qquad (1.)$$

If there be only one driving point, or if the velocities of all the driving points be alike, then P being the total effort, the single product $\mathrm{P}\, v$ may be put in in place of the sum $\Sigma \cdot \mathrm{P}\, v$; reducing the above equation to

$$\mathrm{P}\, v = \Sigma \cdot \mathrm{R}\, v' \qquad (2.)$$

Referring now to Article 398, let the machine be one in which the *comparative* or *proportionate* velocities of all the points at a given instant are known independently of their absolute velocities, from the construction of the machine; so that, for example, the velocity of the point where the resistance R is overcome bears to that of the driving point the ratio

$$\frac{v'}{v} = n;$$

then the condition of uniform speed may be thus expressed:—

$$\mathrm{P} = \Sigma \cdot n\, \mathrm{R}; \qquad (3.)$$

that is, *the total effort is equal to the sum of the resistances reduced to the driving point.*

408 **Energy—Potential Energy.**—*Energy* means *capacity for performing work*, and is expressed, like work, by the product of a force into a space.

The energy of an effort, sometimes called "*potential energy*" (to distinguish it from another form of energy to be referred to in Article 414), is the *product of the effort into the distance through which it is capable of acting.* Thus, if a weight of 100 pounds be placed at an elevation of 20 feet above the ground, or above the lowest plane to which the circumstances of the case admit of its descending, that weight is said to possess potential energy to the amount of $100 \times 20 = 2{,}000$ *foot-pounds;* which means, that in descending from its actual elevation to the lowest point of its course, the weight is *capable of performing work* to that amount.

To take another example, let there be a reservoir containing 10,000,000 gallons of water, in such a position that the centre of gravity of the mass of water in the reservoir is 100 feet above the lowest point to which it can be made to descend while overcoming resistance. Then as a gallon of water weighs 10 lbs., the weight of the store of water is 100,000,000 lbs., which being multiplied by the height through which that weight is capable of acting, 100 feet, gives 10,000,000,000 foot-pounds for the potential energy of the weight of the store of water.

409. **Equality of Energy Exerted and Work Performed, or the Conservation of Energy.**—When an effort actually does drive its point of application through a certain distance, energy to the amount of the product of the effort into that distance is said to be *exerted;* and the potential energy, or energy which remains *capable of being exerted,* is to that amount diminished.

When the energy is exerted in driving a machine at an uniform speed, it is *equal to the work performed.*

To express this algebraically, let t denote the time during which the energy is exerted, v the velocity of a driving point at which an effort P is applied, s the distance through which it is driven, v' the velocity of any working point at which a resistance R is overcome, s' the distance through which it is driven; then

$$s = v\,t;\ s' = v'\,t;$$

and multiplying Equation 1 of Article 407 by the time t, we obtain the following equation:—

$$\Sigma \cdot \mathrm{P}\,v\,t = \Sigma \cdot \mathrm{R}\,v'\,t = \Sigma \cdot \mathrm{P}\,s = \Sigma \cdot \mathrm{R}\,s'; \qquad (1.)$$

which expresses the equality of energy exerted, and work performed, for constant efforts and resistances.

When the efforts and resistances vary, it is sufficient to refer to Articles 400 and 29, to shew that the same principle is expressed as follows:—

$$\Sigma \int \mathrm{P}\,d\,s = \Sigma \int \mathrm{R}\,d\,s'; \qquad (2.)$$

where the symbol $\int$ expresses the operation of finding the work performed against a varying resistance, or the energy exerted by a varying effort, as the case may be; and the symbol Σ expresses the operation of adding together the quantities of energy exerted, or work performed, as the case may be, at different points of the machine.

410. **Various Factors of Energy.**—A quantity of energy, like a quantity of work, may be computed by multiplying either a force into a distance, or a statical moment into an angular motion, or the intensity of a pressure into a volume. These processes have already been explained in detail in Articles 394 and 395, pages 244 to 246.

411. **The Energy Exerted in Producing Acceleration** is equal to the work of acceleration, whose amount has been investigated in Articles 403 and 404, pages 252 to 257.

412. **The Accelerating Effort** by which a given increase of velocity in a given mass is produced, and which is exerted by the *driving body* against the *driven body,* is equal and opposite to the resistance due to acceleration which the driven body exerts against the driving body, and whose amount has been given in Articles

403 and 404. Referring, therefore, to Equations 4 and 8 of Article 403, we find the two following expressions, the first of which gives the accelerating effort required to produce a given acceleration $d\,v$ in a body whose weight is W, when the *time* $d\,t$ in which that acceleration is to be produced is given, and the second, the same accelerating effort, when the *distance* $d\,s = v\,d\,t$ in which the acceleration is to be produced is given:—

$$P = \frac{W}{g} \cdot \frac{d\,v}{d\,t} \quad \text{(1.)}$$

$$= \frac{W}{g} \cdot \frac{v\,d\,v}{d\,s} = \frac{W}{g} \cdot \frac{d\,(v^2)}{2\,d\,s} \quad \text{(2.)}$$

Referring next to Article 404, page 257, we find, from Equation 5, that the work of acceleration corresponding to an increase $d\,a$ in the angular velocity of a rotating body whose moment of inertia is I, is

$$\frac{I \cdot d\,(a^2)}{2\,g} = \frac{I\,a\,d\,a}{g}.$$

Let $d\,t$ be the *time*, and $d\,i = a\,d\,t$ the *angular motion* in which that acceleration is to be produced; let P be the accelerating effort, and l its *leverage*, or the perpendicular distance of its line of action from the axis; then, according as the time $d\,t$, or the angle $d\,i$, is given, we have the two following expressions for the *accelerating couple:*—

$$P\,l = \frac{I}{g} \cdot \frac{d\,a}{d\,t} \quad \text{(3.)}$$

$$= \frac{I}{g} \cdot \frac{a\,d\,a}{d\,i} = \frac{I}{g} \cdot \frac{d\,(a^2)}{2\,d\,i} \quad \text{(4.)}$$

Lastly, referring to Article 405, page 257, Equation 2, we find, that if a train of mechanism consists of various parts, and if W be the weight of any one of those parts, whose velocity v' bears to that of the driving point v the ratio $\frac{v'}{v} = n$, then the accelerating effort which must be applied to the driving point, in order that, during the interval $d\,t$, in which the driving point moves through the distance $d\,s = v\,d\,t$, that point may undergo the acceleration $d\,v$, and each weight W the corresponding acceleration $n\,d\,v$, is given by one or other of the two formulæ—

$$P = \frac{\Sigma\,n^2\,W}{g} \cdot \frac{d\,v}{d\,t} \quad \text{(5.)}$$

$$= \frac{\Sigma\,n^2\,W}{g} \cdot \frac{v\,d\,v}{d\,s} = \frac{\Sigma\,n^2\,W}{g} \cdot \frac{d\,(v^2)}{2\,d\,s} \quad \text{(6.)}$$

both of which are derived from the equation $P\,ds = P\,v \cdot dt = \frac{v \cdot dv}{g} \cdot \Sigma\, n^2 w$.

413. **Work During Retardation—Energy Stored and Restored.**—In order to cause a given retardation, or diminution of the velocity of a given body, in a given time, or while it traverses a given distance, resistance must be opposed to its motion equal to the effort which would be required to produce in the same time, or in the same distance, an acceleration equal to the retardation.

A moving body, therefore, while being retarded, *overcomes resistance* and *performs work;* and that work is equal to the energy exerted in producing an acceleration of the same body equal to the retardation.

It is for this reason that it has been stated, in Article 403, that the work performed in accelerating the speed of the moving pieces of a machine is not necessarily lost; for those moving pieces, by returning to their original speed, are capable of performing an equal amount of work in overcoming resistance; so that the performance of such work is not prevented, but only deferred. Hence energy exerted in acceleration is said to be *stored;* and when by a subsequent and equal retardation an equal amount of work is performed, that energy is said to be *restored.*

The algebraical expressions for the relations between a retarding resistance, and the retardation which it produces in a given body by acting during a given time or through a given space, are obtained from the equations of Article 412 simply by putting R, the symbol for a resistance, instead of P, the symbol for an effort, and $-dv$, the symbol for a retardation, instead of dv, the symbol for an acceleration.

414. **The Actual Energy** of a moving body is the work which it is capable of performing against a retarding resistance before being brought to rest, and is equal to the energy which must be exerted on the body to bring it from a state of rest to its actual velocity. The value of that quantity is the *product of the weight of the body into the height from which it must fall to acquire its actual velocity;* that is to say,

$$\frac{W v^2}{2g} \quad \text{..............................(1.)}$$

The total actual energy of a system of bodies, each moving with its own velocity, is denoted by

$$\frac{\Sigma \cdot W v^2}{2g}; \quad \text{..............................(2.)}$$

and when those bodies are the pieces of a machine, whose velocities bear definite ratios (any one of which is denoted by n) to the velocity of the driving point v, their total actual energy is

$$\frac{v^2}{2g} \cdot \Sigma n^2 W, \quad \text{(3.)}$$

being *the product of the reduced inertia* (or coefficient of steadiness, as Mr. Moseley calls it) *into the height due to the velocity of the driving point.*

The actual energy of a rotating body whose angular velocity is a, and moment of inertia $\Sigma W r^2 = I$, is

$$\frac{a^2 I}{2g}; \quad \text{(4.)}$$

that is, *the product of the moment of inertia into the height due to the velocity,* a, *of a point, whose distance from the axis of rotation is unity.*

When a given amount of energy is alternately stored and restored by alternate increase and diminution in the speed of a machine, the actual energy of the machine is alternately increased and diminished by that amount.

Actual energy, like motion, is *relative* only. That is to say, in computing the actual energy of a body, which is the capacity it possesses of performing work upon certain other bodies by *reason of its motion*, it is the motion *relatively to those other bodies* that is to be taken into account.

For example, if it be wished to determine how many turns a wheel of a locomotive engine, rotating with a given velocity, would make, before being stopped *by the friction of its bearings only*, supposing it lifted out of contact with the rails,—the actual energy of that wheel is to be taken *relatively to the frame of the engine* to which those bearings are fixed, and is simply the actual energy due to the rotation. But if the wheel be supposed to be detached from the engine, and it is inquired *how high it will ascend up a perfectly smooth inclined plane before being stopped by the attraction of the earth*, then its actual energy is to be taken *relatively to the earth*; that is to say, to the energy of rotation already mentioned, is to be added the energy due to the *translation* or forward motion of the wheel along with its axis.

415. **A Reciprocating Force** is a force which acts alternately as an effort and as an equal and opposite resistance, according to the direction of motion of the body. Such a force is the weight of a moving piece whose centre of gravity alternately rises and falls; and such is the elasticity of a perfectly elastic body. The work which a body performs in moving against a reciprocating force is employed in increasing its own potential energy, and is not lost by

the body; so that by the motion of a body alternately against and with a reciprocating force, energy is *stored* and *restored*, as well as by alternate acceleration and retardation.

Let Σ W denote the weight of the whole of the moving pieces of any machine, and h a height through which the common centre of gravity of them all is alternately raised and lowered. Then the quantity of energy—

$$h \Sigma \mathrm{W},$$

is stored while the centre of gravity is rising, and restored while it is falling.

These principles are illustrated by the action of the plungers of a single-acting pumping steam engine. The weight of those plungers acts as a resistance while they are being lifted by the pressure of the steam on the piston; and the same weight acts as effort when the plungers descend and drive before them the water with which the pump barrels have been filled. Thus the energy exerted by the steam on the piston is stored during the up-stroke of the plungers; and during their down-stroke the same amount of energy is restored, and employed in performing the work of raising water and overcoming its friction.

416. **Periodical Motion.**—If a body moves in such a manner that it periodically returns to its original velocity, then at the end of each period, the entire variation of its actual energy is nothing; and if, during any part of the period of motion, energy has been stored by acceleration of the body, the same quantity of energy exactly must have been during another part of the period restored by retardation of the body.

If the body also returns in the course of the same period to the same position relatively to all bodies which exert reciprocating forces on it—for example, if it returns periodically to the same elevation relatively to the earth's surface—any quantity of energy which has been stored during one part of the period by moving against reciprocating forces must have been exactly restored during another part of the period.

Hence *at the end of each period, the equality of energy and work, and the balance of* mean effort *and* mean resistance, *holds with respect to the driving effort and the resistances, exactly as if the speed were uniform and the reciprocating forces null;* and all the equations of Articles 407 and 409 are applicable to periodic motion, provided that in the equations of Article 407, and Equation 1 of Article 409, P, R, and v are held to denote the *mean values* of the efforts, resistances, and velocities,—that s and s' are held to denote spaces moved through in one or more *entire periods*,—and that in Equation 2 of Article 409, the integrations denoted by $\int$ be held to extend to one or more *entire periods.*

These principles are illustrated by the steam engine. The velocities of its moving parts are continually varying, and those of some of them, such as the piston, are periodically reversed in direction. But at the end of each period, called a *revolution*, or *double-stroke*, every part returns to its original position and velocity; so that the *equality of energy and work*, and the *equality of the mean effort to the mean resistance reduced to the driving point*,—that is, the equality of the mean effective pressure of the steam on the piston to the mean total resistance reduced to the piston—hold for one or any whole number of *complete revolutions*, exactly as for uniform speed.

It thus appears that (as stated at the commencement of this Part) there are two fundamentally different ways of considering a periodically moving machine, each of which must be employed in succession, in order to obtain a complete knowledge of its working.

"I. In the first place is considered the action of the machine during one or more whole periods, with a view to the determination of the relation between the mean resistances and mean efforts, and of the EFFICIENCY; that is the ratio which the *useful* part of its work bears to the whole expenditure of energy. The motion of every ordinary machine is either uniform or periodical.

"II. In the second place is to be considered the action of the machine during intervals of time less than its period, in order to determine the law of the periodic changes in the motions of the pieces of which the machine consists, and of the periodic or reciprocating forces by which such changes are produced."

417. **Starting and Stopping.**—The *starting* of a machine consists in setting it in motion from a state of rest, and bringing it up to its proper mean velocity. This operation requires the exertion, besides the energy required to overcome the mean resistance, of an additional quantity of energy equal to the actual energy of the machine when moving with its mean velocity, as found according to the principles of Article 414, page 262.

If, in order to *stop* a machine, the effort of the prime mover is simply suspended, the machine will continue to go until work has been performed in overcoming resistances equal to the actual energy due to the speed of the machine at the time of suspending the effort of the prime mover.

In order to diminish the time required by this operation, the resistance may be increased by means of the friction of a *brake*. Brakes will be further described in the sequel.

418. **The Efficiency** of a machine is a fraction expressing the ratio of the useful work to the whole work, which is equal to the energy expended. The COUNTER-EFFICIENCY is the reciprocal of the efficiency, and is the ratio in which the energy expended is greater than the useful work. The object of improvements in

machines is to bring their efficiency and counter-efficiency as near to unity as possible.

As to useful and lost work, see Article 401. The algebraical expression of the efficiency of a machine having uniform or periodical motion, is obtained by introducing the distinction between useful and lost work into the equations of the conservation of energy, Article 409. Thus, let P denote the mean effort at the driving point; s, the space described by it in a given interval of time, being a whole number of periods of revolutions; R_1, the mean useful resistance; s_1, the space through which it is overcome in the same interval; R_2, any one of the wasteful resistances; s_2, the space through which it is overcome; then

$$P s = R s_1 + \Sigma \cdot R_2 s_2; \quad (1.)$$

and the efficiency of the machine is expressed by

$$\frac{R_1 s_1}{P s} = \frac{R_1 s_1}{R_1 s_1 + \Sigma \cdot R_2 s_2} \quad (2.)$$

In many cases the lost work of a machine, $R_2 s_2$, consists of a constant part, and of a part bearing to the useful work a proportion depending in some definite manner on the sizes, figures, arrangement, and connection of the pieces of the train, on which also depends the constant part of the lost work. In such cases the whole energy expended and the efficiency of the machine are expressed by the equations

$$\left.\begin{array}{l} P s = (1 + A) R_1 s_1 + B; \\ \dfrac{R_1 s_1}{P s} = \dfrac{1}{1 + A + \dfrac{B}{R_1 s_1}} \end{array}\right\} \quad (3.)$$

and the first of these is the mathematical expression of what Mr. Moseley calls the "modulus" of a machine.

The useful work of an intermediate piece in a train of mechanism consists in driving the piece which follows it, and is less than the energy exerted upon it by the amount of the work lost in overcoming its own friction. Hence the efficiency of such an intermediate piece is the ratio of the work performed by it in driving the following piece, to the energy exerted on it by the preceding piece; and it is evident that *the efficiency of a machine is the product of the efficiencies of the series of moving pieces which transmit energy from the driving point to the working point.* The same principle applies to a train of *successive machines*, each driving that which follows it; and to counter-efficiency as well as to efficiency.

419. **Power and Effect—Horse Power.**—The *power* of a machine

is the energy exerted, and the *effect*, the useful work performed, in some interval of time of definite length, such as a second, a minute, an hour, or a day.

The unit of power called conventionally a *horse-power*, is 550 foot-pounds per second, or 33,000 foot-pounds per minute, or 1,980,000 foot-pounds per hour. The effect is equal to the power multiplied by the efficiency; and the power is equal to the effect multiplied by the counter-efficiency. The loss of power is the difference between the effect and the power. As to the French "Force de Cheval," see Article 392, page 244. It is equal to 0·9863 of a British horse-power; and a British horse-power is 1·0139 of a French force de cheval.

420. **General Equation.**—The following general equation presents at one view the principles of the action of machines, whether moving uniformly, periodically, or otherwise:—

$$\int \mathrm{P}\, d\, s = \Sigma \int \mathrm{R}\, d\, s' \pm h\, \Sigma\, \mathrm{W} + \Sigma \cdot \frac{\mathrm{W}\,(v_2^2 - v_1^2)}{2\, g};$$

where W is the weight of any moving piece of the machine;

h, when positive, the elevation, and when negative, the depression, which the common centre of gravity of all the moving pieces undergoes in the interval of time under consideration; v_1 the velocity at the beginning, and v_2 the velocity at the end, of the interval in question, with which a given particle of the machine of the weight W is moving;

g, the acceleration which gravity causes in a second, or 32·2 feet per second, or 9·81 mètres per second.

$\int \mathrm{R}\, d\, s'$, the work performed in overcoming any resistance during the interval in question;

$\int \mathrm{P}\, d\, s$, the energy exerted during the interval in question.

The second and third terms of the right-hand side, when positive, are *energy stored;* when negative, *energy restored.*

The principle represented by the equation is expressed in words as follows:—

The energy exerted, added to the energy restored, is equal to the energy stored added to the work performed.

421. **The Principle of Virtual Velocities,** when applied to the uniform motion of a machine, is expressed by Equation 3 of Article 407, already given in page 259; or in words as follows:—*The effort is equal to the sum of the resistances reduced to the driving point;* that is, each multiplied by the ratio of the velocity of its working point to the velocity of the driving point. The same principle, when applied to reciprocating forces and to re-actions due to varying speed, as well as to passive resistances, is expressed by

means of a modified form of the general equation of Article 420, obtained in the following manner:—Let n denote either the ratio borne at a given instant by the velocity of a given working point, where the resistance R is overcome, to the velocity of the driving point, or the mean value of that ratio during a given interval of time; let n'' denote the corresponding ratio for the vertical ascent or descent (according as it is positive or negative) of a moving piece whose weight is W; let n' denote the corresponding ratio for the mean velocity of a mass whose weight is W, undergoing acceleration or retardation, and $\frac{d\,v'}{d\,t}$ either the rate of acceleration of that mass, if the calculation relates to an instant, or the mean value of that rate, if to a finite interval of time. Then the effort at the instant, or the mean effort during the given interval, as the case may be, is given by the following equation:—

$$\mathrm{P} = \Sigma \cdot n\,\mathrm{R} + \Sigma \cdot n''\,\mathrm{W} + \Sigma \cdot \frac{n'\,\mathrm{W}\,d\,v'}{g\,d\,t}.$$

If the ratio n', which the velocity of the mass W bears to that of the driving point, is constant, we may put $\frac{d\,v'}{d\,t} = \frac{n'\,d\,v}{d\,t}$, where $\frac{d\,v}{d\,t}$ denotes the rate of acceleration of the driving point; and then the third term of the foregoing expression becomes $\frac{d\,v}{g\,d\,t}\,\Sigma \cdot n'^2\,\mathrm{W}$, as in formula 2 of Article 406, page 258.

422. **Forces in the Mechanical Powers, Neglecting Friction—Purchase.**—The mechanical powers, considered as means of modifying motion only, have been considered in Section 6, Part II., pages 107 to 110. When friction is neglected, any one of the mechanical powers may be regarded as *an uniformly-moving simple machine, in which one effort balances one resistance;* and in which, consequently, according to the principle of virtual velocities, or of the equality of energy exerted and work done, *the effort and resistance are to each other inversely as the velocities along their lines of action of the points where they are applied.*

In the older writings on mechanics, the effort is called the *power*, and the resistance the *weight;* but it is desirable to avoid the use of the word "power" in this sense, because of its being very commonly used in a different sense—viz., the rate at which energy is exerted by a prime mover; and the substitution of "resistance" for "weight" is made in order to express the fact, that the principle just stated applies to the overcoming of all sorts of resistance, and not to the lifting of weights only.

The weight of the moving piece itself in a mechanical power may either be wholly supported at the bearing, if the piece is balanced; or if not, it is to be regarded as divided into two parallel components, one supported directly at the bearing, and the other being included in the effort or in the resistance, as the case may be.

The relation between the effort and the resistance in any mechanical power may be deduced from the principles of statics; viz.:—In the case of the LEVER (including the *wheel* and *axle*), from the balance of couples of equal and opposite moments; in the case of the INCLINED PLANE (including the *wedge* and the *screw*), from the parallelogram of forces; and in the case of the pulley, from the composition of parallel forces. The principle of virtual velocities, however, is more convenient in calculation.

The *total load* in a mechanical power is the resultant of the effort, the resistance, the lateral components of the forces acting at the driving and working points, and the weight directly carried at the bearings; and it is equal and directly opposed to the re-action of the bearings or supports of the machine.

By the *purchase* of a mechanical power is to be understood the ratio borne by the resistance to the effort, which is equal to the ratio borne by the velocity of the driving point to that of the working point. This term has already been employed in connection with the pulley.

The following are the results of the principle of virtual velocities, as applied to determine the purchase in the several mechanical powers:—

I. LEVER.—The effort and resistance are to each other in the inverse ratio of the perpendicular distances of their lines of action from the axis of rotation or fulcrum; so that the *purchase* is the ratio which the perpendicular distance of the effort from the axis bears to the perpendicular distance of the resistance from the axis.

Under the head of the lever may be comprehended all turning or rocking primary pieces in mechanism which are connected with their drivers and followers by linkwork.

II. WHEEL AND AXLE.—The purchase is the same as in the case of the lever; and the perpendicular distances of the lines of action of the effort and of the resistance from the axis are the radii of the pitch-circles of the wheel and of the axle respectively.

Under the head of the wheel and axle may be comprehended all turning or rocking primary pieces in mechanism which are connected with their drivers and followers by means of rolling contact, of teeth, or of bands. By the "wheel" is to be understood the pitch-cylinder of that part of the piece which is driven; and by the "axle," the pitch-cylinder of that part of the piece which drives.

III. INCLINED PLANE, and IV. WEDGE.—Here the purchase, or ratio of the resistance to the effort, is the ratio borne by the whole velocity of the sliding body (represented by B C in fig. 76E, and C *c* in fig. 76F, page 109) to that component of the velocity (represented by B D in fig. 76E, and C *e* in fig. 76F, page 109) which is directly opposed to the resistance: it being understood that the effort is exerted in the direction of motion of the sliding body.

The term *inclined plane* may be used when the resistance to the motion of a body that slides along a guiding surface consists of its own weight, or of a force applied to a point in it by means of a link; and the term *wedge*, when that resistance consists of a pressure applied to a plane surface of the moving body, oblique to its direction of motion.

V. SCREW. Let the resistance (R) to the motion of a screw be a force acting along its axis, and directly opposed to its advance; and let the effort (P) which drives the screw be applied to a point rigidly attached to the screw, and at the distance r from the axis, and be exerted in the direction of motion of that point. Then, while the screw makes one revolution, the working point advances against the resistance through a distance equal to the pitch (p); and at the same time the driving point moves in its helical path through the distance $\sqrt{(4\pi^2 r^2 + p^2)}$; therefore the purchase of the screw, neglecting friction, is expressed as follows:—

$$\frac{R}{P} = \frac{\sqrt{4\pi^2 r^2 + p^2}}{p}$$

$$= \frac{\text{length of one coil of path of driving point}}{\text{pitch}}.$$

VI. PULLEY.—In the pulley without friction, the purchase is the ratio borne by the resistance which opposes the advance of the running block to the effort exerted on the hauling part of the rope; and it is expressed by the number of plies of rope by which the running block is connected with the fixed block.

VII. The HYDRAULIC PRESS, when friction is neglected, may be included amongst the mechanical powers, agreeably to the definition of them given at the beginning of this Article. By the resistance is to be understood the force which opposes the outward motion of the press-plunger; and by the effort, the force which drives inward the pump-plunger. The intensity of the pressure exerted between each of the two plungers and the fluid is the same; therefore the amount of the pressure exerted between each plunger and the fluid is proportional to the area of that plunger; so that the purchase of the hydraulic press is expressed as follows:—

$$\frac{R}{P} = \frac{\text{transverse area of press-plunger}}{\text{transverse area of pump-plunger}};$$

and this is the reciprocal of the ratio of the velocities of those plungers, as already shewn in Article 185, page 110.

The purchase of a train of mechanical powers is the product of the purchases of the several elementary parts of that train.

The object of producing a purchase expressed by a number greater than unity is, to enable a resistance to be overcome by means of an effort smaller than itself, but acting through a greater distance; and the use of such a purchase is found chiefly in machines driven by muscular power, because of the effort being limited in amount.

SECTION 3.—OF DYNAMOMETERS.

423. **Dynamometers** are instruments for measuring and recording the energy exerted and work performed by machines. They may be classed as follows:—

I. Instruments which merely *indicate the force* exerted between a driving body and a driven body, leaving the *distance* through which that force is exerted to be observed independently.

II. Instruments which *record* at once the *force, motion,* and *work* of a machine, by drawing a line, straight or curved, as the case may be, whose abscissæ represent the distances moved through, its ordinates the resistances overcome, and its area the work performed (as in fig. 149, page 249).

A dynamometer of this class consists essentially of two principle parts: a spring whose deflection indicates the force exerted between a driving body and a driven body; and a band of paper, or a card, moving at right angles to the direction of deflection of the spring with a velocity bearing a known constant proportion to the velocity with which the resistance is overcome. The spring carries a pen or pencil, which marks on the paper or card the required line. The Steam Engine Indicator is an example of this class of instruments.

III. Instruments called *Integrating Dynamometers,* which record the work performed, but not the resistance and motion separately.

424. **Steam Engine Indicator.**—This instrument was invented by Watt, and has been improved by other inventors, especially M'Naught and Richards. Its object is to record, by means of a diagram, the intensity of the pressure exerted by steam against one of the faces of a piston at each point of the piston's motion, and so to afford the means of computing, according to the principles of Articles 395 and 400, first, the energy exerted by the steam in driving the piston during the forward stroke; secondly, the work lost by the piston in expelling the steam from the cylinder during the return stroke; and thirdly, the difference of those quantities,

which is the *available* or *effective* energy exerted by the steam on the piston, and which, being multiplied by the number of strokes per minute and divided by 33,000 foot-pounds, gives the INDICATED HORSE-POWER.

The indicator in a common form is represented by fig. 150. AB is a cylindrical case. Its lower end, A, contains a smaller cylinder, fitted with a piston, which cylinder, by means of the screwed nozzle at its lower end, can be fixed in any convenient position on a tube communicating with that end of the engine-cylinder where the work of the steam is determined. The communication between the engine-cylinder and the indicator-cylinder can be opened and shut at will by means of the cock K. When it is open, the intensity of the pressure of the steam on the engine-piston and on the indicator-piston is the same, or nearly the same.

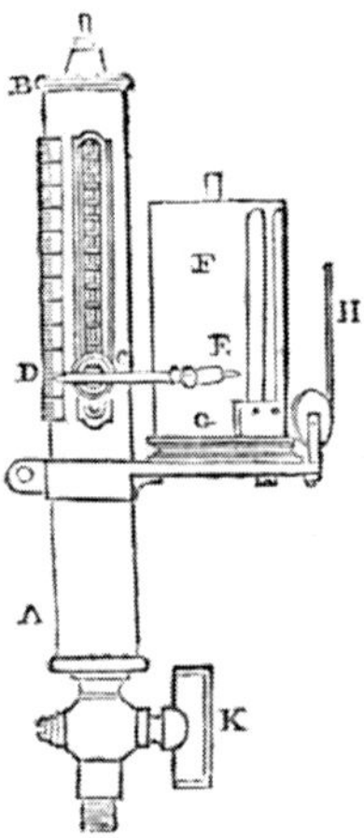

Fig. 150.

The upper end, B, of the cylindrical case contains a spiral spring, one end of which is attached to the piston, or to its rod, and the other to the top of the casing. The indicator-piston is pressed from below by the steam, and from above by the atmosphere. When the pressure of the steam is equal to that of the atmosphere, the spring retains its unstrained length, and the piston its original position. When the pressure of the steam exceeds that of the atmosphere, the piston is driven outwards, and the spring compressed; when the pressure of the steam is less than that of the atmosphere, the piston is driven inwards, and the spring extended. The compression or extension of the spring indicates the difference, upward or downward, between the pressure of the steam and that of the atmosphere.

A short arm, C, projecting from the indicator piston-rod carries at one side a pointer, D, which shews the pressure on a scale whose zero denotes the *pressure of the atmosphere*, and which is graduated into pounds on the square inch both upwards and downwards from that zero. At the other side the short arm has a longer arm jointed to it, carrying a pencil, E.

F is a brass drum, which rotates backward and forward about a vertical axis, and which, when about to be used, is covered with a piece of paper called a "card." It is alternately pulled round in one direction by the cord H, which wraps on the pulley G, and pulled back to its original position by a spring contained within itself. The cord H is to be connected with the mechanism of the steam engine in any convenient manner which shall ensure that the velocity of rotation of the drum shall at every instant bear a constant ratio to that of the steam engine piston: the back and forward motion of the surface of the drum representing that of the steam engine piston on a reduced scale. This having been done, and before opening the cock K, the pencil is to be placed in contact with the drum during a few strokes, when it will mark on the card a line which, when the card is afterwards spread out flat, becomes a straight line. This line, whose position indicates the pressure of the atmosphere, is called the *atmospheric line*. In fig. 151 it is represented by A A.

The cock K is opened, and the pencil, moving up and down with the variations of the pressure of the steam, traces on the card during each complete or double stroke a curve such as B C D E B. The ordinates drawn to that curve from any point in the atmospheric line, such as $\overline{HK}$ and $\overline{HG}$, indicate the differences between the pressure of the steam and the atmospheric pressure at the corresponding point of the motion of the piston. The ordinates of the part B C D E represent the pressures of the steam during the forward stroke, when it is driving the piston; those of the part E B represent the pressures of the steam when the piston is expelling it from the cylinder.

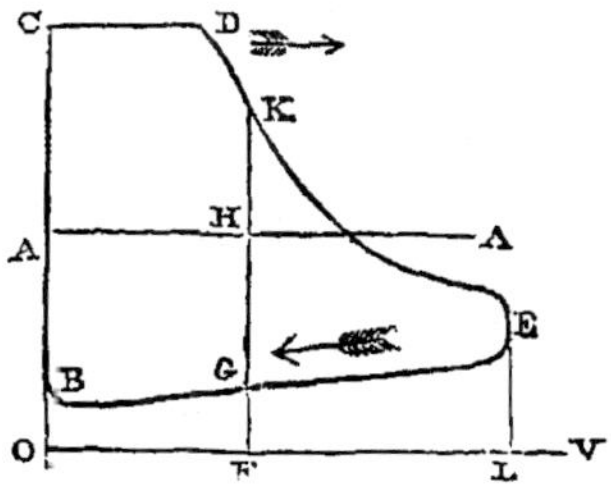

Fig. 151.

To found exact investigations on the indicator-diagrams of steam engines, the atmospheric pressure at the time of the experiment ought to be ascertained by means of a barometer; but this is generally omitted; in which case the atmospheric pressure may be assumed at its mean value, being 14·7 lbs. on the square inch, or 2116·3 lbs. on the square foot, at and near the level of the sea.

Let $\overline{AO} = \overline{HF}$ be ordinates representing the pressure of the atmosphere. Then O F V parallel to A A is the *absolute* or *true* zero line of the diagram, corresponding to *no pressure;* and ordinates drawn to the curve from that line represent the absolute intensities of the pressure of steam. Let O B and L E be ordinates touching the ends of the diagram; then

$\overline{OL}$ represents the *volume* traversed by the piston at each single stroke ($= s$ A, where s is the length of the stroke and A the area of the piston);

The area O B C D E L O represents the energy exerted by the steam on the piston during the forward stroke;

The area O B E L O represents the work lost in expelling the steam during the return stroke;

The area B C D E B, being the difference of the above areas, represents the *effective work* of the steam on the piston during the complete stroke.

Those areas can be found by the rules of Article 34, page 17; and the common trapezoidal rule, D, page 21, is in general sufficiently accurate. The number of intervals is usually ten, and of ordinates eleven.

The *mean forward pressure*, the *mean back pressure*, and the *mean effective pressure*, are found by dividing those three areas respectively by the volume s A, which is represented by $\overline{\mathrm{OL}}$.

Those mean pressures, however, can be found by a direct process, without first measuring the areas, viz.:—having multiplied each ordinate, or breadth, of the area under consideration by the proper multiplier, divide the sum of the products by the sum of the multipliers, which process, when the common trapezoidal rule is used, takes the following form: add together the halves of the endmost ordinates, and the whole of the other ordinates, and divide by the number of intervals. That is, let b_0 be the first, b_n the last, and b_1, b_2, &c., the intermediate breadths; then let n be the number of intervals, and b_m the mean breadth; then

$$b_m = \frac{1}{n}\left(\frac{b_0 + b_n}{2} + b_1 + b_2 + \&c.\right); \text{..............(1.)}$$

and this represents the mean forward pressure, mean back pressure, or mean effective pressure, as the case may be. Let p_e be the mean effective pressure; then the effective energy exerted by the steam on the piston during each double stroke is the product of the mean effective pressure, the area of the piston, and the length of stroke, or

$$p_e \mathrm{A}\, s; \text{.............................(2.)}$$

and if N be the number of double strokes in a minute, the *indicated power in foot-pounds per minute*, in a single-acting engine, is

$$p_e \mathrm{A\,N}\, s; \text{.............................(3.)}$$

from which the *indicated horse-power* is found by dividing by 33,000.

In a *double-acting engine* the steam acts alternately on either side of the piston; and to measure the power accurately, two indicators should be used at the same time, communicating respectively with the two ends of the cylinder. Thus a pair of diagrams will be obtained, one representing the action of the steam on each face of the piston. The mean effective pressure is to be found as above for each diagram separately, and then, if the areas of the two faces of the piston are sensibly equal, *the mean of those two results* is to be taken as the *general mean effective pressure;* which being multiplied by the area of the piston, the length of stroke, and *twice* the number of double strokes or revolutions in a minute, gives the indicated power per minute; that is to say, if p'' denotes the general mean effective pressure, the indicated power per minute is

$$p'' \mathrm{A} \cdot 2 \mathrm{N} s; \qquad (4.)$$

If the two faces of the piston are sensibly of unequal areas (as in "trunk engines"), the indicated power is to be computed separately for each face, and the results added together.

If there are two or more cylinders, the quantities of power indicated by their respective diagrams are to be added together.

The reactions of the moving parts of the indicator, combined with the elasticity of the spring, cause oscillations of its piston. In order that the errors thus produced in the indicated pressures at particular instants may be as small as possible, and may neutralize each other's effects on the whole indicated power, the moving masses ought to be as small as practicable, and the spring as stiff as is consistent with shewing the pressures on a visible scale. In Richard's indicator this is effected by the help of a train of very light linkwork, which causes the pencil to shew the movements of the spring on a magnified scale.

The *friction* of the moving parts of the indicator tends on the whole to make the indicated power and indicated mean effective pressure less than the truth, but to what extent is uncertain.

Every indicator should have the accuracy of the graduation of its scale of pressures frequently tested by comparison with a standard pressure gauge.

The indicator may obviously be used for measuring the energy exerted by any fluid, whether liquid or gaseous, in driving a piston; or the work performed by a pump, in lifting, propelling, or compressing any fluid.

CHAPTER III.

OF REGULATING APPARATUS.

425. **Regulating Apparatus Classed—Brake—Fly—Governor.**—The effect of all regulating apparatus is to control the speed of machinery. A regulating instrument may act simply by consuming energy, so as to prevent acceleration, or produce retardation, or stop the machine if required; it is then called a *brake;* or it may act by storing surplus energy at one time, and giving it out at another time when energy is deficient: in this case it is called a *fly;* or it may act by adjusting the power of the prime mover to the work to be done, when it is called a *governor.* The use of a brake involves waste of power. A fly and a governor, on the other hand, promote economy of power and economy of strength.

SECTION 1.—OF BRAKES.

426. **Brakes Defined and Classed.**—The contrivances here comprehended under the general title of *Brakes* are those by means of which friction, whether exerted amongst solid or fluid particles, is purposely opposed to the motion of a machine, in order either to stop it, to retard it, or to employ superfluous energy during uniform motion. The use of a brake involves waste of energy, which is in itself an evil, and is not to be incurred unless it is necessary to convenience or safety.

Brakes may be classed as follows:—

I. *Block-brakes,* in which one solid body is simply pressed against another, on which it rubs.

II. *Flexible brakes,* which embrace the periphery of a drum or pulley.

III. *Pump-brakes,* in which the resistance employed is the friction amongst the particles of a fluid forced through a narrow passage.

IV. *Fan-brakes,* in which the resistance employed is that of a fluid to a fan rotating in it.

427. **Action of Brakes in General.**—The work disposed of by a brake in a given time is the product of the resistance which it produces into the distance through which that resistance is overcome in a given time.

To *stop* a machine, the brake must employ work to the amount of the whole actual energy of the machine, as already stated in Article 417. To *retard* a machine, the brake must employ work to an amount equal to the difference between the actual energies of the machine at the greater and less velocities respectively.

To *dispose of surplus energy*, the brake must employ work equal to that energy; that is, the resistance caused by the brake must balance the surplus effort to which the surplus energy is due; so that if n is the ratio which the velocity of rubbing of the brake bears to the velocity of the driving point, P, the *surplus effort* at the driving point, and R the resistance of the brake, we ought to have—

$$R = \frac{P}{n}. \quad \ldots\ldots\ldots\ldots\ldots\ldots\ldots\ldots (1.)$$

It is obviously better, when practicable, to store surplus energy, or to prevent its exertion, than to dispose of it by means of a brake.

When the action of a brake composed of solid material is long-continued, a stream of water must be supplied to the rubbing surfaces, to abstract the heat that is produced by the friction, according to the law stated in Article 402, page 252.

428. **Block-Brakes.**—When the motion of a machine is to be controlled by pressing a block of solid material against the rim of a rotating drum, it is advisable, inasmuch as it is easier to renew the rubbing surface of the block than that of the drum, that the drum should be of the harder, and the block of the softer material—the drum, for example, being of iron, and the block of wood. The best kinds of wood for this purpose are those which have considerable strength to resist crushing, such as elm, oak, and beech. The wood forms a facing to a frame of iron, and can be renewed when worn.

When the brake is pressed against the rotating drum, the direction of the pressure between them is obliquely opposed to the motion of the drum, so as to make an angle with the radius of the drum equal to the *angle of repose* of the rubbing surfaces (denoted by φ; see page 154). The component of that oblique pressure in the direction of a tangent to the rim of the drum is the friction (R); the component perpendicular to the rim of the drum is the normal pressure (N) required in order to produce that friction, and is given by the equation

$$N = \frac{R}{f}; \quad \ldots\ldots\ldots\ldots\ldots\ldots\ldots\ldots (1.)$$

f being the coefficient of friction, and the proper value of R being determined by the principles stated in Article 427.

It is in general desirable that the brake should be capable of effecting its purpose when pressed against the drum by means of the strength of one man, pulling or pushing a handle with one hand or one foot. As the required normal pressure N is usually considerably greater than the force which one man can exert, a lever, or screw, or a train of levers, screws, or other convenient mechanism, must be interposed between the brake block and the handle, so that when the block is moved towards the drum, the handle shall move at least through a distance as many times greater than the distance by which the block *directly* approaches the drum, as the required normal pressure is greater than the force which the man can exert.

Although a man may be able occasionally to exert with one hand a force of 100 lbs., or 150 lbs., for a short time, it is desirable that, in working a brake, he should not be required to exert a force greater than he can keep up for a considerable time, and exert repeatedly in the course of a day, without fatigue—that is to say, about 20 lbs. or 25 lbs.

429. **The Brakes of Carriages** are usually of the class just described, and are applied either to the wheels themselves or to drums rotating along with the wheels. Their effect is to stop or to retard the rotation of the wheels, and make them slip, instead of rolling on the road or railway. The resistance to the motion of a carriage which is caused by its brake may be less, but cannot be greater, than the friction of the stopped or retarded wheels on the road or rails under the load which rests on those wheels. The distance which a carriage or train of carriages will run on a level line during the action of the brakes before stopping, is found by dividing the actual energy of the moving mass before the brakes are applied, by the sum of the ordinary resistance and of the additional resistance caused by the brakes; in other words, that distance is as many times greater than the height due to the speed as the weight of the moving mass is greater than the total resistance.

The *skid*, or *slipper-drag*, being placed under a wheel of a carriage, causes a resistance due to the friction of the skid upon the road or rail under the load that rests on the wheel.

Section 2.—Of Fly-Wheels.

430. **Periodical Fluctuations of Speed** in a machine are caused by the alternate excess and deficiency of the energy exerted above the work performed in overcoming resisting forces, which produce an alternate increase and diminution of actual energy, according to the law explained in Article 413, page 262.

To determine the greatest fluctuation of speed in a machine moving periodically, take A B C, in fig. 152, to represent the motion of the driving point during one period; let the effort P of the prime mover at each instant be represented by the ordinate of the curve D G E I F; and let the sum of the resistances, reduced to the driving point as in Article 398, at each instant, be denoted by R, and represented by the ordinate of the curve D H E K F, which cuts the former curve at the ordinates A D, B E, C F. Then the integral,

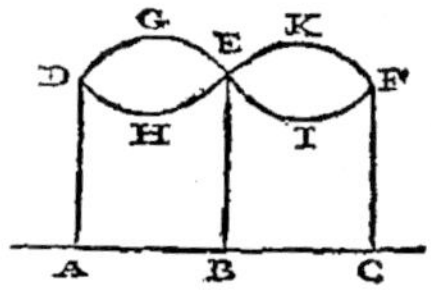

Fig. 152.

$$\int (\mathrm{P} - \mathrm{R})\, d\, s,$$

being taken for any part of the motion, gives the excess or deficiency of energy, according as it is positive or negative. For the entire period A B C, this integral is nothing. For A B, it denotes an *excess of energy received*, represented by the area D G E H; and for B C, an equal *excess of work performed*, represented by the equal area E K F I. Let those equal quantities be each represented by Δ E. Then the actual energy of the machine attains a maximum value at B, and a minimum value at A and C, and Δ E is the difference of those values.

Now let v_0 be the mean velocity, v_1 the greatest velocity, v_2 the least velocity of the driving point, and $\Sigma \cdot n^2$ W the *reduced inertia* of the machine (see Article 405, page 257); then

$$\frac{v_1^2 - v_2^2}{2\,g} \cdot \Sigma \cdot n^2\,\mathrm{W} = \Delta\,\mathrm{E}; \qquad (1.)$$

which, being divided by the *mean actual energy*,

$$\frac{v_0^2}{2\,g} \cdot \Sigma\, n^2\,\mathrm{W} = \mathrm{E}_0,$$

gives

$$\frac{v_1^2 - v_2^2}{v_0^2} = \frac{\Delta\,\mathrm{E}}{\mathrm{E}_0}; \qquad (2.)$$

and observing that $v_0 = (v_1 + v_2) \div 2$, we find

$$\frac{v_1 - v_2}{v_0} = \frac{\Delta \cdot \mathrm{E}}{2\,\mathrm{E}_0} = \frac{g\,\Delta\,\mathrm{E}}{v_0^3\,\Sigma \cdot n^2\,\mathrm{W}}; \qquad (3.)$$

a ratio which may be called the *coefficient of fluctuation of speed* or of *unsteadiness*.

The ratio of the periodical excess and deficiency of energy Δ E to the whole energy exerted in one period or revolution, $\int P\,d\,s$, has been determined by General Morin for steam engines under various circumstances, and found to be from $\frac{1}{10}$ to $\frac{1}{4}$ for single-cylinder engines. For a pair of engines driving the same shaft, with cranks at right angles to each other, the value of this ratio is about one-fourth, and for three engines with cranks at 120°, one-twelfth of its value for single-cylinder engines.

The following table of the ratio, $\Delta E \div \int P\,d\,s$, for *one revolution* of steam engines of different kinds is extracted and condensed from General Morin's works:—

NON-EXPANSIVE ENGINES.

$\frac{\text{Length of connecting rod}}{\text{Length of crank}}$	=	8	6	5	4
$\Delta E \div \int P\,d\,s$	=	·105	·118	·125	·132

EXPANSIVE CONDENSING ENGINES.

Connecting rod = crank × 5.

Fraction of Stroke at which steam is cut off,		$\frac{1}{3}$	$\frac{1}{4}$	$\frac{1}{5}$	$\frac{1}{6}$	$\frac{1}{7}$	$\frac{1}{8}$
$\Delta E \div \int P\,d\,s$	=	·163	·173	·178	·184	·189	·191

EXPANSIVE NON-CONDENSING ENGINES.

Steam cut off at		$\frac{1}{2}$	$\frac{1}{3}$	$\frac{1}{4}$	$\frac{1}{5}$
$\Delta E \div \int P\,d\,s$	=	·160	·186	·209	·232

For double-cylinder expansive engines, the value of the ratio $\Delta E \div \int P\,d\,s$ may be taken as equal to that for single-cylinder non-expansive engines.

For *tools working at intervals*, such as punching, slotting, and plate-cutting machines, coining presses, &c., ΔE is nearly equal to the whole work performed at each operation.

431. **Fly-Wheels.**—A fly-wheel is a wheel with a heavy rim, whose great moment of inertia being comprehended in the

reduced moment of inertia of a machine reduces the coefficient of fluctuation of speed to a certain fixed amount, being about $\frac{1}{32}$ for ordinary machinery, and $\frac{1}{50}$ or $\frac{1}{60}$ for machinery for fine purposes.

Let $\frac{1}{m}$ be the intended value of the coefficient or fluctuation of speed, and $\triangle$ E, as before, the fluctuation of energy. If this is to be provided for by the moment of inertia, I, of the fly-wheel alone, let a_0 be its mean angular velocity; then Equation 3 of Article 430 is equivalent to the following:—

$$\frac{1}{m} = \frac{g \triangle \mathrm{E}}{a_0^2 \mathrm{I}}; \quad (1.)$$

$$\mathrm{I} = \frac{m g \triangle \mathrm{E}}{}; \quad (2.)$$

the second of which equations gives the requisite moment of inertia of the fly-wheel.

The fluctuation of energy may arise either from variations in the effort exerted by the prime mover, or from variations in the resistance, or from both those causes combined. When but one fly-wheel is used, it should be placed in as direct connexion as possible with that part of the mechanism where the greatest amount of the fluctuation originates; but when it originates at two or more points, it is best to have a fly-wheel in connection with each of those points.

For example, let there be a steam engine which drives a shaft that traverses a workshop, having at intervals upon it pulleys for driving various machine-tools. The steam engine should have a fly-wheel of its own, as near as practicable to its crank, adapted to that value of $\triangle$ E which is due to the fluctuations of the effort applied to the crank-pin above and below the mean value of that effort, and which may be computed by the aid of General Morin's tables, quoted in Article 430; and each machine tool should also have a fly-wheel, adapted to a value of $\triangle$ E equal to the whole work performed by the tool at one operation.

As the rim of a fly-wheel is usually heavy in comparison with the arms, it is often sufficiently accurate for practical purposes to take the moment of inertia as simply equal to the weight of the rim multiplied by the square of the mean between its outside and inside radii—a calculation which may be expressed thus:—

$$\mathrm{I} = \mathrm{W} r^2; \quad (3.)$$

whence the weight of the rim is given by the formula—

$$\mathrm{W} = \frac{m g \triangle \mathrm{E}}{a_0^2 r^2} = \frac{m g \triangle \mathrm{E}}{v'^2}, \quad (4.)$$

if v' be the velocity of the rim of the fly-wheel.

In millwork the ordinary values of the product *m g*, the unit of time being the second, lie between 1,000 and 2,000 feet, or approximately between 300 and 600 mètres. In pumping-machinery it is sometimes only about 300 feet, or 90 mètres.

The rim of the fly-wheel of a factory steam engine is very often provided with teeth, or with a belt, in order that it may directly drive the machinery of the factory.

SECTION 3.—OF GOVERNORS.

432. **The Regulator** of a prime mover is some piece of apparatus by which the rate at which it receives energy from the source of energy can be varied; such as the sluice or valve which adjusts the size of the orifice for supplying water to a water-wheel, the apparatus for varying the surface exposed to the wind by windmill sails, the throttle-valve which adjusts the opening of the steam pipe of a steam engine, the damper which controls the supply of air to its furnace, and the expansion gear which regulates the volume of steam admitted into the cylinder at each stroke of the piston.

In prime movers whose speed and power have to be frequently and rapidly varied at will, such as locomotives and winding engines for mines, the regulator is adjusted by hand. In other cases the regulator is adjusted by means of a self-acting instrument driven by the prime mover to be regulated, and called a GOVERNOR.

The special construction of the different kinds of regulators is a subject for a treatise on prime movers. In the present treatise it is sufficient to state that in every governor there is a moving piece which acts on the regulator through a suitable train of mechanism, and which is itself made to move in one direction or in another according as the prime mover is moving too fast or too slow.

The object of a governor, properly so called, is to preserve a certain uniform speed, either exactly or approximately; and such is always the case in millwork. There are other cases, as in marine steam engines, where it may be considered sufficient to prevent sudden variations of speed, without preserving an uniform speed; and in those cases an apparatus may be used possessing only in part the properties of a governor: this may be called a *fly-governor*, to distinguish it from a governor proper.

Governors proper may be distinguished into *position-governors*, *disengagement-governors*, and *differential governors:* a position-governor being one in which the moving piece that acts on the regulator assumes positions depending on the speed of motion, as in the common steam engine governor, which consists of a pair of revolving pendulums acting directly on a train of mechanism which adjusts the throttle-valve: a disengaging-governor being one which, when the speed deviates above or below its proper value, throws the regulator into gear with one or other of two trains of mechanism which move it in contrary directions so as to diminish or increase the speed, as the case may require, as in water-mill governors; and a differential-governor being one which, by means of an aggregate combination, moves the regulator in one direction or in another with a speed proportional to the difference between the actual speed and the proper speed of the engine.

In almost all governors the action depends on the centrifugal force exerted by two or more masses which revolve round an axis. By another classification, different from that which has already been described, governors may be distinguished into *gravity-governors*, in which gravity is the force that opposes the centrifugal force; and *balanced-governors*, in which the actions of gravity on the various moving parts of the governor are mutually balanced, and the centrifugal force is opposed by the elasticity of a spring.

Governors may be further distinguished into those which are truly isochronous—that is to say, which remain without action on the regulator at one speed only; and those which are nearly isochronous—that is to say, which admit of some variation of the permanent or steady speed when the resistance overcome by the engine varies; and lastly, governors may be distinguished into those which are specially adapted to one speed, and those which can be adjusted at will to different speeds.

433. **Pendulum-Governors.**—A pendulum-governor is the simplest kind of gravity-governor. It has a vertical spindle, driven by the engine to be regulated; and from that spindle there hang, at opposite sides, a pair of revolving pendulums, which, by the positions that they assume at different speeds, act on the regulator.

The relation between the height of a simple revolving pendulum and the number of turns which it makes per second has already been stated in Article 336; but for the sake of convenience it may here be repeated:—Let h denote the height or *altitude* of the pendulum ($=$ O H in fig. 153), and T the number of turns per second; then

$$h = \frac{g}{4\pi^2 T^2} = \frac{\cdot 815 \text{ foot}}{T^2} = \frac{9\cdot 78 \text{ inches}}{T^2} = \frac{0\cdot 248 \text{ mètre}}{T^2} \ldots\ldots (1.)$$

If the rods of the revolving pendulums are jointed, as in fig. 154, not to a point in the vertical axis, but to a pair of points, such as C, c, in arms projecting from that axis, the height is to be measured to the point O, where the lines of tension of the rods cut the axis.

In most cases which occur in practice, the balls are so heavy, as compared with the rods, that the height may be measured without sensible error from the level of the centres of the balls to the point O, where the lines of suspension cut the axis. This amounts to neglecting the effects both of the weight and of the centrifugal force of the rods.

The ordinary steam engine governor invented by Watt, which is represented in fig. 153, is a position-governor, and acts on the

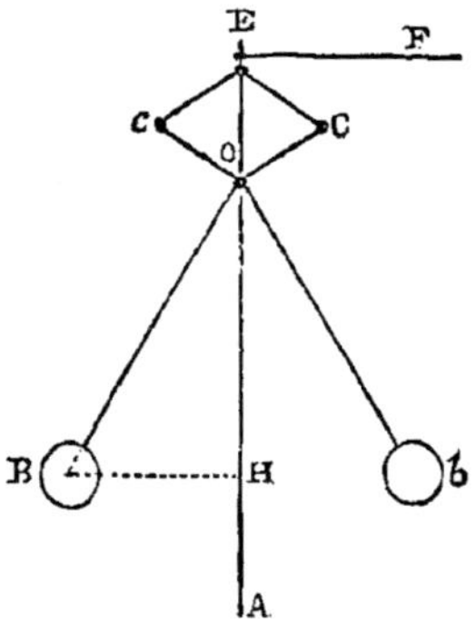

Fig. 153.

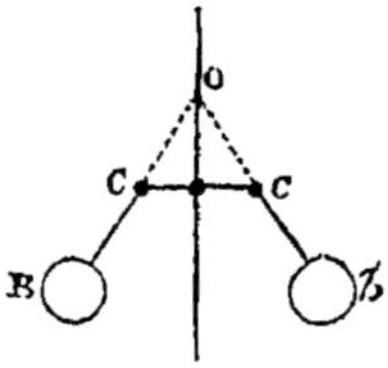

Fig. 154.

regulator by means of the variation of its altitude, through a train of levers and linkwork. That train may be very much varied in detail. In the example shewn in the figure, the lever O C forms one piece with the ball-rod O B, and the lever O c with the ball-rod O b; so that when the speed falls too low, the balls B, b, by approaching the spindle, cause the point E to rise; and when the speed rises too high, the balls, by receding from the spindle, cause the point E to fall. At the point E there is a collar, held in the forked end of the lever E F, which communicates motion to the regulator.

The ordinary pendulum-governor is not truly isochronous; for when, in order to adapt the opening of the regulator to different loads, it rotates with its revolving pendulums at different angles to the vertical axis, the altitude h assumes different values, corresponding to different speeds.

As in Article 431, let the utmost extent of fluctuation of the speed of the engine between its highest and lowest limits be the fraction $\frac{1}{m}$ of the mean speed; let h be the altitude of the governor corresponding to the mean speed; and let k be the utmost extent of variation of the altitude between its smaller limit, when the regulator is shut, and its greater limit, when the regulator is full open. Then we have the following proportion:—

$$1 : \left(1 + \frac{1}{2m}\right)^2 - \left(1 - \frac{1}{2m}\right)^2 :: h : k;$$

and consequently

$$\frac{k}{h}=\frac{2}{m} \dots\dots\dots\dots\dots\dots\dots\dots (3.)$$

434. **Loaded Pendulum-Governor.**—From the balls of the common governor, whose collective weight is (say) A, let there be hung by a pair of links of lengths equal to the ball-rods, a load B, capable of sliding up and down the spindle, and having its centre of gravity in the axis of rotation. Then the centrifugal force is that due to A alone; and the effect of gravity is that due to A + 2 B; for when the ball-rods shift their position, the load B moves through twice the vertical distance that the balls move through, and is therefore equivalent, to a double load, 2 B, acting directly on the balls. Consequently the altitude for a given speed is greater than that of a simple revolving pendulum, in the ratio $1+\frac{2\text{ B}}{\text{A}}$; a given *absolute* variation of altitude in moving the regulator produces a proportionate variation of speed smaller than in the common governor, in the ratio $\frac{\text{A}}{\text{A}+2\text{ B}}$; and the governor is said to be *more sensitive* than a common governor, in the ratio of A : A + 2 B. Such is the construction of Porter's governor.

The links by which the load B is hung may be attached, not to the balls themselves, but to any convenient pair of points in the ball-rods; the links, and the parts of the ball-rods to which they are jointed, always forming a rhombus, or equilateral parallelogram. Let q be the ratio borne by each of the sides of that rhombus to the length on the ball-rods from the centre of a ball to the point where the line of suspension cuts the axis; then in the preceding expressions 2 q B is to be substituted for 2 B.

In the one case 2 B, and in the other 2 q B, is the weight, applied directly at A, which would be *statically equivalent* to the load B, applied where it is.

CHAPTER IV.

OF THE EFFICIENCY AND COUNTER-EFFICIENCY OF PIECES, COMBINATIONS, AND TRAINS IN MECHANISM.

435. **Nature and Division of the Subject.**—The terms *Efficiency* and *Counter-efficiency* have already been explained in Article 418, page 265; and the laws of friction, the most important of the wasteful resistances which cause the efficiency of a machine to be less than unity, have been stated in Articles 261 and 402, pages 153 and 251. In the present chapter are to be set forth the effects of wasteful resistance, and especially of friction, on the efficiency and counter-efficiency of single pieces, and of combinations and trains of pieces, in mechanism. In practical calculations the *counter-efficiency* is in general the quantity best adapted for use; because the useful work to be done in an unit of time, or *effective power*, is in general given; and from that quantity, by multiplying it by the counter-efficiency, of the machine—that is, by the continued product of the counter-efficiencies of all the successive pieces and combinations by means of which motion is communicated from the driving-point to the useful working-point—is to be deduced the value of the expenditure of energy in an unit of time, or *total power*, required to drive the machine. In symbols, let U be the useful work to be done per second; c, c', c'', &c., the counter-efficiencies of the several parts of the train; T, the total energy to be expended per second; then

$$T = c \cdot c' \cdot c'' \cdot \&c. \ldots U \ldots\ldots\ldots\ldots\ldots\ldots\ldots (1.)$$

When the mean effort required at the driving-point can conveniently be computed by reducing each resistance to the driving-point, and adding together the reduced resistances (as in Article 407, page 253, and Article 421, page 267), the ratio in which the actual effort required at the driving-point is greater than what the required effort would be, in the absence of wasteful resistance, is expressed by the continued product of the counter-efficiencies of the parts of the train, as follows: let P_0 be the effort required, in the absence of wasteful resistance; P, the actual effort required; then

$$P = c \cdot c' \cdot c'' \cdot \&c. \ldots P_0; \ldots\ldots\ldots\ldots\ldots\ldots\ldots (2.)$$

and in determining the efficiency or the counter-efficiency of a single piece, the most convenient method of proceeding often consists in comparing together the efforts required to drive that piece, with and without friction, and thus finding the ratios

$$\frac{P_0}{P} = \text{efficiency}; \ \frac{P}{P_0} = \text{counter-efficiency} \ldots\ldots\ldots (3.)$$

In the ensuing sections of this chapter, the efficiency of single primary pieces is first treated of, and then that of the various modes of connexion employed in elementary combinations.

SECTION 1.—EFFICIENCY AND COUNTER-EFFICIENCY OF PRIMARY PIECES.

436. **Efficiency of Primary Pieces in General.**—A primary piece in mechanism, moving with an uniform velocity, is balanced under the action of four forces, viz.:—

I. The re-action of the piece which it drives: this may be called the *Useful Resistance*, and denoted by R;

II. The *weight* of the piece itself: this may be denoted by W;

III. The *effort* by which the piece is driven: this may be denoted by P; and its values with and without friction by P_0 and P_1 respectively;

IV. The resultant pressure at the bearings, or *bearing-pressure*, which may be denoted by Q; and which of course is equal and directly opposed to the resultant of the first three forces.

In the absence of friction, the bearing-pressure would be normal to the bearing surface. The effect of friction is, that the line of action of the bearing-pressure becomes oblique to the bearing-surface, making with the normal to that surface the angle of repose (φ), whose tangent ($f = \tan \varphi$) is the coefficient of friction (see Article 261, page 154); and the amount of the friction is expressed by Q $\sin \varphi$, or very nearly by f Q, when the coefficient of friction is small.

In the class of problems to which this chapter relates, the first two forces—that is, the useful resistance R, and the weight W—are given in magnitude, position, and direction; and in most cases it is convenient to find their resultant, in magnitude, position, and direction, by the rules of statics: that is to say, if the line of action of R is vertical, by Article 226, page 128; and if inclined, by the rules given or referred to in Article 209, page 122. In what follows, the resultant of the useful resistance and weight will be called the *given force*, and denoted by R′.

The third force—that is, the effort required in order to drive the piece at an uniform speed—is given in position and direction; for its line of action is the line of connection of the piece under consideration with the piece that drives it. The magnitude of the effort is one of the quantities to be found.

The fourth force—that is, the bearing-pressure—has to be found in position, direction, and magnitude. The general principles according to which it is determined are the following:—

First, That if the lines of action of the given force and the effort are parallel to each other, the line of action of the resultant bearing-pressure must be parallel to them both; and that if they are inclined to each other, the line of action of the resultant bearing-pressure must traverse their point of intersection.

Secondly, That at the centre of pressure, where the line of action of the resultant bearing-pressure cuts the bearing surfaces, it makes an angle with the common normal of those surfaces equal to their angle of repose, and in such a direction that its tangential component (being the friction) is directly opposed to the relative sliding motion of that pair of surfaces over each other.

Thirdly, That the given force, the effort, and the bearing-pressure, form a system of three forces that balance each other; and are therefore proportional to the three sides of a triangle parallel respectively to their directions.

437. **Efficiency of a Straight-sliding Piece.**—In fig. 155, let A A be a straight guiding-surface, upon which there slides, in the direc-

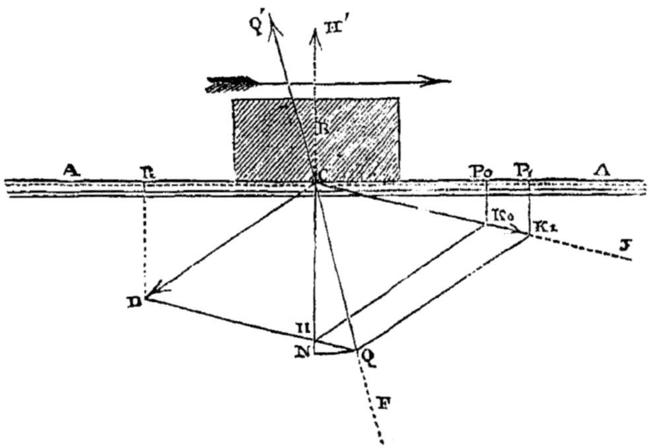

Fig. 155.

tion marked by the feathered arrow, the moving piece B. Let C D represent the *given force*, being the resultant of the useful resistance and of the weight of the piece B. (The figure shews the motion of B as horizontal; but it may be in any direction.) Let C J be the line of action of the effort by which the piece B is driven.

Draw C N perpendicular to A A; and C F making the angle N C F = the angle of repose. Through D, parallel to C J, draw the straight line D H Q, cutting C N in H, and C F in Q; and through H and Q, and parallel to D C, draw H K_0 and Q K_1, cutting C J in K_0 and K_1 respectively. Produce H C to H′, and Q C to Q′, making C H′ = H C, and C Q′ = Q C.

Then, in the absence of friction, C H′ will represent the resultant bearing-pressure exerted upon B by A A; and C K_0 = D H will represent the force in the given direction C J required to drive B at an uniform speed; and when friction is taken into account, C Q′ will represent the resultant bearing-pressure, and C K_1 the actual driving force required; and we shall have

$$\text{the efficiency} = \frac{C K_0}{C K_1}; \text{ and the counter-efficiency} = \frac{C K_1}{C K_0}$$

If from D, K_0, and K_1 there be let fall upon A A the perpendiculars D R, K_0 P_0, and K_1 P_1, C R will represent the direct resistance to the advance of B; C P_0 the direct effort in the absence of friction; and C P_1, the direct effort taking friction into account; so that the distance P_0 P_1 will represent the friction itself; which is also represented by Q N perpendicular to C N.

To express these results by symbols, let C D = R′ (the given force); let the acute angle A C D be denoted by α, and the acute angle A C J by β; and let ϕ denote the angle of repose N C Q.

Then, in the triangle C D H, we have $\angle$ D C H $= \frac{\pi}{2} - \alpha$, and C H D $= \frac{\pi}{2} - \beta$; and in the triangle C Q D, we have $\angle$ D C Q $= \frac{\pi}{2} - \alpha + \phi$, and $\angle$ C Q D $= \frac{\pi}{2} - \beta - \phi$; consequently

$$D H = R' \frac{\cos \alpha}{\cos \beta}; \quad D Q = R' \cdot \frac{\cos(\alpha - \phi)}{\cos(\beta + \phi)};$$

whence it follows that the efficiency and counter-efficiency are given by the following equations:—

$$\text{Efficiency} = \frac{P_0}{P_1} = \frac{D H}{D Q} = \frac{\cos \alpha \cdot \cos(\beta + \phi)}{\cos \beta \cdot \cos(\alpha - \phi)} = \frac{1 - f \tan \beta}{1 + f \tan \alpha} \quad (1.)$$

$$\text{Counter-efficiency} = \frac{P_1}{P_0} = \frac{1 + f \tan \alpha}{1 - f \tan \beta} \quad \ldots\ldots\ldots\ldots\ldots\ldots (2.)$$

It is to be remarked, that the efficiency diminishes to nothing when cotan $\beta = f$; that is to say, when β is the complement of the angle of repose, ϕ. In other words, if the oblique effort is applied in the direction C Q, no force, how great soever, will be sufficient to keep the piece B in motion.

438. **Efficiency of an Axle.**—In fig. 156, let the circle A A A represent the trace of the bearing-surface of an axle on a plane perpendicular to its axis of rotation, O—in other words, the trans-

verse section of that surface. Let the arrow near the letter N represent the direction of rotation. Let C D be the given force; that is, as before, the resultant of the weight of the whole piece that rotates with the axle, and of the useful resistance or re-action exerted on that piece by the piece which it drives; C J, the line of action of the effort by which the rotating piece is driven.

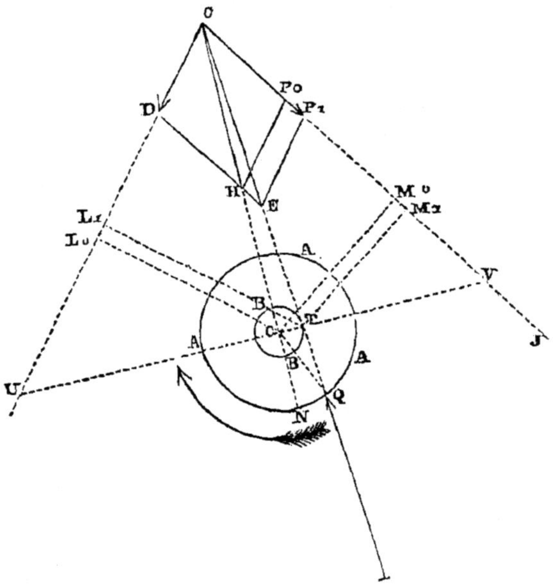

Fig. 156.

Let r denote the radius of the bearing-surface.

About O describe the small circle B B, with a radius $= r \sin \phi = f\,r$, very nearly. Draw the line of action, C T Q, of the resultant bearing-pressure, touching the small circle at that side which will make the bearing-pressure resist the rotation. In the case in which C D and C J intersect each other in a point, C, as shewn in the figure, C T Q will traverse that point also; and in the case in which the lines of action of the given force and the effort are parallel to each other, C T Q will be parallel to both. The centre of bearing-pressure is at Q; and O Q T $= \phi$, the angle of repose.

In the former case the efficiency may be found by parallelograms of forces, as follows:—Draw the straight line C O N; this would be the line of action of the resultant bearing-pressure in the absence of friction, and N would be the centre of bearing-pressure. Through D, parallel to C J, draw D H E, cutting C O N in H, and C T Q in E. Through H and E, parallel to D C, draw H P_0

and E P_1. Then, in the absence of friction, H C would represent the bearing-pressure, and $C P_0 = D H$ the effort; the actual bearing-pressure is represented by E C, and the actual effort by $C P_1 = D E$. Hence the efficiency and counter-efficiency are as follows:—

$$\frac{P_0}{P_1} = \frac{D H}{D E}; \quad \frac{P_1}{P_0} = \frac{D E}{D H} \quad \ldots\ldots\ldots\ldots\ldots\ldots \text{(1.)}$$

Another method, applicable whether the forces are inclined or parallel, is as follows:—From the axis of rotation O, let fall $O L_0$ and $O M_0$ perpendicular respectively to the lines of action of the given force and of the effort. Then, by the balance of moments, the effort in the absence of friction is

$$P_0 = R' \cdot \frac{O L_0}{O M_0}.$$

From a convenient point in the actual line of action, C Q, of the bearing-pressure (such, for example, as T, where it touches the small circle B B), let fall $T L_1$ and $T M_1$ perpendicular respectively to the same pair of lines of action; then the actual effort will be

$$P_1 = R' \cdot \frac{T L_1}{T M_1}.$$

Hence the efficiency and the counter-efficiency have the following value:—

$$\left.\begin{aligned} \frac{P_0}{P_1} &= \frac{O L_0 \cdot T M_1}{O M_0 \cdot T L_1}; \\ \frac{P_1}{P_0} &= \frac{O M_0 \cdot T L_1}{O L_0 \cdot T M_1}; \end{aligned}\right\} \quad \ldots\ldots\ldots\ldots\ldots\ldots \text{(2.)}$$

The same results are expressed, to a degree of approximation sufficient for practical purposes, by the following trigonometrical formulæ:—Let $O L_0 = l$; $O M_0 = m$; $\angle C O L_0 = \alpha$; $\angle C O M_0 =$. Then we have, very nearly,

$$\frac{P_0}{P_1} = \frac{l}{m} \cdot \frac{m - f r \sin \beta}{l + f r \sin \alpha} = \frac{1 - \frac{f r}{m} \cdot \sin \beta}{1 + \frac{f r}{l} \cdot \sin \alpha} \quad \ldots\ldots \text{(3.)}$$

In making use of the preceding formula, it is to be observed that the *contrary algebraical signs* of $\sin \alpha$ and $\sin \beta$ apply to those cases in which the two angles α and β lie at contrary sides of O C. In the cases in which those angles lie at the same side of O C, their algebraical signs are the same; and in the formula they are to be made *both positive* or *both negative*, according as β is *less* or *greater* than α; so that the efficiency may be always expressed by a fraction less than unity. That is to say,

$$\text{If } \beta > \alpha; \ \frac{P_0}{P_1} = \frac{1 - \frac{fr}{m} \sin \beta}{1 - \frac{fr}{l} \sin \alpha}; \quad \text{(3 A.)}$$

$$\text{If } \beta < \alpha; \ \frac{P_0}{P_1} = \frac{1 + \frac{fr}{m} \sin \beta}{1 + \frac{fr}{l} \sin \alpha}; \quad \text{(3 B.)}$$

When the lines of action intersect, let O C be denoted by c; then $l = c \cos \alpha$, and $m = c \cos \beta$; and consequently the three preceding equations take the following form;—

$$\beta \text{ and } \alpha \text{ of contrary signs}; \ \frac{P_0}{P_1} = \frac{1 - \frac{fr}{c} \tan \beta}{1 + \frac{fr}{c} \tan \alpha}; \quad \text{(4.)}$$

β and α of the same sign;

$$\beta > \alpha; \ \frac{P_0}{P_1} = \frac{1 - \frac{fr}{c} \tan \beta}{1 - \frac{fr}{c} \tan \alpha}; \quad \text{(4 A.)}$$

$$\beta < \alpha; \ \frac{P_0}{P_1} = \frac{1 + \frac{fr}{c} \tan \beta}{1 + \frac{fr}{c} \tan \alpha}; \quad \text{(4 B.)}$$

When the lines of action of the forces are parallel, we have $\sin \beta$ and $\sin \alpha = +1$ or -1, as the case may be; and the formulæ take the following shape:—

When l and m lie at contrary sides of O, the piece is a "lever of the first kind;" and

$$\frac{P_0}{P_1} = \frac{1 - \frac{fr}{m}}{1 + \frac{fr}{l}} \quad \text{(5.)}$$

When l and m lie at the same side of O;

If $m > l$, the piece is a "lever of the second kind;" and

$$\frac{P_0}{P_1} = \frac{1 - \frac{f r}{m}}{1 - \frac{f r}{l}} \quad \text{(5A.)}$$

If $m < l$, the piece is a "lever of the third kind;" and

$$\frac{P_0}{P_1} = \frac{1 + \frac{f r}{m}}{1 + \frac{f r}{l}} \quad \text{(5B.)}$$

(As to levers of the first, second, and third kinds, see Article 184, page 108.)

The following method is applicable whether the forces are inclined or parallel; in the former case it is approximate, in the latter exact. Through O, perpendicular to O C, draw U O V, cutting the lines of action of the given force and of the effort in U and V respectively. The point where this transverse line cuts the small circle B B coincides exactly with T when the forces are parallel, and is very near T when they are inclined; and in either case the letter T will be used to denote that point. Then

$$\frac{P_0}{P_1} = \frac{\text{O U}}{\text{O V}} \cdot \frac{\text{T V}}{\text{T U}} \quad \text{(6.)}$$

It is evident that with a given radius and a given coefficient of friction, the efficiency of an axle is the greater the more nearly the effort and the given force are brought into direct opposition to each other, and also the more distant their lines of action are from the axis of rotation.

439. **Efficiency of a Screw.**—The efficiency of a screw acting as a primary piece is nearly the same with that of a block sliding on a straight guide, which represents the development of a helix situated midway between the outer and inner edges of the screw-thread; the block being acted upon by forces making the same angles with the straight guide that the actual forces do with that helix. As to the development of a helix, see Article 160, page 94; and as to the efficiency of a piece sliding along a straight guide, see Article 437, page 288.

Section 2.—Efficiency and Counter-efficiency of Modes of Connection in Mechanism.

440. **Efficiency of Modes of Connection in General.**—In an elementary combination consisting of two pieces, a driver and a

follower, there is always some work lost in overcoming wasteful resistance occasioned by the mode of connection; the result being that the work done by the driver at its working-point is greater than the work done upon the follower at its driving-point, in a proportion which is *the counter-efficiency of the connection;* and the reciprocal of that proportion is *the efficiency of the connection.* In calculating the efficiency or the counter-efficiency of a train of mechanism, therefore, the factors to be multiplied together comprise not only the efficiencies, or the counter-efficiencies, of the several primary pieces considered separately, but also those of the several modes of connection by which they communicate motion to each other.

441. **Efficiency of Rolling Contact.**—The work lost when one primary piece drives another by rolling contact is expended in overcoming the *rolling resistance* of the pitch-surfaces, a kind of resistance whose mode of action has been explained in Article 402, page 251; and the value of that work in units of work per second is given by the expression $a\, b\, \mathrm{N}$; in which N is the normal pressure exerted by the pitch-surfaces on each other; b, a constant arm, of a length depending on the nature of the surfaces (for example 0·002 of a foot = 0·6 millimètre for cast iron on cast iron, see page 252); and a the relative angular velocity of the surfaces.

The useful work per second is expressed by $u\, f\, \mathrm{N}$, in which f is the coefficient of friction of the surfaces, and u the common velocity of the pitch lines. Hence the *counter-efficiency* is

$$c = 1 + \frac{a\, b}{u\, f} \quad \text{(1.)}$$

Let p_1 and p_2 be the lengths of two perpendiculars let fall from the two axes of rotation on the common tangent of the two pitch-lines; if the pieces are circular wheels, those perpendiculars will be the radii. Then the absolute angular velocities of the pieces are respectively $\frac{u}{p_1}$ and $-\frac{u}{p_2}$; and their relative angular velocity is therefore

$$a = u\left(\frac{1}{p_1} + \frac{1}{p_2}\right);$$

which value being substituted in Equation 1, gives for the counter-efficiency the following value:—

$$c = 1 + \frac{1}{f}\left(\frac{b}{p_1} + \frac{b}{p_2}\right) \quad \text{(2.)}$$

It is assumed that the normal pressure is not greater than is necessary in order to give sufficient friction to communicate the motion.

It is evident, from the smallness of b, that the lost work in this case must be almost always a very small fraction of the whole.

442. **Efficiency of Sliding Contact in General.**—In fig. 157, let T be the point of contact of a pair of moving pieces connected by sliding contact. Let the plane of the figure be that containing the directions of motion of the two particles which touch each other at the point T; and let T V be the velocity of the driving-particle, and T W the velocity of the following particle; whence V W will represent the velocity of sliding, and T U, perpendicular to V W, the common component of the velocities of the two particles along their line of connection R T P. C T C, parallel to V W, and perpendicular to R T P, is a common tangent to the two acting surfaces at the point T; the arrow A represents the direction in which the driver slides relatively to the follower; and the arrow B, the direction in which the follower slides relatively to the driver.

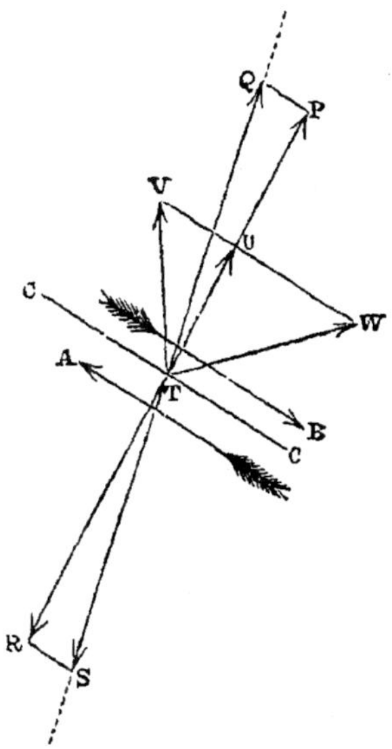

Fig. 157.

Along the line of connection, that is, normal to the acting surfaces at T, lay off T P to represent the effort exerted by the driver on the follower, and T R (= – T P) to represent the equal and opposite useful resistance exerted by the follower against the driver. Draw S T Q, making with R T P an angle equal to the angle of repose of the rubbing surfaces, (see Article 261, page 154), and inclined in the proper direction to represent forces opposing the sliding motion; draw P Q and R S parallel to C C. Then T Q will represent the resultant pressure exerted by the driver on the follower, and T S (= – T Q), the equal and opposite resultant pressure exerted by the follower against the driver, and P Q = – R S will represent the friction which is overcome, through the distance V W, in each second; while the useful resistance, T R, is overcome through the distance T U. Hence the useful work per second is T U · T R; the lost work is V W · R S; and the counter-efficiency is

$$c = 1 + \frac{\mathrm{V\,W} \cdot \mathrm{R\,S}}{\mathrm{T\,U} \cdot \mathrm{T\,R}} \quad \text{(1.)}$$

Let the angle U T V $=\alpha$, the angle U T W $=\beta$, and let f be the coefficient of friction. Then we have—

$$\frac{\text{V W}}{\text{T U}}=\tan\alpha+\tan\beta;\ \frac{\text{R S}}{\text{T R}}=f;$$

and consequently

$$c=1+f(\tan\alpha+\tan\beta)\ldots\ldots\ldots\ldots\ldots\ldots(2.)$$

443. **Efficiency of Teeth.**—It has already been shewn, in Article 148, page 87, that the relative velocity of sliding of a pair of teeth in outside gearing is expressed at a given instant by

$$(a_1+a_2)\,t;$$

where t denotes the distance at that instant of the point of contact from the pitch-point. (In inside gearing the angular velocity of the greater wheel is to be taken with the negative sign.)

The distance t is continually varying from a maximum at the beginning and end of the contact, to nothing at the instant of passing the pitch-point. Its *mean value* may be assumed, with sufficient accuracy for practical purposes, to be sensibly equal to *one-half* of its greatest value; and in the formulæ which follow, the symbol t stands for that mean value.

Let P be the mutual pressure exerted by the teeth; f, the coefficient of friction; then the work lost per second through the friction of the teeth is

$$(a_1+a_2)\,tf\,\text{P}.$$

Let u be the common velocity of the two pitch-circles; θ, the *mean obliquity* of the line of connection to the common tangent of the pitch-circles; then $u\cos\theta$ is the mean value of the common component of the velocities of the acting surfaces of the teeth along the line of connection; and the useful work done per second is expressed by

$$\text{P}\,u\cos\theta.$$

so that the counter-efficiency is

$$c=1+\frac{(a_1+a_2)\,tf}{u\cos\theta}\ldots\ldots\ldots\ldots\ldots\ldots(1.)$$

Let r_1 and r_2 be the radii of the two pitch circles; then we have

$$a_1=\frac{u}{r_1};\ a_2=\frac{u}{r_2};$$

and consequently

$$c=1+ft\sec\theta\left\{\frac{1}{r_1}+\frac{1}{r_2}\right\}\ldots\ldots\ldots\ldots\ldots\ldots(2.)$$

If two pairs of teeth at least are to be in action at each instant (as in the case of involute teeth, and of some epicycloidal teeth), and if the pitch be denoted by p, we have † see $\theta = \frac{p}{2}$; and therefore

$$c = 1 + \frac{fp}{2}\left\{\frac{1}{r_1} + \frac{1}{r_2}\right\} = 1 + \pi f\left\{\frac{1}{n_1} + \frac{1}{n_2}\right\}; \text{.......(3.)}$$

where n_1 and n_2 are the number of teeth in the two wheels.

In many examples of epicycloidal teeth, especially where small pinions are used, the duration of the contact is only $\frac{2}{3}$ or $\frac{3}{4}$ of that assumed in Equation 3; and the work lost in friction is less in the same proportion.

444. **Efficiency of Bands.**—A band, such as a leather belt or a hempen rope, which is not perfectly elastic, requires the expenditure of a certain quantity of work—first to bend it to the curvature of a pulley, and then to straighten it again; and the quantity of work so lost has been found by experiment to be nearly the same as would be required in order to overcome an additional resistance, varying directly as the sectional area of the band, directly as its tension, and inversely as the radius of the pulley. In the following formulæ for leather belts, the stiffness is given as estimated by Reuleaux (*Constructionslehre für Maschinenbau*, § 307).

Let T be the mean tension of the belt; S, its sectional area; r, the radius of the pulley; b, a constant divisor determined by experiment; R′, the resistance due to stiffness; then

$$R' = \frac{S\,T}{b\,r} \text{..............................(1.)}$$

b (for leather) = 3·4 inch = 87 millimètres.

To apply this to an endless belt connecting a pair of pulleys of the respective radii r_1 and r_2, let T_1 and T_2 be the tensions of the two sides of the belt. Then the useful resistance is $T_1 - T_2$, the mean tension is $\frac{T_1 + T_2}{2}$; and the additional resistance due to stiffness is

$$\frac{T_1 + T_2}{2} \cdot \frac{S}{b}\left\{\frac{1}{r_1} + \frac{1}{r_2}\right\};$$

consequently the counter-efficiency is

$$\left.\begin{aligned} c &= 1 + \frac{T_1 + T_2}{2(T_1 - T_2)} \cdot \frac{S}{b}\left\{\frac{1}{r_1} + \frac{1}{r_2}\right\}; \\ &= 1 + \frac{N + 1}{2(N - 1)} \cdot \frac{S}{b}\left\{\frac{1}{r_1} + \frac{1}{r_2}\right\}; \end{aligned}\right\} \text{.............(2.)}$$

N denoting $\frac{T_1}{T_2}$. The sectional area, S, of a leather belt is given by the formula

$$S = \frac{T_1}{p}; \quad \text{(3.)}$$

where p denotes the safe working tension of leather belts, in units of weight per unit of area; its value being, according to Morin,

0·2 kilogramme on the square millimètre, or
285 lbs. on the square inch.

The ordinary thickness of the leather of which belts are made is about 0·16 of an inch, or 4 millimètres; and from this and from the area the breadth may be calculated. A double belt is of double thickness, and gives the same area with half the breadth of a single belt.

When a band runs at a high velocity, the *centrifugal tension*, or tension produced by centrifugal force, must be added to the tension required for producing friction on the pulleys, in order to find the total tension at either side of the band, with a view to determining its sectional area and its stiffness. The centrifugal tension is given by the following expression :—

$$\frac{w\,S\,v^2}{g}; \quad \text{(4.)}$$

in which w is the heaviness (being, for leather belts, nearly equal to that of water); S, the sectional area; v, the velocity; and g, gravity (= 32·2 feet, or 9·81 mètres per second).

When centrifugal force is taken into consideration, the following formula is to be used for calculating the sectional area; T_1 being the tension at the driving-side of the belt, *exclusive of centrifugal tension* :—

$$S = \frac{T_1}{p - \frac{w\,v^2}{g}}; \quad \text{(5.)}$$

and the following formula for the counter-efficiency :—

$$c = 1 + \frac{T_1 + T_2 + \frac{2\,w\,v^2}{g}}{2\,(T_1 - T_2)} \cdot \frac{S}{b} \cdot \left\{ \frac{1}{r_1} + \frac{1}{r_2} \right\} \quad \text{(6.)}$$

For calculating the efficiency of hempen ropes used as bands, it is unnecessary in such questions as that of the present article to use a more complex formula than that of Eytelwein—viz.,

$$R' = \frac{D^2\,T}{b'\,r}; \quad \text{(7.)}$$

where D is the diameter of the rope, and $b' = 54$ millimètres = 2·125 inches.

In all the formulæ, $\frac{D^2}{b'}$ is to be substituted for $\frac{S}{b}$. The proper value of D^2 is given by the formula

$$D^2 = \frac{T_1}{p'}; \qquad (8.)$$

where $p' = 1,000$ for measures in inches and lbs.; and
$p' = \ \ 0·7$ for measures in millimètres and killogrammes.

445. **Efficiency of Linkwork.**—In fig. 158, let $C_1 T_1$, $C_2 T_2$ be two levers, turning about parallel axes at C_1 and C_2, and connected with each other by the link $T_1 T_2$; T_1 and T_2 being the connected points.

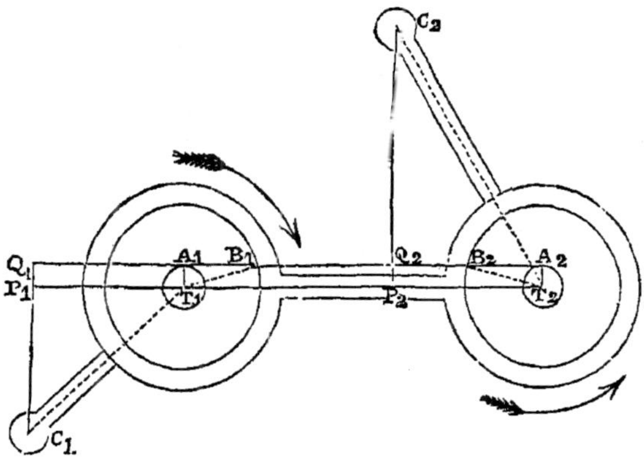

Fig. 158.

The pins, which are connected with each other by means of the link, are exaggerated in diameter, for the sake of distinctness. Let $C_1 T_1$ be the driver, and $C_2 T_2$ the follower, the motion being as shewn by the arrows. From the axes let fall the perpendiculars $C_1 P_1$, $C_2 P_2$, upon the line of connection. Then the angular velocities of the driver and follower are inversely as those perpendiculars; and, in the absence of friction, the driving moment of the first lever and the working moment of the second are directly as those perpendiculars; the driving pressure being exerted along the line of connection $T_1 T_2$. Let M_2 be the working moment; and let M_0 be the driving moment in the absence of friction; then we have

$$M_0 = \frac{M_2 \cdot C_1 P_1}{C_2 P_2}.$$

To allow for the friction of the pins, multiply the radius of each pin by the sine of the angle of repose; that is, very nearly by the coefficient of friction; and with the small radii thus computed, $T_1 A_1$ and $T_2 A_2$, draw small circles about the connected points. Then draw a straight line, $Q_1 A_1 B_1 Q_2 A_2 B_2$, touching both the small circles, and in such a position as to represent the line of action of a force that resists the motion of both pins in the eyes of the link. This will be the line of action of the resultant force exerted through the link. Let fall upon it the perpendiculars $C_1 Q_1$, $C_2 Q_2$; these will be proportional to the actual driving moment and working moment respectively; that is to say, let M_1 be the driving moment, including friction; then

$$M_1 = \frac{M_2 \cdot C_1 Q_1}{C_2 Q_2}.$$

Comparing this with the value of the driving moment without friction, we find for the counter-efficiency

$$c = \frac{M_1}{M_0} = \frac{C_1 Q_1 \cdot C_2 P_2}{C_2 Q_2 \cdot C_1 P_1}; \qquad (1.)$$

and for the efficiency

$$\frac{1}{c} = \frac{M_0}{M_1} = \frac{C_2 Q_2 \cdot C_1 P_1}{C_1 Q_1 \cdot C_2 P_2} \qquad (2.)$$

446. **Efficiency of Blocks and Tackle.**—(See Articles 181, 182, pages 105 and 106.)—In a tackle composed of a fixed and a running block containing sheaves connected together by means of a rope, let the number of plies of rope by which the blocks are connected with each other be n. This is also the collective number of sheaves in the two blocks taken together, and is the number expressing the *purchase*, when friction is neglected.

Let c denote the counter-efficiency of a single sheave, as depending on its friction on the pin, according to the principles of Article 373, page 290. Let c' denote the counter-efficiency of the rope, when passing over a single sheave, determined by the principles of Article 444, the tension being taken as nearly equal to $\frac{R}{n}$; where R is the useful load, or resistance opposed to the motion of the running block. $R \div n$ is also the effort to be exerted on the hauling part of the rope, in the absence of friction. Then the counter-efficiency of the tackle will be expressed approximately by

$$(c c')^n; \qquad (1.)$$

so that the actual or effective purchase, instead of being expressed by n, will be expressed by

$$n (c c')^{-n} \qquad (2.)$$

447. **Efficiency of Connection by means of a Fluid.**—When motion is communicated from one piston to another by means of an intervening mass of fluid, as described in Articles 185 to 188, pages 110 and 111, the efficiencies and counter-efficiencies of the two pistons have in the first place to be taken into account; that is to say, with ordinary workmanship and packing, the efficiency of each piston may be taken at 0·9 nearly; while with a carefully made cupped leather collar the counter-efficiency of a plunger may be taken at the following value:—

$$1 - \frac{4b}{d}; \qquad (1.)$$

in which d is the diameter of the plunger; and b a constant, whose value is from 0·01 to 0·015 of an inch, or from 0·25 to 0·38 of a millimètre. For if c be the circumference of the plunger, and p the effective pressure of the liquid, the whole amount of the pressure on the plunger is $\frac{pcd}{4}$; and the pressure required to overcome the friction is pcb.

The efficiency and counter-efficiency of the intervening mass of fluid remain to be considered; and if that fluid is a liquid, and may therefore be regarded as sensibly incompressible, these quantities depend on the work which is lost in overcoming the resistance of the passage which the liquid has to traverse.

To prevent unnecessary loss of work, that passage should be as wide as possible, and as nearly as possible of uniform transverse section; and it should be free from sudden enlargements and contractions, and from sharp bends, all necessary enlargements and contractions which may be required being made by means of gradually tapering conoidal parts of the passage, and all bends by means of gentle curves. When those conditions are fulfilled, let Q be the volume of liquid which is forced through the passage in a second; S, the sectional area of the passage; then,

$$v = \frac{Q}{S}, \qquad (2.)$$

is the velocity of the stream of fluid. Let b denote the wetted border or circumference of the passage; then,

$$m = \frac{S}{b}, \qquad (3.)$$

is what is called the *hydraulic mean depth* of the passage. In a cylindrical pipe, $m = \frac{1}{4}$ diameter. Let l be the length of the passage, and w the heaviness of the liquid. Then the loss of pressure in overcoming the friction of the passage is

$$p' = \frac{fl}{m} \cdot \frac{w v^2}{2g}; \quad \text{.......................(4.)}$$

in which g denotes gravity, and f a coefficient of friction whose value, for water in cylindrical cast-iron pipes, according to the experiments of Darcy, is

$$f = 0{\cdot}005 \left(1 + \frac{1}{12\,d}\right);^{*} \quad \text{..................(5.)}$$

d being the diameter of the pipe in feet.

Let p be the pressure on the driven or following piston; then the pressure on the driving piston is $p + p'$; and the *counter-efficiency of the fluid* is

$$1 + \frac{p'}{p}; \quad \text{...........................(6.)}$$

which, being multiplied by the product of the counter-efficiencies of the two pistons, gives the *counter-efficiency of the intervening liquid.*

When the intervening fluid is *air*, there is a loss of work through friction of the passage, depending on principles similar to those of the friction of liquids; and there is a further loss through the escape by conduction of the heat produced by the compression of the air.

The friction which has to be overcome by the air, and which causes a certain loss of pressure between the compressing pumps and the working machinery, consists of two parts, one occasioned by the resistance of the valves, and the other by the friction along the internal surface of pipes.

To overcome the resistance of valves, about *five per cent.* of the effective pressure may be allowed.

The friction in the pipes depends on their length and diameter, and on the velocity of the current of air through them. It is nearly proportional to the square of the velocity of the air.

A velocity of about *forty feet per second* for the air in its compressed state has been found to answer in practice. The diameter of pipe required in order to give that velocity can easily be computed, when the dimensions of the cylinders of the machinery to be driven, and the number of strokes per minute, are given.

When the diameter of a pipe is so adjusted that the velocity of the air is 40 feet per second, the pressure expended in overcoming its friction may be estimated at *one per cent. of the total or absolute*

* When the diameter is expressed in millimètres, for $\frac{1}{12\,d}$ substitute $\frac{25{\cdot}4}{d}$.

pressure of the air, for every five hundred diameters of the pipe that its length contains.

Although the abstraction from the air of the heat produced by the compression involves a certain sacrifice of motive power (say from 30 to 35 per cent.) still the effects of the heated air are so inconvenient in practice, that it is desirable to cool it to a certain extent during or immediately after the compression. This may be effected by injecting water in the form of spray into the compressing pumps; and for that purpose a small forcing pump of about $\frac{1}{100}$th of the capacity of the compressing pumps has been found to answer in practice. The air may thus be cooled down to about 104° Fahr. or 40° Cent.

The factor in the counter-efficiency due to the loss of heat expresses the ratio in which the volume of air as discharged from the compressing pump at a high temperature is greater than the volume of the same air when it reaches the working machinery at a reduced temperature; which ratio may be calculated approximately by taking *two-sevenths of the logarithm of the absolute working pressure of the compressed air in atmospheres, and finding the corresponding natural number.* That is to say, let p_0 denote one atmosphere (= at the level of the sea 14·7 lbs. on the square inch, or 10,333 kilogrammes on the square mètre); let p_1 be the absolute working pressure of the air, so that $p_1 - p_0$ is the effective pressure; then the counter-efficiency due to the escape of heat is,

$$c = \left(\frac{p_1}{p_0}\right)^{\frac{2}{7}} \text{..........................} (7.)$$

From examples of the practical working of compressed air, when used to transmit motive power to long distances, it appears that in order to provide for leakage and various other imperfections in working, the capacity of the compressing pumps should be very nearly double of the net volume of uncompressed air required; and it has also been found necessary, in working the compressing pumps, to provide from three to four times the power of the machinery driven by the compressed air.

INDEX.

THE END.

PRINTED BY BELL AND BAIN, GLASGOW

LIST OF STANDARD BOOKS

PUBLISHED BY

CHARLES GRIFFIN & COMPANY

10, STATIONERS' HALL COURT, LONDON.

Aitken's (William, M.D., Prof. of Pathology in The Army Medical School, &c., &c., &c.) THE SCIENCE and PRACTICE of MEDICINE, in two vols., 8vo., bound in cloth, with a Steel Plate, Map, and nearly 200 Woodcuts. *Sixth edition.* Price 38s.

"The present work contains information that will not be found in any other manual of medicine."—*Athenæum.*

"The book is the most comprehensive of any that have in late years been published on the Practice of Medicine."—*British Medical Journal.*

"We know of no work that contains so much, or such full and varied information on all topics connected with the Science and Practice of Medicine."—*Lancet,* Feb. 17, 1872.

"These two volumes are a library in themselves."—*Medical Press,* March 26, 1872.

"The extraordinary merit of Dr. Aitken's book....The author has unquestionably performed a service to the profession of the most valuable kind."—*Practitioner,* March, 1872.

Aitken's (William, M.D.) The Growth of the Recruit, and the Young Soldier, with a view to the Selection of "Growing Lads," and the training of them. By WILLIAM AITKEN, M.D. 12mo., cloth, 2s. 6d.

"This little work should be in the hands of all instructors of youth, and all employers of youthful labour."—*Lancet.*

Aitken's (William, M.D.) Outline Figures of the Trunk of the Human Body. On which to indicate the areas of physical signs in the Clinical Diagnosis of Disease. For the use of Students and Practitioners of Medicine. By WILLIAM AITKEN, M.D. 1s. 6d., by post, 1s. 8d.

Altar of the Household (The); a Series of Prayers, and Selections from the Scriptures, for Domestic Worship, for every Morning and Evening in the year. Edited by the Rev. Dr. HARRIS, with an Introduction by the Rev. W. LINDSAY ALEXANDER, D.D. Royal 4to., cloth, gilt edges, 21s.; half bound calf, marbled edges, strong, 25s.; whole bound, calf, gilt edges, 42s., best levant morocco, antique, gilt edges, 50s.

Altar of the Household, Illustrated with a series of First-class Engravings on Steel, descriptive of the most important events in the Old and New Testaments. Royal 4to., cloth, gilt edges, 25s.; half-bound calf, marble edges, 31s. 6d.; full calf, gilt edges, 47s.; best levant morocco, 55s.

A Manual of Instruction for Attendants on Sick and Wounded in War. By Staff Assistant-Surgeon A. MOFFITT. Published under the sanction of the National Society for Aid to the Sick and Wounded in War. With numerous illustrations, post 8vo, cloth, 5s.

"A work by a practical and experienced author. After an explicit chapter on the Anatomy of the Human Body, directions are given concerning bandaging, dressing of sores, wounds, &c., assistance to wounded on field of action, stretchers, mule litters, ambulance, transport, &c. All Dr. Moffitt's instructions are assisted by well executed illustrations."—*Public Opinion*.

"A well written volume. Technical language has been avoided as much as possible, and ample explanations are afforded on all matters on the uses and management of the field hospital equipment of the British Army."—*Standard*.

A Novel with Two Heroes; By ELLIOTT GRÆME, Author of "Beethoven; a Memoir," in two vols., post 8vo., 21s. *Second Edition.*

"A decided literary success."—*Athenæum*.

"Clever and amusing . . . above the average even of good novels . . . free from sensationalism, but full of interest . . . touches the deeper chords of life . . . delineation of character remarkably good."—*Spectator*.

"Superior in all respects to the common run of novels."—*Daily News*.

"A story of deep interest......The dramatic scenes are powerful almost to painfulness in their intensity."—*Scotsman*.

Ansted's (Professor) Natural History of the Inanimate Creation, recorded in the Structure of the Earth, the Plants of the Field, and the Atmospheric Phenomena. With numerous Illustrations. Large post 8vo., 8s. 6d., cloth.

Astronomy, Elements of, by FREDERICK SCHOEDLER, Ph.D., Professor of the Natural Sciences at Worms. Translated and Edited by HENRY MEDLOCK, Ph.D., F.C.S. New Edition, *numerous engravings*, crown 8vo., cloth, 1s. 6d.

Applied Mechanics, a Manual of; comprising the Principles of Statics and Cinematics, and Theory of Structures, Mechanism, and Machines. By W. J. MACQUORN RANKINE, C.E., LL.D., F.R.SS., late Regius Prof. of Civil Engineering in the University of Glasgow. With numerous diagrams. *Sixth edition*, revised. Cloth, 8vo., 12s. 6d.

"Cannot fail to be adopted as a text-book........The whole of the information is so admirably arranged, that there is every facility for reference."—*Mining Journal*.

Baird's (W., M.D., F.L.S.) Student's Natural History: being a Dictionary of the Natural Sciences. With a Zoological Chart, showing the Distribution and Range of Animal Life. Numerous Illustrations. Demy 8vo., cloth gilt, red edges, 10s. 6d.

"The work is a very useful one, and will contribute, by its cheapness and comprehensiveness, to foster the extending taste for natural science."—*Westminster Review*.

Beethoven. A Memoir. By ELLIOTT GRAEME. With an Essay (Quasi Fantasia) "on the Hundredth Anniversary of his Birth," and remarks on the Pianoforte Sonatas, with hints to Students. By Dr. FERDINAND HILLER, of Cologne. Beautifully printed, and handsomely bound in cloth, gilt, 3s. 6d.

"This elegant and interesting Memoir.........The newest, prettiest, and most readable sketch of the immortal Master of Music."—*Musical Standard.*

"A gracious and pleasant Memorial of the Centenary."—*Spectator.*

"This delightful little book—concise, sympathetic, judicious."—*Manchester Examiner.*

"We can, without reservation, recommend it as the most trustworthy and the pleasantest Memoir of Beethoven published in England."—*Observer.*

"A most readable volume, which ought to find a place in the library of every admirer of the great Tone-Poet."—*Edinburgh Daily Review.*

Biblical Cyclopædia; or, Dictionary of Eastern Antiquities, Geography, Natural History, Sacred Annals and Biography, Theology, and Biblical Literature, Illustrative of the Old and New Testaments. By the Rev. JOHN EADIE, D.D., LL.D. With Maps, prepared expressly by W. and A. K. JOHNSTON, and numerous Pictorial Illustrations. *Fourteenth edition*, embracing all the latest Discoveries and Explorations. Large post 8vo., 700 pages, handsome cloth, price 7s. 6d.

"It gives within a moderate compass a great amount of information, accurate and well put together. The article on 'Creation,' with its survey of the question as it stands between science and the Mosaic Cosmogony, may be cited as a specimen of the candour and liberality with which the editor has done his work."—*Spectator.*

"We must regard this Bible Dictionary of Dr. Eadie's as decidedly the most adapted for popular use, and have always found it a reliable authority. To the Clergy not possessed of large libraries, and to whom the price of books is important, we can cordially recommend the present volume."—*Clerical Journal.*

Book-keeping, a Complete System of Practical, exemplified in Five Sets of Books, arranged by Single Entry, Double Entry in present practice, and a new method of Double Entry by Single. And an Appendix, containing a variety of Illustrations and Exercises, with a Series of Engraved Forms of Accounts. By C. MORRISON, Accountant. *Ninth edition*, with valuable additions and improvements. 8vo., half bound, 8s.

Botany; Elements of. With *Numerous Illustrations of Fruits, Flowers, &c.* By Professors SCHOEDLER and MEDLOCK. Crown 8vo., cloth, 1s. 6d.

Bowdler's (Thomas, F.R.S.) Family Shakspeare. The Dramatic Works of Shakspeare. Edited and adapted for Family Reading. Small 8vo., cloth, gilt. *New edition in preparation.*

*** This unique Edition of the great Dramatist is admirably adapted for a Gift, or Prize-book for young people.

Brougham, the Life of Lord. His Career as a Statesman, a Lawyer, and a Philanthropist, from Authentic sources. By JOHN MCGILCHRIST. Foolscap 8vo., with a fine Portrait on Steel. Cloth, 2s. 6d.

Bryce's (Archibald Hamilton, B.A., LL.D.) The Works of Virgil. Text based on Heyne and Wagner; Notes original, and selected from the best German, American, and English Commentators; Illustrations from the Antique. In three Parts, fcap. 8vo., cloth, 2s. 6d. each.

Part I. Bucolics and Georgics.

Part II. The Æneid. Liber I—VI.

Part III. The Æneid. Liber VII—XII.

Or complete in one volume, cloth, 6s.

"Contains the pith of what has been written by the best scholars on the subject."—*Athenæum.*

"The most complete, as well as elegant and correct edition of Virgil ever published in this country."—*Educational Times.*

"The best Commentary on Virgil which the English Student of his language can obtain."—*Scotsman.*

Bunyan's Pilgrim's Progress. With Life and Notes, Experimental and Practical, by William Mason. Printed in large type, and illustrated with full page Woodcuts, crown 8vo., handsomely bound in gilt cloth, gilt edges, price 3s. 6d.

Bunyan's Pilgrim's Progress. With Expository Lectures, by the Rev. Robert Maguire, Incumbent of Clerkenwell. With Steel Engravings, *Second edition*, imperial 8vo., cloth, 10s. 6d.

Burns' Songs and Ballads. With an Introduction on the Character and Genius of Burns, by Thomas Carlyle. Carefully printed in antique type, and illustrated with beautiful Engravings on Steel. Foolscap 8vo., elegantly bound in cloth and gold, 3s.; malachite, 10s. 6d.

Byron's Childe Harold's Pilgrimage. With a Memoir by Prof. Spalding, of St. Andrew's. Illustrated with Engravings on Steel by Greatbach, Miller, Lightfoot, &c., from Paintings by Cattermole, Sir T. Lawrence, H. Howard, and Stothard. Beautifully printed on Toned Paper. Foolscap 8vo., cloth elegant, 3s.; malachite, 10s. 6d.

Campbell's Pleasures of Hope, and other Poems; Including some Verses never before published. With an original Memoir by the Rev. Charles Rogers, LL.D. Beautifully embellished with fine Portrait and several Steel Engravings. Cloth and gold, 3s.; morocco, 8s.; malachite, 10s. 6d.

"Well adapted for a choice present."—*Public Opinion.*

"The illustrations engraved on steel . . . . are very beautiful."—*Bookseller.*

"Dr. Rogers' memoir of the poet is exceedingly well written, and contains several facts illustrative of Campbell's character and genius which are not to be found in any other edition."—*North British Daily Review.*

Celestial Scenery; or, the Wonders of the Planetary System Displayed, including all new Discoveries. This Work is intended for general readers, presenting to their view, in an attractive manner, sublime objects of contemplation. By THOMAS DICK, LL.D. *New edition.* Printed on Toned Paper, handsomely bound, with gilt edges, price 5s.

Characteristics of Eminent Men, by John Timbs. Bound in neat cloth, price 1s. (one of the series entitled "Griffin's Shilling Manuals," edited by JOHN TIMBS.)

"It is impossible to dip into this booklet without finding something that is new or that will bear repeating."—*Sheffield Daily Telegraph.*

Chatterton's Poetical Works. With an Original Memoir by FREDERICK MARTIN. Beautifully Illustrated *by First-class Engravings and a fine portrait on steel,* and elegantly printed. Fcap. 8vo., cloth and gold, 3s.; malachite, 10s. 6d.

Chaucer's Poetical Works, with Introduction, Notes, Memoir, and Glossary, by ROBERT BELL. Eight volumes. Foolscap 8vo., cloth gilt, 10s.

Chemistry, Elements of, by Friedrich Schoedler, Ph.D. New Edition, translated from the Sixth German Edition by HENRY MEDLOCK, F.C.S. Illustrated by numerous Engravings. Crown 8vo., cloth, 1s. 6d.

Christian Philosopher, The; or, The Connection of Science and Philosophy with Religion. By THOMAS DICK, LL.D. *Twenty-seventh edition,* revised and enlarged. Illustrated with 150 Engravings on Wood. Crown 8vo., cloth, printed on Toned Paper, handsomely bound, with gilt edges, 5s.

Christison's (John) Complete System of Interest Tables at 3, 4, 4½, and 5 per cent.; Tables of Exchange or Commission, Profit and Loss, Discount, Clothiers', Malt, Spirit, and various other useful Tables. To which is prefixed the Mercantile Ready Reckoner, containing Reckoning Tables from one thirty-second part of a penny to one pound. *New edition,* greatly enlarged, 12mo., bound in leather, 4s. 6d.

Christison's New Tables for computing the Weight of Cattle by Measurement, the quantity of Hay in Stacks or Ricks, the Valuing of Land, with other Practical Tables, and directions for using them. *Tenth edition,* bound, 2s.

Circle of the Sciences, by Owen, Ansted, Latham, etc., etc. Each Vol. 5s., cloth binding.

Vol. I.—Organic Nature, Part 1—Physiology.
Vol. II.—Organic Nature, Part 2—Botany, &c.
Vol. III.—Organic Nature, Part 3—Zoology.
Vol. IV.—Inorganic Nature—Geology, &c.

CIRCLE OF THE SCIENCES.—(*Continued.*)

Vol. V.—Navigation, Astronomy, &c.
Vol. VI.—Elementary Chemistry, Light, Heat, &c.
Vol. VII.—Practical Chemistry.
Vol. VIII.—The Mathematical Sciences.
Vol. IX.—Mechanical Philosophy.

The Treatises separately bound in cloth

	£	s.	d.
Ansted's Geology - - - -	0	2	6
Breen's Practical Astronomy - - -	0	2	6
Bronner and Scoffern's Food and Diet - -	0	1	6
Bushnan's Physiology - - -	0	1	6
Gore's Electro-Deposition - - -	0	1	6
Imray's Practical Mechanics - - -	0	1	6
Jardine's Practical Geometry - - -	0	1	6
Latham's Human Species - - -	0	1	6
Martin's Photographic Art - - -	0	2	6
Mitchell and Tennant's Mineralogy - -	0	3	0
Mitchell's Properties of Matter - -	0	1	6
Owen's Principal Forms of the Skeleton - -	0	1	6
Primary Atlas of Geography - - -	0	2	6
Primary Atlas of Geography, coloured - -	0	3	6
Scoffern's Light, Heat, &c. - - -	0	3	0
Scoffern's Inorganic Bodies - - -	0	3	0
Scoffern's Artificial Light - - -	0	1	6
Scoffern and Lowe's Meteorology - -	0	1	6
Smith's Botany - - - -	0	2	0
Twisden's Trigonometry - - -	0	1	6
Twisden on Logarithms - - -	0	1	0
Young's Elements of Algebra - - -	0	1	0
Young's Solutions of Questions in Algebra -	0	1	0
Young's Navigation and Nautical Astronomy -	0	2	6
Young's Plane Geometry - - -	0	1	6
Young's Simple Arithmetic - - -	0	1	0
Young's Elementary Dynamics - -	0	1	0

Civil Engineering, a Manual of, by Professor W. J. MACQUORN RANKINE. Comprising Engineering, Surveys, Earthwork, Foundations, Masonry, Carpentry, Metal Work, Roads, Railways, Canals, Rivers, Water Works, Harbours, etc. With numerous Tables and Illustrations. *Ninth edition.* Cloth, 16s.

"A work comprising much original research, as well as comprehensive study. Its pages contain a large amount of instructive matter, very clearly arranged and put into a shape readily available, both to scientific and practical students."—*Civil Engineer and Architect's Journal.*

"Far surpasses in merit every existing work of the kind. As a 'Manual' for the hands of the professional Civil Engineer it is sufficient and unrivalled, and even when we say this we fall short of that high appreciation of Dr. Rankine's labours which we should like to express."—*The Engineer.*

"The 'Manual of Civil Engineering' might without any impropriety be termed an Encyclopædia of the science, for it touches, and that with a master hand, every branch of it."—*Mechanics' Magazine.*

"A compact compression of the science of Engineering."—*Builder.*

WILLIAM COBBETT'S WORKS.

The only Correct and Authorised Editions.

Cobbett's (William) Advice to Young Men, and (incidentally) to Young Women, in the Middle and Higher Ranks of Life. In a Series of Letters addressed to a Youth, a Bachelor, a Lover, a Husband, a Father, a Citizen, and a Subject. *New Edition.* Foolscap 8vo., cloth, 2s. 6d.

Cobbett's (William) Legacy to Labourers; an Argument showing the Right of the Poor to Relief from the Land. With a Preface by the Author's son, J. M. Cobbett, M.P. for Oldham. Cloth, 1s. 6d.

"The book cannot be too much studied."—*Nonconformist.*

Cobbett's (William) Legacy to Parsons; or, Have the Clergy of the Established Church an Equitable Right to the Tithes and Church Property? With a New Preface by the Author's son, Mr. W. Cobbett. Now ready, a *New Edition.* Cloth, 1s. 6d.

"The most powerful work of the greatest master of political controversy this country has ever produced."—*Pall Mall Gazette.*

Cobbett's (William) English Grammar. With an additional Chapter on Pronunciation. By J.P. Cobbett. Cl., 1s. 6d.

Cobbett's (William) French Grammar. No better extant. Cloth, 3s. 6d. *Fourteenth edition.*

"'Cobbett's French Grammar' comes out with perennial freshness. There are few grammars equal to it for those who are learning, or desirous of learning, French without a teacher. The work is excellently arranged, and in the present edition we note certain careful and wise revisions of the text."—*School Board Chronicle,* Feb. 18, 1871.

Cobbett's (William) Cottage Economy; containing Information relative to the Brewing of Beer, Making of Bread, Keeping of Cows, Pigs, Bees, Poultry, &c.; and relative to other matters deemed useful in conducting the affairs of a Poor Man's Family. *New edition,* revised by the Author's son. Foolscap 8vo., cloth, 2s. 6d.

Cobbett's (William) Poor Man's Friend. A Defence of the Rights of those who do the Work and Fight the Battles. Foolscap 8vo., limp, 1s.

Cobbett's (William) Rural Rides in Twenty-eight English Counties: with Economical and Political Observations. With Notes by J. P. Cobbett. 12mo., cloth, 3s. 6d.

Cobbett's (James Paul) A Latin Grammar. Cloth, 2s.

Cobbett's (Miss) The English Housekeeper. Cloth, 3s. 6d.

Cobbett's (Miss S.) French Verbs and Exercises. Cloth, 1s. 6d.

Classified Bible, The: an Analysis of the Holy Scriptures. By the Rev. JOHN EADIE, D.D., LL.D., &c. Illustrated with Maps. *Third Edition.* Post 8vo., handsome cloth antique, 8s. 6d.; morocco, 16s.

"It is an attempt so to classify Scripture under separate heads as to exhaust its contents. The reader will find, under the respective Articles or Sections, what the Bible says on the separate subjects in relation to Doctrine, Ethics, and antiquities."—*Preface.*

"We have only to add our unqualified commendation of a work of real excellence to every Biblical student."—*Christian Times.*

"This massive and valuable volume requires no notice at our hands."—*Edinburgh Mercury.*

Coleridge (Samuel Taylor), A Dissertation on the Science of Method. (*Encyclopædia Metropolitana.*) With a Synopsis. *Ninth edition.* Crown 8vo., cloth, 2s.

Cowdy's (Rev. Samuel, F. R. Hist. Soc.) Heavenward Ho; or, Story Coxen's Log. Crown 8vo., cloth, 5s.

"The style is simple and pointed, characterised by a nervous energy which never slackens, and there is scarcely a sentence which does not felicitously embody a great truth of Scripture."—*Literary World.*

"A very effective and cleverly-written allegory, in which the progress of the soul in religious experience is portrayed."—*The Rock.*

Craik's English Literature. A Compendious History of English Literature, and of the English Language from the Norman Conquest. With numerous Specimens. By GEORGE L. CRAIK, LL.D. (*late Professor of History and English Literature, Queen's College, Belfast*). Now ready, a *new edition,* in two large 8vo. vols., handsomely bound in cloth, £1 5s.; in tree calf, £1 17s. 6d.

"Professor Craik's book going, as it does, through the whole history of the language, probably takes a place quite by itself. The great value of the book is its thorough comprehensiveness. He is always clear and straightforward, and deals not in theories but in facts."—*Saturday Review.*

"Professor Craik has succeeded in making a book more than usually agreeable."—*The Times.*

Craik's Manual of English Literature, for the Use of Colleges, Schools, and Civil Service Examinations. Selected from the larger work. By Professor CRAIK. *Fifth Edition.* Crown 8vo., 7s. 6d., cloth.

"A manual of English literature from so experienced and well read a scholar as Professor Craik needs no other recommendation than the mention of its existence."—*Spectator.*

"This augmented effort will be, we doubt not, received with decided approbation by those who are entitled to judge, and studied with much profit by those who want to learn. . . . . If our young readers will give healthy perusal to Dr. Craik's work, they will greatly benefit by the wide and sound views he has placed before them."—*Athenæum.*

Currie's (Joseph, formerly Head-Classical Master of Glasgow Academy), THE WORKS OF HORACE. Text from Orellius. Notes, original and selected from the works of the leading Commentators. Illustrations from the antique. In two parts. Fcap. 8vo., cloth, price 3s. each.

Part I. Carmina ... ... 3s.
Part II. Satires ... ... 3s.

Or, complete in one volume, cloth 5s.

"The notes are excellent and exhaustive."—*Quarterly Journal of Education.*

Currie's (Joseph) Extracts from Cæsar's Commentaries; containing his Description of Gaul, Britain, and Germany. Selected, with Notes and Vocabulary for the Use of Young Students. *Fourth edition.* 18mo., cloth, 1s. 6d.

Cook's Voyages. Voyages Round the World, by Captain COOK. Illustrated with Maps and numerous Engravings. 2 vols., super-royal 8vo., cloth, 30s.

Creation's Testimony to its God: the Accordance of Science, Philosophy, and Revelation. A Manual of the Evidences of Natural and Revealed Religion; with especial Reference to the Progress of Science and Advance of Knowledge. By the Rev. THOMAS RAGG. *Twelfth edition*, revised and enlarged. In handsome cloth, bevelled boards, 5s.

"We are not a little pleased again to meet with the author of this volume in the tenth edition of his far-famed work. Mr. Ragg is one of the few original writers of our time to whom justice is being done."—*British Standard.*

This work, originally published in 1855, has now, almost without effort or advertisement, reached its *eleventh* edition. It has been pronounced "The book of the age," "The marvel of the age," "The best popular text book of the sciences," and "The only complete manual of religious evidence natural and revealed."

Cruden's Complete Concordance to the Old and New Testaments, and the Books called Apochryphal. Edited and corrected by WILLIAM YOUNGMAN. With fine Portrait of Cruden. New edition. Imperial 8vo., cloth, handsome gilt top, 7s. 6d.

Cruden's Concordance to the Holy Scriptures. By the Rev. JOHN EADIE, D.D., LL.D. With an introduction by the Rev. Dr. KING. This has long and deservedly borne the reputation of being the completest and best edition extant, and the present reduction in price will also cause it to be by far the cheapest published. *Thirty-fourth edition.* Cloth, 3s. 6d.; whole calf, 8s.; whole morocco, 10s. 6d.

Curiosities of Animal and Vegetable Life. By JOHN TIMBS. Neat cloth, 1s. (one of a Series entitled "Griffin's Shilling Manuals," edited by John Timbs).

"Has the charm of freshness, besides containing much curious information."—*Sheffield Telegraph.*

Dalgairns' Cookery: The Practice of Cookery, adapted to the Business of Every-day Life. By Mrs. DALGAIRNS. *The best book for Scotch dishes.* About 50 new recipes have been added to the present edition, but only such as the Author has had adequate means of ascertaining to be valuable. Foolscap 8vo., cloth, 3s. 6d.

"This is by far the most complete and truly practical work which has yet appeared on the subject. It will be found an infallible 'Cook's Companion,' and a treasure of great price to the mistress of a family."—*Edinburgh Literary Journal.*

"We consider we have reason strongly to recommend Mrs. Dalgairns' as an economical useful, and practical system of cookery, adapted to the wants of all families, from the tradesman to the country gentleman."—*Spectator.*

Dallas' (W. S.) Popular History of the Animal Creation: being a systematic and popular Description of the Habits, Structure, and Classification of Animals. *New edition*, with many hundred Illustrations. Crown 8vo., cloth, 8s. 6d.

Dictionary of the Holy Bible. Designed chiefly for the Use of Young Persons. By the Rev. Prof. Eadie. With numerous Illustrations. *Thirteenth edition.* Small 8vo., cloth antique, red edges, 2s. 6d.; morocco, 7s. 6d.

"This is a most useful compendium of Bible lore. It is portable, yet comprehensive." —*Aberdeen Banner.*

"Parents and tutors will unanimously thank the author for this result of a labour of love."—*Critic.*

"A very good and useful compilation for youth."—*Literary Gazette.*

Domestic Medicine: a Dictionary of, and Household Surgery. By Spencer Thomson, M.D., L.R.C.S., Edinburgh. Thoroughly revised and brought down to the present state of Medical Science. With an additional Chapter on the Management of the Sick Room. *Tenth edition.* With Illustrations. Large 8vo., 750 pages, cloth 8s. 6d.

"The best and safest book on Domestic Medicine and Household Surgery which has yet appeared."—*London Journal of Medicine.*

"Dr. Thomson has fully succeeded in conveying to the public a vast amount of useful professional knowledge."—*Dublin Journal of Medical Science.*

"Worth its weight in gold to families and the clergy."—*Oxford Herald.*

Earth Delineated with Pen and Pencil, The: an Illustrated record of Voyages, Travels, and Adventures all round the World. Illustrated with more than 200 Engravings in the first style of Art, by the most eminent Artists, including several from the master pencil of Gustave Dore. Demy 4to., 750 pages, profusely illustrated, very handsomely bound, price £1 1s.

Education of Man, The. A Suggestive Dissertation on the Soul: What it is, and how trained. With an Appendix, "Have Savages Souls?" By a Member of the New Zealand Bar. Crown 8vo., cloth, 10s. 6d.

English Woman's Library; a Series of Moral and Descriptive Tales. By Mrs. Ellis. Cloth, each volume, 2s. 6d.; with gilt back and edges, 3s.

1.—The Women of England: their Social Duties and Domestic Habits. 39th thousand.

2.—The Daughters of England: their Position in Society, Character, and Responsibilities. 20th thousand.

3.—The Wives of England: their Relative Duties, Domestic Influence, and Social Obligations. 18th thousand.

4.—The Mothers of England: their Influence and Responsibility. 20th thousand.

ENGLISH WOMAN'S LIBRARY.—(*Continued*).

5.—FAMILY SECRETS; or, Hints to Make Home Happy. 3 vols. 23rd thousand.

6.—SUMMER AND WINTER IN THE PYRENEES. 10th thousand.

7.—TEMPER AND TEMPERAMENT; or, Varieties of Character. 2 vols. 10th thousand.

8.—PREVENTION BETTER THAN CURE; or, the moral Wants of the World we Live in. 12th thousand.

9.—HEARTS AND HOMES; or, Social Distinction. 3 vols. 10th thousand.

Eventful Life of a Soldier in the Peninsular. By JOSEPH DONALDSON, Sergeant in the Ninety-fourth Scots Regiment. Fcap. 8vo., cloth, 3s. 6d.; with gilt sides and edges, 4s.

Gilmer's Interest Tables: Tables for the Calculation of Interest, on any sum, for any number of days, at ½, 1, 1½, 2, 2½, 3, 3½, 4, 4½, 5, and 6 per cent. By ROBERT GILMER. Sixth Edition, corrected and enlarged. 12mo., roan, lettered, 5s.

Golden Leaves. From the Works of the Poets and Painters. Edited by ROBERT BELL. Illustrated by 64 superb Engravings on Steel, after Paintings by David Roberts, Stanfield, Leslie, Stothard, Haydon, Howard, Levaint, Cattermole, Nasmyth, Sir Thomas Lawrence, and many others, and engraved in the first style by Finden, Greatbach, Lightfoot, &c. In 4to., elegantly bound in cloth and gold, with gilt edges, One Guinea; also, in morocco antique, handsome, 35s. *Second Edition.*

"'Golden Leaves' is by far the most important book of the season. The illustrations are really works of art, and the volume does credit to the arts of England."—*Saturday Review.*

"The Poems are selected with taste and judgment."—*Times.*

"The engravings are from drawings by Stothard, Newton, Danby, Leslie, and Turner, and it is needless to say how charming are many of the above here given."—*Athenæum.*

"Mr. Bell has preserved a judicious mean between too wide a range of poets and too imperfect a collection of poetry; altogether 'Golden Leaves' is one of the most attractive of gift-books."—*Spectator.*

Goldsmith's (Oliver) Poetical Works. With a Memoir by WILLIAM SPALDING, A.M., and numerous Illustrations on Steel and Wood. Foolscap 4to., most elaborately gilt, cloth, 5s.

Griffin's (John Joseph, F.C.S.) Chemical Recreations: A Popular Manual of Experimental Chemistry. *Tenth edition.* With 540 engravings of Apparatus. Crown 4to., cloth.

Part I. Elementary Chemistry, price 2s.

Part II. The Chemistry of the Non-Metallic Elements, including a Comprehensive Course of Class Experiments, price 10s. 6d.

Or, complete in one volume cloth, gilt top, 12s. 6d.

WORKS by the Rev. JOHN EADIE, D.D., LL.D.

I Ecclesiastical Cyclopædia: a Dictionary of Christian Antiquities, Sects, Denominations, and Heresies; History of Dogmas, Rites, Sacraments, Ceremonies, etc., Liturgies, Creeds, Confessions, Monastic and Religious Orders, Modern Judaism, etc., etc. By the Rev. JOHN EADIE, D.D., LL.D., assisted by numerous Contributors. *Fourth edition,* post 8vo., cloth, 8s. 6d.; morocco, 15s.

"This Cyclopædia will prove acceptable both to the clergy and laity of Great Britain. A great body of curious and useful information will be found in it; the aim has been to combine popularity with exactness."—*Athenæum.*

"We very warmly commend a book prepared with so much fulness of knowledge and conscientious care; and we especially press it on the attention of ordinarily educated persons in our general congregations, of Sunday School teachers, and of ministers not possessed of large libraries."—*Nonconformist.*

"Our readers will not need to be told that this is a 'comprehensive' work, and we may add that it is one that will be found useful and convenient to a large number of both clergy and laity."—*English Churchman.*

II. Rev. Professor Eadie's Commentary on the Greek Text of the Epistle of St. Paul to the Ephesians. *Second edition,* revised throughout and enlarged. 8vo., cloth, 14s.

"The book is one of prodigious learning and research. The author seems to have read all, in every language, that has been written upon the Epistle; it is also a work of independent criticism, and casts much new light upon many passages."—APPOLODORUS *in Literary Gazette.*

III. Eadie's Biblical Cyclopædia. Large post 8vo., 700 pp., handsome cloth, price 7s. 6d. *Fourteenth edition.*

IV. Eadie's Classified Bible. Post 8vo., handsome cloth, antique, price 8s. 6d. *Third edition.*

V. Eadie's Cruden's Concordance. Large post 8vo., 576 pp., price 3s. 6d. *Thirty-fourth edition.*

VI. Eadie's Bible Dictionary. Small 8vo. Price 2s. 6d. *Thirteenth edition.*

Hallowed Spots of Ancient London, The. Historical, Biographical, and Antiquarian. *With* 40 *Illustrations of Remarkable Places, Objects, and Incidents as they appeared and occurred in the Olden Time.* By ELIZA METEYARD (Silverpen). As the City of their Empire, London must ever have a peculiar interest to Englishmen. In this volume its long past scenes are clothed anew with vivid picturesqueness, rendering it a work most curious and entertaining. *A new edition in preparation.*

Henry's (Matthew) Commentary on the Holy Bible. *New edition,* 3 vols., super-royal 8vo., strongly bound in cloth, 50s., calf, marbled edges, 67s. 6d., levant morocco, antique, gilt edges, 84s.

Herbert's Poetical Works. With Memoir by J. NICHOL, B.A., Oxon. Edited by CHARLES COWDEN CLARKE. Illustrations to each page. Foolscap 8vo., cloth and gold, 3s.; malachite, 10s.

FINDEN'S FINE ART WORKS.

Gallery of Modern Art; a Series of 31 highly-finished Steel Engravings, with Descriptive Tales by Mrs. S. C. HALL, MARY HOWITT, and others. Folio, cloth extra, gilt edges, 21s.

Beauties of Moore; being a Series of Portraits of his principal Female Characters, from Paintings by eminent Artists, engraved in the highest style of Art, by Mr. EDWARD FINDEN, with a Memoir of the Poet, and Descriptive Letter-press. Folio, cloth extra, gilt edges, 42s.

Drawing-room Table Book; a Series of 31 highly-finished Steel Engravings, with Descriptive Tales, by Mrs. S. C. HALL, MARY HOWITT, and others. Folio, cloth extra, gilt edges, 21s.

Hogarth—The Works of William Hogarth, in a Series of 150 Steel Engravings by the First Artists, with Descriptions by Rev. JOHN TRUSLER, and Introductory Essay on the Genius of Hogarth, by JAMES HANNAY. Small folio, cloth, gilt edges, 52s. 6d.

Kitto's (Dr.) Holy Land. The Mountains, Valleys, and Rivers of the Holy Land; being the Physical Geography of Palestine. By JOHN KITTO, D.D., F.S.A. *New Edition*, fcap. 8vo., handsomely bound, 2s. 6d.; in bevelled boards, red edges, 3s.

Lamb's (Charles and Mary) Tales from Shakespeare. New Edition. To which are now added Scenes illustrating each Tale. With numerous Woodcuts from Designs by HARVEY. Edited by CHARLES KNIGHT. Small 8vo. cloth, bevelled boards, 2s. 6d.; with gilt edges, 3s.; morocco antique, 7s.

Language of Flowers, The; or, The Pilgrimage of Love. By THOMAS MILLER. With eight beautifully coloured Plates. Just published, a *New Edition*. Small 8vo., cloth, gilt edges, 3s. 6d.; handsomely bound in silk and in morocco, 8s.

Language of Flowers (The). Cheap Edition. Abridged from the larger edition. By THOMAS MILLER. With Coloured Frontispiece. Limp cloth, 6d.

Leared (Dr.) On Imperfect Digestion, Its Causes and Treatment, by ARTHUR LEARED, M.D., F.R.C.P., Senior Physician to the Great Northern Hospital. *Fifth edition*, post 8vo., cloth, 4s. 6d.

"It now constitutes about the best work on the subject."—*Lancet.*

"Dr. Leared has treated a most important subject in a practical spirit and popular manner."—*Medical Times and Gazette.*

"A useful manual of the subject upon which it treats, and we welcome it as an addition to our Medical Literature."—*Dublin Quarterly Journal of Medical Science.*

GRIFFIN'S EMERALD GEMS.

Gray's Poetical Works. With Life by the Rev. JOHN MITFORD, and Essay by the EARL OF CARLISLE. With Portrait and numerous Engravings on Steel and Wood. *Eton edition, with the Latin Poems.* Elegantly printed on toned paper, foolscap 8vo., richly bound in cloth and gold, 5s.; malachite, 12s. 6d.

Goldsmith's Poetical Works. With Memoir by WILLIAM SPALDING, A.M. Exquisitely Illustrated with Steel Engravings. *New Edition.* Printed on Superior toned paper. Foolscap 8vo., cloth and gold, 3s.; malachite, 10s. 6d.

Burns' Songs and Ballads. With an Introduction on the Character and Genius of Burns. By THOMAS CARLYLE. Carefully printed in antique type, and illustrated with beautiful Engravings on Steel. Foolscap 8vo., elegantly bound in cloth and gold, 3s.; malachite, 10s. 6d.

Poe's Poetical Works, Complete. Edited by J. HANNAY. Illustrations after WEHNERT, WEIR, &c. Toned paper. Foolscap 8vo., cloth elegant, 3s.; malachite, 10s. 6d.

Byron's Childe Harold's Pilgrimage. With Memoir by Prof. SPALDING. Illustrated with Engravings on Steel by GREATBACH, MILLER, LIGHTFOOT, &c., from Paintings by CATTERMOLE, Sir T. LAWRENCE, H. HOWARD, and STOTHARD. Beautifully printed on toned paper, Foolscap 8vo., cloth elegant, 3s.; malachite, 10s. 6d.

Chatterton's Poetical Works. With an Original Memoir by FREDERICK MARTIN. Beautifully Illustrated, and elegantly printed. Foolscap 8vo., cloth and gold, 3s.; malachite, 10s. 6d.

Herbert's Poetical Works. With Memoir by J. NICHOL, B.A., Oxon. Edited by CHARLES COWDEN CLARKE. *Beautiful Illustrations to each page.* Foolscap 8vo., cloth and gold, 3s.; malachite, 10s. 6d.

Campbell's Pleasures of Hope. With Introductory Memoir by Rev. CHARLES ROGERS, LL.D. Illustrated with splendid Steel Engravings. Price 3s.; malachite, 10s. 6d.

Other Volumes will be added from time to time.

Machinery and Millwork: comprising the Geometry, Motions, Work, Strength, Construction, and Objects of Machines, &c., by Professor W. J. MACQUORN RANKINE, LL.D., &c. Illustrated with nearly 300 Woodcuts. Crown 8vo., cloth, 12s. 6d.

"Professor Rankine's 'Manual of Machinery and Millwork' fully maintains the high reputation which he enjoys as a scientific author; higher praise it is difficult to award to any book. It cannot fail to be a lantern to the feet of every engineer."—*The Engineer.*

GRIFFIN'S SHILLING MANUALS.

Edited by JOHN TIMBS. Foolscap 8vo., neat cloth, price 1s. each.

I. Popular Science.
II. One Thousand Domestic Hints.
III. Oddities of History.
IV. Thoughts for Times and Seasons.
V. Characteristics of Eminent Men.
VI. Curiosities of Animal and Vegetable Life.

"It is difficult to determine which of these volumes is the most attractive. Will be found equally enjoyable on a railway journey, or by the fireside."—*Mining Journal.*

"These additions to the Library, produced by Mr. Timbs' industry and ability, are useful, and in his pages many a hint and suggestion, and many a fact of importance, is stored up that would otherwise have been lost to the public."—*Builder.*

"Capital little books of about a hundred pages each, wherein the indefatigable Author is seen at his best."—*Mechanics' Magazine.*

"Extremely interesting volumes."—*Evening Standard.*

"Amusing, instructive, and interesting. . . . . As food for thought and pleasant reading, we can heartily recommend the 'Shilling Manuals."—*Birmingham Daily Gazette.*

Mackey's Freemasonry: a Lexicon of Freemasonry; containing a Definition of all its Communicable Terms, Notices of its History, Traditions, and Antiquities, and an Account of all the Rites and Mysteries of the Ancient World. By ALBERT G. MACKEY, M.D., Secretary-General of the Supreme Council of the U.S., &c. Handsomely bound in cloth, price 5s.

Magic (The) of Science: a Manual of Easy and Instructive Scientific Experiments. By JAMES WYLDE, formerly Lecturer on Natural Philosophy at the Polytechnic. With Steel Portrait of Faraday and many hundred Engravings. Crown 8vo., cloth, gilt, 5s.

"Among the many books written to make science 'pleasant,' this is one of the very best. It is simple, free from technical terms, as far as possible, and yet the experiments are described with scientific accuracy."—*Spectator.*

"To those who need to be allured into the paths of natural science by witnessing the wonderful results that can be produced by well-contrived experiments, we do not know that we could recommend a more useful volume."—*Athenæum.*

Mangnall: Historical and Miscellaneous Questions, for the Use of Young People. By RICHMAL MANGNALL. *New Illustrated Edition*, greatly enlarged and corrected, and continued to the present time, by INGRAM COBBIN, M.A. 12mo., bound, 4s. *Forty-eighth thousand.*

McBurney's (Isaiah, LL.D.) Extracts from Ovid's Metamorphoses. With Notes and Vocabulary for the use of young Students. 18mo., cloth, 1s. 6d. *Third edition.*

London Labour and the London Poor:

A Cyclopædia of the Condition and Earnings of those that Will Work, those that Cannot Work, and those that Will Not Work. By HENRY MAYHEW. With Illustrations from Photographs. In 3 vols, 8vo., cloth, 4s. 6d. each.

"Every page of the work is full of valuable information, laid down in so interesting a manner that the reader can never tire."—*Illustrated News.*

"Mr. Henry Mayhew's famous record of the habits, earnings, and sufferings of the London poor."—*Lloyd's Weekly London Newspaper.*

"This remarkable book, in which Mr. Mayhew gave the better classes their first real insight into the habits, modes of livelihood, and current of thought of the London poor." —*The Patriot.*

London Labour and the London Poor:

The Extra Volume: Those that Will Not Work; Comprising the Non-workers, by HENRY MAYHEW; Prostitutes, by BRACEBRIDGE HEMYNG; Thieves, by JOHN BINNY; Beggars, by ANDREW HALLIDAY. With an Introductory Essay on the Agencies at Present in Operation in the Metropolis for the Suppression of Crime and Vice, by the Rev. WILLIAM TUCKNISS, B.A., Chaplain to the Society for the Rescue of Young Women and Children. With Illustrations of Scenes and Localities. 1 large vol., 8vo., cloth, 7s. 6d.

"The work is full of interesting matter for the casual reader, while the philanthropist and the philosopher will find details of the greatest import."—*City Press.*

Companion Volume to the above.

Criminal Prisons of London,

And Scenes of Prison Life. By HENRY MAYHEW and JOHN BINNEY. Illustrated by nearly 200 Engravings on Wood, principally from Photographs. 1 large vol., 8vo., cloth, 7s. 6d.

CONTENTS:—General View of London, its Population, Size, and Contrasts — Professional London—Criminal London — Pentonville Prison—The Hulks at Woolwich—Millbank Prison—The Middlesex House of Detention—Coldbath Fields—The Middlesex House of Correction, Tothill Fields—The Surrey House of Correction, Wandsworth—Newgate—Horsemonger Lane—Clerkenwell.

"This volume concludes Mr. Henry Mayhew's account of his researches into the crime and poverty of London. The amount of labour of one kind or another, which the whole series of his publications represents, is something almost incalculable."—*Literary Budget.*

Mineralogy and Geology, Elements of. By Prof.

F. SCHOEDLER, Ph.D. *New edition*, translated from the Sixth German Edition by Dr. MEDLOCK. *Numerous Engravings.* Crown 8vo., cloth, 1s. 6d.

Natural Philosophy, Elements of. By FRIEDRICH

SCHOEDLER, Ph.D. Translated from the Sixth German Edition by HENRY MEDLOCK, F.C.S. *New edition, Illustrated by above 100 Engravings.* 8vo., cloth, 1s. 6d.

Many Thoughts of Many Minds: being a Treasury of Reference, consisting of Selections from the Writings of the most celebrated Authors. Compiled and analytically arranged by HENRY SOUTHGATE. *Twenty-second thousand.* Square 8vo., printed on toned paper, elegant binding, 12s. 6d.; morocco, £1 1s.

"The produce of years of research."—*Examiner.*
"Destined to take a high place among books of this class."—*Notes and Queries.*
"A treasure to every reader who may be fortunate enough to possess it."—*English Journal of Education.*
"The accumulation of treasures truly wonderful."—*Morning Herald.*
"This is a wondrous book."—*Daily News.*
"Worth its weight in gold to literary men."—*Builder.*

Many Thoughts of Many Minds. Second Series. By HENRY SOUTHGATE. Square 8vo., printed on toned paper, and elegantly bound in cloth and gold, 12s. 6d. *Second Edition.*
The same, handsome morocco antique, 21s.

"Few Christmas Books are likely to be more permanently valuable."—*Scotsman.*
"Fully sustains the deserved reputation achieved by the First Series."—*John Bull.*

Mechanical Text Book, A: or Practical Introduction to the Study of Mechanics, by W. J. MACQUORN RANKINE, LL.D., late Prof. of Civil Engineering in the University of Glasgow, and EDWARD FISHER BAMBER, C.E. *In preparation.*

Mental Science. SAMUEL TAYLOR COLERIDGE'S celebrated Essay on Method; ARCHBISHOP WHATELEY'S Treatises on Logic and Rhetoric. Crown 8vo., cloth, 5s. *Tenth Edition.*

Moral and Metaphysical Philosophy: Systems of Philosophy Anterior to the Time of Christ. By the Rev. F. D. MAURICE, M.A. *Fourth edition*, re-written, 5s.

"Complete and satisfactory."—*Literary Gazette.*

Morrison's (Dr. J.) Family Worship. Prayers for every Morning and Evening throughout the Year. 4to., cloth, gilt edges, 12s.

Nichol's (Professor) Cyclopædia of the Physical Sciences; comprising Acoustics, Astronomy, Dynamics, Electricity, Heat, Magnetism, Meteorology, &c., &c. *Third edition*, enlarged. Maps and Illustrations. Large 8vo., half-bound, Roxburghe, 21s.

"It takes its place at once, and of course among standard works. . . . . . The ground of our opinion is the excellence of the matter, the freshness of the articles, and the attention which has been paid to bringing in the most recent views and discoveries."—*Athenæum.*
"Well printed and illustrated, and most ably edited."—*Examiner.*
"A most useful book of reference, deserving high commendation."—*Westminster Review.*

One Thousand Domestic Hints on the Choice of Provisions, Cookery, and Housekeeping; New Inventions and Improvements; and various branches of Household Management. Written and compiled by JOHN TIMBS. Crown 8vo., cloth, 1s. *Second edition.*

"No money is better spent than what is laid out for domestic satisfaction."—*Dr. Johnson.*

Occult Science, Compendium of the. Containing the Traditions and Superstitions of Past Times, and the Marvels of the Present Day, in a Series of Historical Sketches. By Rev. EDWARD SMEDLEY, M.A., E. RICH, Esq., and other Contributors, 6s.

"A repository of curious information . . . . most instructive and interesting reading."—*Literary Gazette.*

Oddities of History and Strange Stories for all Classes. Selected and compiled by JOHN TIMBS. (*Griffin's Shilling Manuals.*) Crown 8vo., cloth, neat, 1s.

"The reader is presented with many curious tit-bits of history, interesting and valuable in themselves . . . . interesting companion for spare minutes."—*Edinburgh Daily Review.*

Oldham's Poetical Works. With Notes and Memoir. By ROBERT BELL. Foolscap 8vo., cloth, gilt, 1s. 3d.

Paley's Natural Theology; or, the Evidences of the Existence and Attributes of the Deity. With illustrative Notes and Dissertations. By LORD BROUGHAM and Sir C. BELL. *Cheap edition.* In 1 vol., 16mo., cloth, 3s. 6d.

"When Lord Brougham's eloquence in the Senate shall have passed away, and his services as a statesman shall exist only in the free institutions which they have helped to secure, his discourse on Natural Theology will continue to inculcate imperishable truths, and fit the mind for the higher revelations which these truths are destined to foreshadow and confirm."—*Edinburgh Review.*

Phillips' (John Arthur, C.E., F.G.S., F.C.S., &c.) MANUAL OF METALLURGY. The *fourth edition*, thoroughly revised and almost entirely re-written. Large post 8vo., with nearly 200 Illustrations. (*In preparation.*)

Pictorial Gallery of the Useful and Fine Arts. Illustrated by numerous beautiful Steel Engravings, and nearly Four Thousand Woodcuts. Edited by CHARLES KNIGHT. 2 vols. folio, cloth gilt, 34s.

Pictorial Museum of Animated Nature. By CHARLES KNIGHT. Illustrated with Four Thousand Woodcuts. Folio, 2 vols., cloth, gilt, 34s.

Pictorial Sunday Book. Edited by Dr. JOHN KITTO. Cloth, gilt, 28s.

Poe's Poetical Works, Complete. Edited by J. HANNAY. Illustrations after WEHNERT, WEIR, &c. Toned paper. Foolscap 8vo., cloth, elegant, 3s.; malachite, 10s. 6d.

Popular Science. Recent Researches on the Sun, Moon, Stars, and Meteors; The Earth; Phenomena of Life; Sight and Sound; Inventions and Discoveries. Familiarly Illustrated. Written and Compiled by JOHN TIMBS. Crown 8vo., cloth, 1s.

"A very useful little manual."—*Sheffield Daily Telegraph.*

"Will be found abundant in much out-of-the-way information on almost every conceivable topic, where the popular mind is for the most part in error."—*Evening Standard.*

PROFESSOR RAMSAY'S WORKS.

A Manual of Roman Antiquities. By WILLIAM RAMSAY, M.A., Trinity College, Cambridge, late Professor of Humanity in the University of Glasgow. With Map, numerous Engravings, and very copious Index. *Ninth edition*, revised and enlarged. Crown 8vo., cloth, 8s. 6d.

"Comprises all the results of modern improved Scholarship within a moderate compass."—*Athenaeum.*

"A useful work, evidencing much thought in compilation and arrangement. Students of Latin literature will find this volume handy in form and matter. There is a perfect index and numerous illustrations, with a chapter on Roman Agriculture."—*School Board Chronicle.*

Ramsay's (Professor) Elementary Manual of Roman Antiquities. Adapted for Junior Classes. Numerous Illustrations. *Fifth edition.* Crown 8vo., cloth, 4s.

Ramsay's (Professor) Manual of Latin Prosody. *Fourth edition,* revised and greatly enlarged. Crown 8vo., cloth, 5s.

Ramsay's (Professor) Elementary Manual of Latin Prosody. Adapted for Junior Classes. Crown 8vo., cloth, 2s.

Ready Reckoner

(Fisher's). The best in the world. *New edition,* 18mo., bound, 1s. 6d.

Religions of the World, The. Being Confessions of Faith, contributed by eminent members of each Denomination of Christians, also of Mahometanism, the Parsee Religion, the Hindoo Religion, Mormonism, &c., &c. Cloth, 8vo., bevelled boards, 3s. 6d.

"In this volume each denomination, through some leading member, has expressed its own opinions. There is no book in the language on the same plan. All other works on the subject being written by one individual, are necessarily incorrect and unauthentic.'

Religious and Moral Anecdotes. With an Introductory Essay by the Rev. GEORGE CHEEVER, D.D. *Fourteenth Thousand.* Crown 8vo., cloth, 3s. 6d.

"These Anecdotes relate to no trifling subjects, and are invaluable to those engaged in the instruction of the young."

Shakespeare's Dramatic and Poetical Works. Revised from the Original Editions, with a Memoir, and Essay on his Genius, by BARRY CORNWALL. Also, Annotations and Introductory Remarks on his Plays, by R. H. HORNE. With numerous Engravings by Kenny Meadows. 3 vols., super-royal 8vo., cloth, 42s.

Shakspeare's Works. Edited by T. O. HALLIWELL, with a Series of Steel Portraits. 3 volumes, royal 8vo., cloth gilt, 50s.

PROFESSOR RANKINE'S WORKS.

Machinery and Millwork, by W. J. MACQUORN RANKINE, LL.D., late Professor of Civil Engineering in the University of Glasgow. Illustrated with nearly 300 Woodcuts. Crown 8vo., cloth, 12s. 6d.

Rankine's Civil Engineering. *Ninth edition.* Cloth, 16s.

Rankine's Applied Mechanics. *Sixth edition.* Cloth, 12s. 6d.

Rankine's The Steam Engine and other Prime Movers. *Sixth edition.* Cloth, 12s. 6d.

Rankine's Useful Rules and Tables. *Fourth edition.* Cloth, 9s.

Rankine's Mechanical Text-Book : a Practical Introduction to the Study of Mechanics. By Prof. RANKINE and E. F. BAMBER, C.E. (*In preparation.*)

Shakspeare, The Family. Edited by THOMAS BOWDLER, F.R.S., and expressly adapted for Family Reading. (*New Edition in preparation.*)

Scott—A Commentary on the Bible ; Containing the Old and New Testaments, according to the Authorized Version. *New edition,* with Explanatory Notes. By Rev. THOMAS SCOTT. With Practical Observations, copious Marginal References, Indexes, etc. 3 vols. royal 4to., cloth, 63s.; calf, marble edges, 90s.; Levant morocco, antique, gilt edges, 130s.

Senior's (Professor) Treatise on Political Economy: the Science which treats of the Nature, the Production, and Distribution of Wealth. *Fifth edition.* Crown 8vo., cloth, 4s.

Sidereal Heavens, The, and other Subjects connected with Astronomy, as illustrative of the Character of the Deity, and of an Infinity of other Worlds. By THOMAS DICK, LL.D., author of the "Christian Philosopher," &c. *New edition.* Printed on toned paper, handsomely bound, with gilt edges, price 5s.

The School Board Readers.

A New Series of Standard Reading Books for Elementary Schools. Edited by a Former H.M. Inspector of Schools. Recommended by the London School Board, and adopted by many School Boards throughout the country. The prices are as follows:—

ELEMENTARY READING BOOK, Part I., containing lessons in all the short vowel sounds. Demy 18mo., 16 pages, in stiff wrapper, price 1d.

ELEMENTARY READING BOOK, Part II., containing the long vowel sounds, and other monosyllables. Demy 18mo., 48 pages, in stiff wrapper, price 2d.

STANDARD I., containing Reading, Dictation, and Arithmetic. Demy 18mo., 96 pages, neat cloth, price 4d.

STANDARD II., Ditto demy 18mo., 128 pages, neat cloth, price 6d.

STANDARD III., Ditto fcap. 8vo., 160 pages, neat cloth, price 9d.

STANDARD IV., Ditto fcap. 8vo., 192 pages, neat cloth, price 1s.

STANDARD V., Ditto crown 8vo., 256 pages, neat cloth, price 1s. 6d.

STANDARD VI., containing Selections from the Best English Authors, chronologically arranged, Hints on Composition, and Lessons on Scientific Subjects. Crown 8vo., 320 pages, neat cloth, price 2s.

KEY to the "Questions in Arithmetic," in STANDARDS I., II., and III., price 6d.

Ditto, in IV., V., VI., price 6d.

*** Each Book of this Series contains within itself all that is necessary to fulfil the requirements of the Revised Code, viz., Reading, Spelling, and Dictation Lessons, together with Exercises in Arithmetic for the whole year. The paper, type, and binding are all that can be desired.

"The books generally are very much what we should desire."—*Times.*

"The choice of matter is excellent, and so are the method and style."—*School Board Chronicle.*

"There are no better reading books published . . The advanced books are gems."—*Educational Reporter.*

"Remarkably fresh; most of the old stereotyped forms have been abandoned. . . . The Fifth and Sixth Books are capital productions, and form a valuable epitome of English literature."—*Leeds Mercury.*

School Board Manuals, The: on the Specific

Subjects of the Revised Code. By a Former H.M. Inspector of Schools, Editor of the "School Board Readers."

I. ALGEBRA. 64 pages, stiff wrapper, 6d.; neat cloth, 7d.
II. ENGLISH HISTORY " " "
III. GEOGRAPHY " " "
IV. PHYSICAL GEOGRAPHY " "
V. ANIMAL PHYSIOLOGY, " "
(Well Illustrated with good Engravings.)
VI. BIBLE HISTORY. " "
(Entirely free from any denominational bias.)

*** The SCHOOL BOARD MANUALS contain all that is necessary for passing in Extra subjects in Standards IV., V., and VI.

Spelling by Dictation: Progressive Exercises in English Orthography for Schools and Civil Service Examinations. By the Rev. A. J. D. D'ORSEY, B.D., M.C.P., of King's College. *New edition.* 18mo., cloth, 1s. *Fourteenth thousand.*

Spenser and His Poetry. By ROBERT BELL. Three vols in one. Fcap. 8vo., cloth gilt, 3s. 6d.

Steam Engine and Other Prime Movers (a Manual of, The). By W. J. MACQUORN RANKINE, LL.D., F.R.SS., &c., &c., late Regius Professor of Civil Engineering in the University of Glasgow. With numerous Tables and Illustrations. *Sixth edition.* Crown 8vo., cloth, 12s. 6d.

Surrey, Poetical Works of the Earl of, of Minor Contemporaneous Poets, and of SACKVILLE, Lord BUCKHURST. With Notes and Memoir by ROBERT BELL. Foolscap 8vo., cloth gilt, 1s. 3d.

Thoughts for Times and Seasons. Selected and compiled by JOHN TIMBS. Crown 8vo., cloth neat, 1s. 3d. (*Griffin's Shilling Manual.*)

"These little volumes contain a very great amount and variety of out-of-the-way extracts from modern and old writings."—*Mechanics' Magazine.*

"In a neat and concise form, are brought together striking and beautiful passages from the works of the most eminent divines and moralists, and political and scientific writers of acknowledged ability."—*Edinburgh Daily Review.*

Thomson's (Robert Dundas, M.D., F.R.S.) Popular Dictionary of Chemistry, Practical and Theoretical, with its applications to Mineralogy, Physiology, and the Arts. With numerous Illustrations. Post 8vo., cloth, 8s. 6d.

"Its limited compass, its completeness and correctness, recommend this cyclopædia of Chemistry to all who feel an interest in the science."—*Athenæum.*

"In every respect worthy of praise. It is within a limited space by far the best compendium of chemistry which we have yet seen."—*Lancet.*

Thomson's Seasons. *School edition.* Edited with Notes. By ROBERT BELL. Foolscap 8vo., cloth, 1s. 6d.

"An admirable introduction to the study of our English classics."

Timbs' (John, F.S.A., Author of "The Curiosities of London," &c.) PLEASANT HALF-HOURS FOR THE FAMILY CIRCLE. Containing Popular Science, Thoughts for Times and Seasons, Oddities of History, Characteristics of Great Men, and Curiosities of Animal and Vegetable Life. Crown 8vo., cloth, gilt, 5s.

Useful Rules and Tables. For Architects, Builders, Carpenters, Coachbuilders, Engravers, Engineers, Founders, Mechanics, Shipbuilders, Surveyors, Wheelwrights, &c. By W. J. MACQUORN RANKINE, LL.D., F.R.SS., &c., &c., late Regius Professor of Civil Engineering in the University of Glasgow. *Fourth edition.* 8vo., cloth, price 9s.

"Undoubtedly the most useful collection of engineering data hitherto produced."—*Mining Journal.*

"A necessity of the engineer . . . will be useful to any teacher of mathematics." —*Athenæum.*

Voices of the Year, The, or the Poet's Kalendar. Containing the choicest Pastorals in our Language. Profusely illustrated by the best Artists. In bevelled boards, elaborately ornamented and gilt, 12s. 6d.

Wanderings in every Clime; or Voyages, Travels, and Adventures All Round the World. Edited by W. F. AINSWORTH, F.R.G.S., F.S.A., &c., and embellished with upwards of Two Hundred Illustrations by the first Artists, including several from the master pencil of GUSTAVE DORE. Just published, large 4to., 800 pages, cloth and gold, bevelled boards, 21s.

Waller's Poetical Works. With Notes and Memoir by ROBERT BELL. Foolscap 8vo., cloth, gilt, 1s. 3d.

Whately's (Archbishop) Treatise on Logic. The *Original edition*, with Synopsis and Index. Crown 8vo., 3s., cloth.

Whately's (Archbishop) Treatise on Rhetoric. The *Original edition*, with Synopsis and Index. Crown 8vo., 3s. 6d., cloth.

Wyatt's Poetical Works. With Notes and Memoir by ROBERT BELL. Foolscap 8vo., cloth gilt, 1s. 3d.

Zoology and Physiology, Elements of, by Friedrich SCHOEDLER. *Tenth edition*, translated by Dr. HENRY MEDLOCK. Illustrated by above One Hundred Engravings. Small 8vo., cloth, 1s. 6d.

Specimen Copies of all the Educational Works published by Messrs. Griffin and Company may be seen at the Libraries of the College of Preceptors, South Kensington Museum, and Crystal Palace; also at the depôts of the chief Educational Societies.

Made in United States
North Haven, CT
24 September 2026

19102540R00191

Between Militarism and Technocratic Governance

State Formation in Contemporary Uganda

Anders Sjögren

FOUNTAIN PUBLISHERS
www.fountainpublishers.co.ug

THE NORDIC AFRICA INSTITUTE
www.nai.uu.se

Fountain Publishers
P.O. Box 488
Kampala
E-mail: sales@fountainpublishers.co.ug
publishing@fountainpublishers.co.ug
Website: www.fountainpublishers.co.ug

In cooperation with
The Nordic Africa Institute
P.O. Box 1703
SE-75147, Uppsala, Sweden
Website: www.nai.uu.se

Distributed in Europe and Commonwealth countries outside Africa by:
African Books Collective Ltd,
P.O. Box 721,
Oxford OX1 9EN, UK.
Tel/Fax: +44(0) 1869 349110
E-mail: orders@africanbookscollective.com
Website: www.africanbookscollective.com

First published 2013

ISBN: 978-9970-25-150-6

Contents

Abbreviations

AGOA	African Growth Opportunity Act
BUCADEF	Buganda Cultural and Development Foundation
BUFA	Bulando United Farmer's Association
CA	Constituent Assembly
CBO	Community Based Organisation
CCM	*Chama Cha Mapinduzi*
COFTU	Confederation of Free Trade Unions in Uganda
CP	Conservative Party
DA	District Administrator
DFID	Department for International Development
DP	Democratic Party
DRC	Democratic Republic of Congo
ERP	Economic Recovery Programme
FDC	Forum for Democratic Change
FEDEMU	Federalist Democratic Movement of Uganda
FOBA	Force Obote Back Again
HIPC	Highly Indebted Poor Countries
HSM	Holy Spirit Movement
HSMF	Holy Spirit Mobile Forces
IBEAC	Imperial British East African Company
ICC	International Criminal Court
ICFTU	International Confederation of Free Trade Unions
IFI	International Financial Institutions
IMF	International Monetary Fund
KANU	Kenya Africa National Union
KY	*Kabaka Yekka*
LC	Local Council
LRA	Lord's Resistance Army
MADDO	Masaka Diocese Development Organisation

MoFPED	Ministry of Finance, Planning and Economic Development
MoH	Ministry of Health
MoPS	Ministry of Public Service
NEC	National Executive Committee
NGO	Non-Governmental Organisation
NOM	Ninth October Movement
NOTU	National Organisation of Trade Unions
NRA	National Resistance Army
NRC	National Resistance Council
NRM	National Resistance Movement
PAPSCA	Programme to Alleviate Poverty and the Social Costs of Adjustment
PEAP	Poverty Eradication Action Plan
PHC	Public Health Care
PMB	Produce Marketing Board
PNDC	People's National Defence Council
PRA	Participatory Rural Appraisal
PRA	People's Redemption Army
PRA	Popular Resistance Army
PRS	Poverty Reduction Strategy
RC	Resistance Council
SAP	Structural Adjustment Programme
SPLA	Sudan People's Liberation Army
SWAp	Sector Wide Approach
UCC	Uganda Constitutional Commission
UCMB	Uganda Catholic Medical Bureau
UDC	Uganda Development Corporation
UFL	Uganda Federation of Labour
UFM	Uganda Freedom Movement
UMA	Uganda Medical Association
UMWU	Uganda Medical Workers Union
UNC	Uganda National Congress
UNICEF	United Nations Children's Fund

UNLA	Uganda National Liberation Army
UNRF	Uganda National Rescue Front
UPA	Uganda People's Army
UPC	Uganda People's Congress
UPM	Uganda People's Movement
UPMB	Uganda Protestant Medical Bureau
UPDA	Uganda People's Democratic Army
UPDF	Uganda People's Defence Forces
UPE	Universal Primary Education
USAID	United States Agency for International Development
UWESO	Uganda Women's Effort to Support Orphans
WNBF	West Nile Bank Front

Glossary

Busulu	Ground tax
Chaka mchaka	pro-NRM political training courses
Envujjo	Portions of the harvest that the tenant would give to the landlord in order to keep in his favour.
Kabaka	King of Buganda
Lukiiko	Parliament of Buganda kingdom
Mafutamingi	A stratum of state created proprietors
Magendo	Underground (economy)
Munno mukabi	"a friend in need", Community Based Organisation (CBO)
Shamba	Plot for cultivation
Uganda n'Eddagala Lyayo	Uganda Herbalist Association

Acknowledgements

This book grew out of my doctoral thesis, written over a long period. Consequently, my gratitude extends to a great number of individuals whose invaluable support, over the years, has enabled me to finish it.

Two persons have been particularly important to me when working on this book. Professor Björn Beckman has been my supervisor ever since my undergraduate days and I cannot really express how thankful I am to have benefited from his wisdom, sharp intellect and generosity as a mentor, colleague and friend over the years. It has been a wonderful learning experience in the best sense.

Haruna Bukenya in Masaka, Uganda, has become much more than a research assistant, as the term goes. He is a colleague and a friend, with whom I have shared all the practical and intellectual ups and downs of fieldwork as well as wonderful meals in his home.

Many people have commented on this work at various stages of its development. In the Department of Political Science, Stockholm University; special thanks go to Henrik Berglund, Drude Dahlerup, Eva Hansson and Magnus Lembke. During the final stages, I received very valuable comments from Henrik Angerbrandt, Maud Eduards, Lasse Lindström and Rune Premfors.

Other people whose comments and criticisms have strengthened the arguments in this book immensely are: Emmanuel Akwetey, Gunilla Andræ, Yusuf Bangura, Helena Bjuremalm, Jeremy Gould, Ben Jones, Raufu Mustapha, Cyril

Obi, Lars Rudebeck, K. Sivaramakrishnan, Olle Törnquist, and the late Mariken Vaa.

Also in the department, Lena Helldner and Claes Linde offered invaluable support. Merrick Tabor skilfully copyedited the manuscript into readable English. Peter Mungai provided formatting expertise towards the very end.

The research was generously supported by Sida/Sarec, for which I am most grateful. An early travel grant from the Nordic Africa Institute in Uppsala was also very valuable.

In Kampala, the Centre for Basic Research (CBR) became a second home. Thanks to everyone at CBR for the assistance and friendship. During my first visits, the then executive director, the late Nyangabyaki Bazaara, with typical generosity supported me with practical assistance, intellectual discussions and enjoyable music. May you rest in peace.

For comments on various drafts at CBR seminars, I wish to thank Godfrey Asiimwe, John-Jean Barya, Nyangabyaki Bazaara, Nicholas de Torrenté, Frederick Golooba-Mutebi, David Kibikyo, Mwambutsya Ndebesa, Joe Oloka-Onyango, Samson Opolot and Edward Rubanga. Simon Rutabajuuka read and commented upon the entire manuscript. Sallie Simba Kayunga, Richard Ssewakiryanga, David Ouma Balikowa, Anders Jeppsson, and Sam Okuonzi were all very helpful at different stages. Other friends in Kampala and Nairobi who have always given me of their precious time and immense knowledge are Andrew Mwenda, Ibrahim Ssemujju Nganda, Charles Onyango-Obbo and Sylvia Tamale.

In the past, I was repeatedly treated to hospitality and friendship in Nairobi by Per and Lilian Karlsson. I am happy to be able to pay back that particular debt now! Ros-Mari Bålöw, Mia Haglund-Heelas and Paul Heelas, and Pernilla Trägårdh and Fredrik Malmberg at various times opened their homes for me in Kampala. Thank you all.

In Masaka, I would particularly like to thank Michael Ssali, Eddie Ssejjoba, Vincent Basajja, John Kinsman, Sabina Bergsten and Nils Fagerberg, who all did their very best to make me feel at home. There are also innumerable other people in Masaka who assisted me in different ways, many without knowing it. And above all: to everyone in Bulando who took their time and shared their experiences with me – *mwebale nnyo*!

A special thank you extends to Per Karlsson and Sverker Finnström for our ongoing informal seminar on Ugandan politics and society. It has been great – long may it continue!

My parents, Eva and Ragnar Sjögren have always supported me to the full in every way possible. I am immensely thankful to you. Finally, my deepest and most heartfelt gratitude goes to Helena and our children Mikael and Rebecka, whose love, support, endurance and shared interest in Ugandan politics have exceeded all I could ever hope for.

Anders Sjögren
Stockholm
October 2012

Map of Uganda

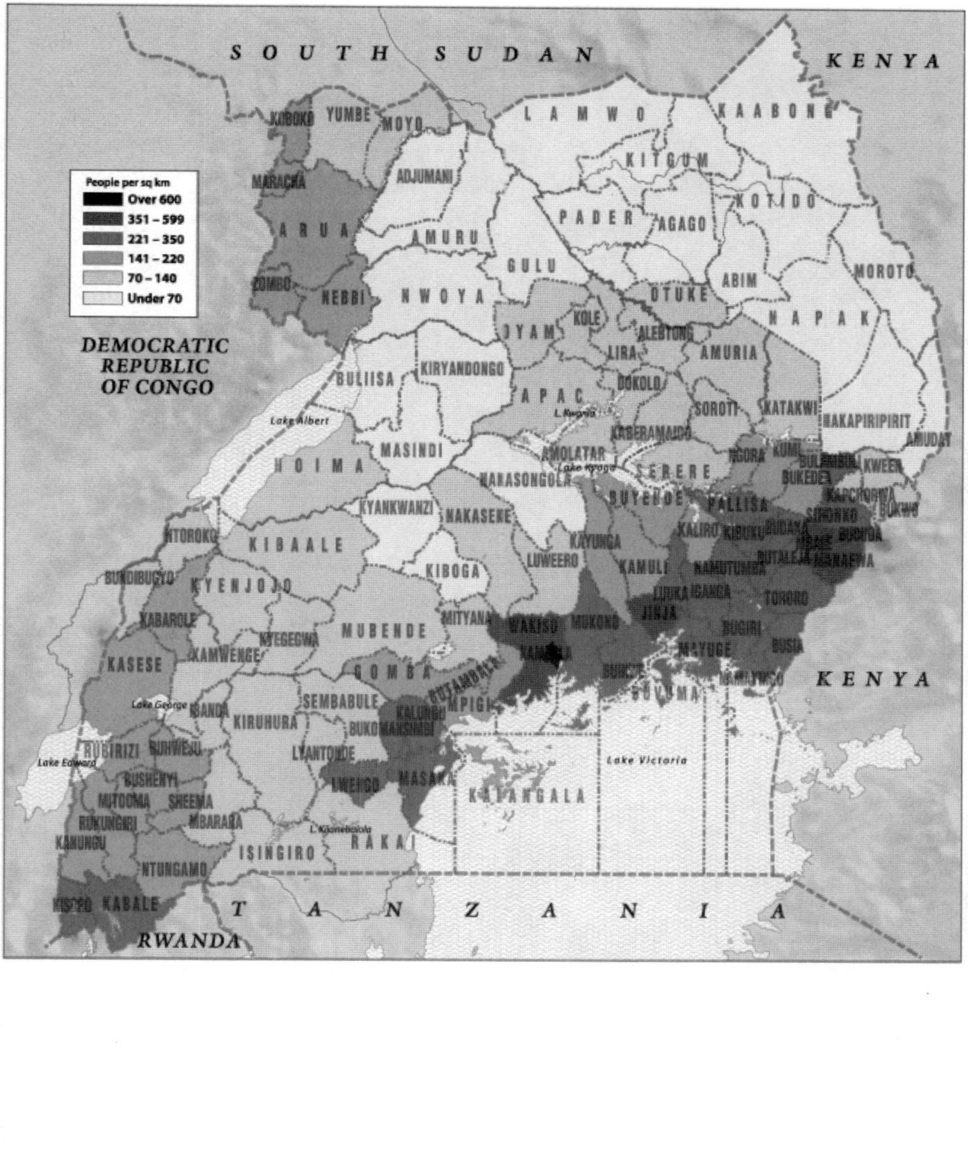

Chapter 1

Introduction

In the wake of the socio-economic crisis of the 1980s, African states and societies have been reshaped in the context of global patterns of economic liberalisation and by specific forms of intervening reform programmes designed to transform state-society relations. Externally imposed economic liberalisation and state reform packages soon came to be discussed under the rubric of "governance reforms", promoted with the stated aim of transforming the "problematic" nature of African state-society relations. Governance reforms were closely related – at the discursive level – to the promotion of greater scope of influence for civil society groups, as well as the institutional recomposition of civil society. During the same period, governments lacking in developmental and democratic credibility have constantly been challenged by social forces. The interaction between these external and internal sets of processes has restructured state-society relations along a wide spectrum of political regimes, ranging from intensified authoritarianism, to limited political liberalisation and in some cases to democratisation of the state and of social relations in a broader sense.

The diverse political dynamics of changes associated with governance reforms warrant investigations of the shifting bases of political rule in different political arenas. This study attempts to analyse such changes through the case of Uganda. It sets out from Bangura's (1994) suggestion that while governance reforms have certainly changed African states, they have not done so merely in line with proclaimed intentions. Reforms were from the start beset by inherent contradictions between the technocratic orientation of the stated political aims of institutional remodelling and the deeply contested political nature of social processes and struggles, reflecting the different political agendas of the actors involved. The fact that political struggles have not conformed to the expectations of governance reforms can be understood as their being shaped by state withdrawal and resulting informalisation in relation to the contestation of authoritarian rule by manifold actors in society, making claims for autonomy and participation in decision making.

Uganda's economic and political crisis during the 1980s was deeper than most. Following years of military rule, economic decay and civil war, the central state had all but collapsed by the mid-1980s. Only a decade later, in the late 1990s, was Uganda frequently mentioned as a prime example of state resurgence, exhibiting stabilised institutions, a booming economy and a vital civil society participating in relatively successful poverty reduction policies – all in all, an exponent of the "African renaissance" and a governance showcase attracting attention and resources. Yet, a series of rebellions had kept war raging in the northern parts of the country since 1986. Uganda's military had become deeply involved in the DR Congo civil war, with high-ranking officers and politicians accused of human rights abuses and also of looting natural resources. At home, the government faced allegations of authoritarianism and corruption.

This book is about the formation of the Ugandan state and civil society in this conjuncture. It addresses the scope for political forces to democratise the state in a context characterised by seemingly contradictory expressions of state-society relations – a "participatory governance" success story on the one hand, lingering militarism on the other – by tracing the historical roots of contemporary formation of state and civil society. In so doing, the study seeks to analyse state-society relations as these are embedded in the wider political economy. The overarching theoretical problem of the study concerns the social basis of political rule: how social relations block or promote political democracy. More specifically, the aim is to explore state-civil society relations in contemporary Uganda in order to illustrate and explain the scope for and capacity of different social forces to create access to and democratise the state.

The study interrogates state-civil society relations as these are expressed through forms of interest representation and conflict regulation in different political arenas. This takes place in the contemporary context of economic liberalisation and state reform. How have these changes affected state-society relations? How does the state attempt to shape its relations to different social forces? How have social forces sought to influence the state? What explains the scope for interest groups to intervene and influence policy making and ultimately either open up the state to make it more responsive to wider sections of the population, or to block the access to it for others?

The study addresses these issues through the lens of the political regulation of the health sector, and more specifically by analysing the relations between the state and a selection of interest groups that operate in this sector. Health sector politics illustrates a central feature of liberalisation and state reform – public sector reform and semi-privatised social development.

State-society relations

A useful starting point for analysing the research problem, the scope for social forces to democratise the state, is that these issues essentially concern the organisation and distribution of power. The main argument of the study is that the democratisation of the state hinges upon relations of domination in society, within the state itself and among international actors. Political forces at all these levels attempt to access, control and shape the state for very different purposes, and the state promotes some social forces at the expense of others. In that sense, the state is rooted in society, and the form as well as content of state institutions, including their capacity and legitimacy, is structured by relations of power in society as reflected in and shaped by the state itself.

Such power relations are of many kinds, based on for instance; property, status and hegemony and are articulated along lines of class, gender, ethnicity, religion and so on. This is obviously not to claim that such forms of domination in society are translated into state power in a direct sense. The state wields enormous resources, and relative institutional autonomy allows state managers the potential scope and capacity to reshape and institutionalise social relations. Nevertheless, where deep-rooted economic and social inequalities prevail, as they do in Uganda, they can be predicted to correspond fairly closely to political inequality in a real as opposed to a formal sense.

A preliminary analytical point of departure, then, is the rather established perspective that views the state as a relatively permanent structure of domination and coordination of social relations. The state is both an arena of struggles between competing social forces, and an actor, involved in such struggles either directly as a player or more indirectly as regulator of social relations, creating and upholding rules for resolving conflicts. Such a perspective means viewing state formation as an ongoing overlapping process that occurs in many arenas and on many levels. This general remark is vividly illustrated by the Ugandan case, where political instability has made state formation a very much unfinished and fluid process indeed. A crucial dimension of this is that the historical development of the Ugandan political economy has entrenched deep regional inequalities. One can therefore expect dramatic regional shifts with regard to both state institutions and the composition of social forces.

What is then meant by the democratisation of the state? Beyond the obvious meaning of the constitutionalised regulation of national politics capable of safeguarding the establishment of democratic institutions and procedures at that level, this study the forms of regulation of interest representation and conflicts in numerous institutions at different levels of the state. Economic liberalisation and state reform have all over Africa transformed the role of the state, including downsizing and decentralisation, and have set off a proliferation of associational life. Large sections of society have developed a multiplicity of strategies to cope with crisis and restructuring. Even where the state has largely withdrawn from its previous role, as in the field of welfare provision, it still intervenes and regulates development. Different sections of civil society, including community groups, social movements and local and foreign NGOs, all seek to engage the state, hoping to influence the direction of policy. According to one view, civil society groups can, in the face of fragile political institutions, be expected to hold political power-holders or bureaucrats accountable on behalf of the citizenry. Another more pessimistic version of the political content of state-society relations in the current conjuncture holds that appeals to civil society and good governance largely function as token devises, employed by political elites who seek to conceal persistent patterns of patronage politics.

Do restructured relations between state and society open up for democratic interest representation? Or do they reinforce authoritarian relations? The study sets out from the assumption that the interplay between institutional restructuring and wider social processes contributes to a recomposition of state-society relations, immediately visible as a shift in the division of labour with regard to, for instance, social service provision, but also in terms of how this restructuring affects the way social interests are given political expression. The research problem will be approached within a framework that stresses the interaction between the balance of power among various social forces and the strategic intervention of the state to regulate conflicts in different arenas in society. Ruling groups seek to enforce control and support within structural constraints of resource extraction and in relation to demands from contending social forces. This results in a variety of regulatory forms that may be formal or informal, arbitrary or predictable, legitimate or illegitimate, authoritarian or democratic, and so on. Therefore, the study analyses the degree to which different social forces are capable of shaping forms of regulation in the direction of, for instance fairness and access in form and substance and to what extent they may defend or expand political space. The latter notion may be thought of as the institutional and discursive scope for exercising politics.

Most African countries, including Uganda, were from the early 1980s subjected to economic reforms by the international financial institutions as part of structural adjustment programmes. This followed upon economic stagnation due to a combination of international recession and domestic political mismanagement. Reforms typically included devaluation, price stabilisation, privatisation and reduction of public expenditures. After nearly a decade of continued dismal performance under market policies, institutional reforms were introduced in order to make neo-liberal economic adjustment more "embedded". Governance reforms included measures such as new public management techniques, civil service reforms and decentralisation. There were protests in many countries against worsening economic conditions that followed upon withdrawn subsidies to or privatisation of health and education, as well as the often heavy handed imposition of these policies. Hence, demands for political reforms were made in the context of structural adjustment. In Uganda, state withdrawal was not primarily a consequence of economic liberalisation; it had already occurred in a dramatic fashion. Rather, economic liberalisation came to be associated with state restructuring and growth. This proved to be of significance for the way interest groups responded.

In Uganda, the constraints of resource extraction are very much set by the high level of dependence on foreign development aid. Economic growth and institutional restructuring remain donor-dependent. For many years, between 40 and 50 per cent of the state budget has been financed by externally provided resources. Donors have offered these resources in exchange for the government's continued commitment to donors' preferred direction of economic policy making and state reform. In this sense, Uganda's development model seems to be fundamentally shaped by external requirements. Yet, to stop there would be simplistic. Clearly, the Ugandan government seeks to adapt to the requests of donors and ultimately to the imperatives of globalised capitalism. As Harrison (2004: 6) suggests, the retooling of the state through administrative reforms and management techniques should, in a wider sense, be thought of as a strategy for embedding liberalism, or as "the politics of the encounter between the institutions of global capitalism and African nation states." To follow Harrison further, the politics of this encounter cannot however be reduced to a model of foreign domination. For one, the main dividing line is not between "national" and "international" interests, even though such conflicts do exist. Just as the state is shaped by competing political projects in Ugandan society, this applies to external intervention as well. A plethora of external actors, from bilateral donors to NGOs, from the UN system to religious organisations, seek to shape the Ugandan state, in alliance with domestic forces within or outside the state, although the government has to a large extent been able to skilfully mediate these interventions for its own purposes.

There is more mutual dependence between donors and Uganda than a cursory presumption would suggest. Donors have invested much commitment and prestige in Uganda as a success story. Hence, the content of governance reforms is mediated and shaped by struggles between competing constellations of domestic and foreign political interests that pull and push the state in different directions. This holds for state formation anywhere, of course, but Uganda's heavy reliance on foreign funding clearly highlights the ways in which external interests do not merely push the state from outside but are internalised into it at all levels – and how they are in the process remoulded by relations of domination in Uganda itself.

State and society in Uganda

State formation in Uganda has been, and still is, a complex and tortuous process with regard to its institutional, social and territorial dimensions. As will be shown in greater detail in Chapter 2, colonial rule was built on a diversity of social orders and perpetuated the fragmentation of these within the colonial protectorate, alongside promoting an economy that was deeply unevenly developed among regions. Those structural features created the conditions for inherent political destabilisation, which after twenty-five years of independence had resulted in two military coups d'etat; one war with a neighbouring country and a bloody civil war.

The conventional story told of contemporary Uganda is one in which the process of institutional collapse was followed by subsequent resurrection under the incumbent National Resistance Movement (NRM) government, in power since 1986.

After the NRM had come to power, its leaders pledged to break with the political past. This past was said to be characterised by a corrupt and repressive neo-colonial state ruled either by bourgeois sectarian politicians or by the military, both groups unable and unwilling to overcome economic dependence and underdevelopment.

The new government promised a "fundamental change" with a nationalist – and initially socialist – orientation.[1] Socialism was soon substituted for capitalism, but growth and stability lay the ground for economic recovery in the southern part of the country, something that attracted the attention of donors and creditors as well as foreign investors. Uganda, or at least its ruling strata, was incorporated into the circuits of global capitalism. Governance reforms such as decentralisation, new public management techniques and eventually debt and poverty reduction programmes were often tested first in Uganda, with its perceived farsighted and competent leadership, before they were applied elsewhere. One good thing, it seemed, led to another in a virtuous circle of converging aspects of modernisation.

This view of Uganda overlooked the ongoing strife and biting poverty in the north and parts of the east. This was for a long time neglected or seen as a temporary aberration by the international community. Reality was more complicated, of course. The Ugandan state, just like all states, but manifested in a very pronounced way during a period of rather dramatic shifts, is the meeting ground of contradictory forces and demands in

[1] The NRM's "Ten-Point Political Programme" included the following components: restoration of democracy, restoration of security, consolidation of national unity and elimination of all forms of sectarianism, defending and consolidating national independence, building an independent, integrated and self-sustaining national economy, restoration and improvement of social services and rehabilitation of war-ravaged areas, elimination of corruption and misuse of power, redressing errors that have resulted in the dislocation of some sections of the population, co-operation with other African countries, and following an economic strategy of a mixed economy (Amaza 1998: 242-245).

society, its institutions shaped by tendencies of continuity and change, authoritarianism and democratisation, and with great institutional and regional unevenness. Certainly, the NRM government had brought about a new sense of purpose and cohesion. Yet, instances of erratic decision making, corruption and ruling group's continued reliance on military force suggested a lingering inherent instability underneath the surface of change.

This study seeks to bring out this more complex story by investigating state-society relations as they have evolved in shaping institutions of the state at various levels. The historical legacy, reinforced in the present, of uneven patterns of development and of state-society relations suggests that an analytical disaggregation of both state and civil society is required. It is necessary to assume that state-society relations as well as the composition of the state and the forms of exercising state power have taken very shifting forms in different parts of the country. The most obvious example is how the composition and effective presence of bureaucratic, welfare and coercive state institutions diverge dramatically between the southern and northern parts of Uganda.

During the years of turmoil and institutional dilapidation in the 1970s and 1980s, the health status of Ugandans worsened disastrously. War and displacement rendered large sections of the population very vulnerable. Poor water and sanitation conditions in the context of the near collapse of the public health care system caused people to die in large numbers from preventable diseases. In addition to this, the HIV/AIDS epidemic struck hard from the early 1980s and tore existing family structures apart. People turned to whatever solutions there were: remaining health units run by religious organisations, relief from various agencies of the UN system, international NGOs, formal or informal private practitioners, traditional medicine and self-medication. Needs were desperate and state capacity to respond to them and to restructure the health sector was seriously limited.

The social crisis produced a complex institutional landscape inhabited by formal and informal public and private providers, poorly coordinated and regulated. Even with political stabilisation in the south from the late 1980s, forms of provision were fragmented and access was uneven. In the north, the situation worsened in the wake of conflicts. Throughout the 1990s, donors entered, but institutional reforms and international intervention are, as previously stated, mediated by coping strategies and political struggles. By approaching state-society through an of one sector, health, at national, district and village levels, it is possible to bring out in a detailed manner how different forms of state-society relations have evolved – and how they are continuously reshaped, as state formation in Uganda is very much an unsettled and an ongoing process.

Scope and organisation

The book is organised as follows: In Chapter 1, the research problem is situated in its empirical and theoretical contexts of recent changes in state-society relations in Sub-Saharan Africa and of theoretical perspectives on this. I develop the theoretical argument about state formation analysed as the regulation of power relations in various arenas in society and suggest how such a perspective is able to make sense of the contradictory tendencies of Ugandan state-society relations.

It may appear superfluous to stress that states and societies are fundamentally shaped in complex ways by their history. However, much prescriptive literature and intervening policy making – and this is very much so with regard to contemporary Uganda – is curiously ahistorical in its prescriptions of getting institutions right, in spite of historical institutionalists' reminders to the contrary. Therefore, a historical is undertaken in Chapter 2, where I outline the main features of state formation from colonial times until 1986. It relates the historical trajectory of the regulation of health and welfare to these main features. To take only one example, just like most other social relations during the 1970s and 1980s, health provision was subject to thoroughgoing informalisation. The specific forms of informalisation were in turn shaped by existing forms of health provision. This legacy was not easily changed by institutional reforms, and it lingers on to this day, as is discussed in chapters 3 and 4.

These chapters review shifts in the formation of the Ugandan state and civil society under the NRM government, that is to say, from 1986 to the mid-2000s. Chapter 3 discusses the period from 1986 to the early 1990s, during which the NRM, from a position of political vulnerability and having inherited a very weak state, sought to establish control over society. Within these constraints, what strategies did state managers employ, and how did different social forces respond? How did overall changes in state-society relations affect health provision? Chapter 4 moves on to analyse developments after the early 1990s, a period characterised by a high degree of influx of foreign development assistance and one often thought of as marked by consolidation of state power – yet, as shall be shown, the particular mode of consolidation generated strong undercurrents of destabilisation, which would later become manifest. This chapter also incorporates international actors into the analysis.

External influence has not been limited to the expansion of monetary resources from overseas. As suggested above, the Ugandan state has become a testing ground for many governance reforms and, again as already hinted, it has been referred to as a leading example of success. This rising status gradually gave the government more leverage and leeway in relation to foreign providers of resources. In what ways did the political constellations brought about by external influence and resources shape the Ugandan state and civil society? How were conditions for democratisation of the state affected by economic liberalisation and state reform?

Another prominent feature of state reform throughout the period under study is decentralisation. Local politics often carries specific (sub-) regional traits which shape the local state, in its real as opposed to formal sense in rather divergent directions, depending among other things on the local political economy and relations between the central and the local state. This is analysed in Chapter 5 through the case study of Masaka District, and an examination of state-society relations there. The Masaka case illustrates how sources of formal and informal power converge and shape local politics in locally specific ways that give specific content to general forms. How has decentralisation affected the scope for participation in and influence on politics? In the concluding chapter, the main findings are summarised and related to the theoretical argument. Some questions nevertheless linger on. To what extent do state-society relations in the health sector differ from those in other political arenas? In what ways does state formation differ between regions? How are different state institutions connected to or detached from one another? A final section of the concluding chapter broadens the scope of the in order to address these issues and to incorporate them into a discussion of comparative experiences of state formation within and beyond Uganda.

Chapter 2

State and Society: Theoretical Arguments and Empirical Context

States and societies in Africa have long been strained by enduring socio-economic stress and political tension. Relatively successful economic development in many African nations during the 1960s and early 1970s slowed down from the late 1970s due to a combination of international recession and domestic structural weaknesses. Economic stagnation resulted in budget, balance of payment and debt crises, in many cases presided over by authoritarian governments with waning legitimacy, which in turn typically created a political crisis. Most governments turned to the international financial institutions (IFIs), the World Bank and the International Monetary Fund (IMF) for credits, which were granted on the condition that governments pursued Structural Adjustment Programmes (SAPs) along the lines of economic liberalisation.

These programmes in most cases did not produce the intended outcomes and were from the late 1980s complemented by institutional reforms. A common denominator of these two generations of programmes was the assumption that state-society relations in Africa exhibited features that impeded growth and development, and that the reform programmes would correct these distortions.

Furthermore, during the early 1990s, in the midst of a global trend of transitions to democracy, expectations were advanced among donors and creditors of a convergence between economic and political liberalisation. Indeed, popular and parallel elite demands for political democracy had for some time been imposed on ruling elites, though just as often by way of protests against economic liberalisation as linked to economic decline and mismanagement and political authoritarianism more broadly. Also, in some cases ruling elites made use of the return to democracy to reinvent themselves and their hold on power.

The outcome of economic and political liberalisation has been mixed. Growth has often been unequally distributed, comes with high social costs and has typically not translated into substantial economic and social development. In many cases where liberal democracy was established, its content has remained shallow with limited scope for real access to the state or influence on decision making for most sections of society.

The manifestations of the crisis extended from relatively modest economic stagnation to deeper institutional decay, or what some authors refer to as "state collapse" (Zartman 1995). By any standard, Uganda experienced an acute crisis of statehood in the mid-1980s. The Uganda Peoples Congress (UPC) government that had been fighting a guerrilla army for four years was overthrown in a coup d'etat in 1985, and Kampala was effectively divided into zones controlled by factions of the military or various militias. Routine bureaucratic functions of the state had long since been undercut by informal exit strategies of its cadres, reflecting the informalisation of the wider commercial economy. State welfare institutions were reduced to shadows of their former selves. All sections of society endured through various kinds of coping strategies. Nationhood was, if not falling apart, so at least under serious stress after years of factional rivalry, not least within the armed forces. Such a dramatic predicament with regard to the institutional and territorial dimensions of the state raises fundamental questions about underlying conditions for the formation of state and society.

More than twenty years later, while not pushed to the very edge of survival, Uganda still faces serious challenges of state and nationhood. The country, and particularly so its northern parts, has suffered from a continuous series of armed insurgencies and civil wars, dividing the country to the extent that issues of future national disintegration were raised by political leaders in the north. But even more peaceful parts of the country are marked by underlying tensions and deep mistrust between government and opposition. Recent years have seen recurring conflicts between different state agencies, including a tendency towards a threatening militarisation of the state. In between, though, Uganda has been declared a success story. Its economic recovery and relative political stabilisation is much due to a reconstructed state. While some of the praise heaped upon Uganda by donors and creditors can easily be dismissed as exaggerated rhetoric, many achievements are undoubtedly real.

However, just as these achievements partly have their roots in pre-1986 history, so do some of today's problems. Unlike what is claimed in semi-official history writing in contemporary Uganda (Amaza 1998), there was no mythical complete break with the past or political revolution in 1986; the crisis in northern Uganda testifies to that.

The more general theoretical point is that state formation is a complex process of intertwined and often contradictory dimensions with different historical roots in society, and hence, characterised by both continuity and change. It is precisely the contradictory manifestations of state power in Uganda that call for a deeper of how they are related to the overall complexity of state-society relations. Furthermore, just as general forms of intervention, such as economic liberalisation and state reform as imposed on most countries in Sub-Saharan Africa, take on their specific content shaped by political struggles in each setting, so are the particular trajectories of these settings in turn shaped by changes in the broader conjuncture. This means that it is necessary to analyse both how, for instance, governance reforms are internalised by political forces in Uganda as well as how economic liberalisation under global capitalism spells out new conditions for those forces to do so. I shall return to the analytical implications of this admittedly abstract argument below by discussing how state managers, external actors and various social forces all seek to shape state policies and institutions to promote them.

At this stage, a conceptual clarification is needed. When I in this study apply the term "technocratic governance" in order to capture some of the characteristics of the contemporary Ugandan state, I do so, as the attribute "technocratic" clearly reveals, with a narrow policy related conception of governance in mind, well aware that the concept is also sometimes used in a generic and theoretically more ambitious ways to designate "any form of coordination of interdependent social relations" (Jessop 2002: 52). The policy related concept was developed in the early 1990s. To quote the World Bank (1992: 1), governance in their view is, "the manner in which power is exercised in the management of a country's economic and social resources for development", and incorporates public sector management, accountability, a legal framework for democracy and information and transparency (World Bank 1992: 12).

It should be noted that while the World Bank concentrated on managerial aspects, bilateral donors who incorporated the concept into their thinking were more explicitly political and beyond supporting institutional reforms also in some cases demanded multi-party democracy and respect for human rights as conditions for foreign aid (Leftwich 1993: 606). It is noteworthy, however, that Uganda, although occasionally criticised for human rights abuses, was for long exempted from such pressures (Haynes 2001). The reason for using the narrower and no doubt ideological sense of the concept "good governance" is that it is roughly that conception which, put to practice, has been central in restructuring the Ugandan state and in justifying this project as a success (Nsibambi 1998a). In that sense, "governance" figures in this study both as a descriptive characterisation and, as in "governance state" (Harrison 2004), as an object of analysis; more on this below.

This chapter deepens the probing of the research problem by relating it to, first, the comparative context of crisis, economic restructuring and state reform in Sub-Saharan Africa and, secondly, to theoretical debates about state-society relations in this context. The third section elaborates the theoretical perspective and argument while the fourth situates the preceding three into the Ugandan context. The chapter concludes with a section on methodology.

The comparative context: Economic liberalisation, state reform and pluralised development

The post-colonial accumulation model dominant in most African countries during the first decade and a half after independence was characterised by strong degrees of state intervention in the economy, import-substitution industrialisation and public sector expansion. State-led development models typically incorporated social forces into loyalty and submission, and the political survival of states to a large degree rested on their capacity for delivery of economic development and social welfare. Such "social contracts" and the various political alliances that underpinned them began to disintegrate as their basis for accumulation were undermined through intertwined processes of declining terms of trade and internal structural problems, resulting in fiscal crisis.

Since the early 1980s, most countries in Sub-Saharan Africa have applied programmes for economic liberalisation; Uganda embarked on its first Economic Recovery Programme in 1981. These programmes, whether "home-grown" or explicitly and formally imposed from outside by the IFIs, have involved measures for securing financial stabilisation and promoting export-led growth. Prominent features of the programmes include removal of subsidies, trade liberalisation, devaluation of currencies, and privatisation of the economy (Engberg-Pedersen, Gibbon, Raikes and Udsholt 1996; Mohan *et al* 2000: chapters 1-2).

The transformation from a state-interventionist to a neo-liberal development model has been underpinned by institutional reforms. State and public sector reforms, such as budget cuts, public sector worker retrenchment, user charges for health and education, privatisation and the introduction of new public management methods have been cornerstones of the reforms. Specific programmes were introduced in order to reduce the social cost of economic liberalisation.

A central feature of state reform has been decentralisation, with the stated aim of improving efficiency in service delivery and creating conditions for popular participation in development. The restructuring of state institutions along market lines has reduced the role of the state as well as transforming the foundations for its popular support. These policies have aimed at shaping managerial efficiency and have shifted the basic logic of welfare provision from universal to targeted policies, with NGOs entering as alternative providers (Bangura 2001: 256-259).

One more or less explicit assumption tied to these reforms is that by restructuring the state and allowing for more demand-led service provision, it is also possible to discipline public action away from patrimonialism towards more "rational" ways of behaviour (World Bank 1997). This in turn is based on views of the African state as overloaded and dysfunctional (Bangura 1994: 819).

Breakdown in state capacity of providing basic social services, including health, has both preceded and been deepened by adjustment programmes. Conflicts over priorities, an underlying feature of health care anywhere, have been exposed, and the problems of the health sector have been magnified in the context of the economic and social crisis of the previous two decades. A combined set of institutional reforms have thus contributed to the tendency of transforming social welfare services from social rights, albeit previously imperfectly supplied, to marketised commodities.

What has been the impact of state reform and economic liberalisation on state-society relations? Here, the focus is on the consequences for state-civil society relations in the context of social service delivery. In most African countries, adjustment policies in the shape of privatisation and downsizing of public expenditure have led to some degree of state withdrawal from previously central assignments in this field. Institutional reforms have been affected by, and set off, a multiplication of survival or coping strategies, following upon a drastic – or, as in Uganda, long-standing – deterioration of living standards. Coping strategies are conceptualised by Bangura (1994: 790-92) as the way individuals and groups organise themselves to make a living. They overlap or diverge, take individual or collective form and affect both structures of provision and patterns of consumption. Key features include diversification and informalisation of modes of livelihood and typically include migration, subsistence and self-employment. Individuals and households shift between public sector, private and traditionalist forms of health care, depending on accessibility or level of expenses (Bangura 1994; Mustapha 1992).

At the same time, new actors enter into the providing side, including urban-based indigenous or foreign NGOs, district based home area associations, religious organisations, community based self-help groups, and reactivated ethnic-traditional operators (Kiondo 1995; Rudebeck 1990). In Uganda, there were a few hundred NGOs towards the late 1980s. Today there are considerably more, and an unknown quantity of community based organisations (CBOs). The emergence of these actors is associated with external influences, such as the tendency during the late 1980s and early 1990s to channel development aid through NGOs as a way of "rebuilding civil society". Internal factors include a combination of growing demands of expectations from the population and decreasing financial and institutional capacity of the state (Kanyinga 1995; Therkildsen and Semboja 1995).

It would, however, be misleading to present these changes as a case of privatisation in a conventional sense. Rather, it is more a question of straddling or overlapping between donors, NGOs and states, in terms of funding, staffing and provision. Donors and civil society organisations enjoy a variety of political relations with the state, which has retained and may even have strengthened its close links to some of these actors, although in a new manner (Therkildsen and Semboja 1995: 3-4). As Mkandawire argues, political support can be expected to be closely related to the resolution of development and welfare issues (Mkandawire 2004: 12-13).

These changes are part of the wider paradigm of dismantled corporatist arrangements for regulating relations between state and society outlined above. Institutional reforms and economic liberalisation restructure state-society relations in a number of ways, including the internal composition of both state institutions and civil society as well as of the constituencies that civil society groups claim to represent. At the same time as the state has been downsized, privatised and decentralised, civil society has undergone a recomposition with implications for power relations. Previously predominant interest groups, such as trade unions and co-operatives, have been sidelined to shifting degrees in the process of liberalisation. NGOs have taken centre stage, at least in the field of social development. The overall effects of this recomposition, Bangura suggests (2001: 262), are pluralisation and fragmentation of both state and civil society, as well as of the regulatory mechanisms that mediate between them. The organisation of collective social projects is complicated by the multiplication of strategies and actors.

To sum up, economic and public sector reforms involve a growing role of NGOs and other minor actors in service delivery, with remaining but transformed importance of the state and a growing influence of donors on policy making. The combination of multiple coping strategies and a diversified development arena points to straddling as a defining feature, both of structures of provision and patterns of consumption, in relation to social institutions, where people move between contexts, loyalties and interests. The political implications of these changes will be discussed in the next section.

This section has stressed general contemporary features at the expense of historical particularities. It therefore needs to be reiterated that general or external intervention is given its content by the political struggles that emerge within specific contexts. While state formation in Uganda by and large conforms to the model of economic liberalisation state reform that operates within the structural policy constraints of a very weak domestic revenue base and subsequent aid dependence, the political impact of economic liberalisation does not follow a typical pattern of responses to state contraction and dismantled state corporatism. In Uganda, liberalisation has since 1987 been an integral part of a project for rebuilding the state. The intersection between the specific Ugandan trajectory and the general features of the policy context is analysed below.

State-society relations in Africa: Theoretical perspectives

It was suggested above that the wider political implications of structural adjustment programmes need to be seen from the perspective of the gradual decline of an accumulation model centred on an interventionist developmental state that had set up corporatist relations with a specified range of interest groups. The emergence of the contemporary orthodoxy in development thinking that accompanied the structural changes outlined above is well recognised and has already been suggested. "Good governance" emerged, in slightly different versions among policy makers (World Bank 1992; 1997) and scholars (see Hydén 1992), as shorthand for a development model where a combination of a restructured state, market forces and civil society are expected to bring about political pluralism and accountability and, eventually, democracy and development in a more substantial sense (Potter 2000; World Bank 1997).

The rise of the good governance agenda to dominant status as a development model has been addressed elsewhere (Doornbos 2001, Williams and Young 1994). Its application involves, among other things, an institutional modification of an earlier generation of structural adjustment programmes and their perceived failure. These were distinguished by rigid enforcement of deregulation, privatisation and trade liberalisation. The World Bank looked for institutional explanations and solutions to their shortcomings (Leftwich 1993: 607-608; Moore 1993: 2).

Such rectification in terms of state reform does not necessarily involve a drastically changed view of the restricted role allotted to the state in economic policy making, but it does underline the perceived developmental and democratic promise of civil society groups.

The contemporary model emphasises the potential of collaboration in more fluid forms between NGOs, local government and, possibly, a revitalised private sector to enhance "participatory development and community empowerment", through accumulation and distribution of social capital (Williams and Young 1994: 87; Mohan and Stokke 2000). A reformed and decentralised state and a revived civil society become key mechanisms for the realisation of these and a host of other objectives (Crook and Manor 1998: 2; White 1994). Similar beliefs about the propensity of civil society groups to advance political democracy have been shared by some academics, who propose that a pluralised civil society will function as a counterweight to state power, a transmission belt for popular demands and a school for nourishing a democratic culture (Harbeson 1994).

In many cases, economic reforms were, from the late 1980s, accompanied by political liberalisation. Domestic pressures for reforming and opening up political institutions were incorporated into the governance agenda and codified into political conditionalities such as demands for multi-party elections. The great expectations that first met these changes, as being part of the so-called Third Wave of democracy, eventually evaporated. Some countries fell back into military or one-party rule. In other cases, where liberal democracy has been established, observers have pointed to its shallow character and limited reach (Sandbrook 2000: 23-26). These setbacks suggest that patterns of political authoritarianism are deeply rooted and further enforced by structural characteristics of many African social orders. Significant features that impede the consolidation of democracy in Africa are said to include deep social and political inequalities; fragile and incapacitated state institutions controlled by elite coalitions; factional political parties without a solid social base; and weak civil societies, often characterised by ethno-regional fragmentation and disrupted links between the urban and rural population (Joseph 1997; Sandbrook 2000: chapter 2).

Another interpretation of state-society relations in Africa is even more pessimistic. On this account, the bright visions of "synergy effects" between state, civil society and the market, supposedly beneficial for democracy and development, need to be not only modified, but completely reversed. The main thesis of this critique is that political reforms are almost certain to be captured by prevailing structures of clientilism and that the resources that come with them will instead serve to reinforce patronage networks (Chabal and Daloz 1999: chapters 1-3).

This view is derived from a neo-patrimonialist perspective on state and society in Africa, which sets out from the Weberian distinction between rational-legal and patrimonial modes of authority. The latter, which is said to prevail as the core feature of politics in most African societies, is distinguished by personalised authority, privatisation of state power and informalised networks for distributing services and spoils among loyal clients. These characteristics constitute an informal system of rule that runs alongside and erodes formal legal-rational institutions (Bratton and van de Walle 1997: 61-63; Sandbrook 2000: 59-60). Neo-patrimonialism is said to follow from basic features of the social structure, primarily from particularist political identities along ethnic, regional and religious lines which reinforce clientilism.

The neo-patrimonialist argument echoes well-known propositions by Clifford Geertz on the specificity of state, nation and citizenship in the third world. In his seminal article, Geertz (1973 [1963]) argued that people in developing countries within the foreseeable future would sustain their close ties with ethnic, religious or equivalent kind of primordial communities – primordial basically in the sense of resting on persistent cultural idioms and modes of interaction. Similar arguments can be traced in an influential article by Peter Ekeh (1975), in which the author argues that colonialism left Africa with two publics: one civil, the other primordial, each with its own psychological and political logic.

The interaction between the two publics created a moral vacuum. While the virtues of the civic public were officially upheld, most people neglected its exhortations and remained loyal to their primordial publics. This clash of norms and interests between the two publics eventually generated two distinct features of African politics: tribalism and corruption (1975: 110). Incidentally, the patrimonialist motif reappeared in Ugandan political discourse, as it constituted the NRM's main argument against political pluralism.

The explanatory emphasis is sometimes placed on these structural-cultural logics, which penetrate the state and create systemic effects described as an "economy of affection" (Hydén 1983) or a "politics of the belly" (Bayart 1993); sometimes on cunning instrumental intervention from the state based bureaucratic-economic elite itself, creating vast networks of patronage in exchange for political loyalty (Bates 1981). On this view, state interventionism after independence, for instance, was not primarily a strategy for creating development but amounted to a tapping of public resources for maximising personal or sectional interests, by and large a symptom of an underlying logic of leaders distributing spoils to selected vested interest groups. Thus, while governance theorists claim that proper civil society is stifled by a hostile state, the neo-patrimonialist thesis holds that weak states are overpowered by an excess of communalist forces, or the downside of civil society.

Critique of the governance orthodoxy, and in particular of its perceived shallow conceptualisation of politics and power, is now commonplace (Mohan and Stokke 2000; Harriss 2002). This study shares this critical view and sets out from the assumption that contemporary development models to significant degrees are shaped by unintended but fundamental contradictions between institutional reforms (such as decentralisation and New Public Management), and social (such as informalisation and individual and collective coping strategies) and political processes (such as the intervention of ruling groups in society to reproduce their political base). Against the governance model, this study stresses the seemingly obvious, namely that state reform is an inherently political as opposed to technical process. The argument proposed here sets out from the premise that the realisation of decentralisation and local governance reforms will be shaped by deliberate efforts by the ruling class to structure its relations with other social forces due to central features of the political economy. Such efforts are likely to include pressures or concessions in relation to interest groups or regional or local elites to establish or consolidate the political influence of the regime (Boone 2003).

Against the patrimonialist argument, on the other hand, the proposition advanced here is that there is indeed scope, although often narrow and uneven, for interest groups to open up political space in a more democratic direction. While patronage politics is common – factionalist politics based on ethnicity and religion has certainly been a main feature of Uganda's modern political history – and the risks that institutional reforms will turn into "elite capture" by local powerholders are real, it is suggested here that these features are neither inevitable nor exclusive – a view that follows from treating "patrimonialism" as a fact of life.

Neo-patrimonialist theories have been criticised for reducing all forms of politics and interest group pressure to a question of patronage or "rent-seeking", isolated from other aspects of social relations. Such reductionism, it is argued, makes it difficult to explain the very different paths and effects of economic policy making found in Africa (Mkandawire 2001: 298-299). It further closes off analyses of different kinds of relations between state and society by treating these relations as largely antagonistic and civil society itself as homogenous (Gibbon 1992: 4-5; Beckman 1993: 23-24). Also, it can be added, subsuming all social relations under a catch-all patrimonialism concept conceals what patron-client or any other relations are all about, the contexts that have created these relations and what they promote in terms of regulatory mechanisms as well as policy content.

The concept says very little about the capacity or propensity of ruling or subordinate groupings to translate patronage ties into productive accumulation or conspicuous consumption, elitist exclusion or populist inclusion, tight control or broadened accommodation. It is entirely possible, for instance, that just as some individuals or groups may make use of formally democratic channels in order to bypass them by informal particularist favours, others may take advantage of clientilist networks to promote demands for equal access and fairer forms of conflict regulation. That is to say, while it is not difficult to identify examples of patronage, the argument is conspicuously void of convincing explanations of such examples, explanations which need to be related to a broader understanding of power relations that structure access and influence of social forces in order to make sense of the content of the political projects of either "patron" or "client".

Institutional reforms take root in different ways due to the ways in which social forces seek to appropriate them. Bangura (1994) suggests that informalisation and coping strategies to deal with economic crisis all over sub-Saharan Africa have produced underlying contradictions in relation to the institutional reforms mentioned above. This, according to Bangura, is due to the weak capacity of the state to intervene and address problems in relevant ways and a subsequent destabilisation of institutions.

A range of forces, not necessarily with democratic or egalitarian orientation, including reactivated traditionalist groupings, religious institutions and local and foreign NGOs, enter into the development arena to assert their relevance. This pluralisation of development actors involves a multiplicity of development interventions. "Intervention", of course, is a rather heterogeneous category. Development aid projects for agricultural modernisation, state penetration of rural areas for political support and social movement mobilisation seeking recognition all have their own roots and orientation and set off parallel forms of institutional connections and regulation. The next section deepens the theoretical argument and elaborates a framework for addressing the problem.

Framework and argument

How then do we approach the research problem, the scope for and capacity of social forces to create access to and democratise the state?[2] It is proposed here that in order to go beyond both the prescriptive optimism of the governance agenda and pluralist political science and the reductionist pessimism of neo-patrimonialist theory, state-civil society relations need to be theorised from a perspective that stresses relations of domination in overlapping arenas at the level of both state, civil society and among international actors. The study sets out from the proposition that interest representation and conflict regulation at the level of civil society is critical for the restructuring of state institutions. The institutional composition and political orientation of the state are seen as shaped by the regulation of demands and power struggles among social forces outside and

---

2 For overviews on the massive literature on democratisation, see e.g. Grugel (2002), Mahoney (2003) and Potter *et al*. (1997).

within the state, including transnational structures of power. More specifically with regard to democratisation, the draws on the central argument developed by Rueschemeyer *et al* (1992); Huber *et al* (1997) that a changing balance of class power has historically been crucial in promoting political democracy. It extends their argument to the broader hypothesis that the changing balance of power along different lines in society – not only class ones – and that the way this is interconnected with state structures, will continue to be fundamental for sustained as opposed to shallow democratisation of both state power and social relations.

Changing relations of domination are of crucial importance for democratisation for the simple reason that dominant groups, whether based on class, gender, ethnicity or any other marker of social stratification, are likely to try to preserve their economic, social and political privileges by resisting transitions to or the deepening of democracy. Bangura (1992: 46) argues that:

> the basis for authoritarian rule should be located primarily at the level of material relations, i. e. it expresses a particular resolution of contradictions in particular forms of accumulation. But the dynamics of authoritarian rule and struggles for democratisation develop at the level of civil society.

Hence, any set-up of political institutions is embedded in power relations in the wider social formation, and needs to be analysed in that context. Democratisation of the state involves changes in the balance of power among different political forces and of state-society relations to the effect that the state in the process becomes more autonomous from dominant political classes and social groups and correspondingly more embedded among and responsive to subordinate ones. To continue the argument, it is proposed that states are shaped by struggles between different constellations rooted in the state, in local society and among international actors – for instance international capital, but also international NGOs or providers of development aid. In contemporary Uganda, the transnational dimension is very obvious as reconstruction largely rests on donor-driven capitalist development.

A similar argument is applicable to the of the local state. Catherine Boone (2003) in her study on variations in local state forms and central-local relations in West Africa, emphasises the mutual determination of power relations in rural society and strategic state intervention. Forces in local society make demands on or respond to interventions by central governments. Differences in political capacity and interest in rural society are shaped by political struggles at that level. Ruling groups seek to penetrate and control various forces in society in relation to the different political and economic challenges posed by the latter.

This framework allows for capturing authoritarian, democratic and patrimonial features of state-society relations but does not predetermine either. In contrast to governance and civil society oriented theories, it incorporates power relations and political struggles at the level of civil society, within the state and internationally. It also goes beyond neo-patrimonialist perspectives in that it considers how "neo-patrimonial" features feed into other and contradictory tendencies and interconnections between them. Even where authoritarian rule at the level of state power is firmly anchored among hierarchical social relations, it is transformed by the dynamics of conflicts and demands set off by the contradictions of development. This theoretical argument is explored empirically by analysing different forms of conflict regulation and interest representation in the health sector. In the rest of this chapter, I move from the general theoretical argument to a contextualised analytical framework.

Dimensions of state formation

The argument sets out from a perspective of the state as a historically evolving, yet relatively permanent set of institutions that aspire to stabilise and regulate social relations of domination, institutions that in the process both reflect and reshape the latter (Jessop 2002: 40). Capitalist states are deeply rooted in economic and social relations by for instance the need to extract resources from society in exchange for concessions and by redistributing those resources and setting rules for conflict regulation. Formal and real distribution of power between central and local institutions and different arms of government impact on important dimensions of state structures such as their institutional capacity (as they may rest on internally either relatively coherent or fragmented state agencies) and democratic accountability (as they may be responsive to, for instance, representative state agencies and different forces in society, or not).

All state managers and ruling groups meet challenges of how to accommodate and balance different dimensions of state formation and reproduction, including requirements often deemed fundamental to statehood, such as preserving territorial integrity, establishing monopoly of the legitimate use of violence, extracting and allocating resources, promoting accumulation and penetrating society to establish control and support through coercion and consent; this arguably involves recurrent crises with regard to such material and ideological sources of power. After constructing such general propositions, however, I will in the following section hasten to deconstruct them to some extent. As should be clear, most of the core concepts used above, including "coherence", "stabilise", and "legitimate" are contested, and their content varies depending on the perspective of each actor. For instance, what is considered legitimate about state power for one group in society may be highly illegitimate for another. Likewise, what is considered "fundamental" challenges and how to deal with them is highly contextual and subject to political struggles. What is of concern here, however, is the ways in which a group in control of state power in a particular society, in relation to demands by other social forces on the state, applies strategies for accumulation and legitimation, conceived as attempts to reproduce its political base and sustain political control through material and ideological resources.

In the study, I analyse the various attempts of the NRM government to establish and maintain control in relation to demands from contending political forces, employing sometimes contradictory, sometimes reinforcing strategies to promote accumulation and legitimation. "Control" includes the various ways in which the state seeks to achieve legitimate or otherwise domination over social groups and the state territory through coercion and consent. These processes are evidently related and contradictory in complicated ways. They are related through the ways in which the state is dependent upon support among at least most significant sections of society for extracting resources. This in turn typically requires concessions in terms of security and redistribution in order to promote normative and instrumental legitimation; none of this is new or surprising. The processes are potentially contradictory not only in sofar as social forces are likely to have different and occasionally irreconcilable interests, meaning that what constitutes and creates "state legitimacy" will differ in relation to various constituencies but also in that the strategies chosen for extracting resources and promoting accumulation, more often than not undertaken within serious constraints, may undercut legitimation or vice versa. In pursuing projects for capitalist accumulation and development, conflicts arise between different political interests and in relation to the various sources of legitimacy these interests draw on, and structural contradictions emerge between strategies for example aimed at rapid but potentially destabilising accumulation and conditions necessary for the reproduction of the political and economic order in a longer perspective. Bruce Berman (1990) has successfully highlighted such "dialectics of domination" through the case of colonial Kenya, emphasising the inherent instability of the seemingly solid colonial state. The NRM government's attempts to establish control have taken place within severe constraints created by the preceding crisis and in relation to demands made on the state by different political forces. It has therefore been rather an unstable political project.

The character of state institutions, including the extent and kind of state intervention that characterises economic and social policy, is significantly shaped by the way conflicts and power relations in society are reproduced within the state. Contending forces in civil society attempt to create access to the resources and regulatory intervention of the state. State managers seek to reproduce the effective presence of state agencies at the level of society by mobilising constituencies, crafting alliances and building institutional mechanisms for representation and control – often but not always benefiting dominant social forces and external constituencies at the expense of subordinate ones. These mechanisms are mediated by resistance or compliance, due to the balance of power among social forces; again, social relations of domination are both reflected in and reshaped by the state. The strategies of ruling groups for ensuring control range from coercion to more sophisticated and legitimate forms of the political institutionalisation of social relations.

The argument so far may seem overly structuralist and abstract and needs to be modified and placed in context. It should be clear from the foregoing that these processes are inherently complex and unstable. In cases of fragile states with uneven capacity and weak resource bases, as in Uganda, contradictions between different aims are constantly brought to the fore by demands in different directions on the state by various formal and informal, internal and external constituencies; the latter impinge heavily on the functions of the Ugandan state in view of its dependence on donors and creditors for resources. The latter point suggests that challenges of control and strategies to meet them by among other things accumulation and legitimation are contextually specific through the ways states are inserted in the international system and in the local social formation with its historically developed institutional landscape. And this leads to the analytical proposition that state formation needs to be studied with careful consideration of context and historical sequence.

The balancing of different dimensions of state formation takes on its conjunctural political expression in many African countries in the context of economic liberalisation, decorporatisation and changing state-society relations outlined above. In the Ugandan case, it is in the context of the specific trajectory of aid-driven capitalist development and state reconstruction. In most cases, including in Uganda between 1981 and 1985, adjustment and liberalisation followed from fiscal crisis and weakened state capacity. In the context of declining revenue bases, states enter vicious circles and face interlinked crises of weakened capacity for material reproduction and destabilisation of legitimacy. In order to secure control, state rulers have found it necessary to either repress or incorporate demands from various groups in civil society, according to the strength of the latter. Uganda's first adjustment experience during the early 1980s was characterised precisely by the difficulty for the UPC government to balance contradictory demands. In the context of civil war, the government embarked upon measures which sought to secure loyalty from its shrinking power base but which undercut the conditions for its reproduction. In Uganda since the late 1980s, on the contrary, liberalisation has been associated with state reconstruction as opposed to decay. The main sources of legitimation of the NRM government have been peace and security among relevant constituencies, and growth and distribution of public goods.

Civil society shaping the state

The nature of civil society in Africa has long been the topic of a vast literature, with proponents (Bratton 1989; Chazan 1992; Diamond 1994; Harbeson 1994) as well as critics (see the collections edited by Mamdani and Wamba-dia-Wamba 1995; Kasfir 1998; and Comaroff and Comaroff 1999) of the idea that civil society holds the key to democratic transitions in Africa. For an overview, see Sjögren (2001). In contrast to certain contemporary conceptualisations of civil society that tend to view it as a pluralising social phenomenon with relatively homogeneous core properties and liberal-democratic political implications (Diamond 1994), it is argued here that civil society is better understood as an analytical as opposed to prescriptive shorthand term for a wide range of more or less formal social groups without specific pre-determined political properties. Civil society cannot, for instance, be assumed a priority to be autonomous from or opposing the state. Following from this, the formation of civil society in any specific setting is here viewed as rooted in and transformed by the political economy and its different kinds of social stratification, including class, gender, and ethnic relations, and in response to the forms of ideology and representation such stratification sets off (Gibbon 2001; Sjögren 2001). This process does not, of course, occur in any deterministic manner. The composition and political orientation of civil societies cannot be derived straight off from the social structure or from state or ruling class imposed attempts at domination. Another aspect is that very often the distinction between state and civil society is not only blurred but entirely misleading as individuals and resources move freely between the two.

Finally, while not making claims to any absolute defining demarcation, the conceptual distinction between society and civil society made throughout the study refers roughly to "routine modes of interaction in a social formation" and "organisational forms emerging out of that", respectively. Therefore, it sometimes makes more sense to speak of state-society rather than state-civil society relations; particularly so at the level of local politics, as shall be discussed in Chapter 5, but sometimes also at national levels where the notion of civil society may constrain rather than expand the understanding of, for instance, the networks of elite politics. Consequently, state-civil society relations may be thought of as a subset of state-society relations.

Transnational power structures

Transnational structures of power shape both state structures and social relations in different ways, through changing forms of accumulation, for example international regimes of trade, as well as through direct external involvement with a country's policies and institutional framework. A defining feature of contemporary Ugandan politics has been the deep and sustained external involvement. Donors and creditors have for years financed approximately 50 per cent of Uganda's state budget (Tangri and Mwenda 2006), and this external fiscal basis of the state naturally impacts upon internal state-society relations. It also restricts the Ugandan government's options for policy choices. Furthermore, transnational structures of power also affect both internal class structures and the composition of civil society, as many civil society organisations are either external or externally funded. This again influences state-civil society relations. With regard to contradictions between foreign and national interests, it is clear that the Ugandan political economy remains dependent. It relies on income from primary commodity exports and foreign aid or credit. More advanced production is heavily import-dependent. There is a high degree of penetration of foreign capital. Economic policy making is in a broad sense imposed by external providers of resources in close collaboration with state managers (Harrison 2004). It is also evident that the historical trajectory of the political economy has in crucial ways been shaped by colonialism and imperialism, including the form class and state formation took on in this context, through the promotion of Buganda region and the deliberate underdevelopment of other regions in the eastern and northern parts of the country (Jørgensen 1981; Mamdani 1976). It is furthermore clear that independent Uganda has occasionally suffered from foreign political intervention – most notably Israeli and British backing of the 1971 coup (Furley 1989). Nevertheless, the form and content of capitalist development and state formation that emerged has been crucially shaped by Ugandan social relations of power and resistance, as reproduced in and by the state. In making this claim, I also assume that the contemporary state, in spite of dependence on external revenue, retains relative autonomy as regulating instance, although in a slightly new manner.

At this point, conceptual clarification is required. The shorthand notions "external" / "foreign" and "internal" / "domestic" used throughout this book have limited meaning in themselves: evidently, there are strong linkages between local, national and global economic and political factors, blurring clear-cut demarcations. While there are, in principle, contradictions between foreign and national interests, lines of conflict just as often run between different constellations of external-internal forces that, according to power relations, are differently positioned to shape the Ugandan state. Harrison (2001; 2004: 71-74) discusses Uganda as an example of "governance states" or "post-conditionality regimes", where "[p]ost-conditionality politics is both more interventionist and less starkly coercive" (Harrison 2004: 71). This kind of state – the main features of which are particularly prominent in administrative reform processes – are according to Harrison characterised by the high-profile role of donors in economic and institutional reform. Their significant presence is to a high degree ideologically and institutionally internalised within the state, and supervision is exercised on a basis of trust and with implicit references to a "common sense" and mutual interests.

These regimes are thus said to be differentiated from conditionality relations by the employment of promotion and rewards rather than coercive measures in order to exercise reform, and a blurring of distinction between external and internal interests where instances of the former have turned into a permanent feature of the institutional landscape. That is, donors and creditors were early on taken on board the NRM's broad based coalition, and while not in a strict sense part of the state, they have in important respects been internalised into the state formation process (Harrison 2004: 87-88). At the same time, the Ugandan state has been reorganised and externally incorporated into what has been termed an emerging "transnational state" made up of networks of "supranational economic and political forums", such as the World Bank, the IMF and the UN system (Robinson 2004: 88). External actors, and not only donors but also NGOs and private capital, are allied with various domestic political forces located within and outside the state. I therefore incorporate these actors into the of state-society relations as integral parts of competing political alliances, particularly prominent in the creation of "technocratic governance".

State-civil society relations

The preceding sections support a perspective that emphasises the mutual formation of state and civil society and a manifold of state-civil society relations: an institutional landscape of structurally differentiated constellations competing for access to the state (Migdal *et al* 1995). The main argument with regard to the research problem is that the scope for and capacity of groups to support or indeed block the democratisation of the state is best explained by relations of domination. What may be hypothesised about the politics of state-civil society relations in the context of economic liberalisation and state reform? What strategies do state managers use in order to promote accumulation and legitimation? Do they seek to repress or to incorporate civil society groups? Do civil society groups strive for greater autonomy or incorporation? Do they reproduce or challenge received political identities? What is the scope for participation, accountability and influence? Are channels of influence structured as "transmitters of temporary particularistic favours [or] channels to mobilise citizens into influencing policy formation" (Huber, Rueschemeyer, Stephens 1997: 334)? In contemporary Africa, different strands of civil society are developing. These are shaped by related processes of change in the political economy but are based in and represent dissimilar constituencies, for very different purposes. Ideal-typically, many NGOs are oriented towards service provision, while social movements articulate and mobilise constituencies in relation to socio-economic transformation; combinations occur. In theoretical terms this may be related to the distinctions between "deepening" and "politicising" civil society (Gibbon 1996: 38), or between forms of civil society that pluralise social life, those that expand popular participation and those that strengthen the interests of subordinate groups (Huber *et al*: 1997: 328).

Mkandawire has discussed African state-civil society relations in the adjustment context by distinguishing between merchant and rentier states on the one hand and strong and weak civil societies on the other (Mkandawire 1995: 24-29). The utility of his argument for the purposes of this study stems from his attempt to systematically link "structural imperatives of accumulation and the conscious and purposive actions of social agents – individuals, interest groups or classes which set the boundaries of the political legitimacy of state measures" (Mkandawire 1995: 23), or in other words, to situate the scope for action of both dominating and subordinate political forces limited by the structural constraints within which they operate. The fiscal basis of the state shapes the modes of interaction between different social groups and the state, in that, for instance, rentier states that depend on high degrees of external (including enclave based) revenue are less compelled to engage with domestic interest groups, and particularly in cases of weak civil societies. The overall strength and internal composition of civil society is in turn associated with different levels of political clout in relation to the state in terms of being able to consistently politicise social grievances. This generalised argument needs to be thought of as implying the kind of internal differentiation stressed above so as to allow for various tendencies, cohesion or fragmentation among ruling and subordinate classes and relations of domination and political conflicts within civil societies.

Some analysts have claimed that the emergence of NGOs and community organisations is associated with increasing popular participation and empowerment and links informalisation with new forms of democratic popular political cultures (Bratton 1989; Chazan 1992; Hydén 1995). Others argue that self-help groups and NGOs are typically incorporated into existing power structures and therefore have little political emancipatory value except at an individual level. From such a position, state contraction and the corresponding pluralisation of service provisioning has largely produced a fragmented and individualised civil society containing organised groups with limited incentives or capacity to move beyond their own concerns and with doubtful propensity to advance democratic demands (Kanyinga 1995; Kiondo 1995). NGOs in particular have been said to be limited by the project oriented nature of their concerns and their institutional and financial dependence on donors and the state (Fowler 1991). Such fragmentation is said to stem both from the inclination of NGOs to represent discrete constituencies and from the tendency to enforce distinct and limited agendas.

The argument is that the context of economic liberalisation and coping strategies reinforces the structural differentiation of state-civil society relations in terms of political orientation and the capacity to influence. Therefore, the civil societies that emerge in such contexts could be expected to reproduce rather than to challenge received identities, ideologies and power relations.

Regimes and political space

The contours of state-society relations are shaped through broad processes of state formation, such as the promotion of capitalist development and nation building and the political communities these processes encourage or repress. But, as suggested by Beckman and Jega (1995), more precise linkages also arise around specific instances, as when state agencies intervene strategically among social forces to mediate and regulate by creating institutions for representation and control in different arenas of conflict. These social forces in turn have different capacity to respond and to shape state institutions and the channels for interest representation and mechanisms for decision making, which link state and society and shape the social base and political content of state institutions. The latter cannot of course be reduced to instruments of factional interests, but they are in important respects moulded by social forces through the way the state is drawn into or chooses to enter into conflicts in different arenas. The state may come up against assertive civil society groups that require accommodation and concession, just as it may face a relatively weak and politically insignificant civil society that can easily be co-opted, demobilised, neglected or if necessary repressed. The extent to which interest groups are allowed to, capable of, and inclined to organising around specific and wider interests is highly crucial for the preconditions to engage the state and the constituencies it represents, opening it up to various sections of the population, and even transforming it in democratic directions. This is why the regulation of power relations and conflicting interests in different arenas in society is relevant for the ways in which social forces create organisational autonomy and capacity, bargaining power and access to the institutions of the state.

The argument so far suggests that there is a key linkage between specific issues and more general democratic rights through conflict regulation and the constitution of the state at different levels. One aspect of this linkage is the right of association in its formal and substantial senses. The latter involves a real capacity to become involved in decision making processes and to open up political space for interest representation. Engberg-Pedersen and Webster (2002: 7-12) discuss different dimensions of political space in terms of institutional channels for interest representation, the scope for practice and parameters for discourse.

When analysing political space, it is important to recall the complexity of the concept and the practices it refers to. On the one hand, it is an analytical construct, and on the other, it reflects the different perspectives of various actors. Simply put, a change in for instance legislation over accessing medication, land rights or the right to organise may lead to some social forces succeeding in advancing their positions and opening up space for their own projects – which might close down the space for others. This means that there are competing notions and practices of political space in any social order. Not all of these are democratic or egalitarian. Nevertheless, they point to existing forms for making claims and shaping public authority. Beckman and Jega suggest a range of democratically relevant aspects, from defending the right to organise to articulating wider democratic demands (1995: 171-172). What explains the capacity for civil society organisations to open up political space in any of these senses? Some factors are seemingly evident: the social and material base of the groups in question; their organisational experience and capacity in terms of membership and finances; internal structures of decision making; their bargaining power in their fields of operation; the legal and political framework in place, including the presence or absence of institutionalised forms for demand making and interest representation; their options for striking alliances; and the strategies used by contending social forces and the state. Most of these factors constitute or reflect power relations.

In order to capture the theoretical complexities of mutual determination and the empirical intricacy of messy institutional landscapes in more precise terms, the concept of regimes will be used. This concept refers to the set of institutions, rules and practices, shaped by relations of domination, which regulate interconnected policy arenas at different levels. The empirical arenas under study are structured by formal and informal institutions within the nexus of a variety of regimes such as labour, public sector, welfare and health regimes. Gough (2004) has sought to modify Esping-Andersen's work on welfare regimes for application to research on social policy in developing countries. Under the generic rubric of welfare regimes Gough develops distinctions between welfare state, informal security and insecurity regimes so as to enable the theoretical incorporation of conspicuous features such as informal modes of production, reproduction and forms of provision and, consequently, informal demand making structures (Gough 2004: 26-33). What is more significant than the conceptual distinctions themselves, however, in exploring the many and divergent Ugandan health regimes and how they relate to state-society relations more broadly, is the analytical proposition that stresses how health and welfare regulation must be conceived of and analysed in terms of a series of overlapping institutional regimes. By including both rules and practices, the regime concept opens up to capture precisely the broader features of state-society relations that underpin formal institutional change.

State formation in Uganda: Governance and militarism

Uganda is often considered the prime African successful case of contemporary state reconstruction. After many years of turmoil, civil war and institutional decay, the NRM government set out, from the mid-1980s, to rehabilitate the state through strategies of economic liberalisation in order to construct its fiscal base, and political inclusion of contending forces in a "no-party" government system. The new government suspended all political party activities with the stated aim of putting an end to elite manipulation of ethno-regional and religious sentiments, codified as "sectarianism", something that NRM identified as the root cause of Uganda's problems. The argument was generalised into a doctrine of politics that ruled out multiparty politics in developing societies characterised by vertical communal cleavages. NRM itself cast its role as the driving force behind a "fundamental change" and justified its existence as a political entity in terms of being an all-encompassing "movement" as opposed to a party (Mamdani 1994).[3]

Several means were employed for reshaping institutions of interest representation. At the national level, a broad-based government was formed, including members of antagonistic groups and factions. Locally, institutional channels for linking the rural population to the state were established – a five-tiered system from village to district of so-called Resistance Councils (RCs), later renamed Local Councils (LCs). From the point of

---

[3] The NRM referred to itself as per the 1995 Uganda Constitution as the "Movement" and to the "no-party" system as the Movement Political System, sealed in law with the 1997 Movement Act. With the reintroduction of multiparty politics in 2005, it reassumed the name NRM. Therefore, both "NRM" and " "Movement" are used in the text, depending on the issue under discussion.

view of those who stress the successful reconstruction of the Ugandan state, strategies of accommodation within political structures for representation have overtaken sectarian and extra-constitutional means of competing for and exercising political power (Wunsch and Ottemoeller 2004). This logic of political stabilisation is said to be underpinned by economic growth and the professionalisation of the armed forces.

Among the government's acclaimed accomplishments are the establishment of peace and security in large parts of the country, protection of a basic level of civil liberties and laying the foundations for macro-economic stability and rapid and enduring growth, which was, until recently, claimed to have reduced significantly the level of poverty (Collier and Reinikka 2001). Uganda's Poverty Eradication Action Plan (PEAP) is frequently held up by IFIs as a prototype for poverty reduction strategies (PRS), with regard both to its participatory process that included civil society organisations and its coherent strategy in achieving its aims (Piron and Norton 2004). Furthermore, Uganda's achievements in countering the HIV/AIDS epidemic are widely recognised (for overviews, see Allen 2006; Allen and Heald 2004; Parkhurst 2005). Much of this has been attributed to the enhanced capacity of state institutions in general and the political leadership of President Yoweri Museveni, who was hailed during the 1990s as one of the "New Breed" of African leaders (Oloka-Onyango 2004). State capacity has rested upon and reinforced a "peace-and-growth" dividend. In short, Uganda has long been portrayed as a reconstruction model to learn from in economic adjustment, "good governance" and "post-conflict" senses – save for the lingering conflict in northern Uganda.

While this account of the "Ugandan miracle" has been the dominant one within "the international community", it has not gone unchallenged. The recent shift towards a more hesitant embrace of the NRM government by donors and creditors was preceded from the early 1990s by both critical academic reconsiderations of the contradictions of the Ugandan state (Mamdani 1994; Oloka-Onyango 1991) and accusations from various sections in the Ugandan society of increasing intolerance, nepotism and corruption. Economic analysts have pointed to problems of regional fragmentation, growing aid dependence and over-reliance on coffee as foreign currency earner (Dijkstra and Kees van Donge 2001: 842-845; Ellis and Bahiigwa 2003: 998). A series of armed rebellions have raged in all parts of the country ever since the mid-1980s, the most longstanding one being the conflict in the north between the government and the Lord's Resistance Army (LRA). Political opposition has been dealt with in an increasingly harsh manner; the 2001 Presidential election, where the armed forces were instrumental in coercing opposition candidates and sympathisers, was the most violent in Uganda's history (Ajulu 2001).

Uganda has, directly and by proxy, played a crucial role in breeding and sustaining the conflict in DR Congo (Human Rights Watch 2001 b, 2005; International Crisis Group 2000; International Court of Justice 2005; United Nations 2002). This picture of state-society relations in Uganda as deeply marked by structural continuity of violence and authoritarianism thus contrasts dramatically with the dominant one of institutional change and the stability, growth, poverty reduction and popular participation that this transformation has brought about. This is the context of another term in the title – "militarism". Its meaning is used broadly, as "the pervasiveness in society of symbols, values and discourses validating military power and preparation for war" (Luckham 1994: 24), and evidently closely related to militarisation, referring to:

> a multidimensional process through which a number of elements – such as military coups and regimes, authoritarian government, the dominance of patriarchy, powerful military and repressive state apparatuses, war and armed conflict,

> rising military spending and arms imports, and external military intervention – become dynamically linked, both to each other and more widely to capital accumulation and projects for national and international hegemony (Luckham 1994: 24).

What do these apparently contradictory developments tell about the logic of state formation? After a year in power the NRM abandoned its initial socialist-nationalist programme and, in a context of near state bankruptcy, struck an agreement with the IFIs over financial support in exchange for economic liberalisation. Following a phase of internal struggles within the government over which long-term directions that should follow upon the "tactical concessions" of the IMF agreement, the government embarked on liberalisation reforms in the early 1990s. A fairly typical package of economic adjustment was complemented by institutional reforms, including civil service restructuring and decentralisation, intended to provide a stabilising political framework so as to ensure "good governance". Economic liberalisation has been one central dimension of state formation.

Another has been the considerable degree of external involvement. Ever since the IMF agreement in 1987, it has been evident that state reconstruction and capitalist development has been profoundly donor-driven in terms of resources and its broader policy framework. Economic policy making is distinctly and faithfully applied within the parameters set by international neo-liberal development orthodoxy, rendering Uganda an IFI showcase status during the 1990s (Harrison 2004: 39). Donors and creditors have for years contributed a vast percentage of the state budget. International NGOs are prominent in the development arena. It would be one-sided, of course, to portray institutional and policy changes as entirely externally imposed. The decentralisation programme, for instance, had some of its roots in the Resistance Councils that were created during the guerrilla war in the early 1980s, and even the most typical of economic reforms, such as devaluation, take on their specific form and content mediated by Ugandan social relations. Nevertheless, these and other forms of external influence are conspicuous and significant and their causes, content and impact on for example state revenue and political support beg for analysis. This book discusses the consequences of the heavy presence of foreign actors and resources, be it donors or agents of international capital or a combination of the two, for state-society relations. It pays particular attention to the prospects for democratising the state.

Contrary to the case in many other countries in Sub-Saharan Africa, economic liberalisation in Uganda was associated with state reconstruction rather than with state withdrawal. Needless to say, the real extent to which the state has been rebuilt varies dramatically between regions and between levels and types of state institutions. Nonetheless, economic liberalisation became associated with rapid growth rates and a wider process of political stabilisation and contrasted favourably to past encounters with state intervention and its connotation of elite manipulation, corruption and inefficiency. Furthermore, the crisis that preceded adjustment was so severe that any stabilisation was considered an advance, and according to some observers this experience contributed to the prevention of effective protests against reforms (Brett 1996: 342; Kjær 2001: 154-161) and shaped state-society relations in important ways, as shall be explored later in this study.

Potential resistance to reforms was also weakened as part of the erosion of collective organisation more broadly. During the years of dictatorship and civil war, civil society practically disappeared. The period after 1986 has seen its gradual resurgence and recomposition. Previously predominant forms of collective organisation, such as trade unions and co-operatives, were weakened while local and foreign NGOs and community based organisations (CBOs) mushroomed (Dicklich 1998; Bazaara and Nyago 1999). How does this recomposition of both state and society affect power relations in society and state-society relations? In what ways does the state seek to relate to different civil society groups? To what extent are these inclined and capable of expanding political space? What kinds of regulatory regimes emerge?

Rather than seeking to confirm or reject clear-cut models, the aim of this study is to bring out the complexity of the social basis of political rule in Uganda from the theoretical perspective outlined in this chapter.

Some notes on methodology

Design

Some issues of design have already been discussed above. In approaching the research problem and the theoretical argument analytically, I seek to analyse state-civil society relations through the concept of regimes. This is then applied to the health sector and a range of state-civil society relations thereof. Attention is paid to power relations between state and civil society organisations as well as within both state and civil society. The argument proposes that it is power relations between different political forces that determine the scope for the democratisation of the state, since political institutions are embedded within and are given shape and content by the wider political economy. The argument further posits that both power relations and political institutions are shaped by historical trajectories. These propositions give explanatory emphasis to structural factors shaped by historical sequence; it is taken for granted that structural factors are reshaped by institutions and actors. In order to go beyond the manifest level of power relations, the cases studied will be situated within an argument of structural change of the Ugandan political economy.

Cases, sources and methods

In more ways than one, Uganda is, to borrow a phrase, "extreme but not exceptional" (Mamdani 1987). Its recent history is impaired by dramatic swings that effectively highlight tendencies general to Sub-Saharan Africa – state crisis, donor intervention and some degree of political liberalisation amidst persistent authoritarianism. The Ugandan state has undergone radical reconstruction since the years of turmoil and civil war and stands out as a prime case of privatisation, deregulation, and decentralisation. The fact that the NRM government has implemented far-reaching economic liberalisation and public sector reform makes the health sector in Uganda an advantageous case for investigating state-society relations in the context of the contemporary development model.

In Chapter 2, the study begins to analyse how contemporary state-society relations are shaped by historical trajectories. This chapter situates the development of different health regimes in the context of state-society relations at large. The narrative relies on a rich body of literature and brings together this material through the empirical focus on the health sector with the theoretical argument as the organising principle. This continues in Chapter 3, which covers the first years of the present government in power, and is based on similar material.

In order to approach the theoretical problem in greater empirical detail, the study examines a cross-section of interest groups in the health sector in Uganda and their relationship to the state: Uganda Medical Workers Union (UMWU); World Vision, an international Christian NGO; and Buganda Cultural and Development Foundation (BUCADEF), the development organisation of the revived Buganda kingdom. The study analyses state-civil society relations at both national and local levels. At both levels, the study moves from interrogating the organisations and their respective relations to the state one by one to an of the overall characteristics of the sector. The specific trajectories of these groups have consequences for the kind of political space that is created in relation to them. They embody different conceptions of how to shape social development and structure state-society relations and are vehicles for highly divergent social and political projects.

The present conjuncture, it is argued, is characterised by fragmentation of both state and civil society. The cross-section design depicts the empirical tendency towards a pluralisation of development actors and their relations to different state organs, as well as of forms of regulation in the sector at large. The diverging experiences of the groups under study also offer an illustration to the theoretical argument that state and civil society are mutually constituted through the ways different social forces seek access to the state based on their shifting capacity and orientation and the way the state chooses to structure its relations to these social forces. To capture the wider significance of such diverse organisations, they need to be placed within the contexts that have enabled them. In this case the structuring of state-labour relations, the socio-economic context of the rise of NGOs, and relations between the central government and the Buganda kingdom, respectively. These contexts – further elaborated in chapters 4 and 5 – have separate legacies, and have generated different models for interest representation and conflict regulation.

Chapter 4 relies on primary sources such as newspapers and official documents, secondary literature and interviews. The latter were conducted, repeatedly over time, with representatives of the groups under study but also with Ministry of Health staff and members and representatives of other civil society organisations. In order to obtain an improved understanding of the broader tendencies of Ugandan politics, interviews and informal talks were continuously held with observers of the political scene. I spent altogether seven months between 1999 and 2005 doing fieldwork in Kampala and Masaka, with roughly four months in total in Masaka. A first preliminary visit was made to Kampala in December 1999, followed by a first session of fieldwork during two months in late 2000. Fieldwork was again carried out for two months during February and March 2003, for one month in February 2004 and for another few weeks in February 2005. This drawn out process means that politics did not only change dramatically, particularly on the national scene, but also that I have been able to witness a sequence and get a broader perspective on these changes.

At the level of local politics, the topic of Chapter 5, one dimension of the investigation offers an overview of the district development arena, with a focus on some selected actors. Another dimension offers a slice of the same reality as seen from the community angle, through sub-county and district, which integrates the development arena into a broader picture of local politics. The bulk of the data for this chapter was gathered from semi-structured interviews and group discussions and also from visits to and observations of relations at health centres. With regard to the very local study, I make no claims to serious ethnographic work. I have paid only brief, if numerous, visits to Bulando village. There is rather little of primary documentation at the sub-county and village levels, but I have gone through what was available. I conducted interviews with representatives of civil society organisations, and not only of the three organisations studied, but also with politicians and civil servants at different levels and ordinary citizens.

At the village level, discussions were held in Luganda by my colleague Haruna Bukenya and he then translated them into English. Group discussions were held separately with different categories: men, women and youth, in order to as far as possible enable conditions for frank, open, and fruitful discussions for all.

On other occasions, and typically when meeting representatives of civil society organisations and of the central and local state, interviews were conducted directly in English. Both interviews and group discussions were conducted repeatedly with the same interviewees and groups over the years. Interviews lasted between 30 minutes and two hours, and normally around one hour. Needless to say, there are a great many methodological problems associated with doing research based on interviews in a social context rather different from one's own and occasionally through interpretation. Complex issues of reliability present themselves through translations, memories of those interviewed, their intentions vis-à-vis me and the impact of my presence in the locality – it was often assumed that I represented an NGO, with all that would imply. I sought to counter such difficulties as far as possible by, for instance, conducting repeated interviews, cross-checking information with different respondents and discussing the meaning of particular statements and overall contexts. I shall return to some specific methodological considerations emerging from fieldwork in Chapter 5.

Even more than with national politics, my emerging understanding of the complexities of local politics in Masaka has depended on numerous visits, repeated interviews, constant informal discussions with a wide selection of observers of the local political scene, and on ongoing reflections on all these exercises. That is, many interviews and discussions have been used to build up knowledge of the wider context rather than targeting the specific object of inquiry. For reasons of integrity and sometimes security, not all of the observers mentioned above can be named. The incessant movements I experienced during fieldwork back and forth between theoretical reflection "that embodies a wide range of previous observations and analysis" (Rueschemeyer 2003: 332) and empirical evidence which led to continuous revisions of both, call it "analytical induction" (Rueschemeyer *et al* 1992: 36-38) or something else, presents perhaps no great theoretical revelation since it is what all researchers are engaged in, but I wish to stress this point since it has been a profound learning experience in this case.

Finally, some words of caution are warranted. This is a study of a very much ongoing process, something that makes it complicated to assess, because of the messy and rapidly changing reality and because of the limited time perspective. In many ways, doing this research has been like aiming at a rapidly moving target from a moving platform. It also goes without saying that the reach of the findings are in one sense limited to the immediate objects of study – to use a technical expression. It is also difficult to know whether the cases are representative in a statistical sense. It is however possible to discuss their theoretical representativity by relating the cases to other empirical studies and to theoretical overviews of Uganda's politics, something that is done throughout this work.

Chapter 3

Formation of State and Society 1900-1986: The Roots of Destabilisation and Authoritarianism

In order to understand contemporary strategies for regulating health provision, it is necessary to trace their historical roots. This chapter outlines central features of the historical development of state-society relations in Uganda, interwoven with of health and welfare regimes. In exploring the political regulation of welfare and health regimes, the focal point will be on structures for interest representation and conflict regulation created by different forms of state intervention in relation to demands from groups in society, seeking to enforce control.

What were the demands from different sections of society on the colonial and post-colonial states? What strategies did state managers develop for meeting those demands and challenges of accumulation, legitimation and control? What has characterised the frameworks of interest representation and conflict regulation of various health regimes? How have the ways in which these frameworks evolved shaped conditions for contemporary institutional reforms?

This chapter seeks to place the theoretical issues discussed in the first chapter more firmly in a historical context. It offers the necessary background to the study of contemporary state-society relations and welfare regimes in the following chapters. The relies on a rich body of literature on the political and social history of Uganda. The structure of the chapter is chronological, starting with accounts of the pre-colonial and colonial periods before analysing post-independence politics.

Pre-colonial and early colonial politics

The area that in 1894 was to turn into the British Protectorate Uganda encompassed a great diversity of social orders, with different political systems, cultures and ethnic groups. The dominant form of political order in the south consisted of kingdoms with long histories of rather centralised systems of rule, including Bunyoro, Ankole, Tooro and Buganda.

The northern parts were characterised by less hierarchical societies and more decentralised political structures, including those of the Acholi, the Langi and the Iteso. During the 17th and 18th centuries Bunyoro was the most powerful kingdom, only to be superseded by Buganda during the 19th century in the competition for regional supremacy. Trade links with Zanzibar and the introduction of firearms was one source of Buganda's expansion.

Another form of external influence came from Catholic and Protestant missionaries, whose presence and fierce competition from the 1870s introduced religious sectarianism to Buganda court politics (Jørgensen 1981: 38-39), a persistent legacy that was to remain a dominant feature of Ugandan politics well after independence.

British colonial policy set out from the agreement of protection struck with the Buganda kingdom in 1900, which was to fundamentally shape Uganda's socio-economic and political structure. In exchange for protection against neighbouring kingdoms and granted a special status among the conquered territories, Buganda agreed to assist Britain in colonising the remaining parts of Uganda through military conquest and the system of indirect rule. This conquest met with stiff resistance in most parts of the protectorate, but all in all, colonial rule was established by 1920.

Indirect rule involved the appointment of Baganda chiefs in the conquered areas as administrators in the colonial system. The 1900 Agreement also restructured social relations within Buganda, where feudal lords were converted into landlords by a reshuffling of land ownership (Mamdani 1976: 40-44; Uzoigwe 1982: 66-73). This was not however an entirely externally imposed transformation. From the late 1880s Baganda chiefs had gradually manoeuvred to commercialise land tenure, and in 1888 they deposed the *Kabaka* (Jørgensen 1981: 47).

Mamdani argues, though, that the agreement and the subordination to imperial interests transformed the character of this class from potentially productive mercantilists to a rentier class (1976: 42). The rise of Baganda chiefs, manifested in their alliance with the British, resulted in a "political dyarchy of the colonial state and collaborative chiefs" (Jørgensen 1981: 77). These chiefs struggled for greater autonomy – successfully, initially, as they played an indispensable role for the colonial state in transforming it from a purely coercive apparatus upholding law and order to encompassing more complex functions such as expanding production and infrastructure.

The colonial state gradually moved to block the political and economic aspirations and growing power of the chiefs. This political struggle was resolved during the 1920s, when the colonial state in several steps dismantled the position of the landlords. Politically, these were replaced by appointed chiefs. Economically, the state intervened to undercut their legal basis for accumulation (Jørgensen 1981: 82-87).

This period also witnessed the replacement of Baganda chiefs by local counterparts in other regions, mainly as a result of the need to ensure a degree of local acceptance for chiefs (Bazaara 1997: 34). This shift sealed the process of ethnicisation of the local state within an overall racialised political order. This racial hierarchy consisted not only of British colonialists and African subjects but also of a middle stratum of Asians – the popular term for immigrants from the Indian sub-continent. The effect of their presence was to block as opposed to mediate social mobility. Most Asians had come to East Africa as either forced imported labourers or as commercial workers.

The colonial state made further use of their presence. It allotted a section of them the function as an intermediate bourgeoisie in order to promote capitalist development, while at the same time obstructing African aspirations. The economic privileges granted to a racially alien minority by the colonial state would eventually politicise social tensions along racial lines (Mamdani 1976: Chapter 3).

The main function of Uganda in the wider colonial context was to supply Britain with cotton and from the 1930s also coffee. These crops were grown mainly in the south, where a wealthy class of African landlords and large-scale farmers gradually emerged (Lwanga-Lunyiigo 1989: 28-31). The landlords were also administrators or "chiefs" and used their dual capacities to order and control the cultivation of cotton by peasants-turned-tenants; during the first decades of colonial rule, capitalist relations of production coexisted with semi-feudal ones, including tribute in kind or labour. In the northern areas, cash crops were grown only on a modest scale. These parts of Uganda essentially served as a "labour reservoir for the cash crop economy of the south" (Mamdani 1976: 52), and also for the police and the armed forces. Importantly, Uganda was incorporated into the world economy on the basis of smallholder peasant production and not as a plantage or a settler economy. This reinforced the need for decentralised supervision and indirect rule. Peasant production competed with plantations until the early 1920s, when Britain, after pressure from metropolitan capital and seeking to avoid proletarianisation and political unrest (Bazaara 1997: 49-50) decided to favour the former model.

An alliance between peasants and clan leaders waged political protests against landlords and prevailing agrarian relations. This protest from below was reinforced from within the colonial state, codified in the 1928 *busulu* (ground rent) and *envujjo* (commodity rent) law, which guaranteed security of tenure to peasants. According to some observers, these legal reforms undercut the economic power of the landlords, and transformed them "from semi-autonomous political entrepreneurs into appointed servants of the colonial state" (Jørgensen 1981: 86). Others argue however, that although landlords were weakened, they still managed to block agrarian reform and, through alliances with traders, were able to preserve a great deal of their economic and political power throughout the colonial period (Bazaara 1997: 69, 79-80).

The uneven inclusion of regions into the protectorate and the world economy shaped social relations in colonial Uganda in a critical way. In order to address emerging structural contradictions between demands for a peasant based growing economy and perceived social disruption on the one hand and political control and stability on the other, the colonial state pursued a strategy to promote migrant labour as a stable yet dependent labour force. It did so by initially discouraging the production of cash crops in peripheral regions, designed to be labour reservoirs.

From the late 1920s, however, cash crops were grown on a larger scale also in the north (Jørgensen 1981: 98-106). The creation of a regionally divided economy was evidently a complex process, as the colonial administration itself acted within limitations. Within the constraints following from its place within the British Empire, and in turn the world economy, the colonial administration became enmeshed in domestic social and political struggles, and increasingly so the more complex its tasks. In order to balance its functions and exercise control, the state had to root itself deeper in society, craft political alliances among domestic social groupings and respond to conflicting demands from for instance peasants, chiefs, and different factions of metropolitan capital.

The geographic and social division of labour, including regionally distinct and uneven patterns of recruitment of labour force, and of trade, education and access to the political centre had important political consequences as it delayed the emergence of nationalist movements and shaped their class based and regional composition. Low levels of urbanisation and industrialisation weakened formalised and urban-based forms of political organisation including trade unions and professional associations and their scope for alliances.

Politics during the first decades of colonial rule was confined to the dominant classes, mainly in Buganda, and revolved around the control of land (Mamdani 1976: 171). Political demands advanced from outside these narrow strata were rejected. Legislative Councils were introduced in 1921, with reserved representation for the British and Asians. Nevertheless, economic expansion gave rise to social change, which gradually matured into new expressions of political conflicts. Social movements and interest groups of many kinds emerged from the 1920s, and political opposition to the Buganda government, and ultimately to the colonial state, which grew in strength. Parameters for popular politics, which co-existed with and challenged the aristocracy, were broadened through alliances between peasants, workers and traders; "the politics of the court were giving way to the politics of the marketplace" (Mamdani 1976: 173). However, not until the 1940s would different sources of discontent merge into a stronger form of politicisation that reached out beyond the ruling classes or the urban intelligentsia.

Health and welfare under early colonialism

Pre-colonial health care in Africa was provided in a number of ways, and within each social formation, there were institutional frameworks for regulating public health. These welfare systems were closely connected to forms of political authority, and many forms of healing survived the colonial conquest (Feierman 1985: 118-120). In Uganda, traditional medicine – a sweeping category that included birth attendants, healers, herbalists and psychotherapists – took on great relevance as formal systems collapsed in the early 1980s, and continues to co-exist with the modern and formal sector to this day. Nevertheless, colonialism fundamentally transformed the systems for regulating social needs. Generally speaking, social services in colonial territories were given low priority during the first decades of the 20th century.

Initially, health services were introduced primarily to protect the workforce of various colonial agencies such as soldiers or, in some colonies, settlers. Doctors, just like other colonial employees, were expected to contribute to the upholding of racial dividing lines and supremacist ideologies (Feierman 1985: 120). The Imperial British East African Company (IBEAC) introduced a western type health care in Uganda in 1878 in order to cater for its staff. IBEAC was taken over by the British Colonial office in 1894.

During the colonial era, Christian missions were the main providers of health care and education to Africans, and the Colonial office made use of their presence for purposes of administrative control. A relatively well-developed system of church-run health centres and hospitals was created in Uganda (Iliffe 1998: 41). Despite a dense network, social services were delivered to Africans only on a modest scale, and were concentrated in Buganda. Only about 100,000 African outpatients were handled annually (Jørgensen 1981: 62). Apart from its inbuilt racial segregation, colonial education and medicine had an urban and elitist bias, and the strong reliance on religious institutions as providers further stratified access along lines of class, region, ethnicity and religious affiliation (Mamdani 1976: 161-162).

Colonial rule sought, as Berman and Lonsdale (1992) have argued, to balance challenges posed by accumulation and control. In the field of service provision, colonial powers tried to minimise their obligations and expenses. To the extent that they did provide services, the scope and content of these were mainly predicated upon instrumentalist calculations, concerned with the necessary conditions for the reproduction of the labour force. There were fears within the colonial state that a weakened and stagnating or even declining population would undercut the economic basis of colonialism itself. Summers (1991) analyses in detail how British concerns over population decrease during the first decades of the century prompted a set of ideological and institutional interventions with respect to reproductive health in general and the reproductive behaviour of Baganda women in particular. While the colonial administration mainly responded medically from the perspective of a perceived public health and economic crisis, missionaries addressed the issue as a moral crisis, and these modes of intervention became interlinked (Summers 1991: 796, 798).

This example illustrates how, as the economy expanded and set off more complex social structures, colonial authorities were pushed to respond to new challenges and how these responses in the process transformed the structures of the state. In order to meet demands for accumulation and administration, new strata of more educated and healthy Africans were needed. In Uganda, the expansion of the peasant economy in the 1920s necessitated an expansion of Africans into the state as administrators in the civil service. This in turn required the growth of an education system, financed by the state and run by the missions, which offered the colonial state a possibility of meeting challenges of accumulation and legitimation (Mamdani 1976: 161).

After a decade of conflict between the colonial state and missions over the direction of education policies, the state conceded to the missions' pressure for a more academic oriented education model that created an educated bourgeoisie with elitist outlook (Jørgensen 1981: 164-165). From the 1920s, the expenditure pattern of the colonial state had shifted, and expenses on social services had in quantitative terms overtaken those on the coercive arms of government. Health expenditures rose slowly. A main factor that brought attention to the poor status of health services was the condition of the migrant labour force. From the 1920s, the coffee plantations in Buganda confronted a shortage of labour.

Colonial labour policy oscillated in its strategies between strict imposition of recruitment and a more open policy toward labour migration, from within and outside Uganda. Migrants were mainly recruited from the West Nile region in northwestern Uganda, and from Rwanda and Burundi. Again, the colonial state faced conflicting demands with regard to accumulation based on an inflow of labour migration and control of it. The difficulties in regulating migration flows led to concerns about public health risks. Groups of migrants were seen as unhealthy and certain ethnicities were stigmatised (Lyons 1996). In order to counter marginalisation, these groups developed different strategies such as go-slows and absenteeism, which increased the colonial authorities' awareness of health problems (Opolot 2002: 87-89). Hospitals and health centres were set up along migration routes (Jørgensen 1981: 166).

The dilemmas, from the colonialists' point of view, created by a vulnerable work force were recognised already in the 1930s in the wake of the worldwide economic depression. It would, however, take until the post-war years to initiate more ambitious reforms (Packard 1997: 95). Throughout the British and French empires, indigenous populations had begun protesting against poor living conditions and occasionally connected social demands to political ones. The British government sought to respond to these challenges with the Colonial Development and Welfare Acts of 1939 and 1945 (Cooper 1996: 67); the Ugandan offshoot was the Worthington Plan for industrialisation of 1947 (Mamdani 1976: 253). According to Cowen and Shenton (1996: 296-298), the development doctrine of late colonialism codified in these Acts was an expanded reinvention of development that encompassed a welfarist dimension, with comprehensive state intervention into agrarian production, social services and infrastructure.

These policies were accompanied by limited political reforms. With these reforms, the colonial state sought to improve agricultural productivity through the making of an African bourgeoisie and in the process attempted to craft good relations with it in the context of militant nationalism. This would, it was estimated, pre-empt political demands among these groups and forge ties with them so as to provide the colonial state with support and meet the mounting challenges of control (Berman and Lonsdale 1992: 149; Young 1994: 169).

Late colonialism: Protest and reform

In 1945, a wave of strikes and uprisings comprised of both peasants and workers took place, mainly in Buganda. The protests were primarily targeted at institutions regulating land ownership and wages, but also turned against colonial rule as such. Grievances surfaced again in 1949, when widespread riots shook the ruling coalition consisting of the colonial state, the Buganda government and non-indigenous capitalists. The uprising revolved around demands for political representation and access to trade.

The colonial state offered concessions, but in the process blunted the radical edge of the protests. This is not to say that protests failed. By joining hands in the 1945 and 1949 uprisings and by linking socio-economic to political demands, co-operatives, trade unions and other social movements had pushed the colonial state to make concessions regarding both policy content and legal recognition.

Electoral participation and African representation in the Legislative Councils was expanded from 1949; access to trade was improved. The strength of the demands also forced the colonial state to employ more sophisticated means in controlling discontent. According to Mamdani, at the same time as co-operatives, trade unions and ethnic and religious organisations were legalised, they also became subject to government control through an array of regulatory mechanisms such as registration and supervision. Most aspects, from membership and leadership to finances and overall scope of operation, were regulated by the government (Mamdani 1994). Thus, this move towards legalisation of organised interests also involved their relative depoliticisation (Mamdani 1990a: 61; Mudoola 1993: 17-18).

State control of popular movements was made possible in a context of the political rise of a new middle-class that principally challenged the traditional oligarchy. The coalitions that had upheld the uprisings consisted of a wide range of forces, all with rather shifting motives for their participation. The colonial state related in different ways to the various strands inherent in the emerging nationalist movement and cultivated close relations to nascent elites so as to pre-empt more radical demands from other groups. One central dimension of this strategy, as mentioned above, was to encourage co-operatives as a vehicle to incorporate Africans into crop processing and marketing. Another was the creation of Uganda Development Corporation (UDC) to promote state-regulated industrialisation with a degree of African participation (Jørgensen 1981: 188).

Political fragmentation was not only the result of socio-economic divisions, but also the outcome of the institutional forms of the local state. Mamdani (1996) has argued that the fundamental consequence of indirect rule was the institutionalisation of a "bifurcated state", with dichotomous types of state structures and modes of power.

Urban areas were governed by civil law, with racial exclusion from civil society; rural areas by customary law, and marked by authoritarian inclusion into closed-off worlds of ethnic tradition. On this view, civil society only developed in the urban areas of colonial Africa. As suggested above, the colonial state introduced local chiefs in order to soften resistance. "The customary" was reified, compartmentalised and made absolute by intertwined processes: authoritarian aspects of tradition were selectively derived and deemed fit for rule; each "tribe" was declared a homogeneous cultural (and by implication administrative) unit; customary law was invested in the singular authority of the chief. This is what Mamdani calls "the fused power" of "decentralised despotism" (Mamdani 1996: 23).

Evidently, chiefs, in their day-to-day exercise of power, were embedded in local society and needed to balance the imposition of colonial control against social acceptance. An essential part of their effective authority depended on their capacity to mobilise a local following, mediated along lines of religion and ethnicity (Bazaara 1997: 37). According to Mamdani, post-war reforms of local government, including partial introduction of electoral politics as in the case of the 1949 Legislative Councils in Uganda, did not significantly alter the fact that political power in rural Africa was based on coercion (Mamdani 1996).

Political parties emerged in large numbers in the 1950s. Just like most previous expressions of political discontent, much party-political organisation was fragmented along lines of region, religion and class. Uneven regional capitalist development resulted in very different political demands. The first party with a national base was Uganda National Congress (UNC), set up in 1952. Initially, UNC had its base outside Buganda, but with the colonial government's expulsion of the *Kabaka* in 1953, the party made inroads in the region, riding on a wave of anti-colonial sentiments.

The return of the *Kabaka* in 1955 resulted in a rapid evaporation of UNC support in Buganda. However, its leadership found other avenues for its interests, primarily by taking control of the *Lukiiko* (the Buganda legislative assembly). In response to the perceived use of the *Lukiiko* as a vehicle for Protestant privileges, the Democratic Party (DP) was formed, with its base among Baganda Catholic professionals, chiefs and landlords.

The most influential party would turn out to be the mainly northern and Protestant based Uganda People's Congress (UPC), led by Milton Obote. Although all the main parties had political bases among the peasantry, popular influence on their leadership within or from outside parties was limited. Yet, the political interests of economic elites were mediated by their need to ensure popular consent.

Parties operated in a context of uneven economic development, politically expressed in terms of ethno-regional cleavages, and were for that reason largely reluctant to or incapable of transcending their ethnic or religious affiliations. Most of them can be described as confederations of local strongmen and their respective following (Mamdani 1976: 210-223).

In many areas, ethnic fronts were created by alliances between landlords and tenants for the purpose of blocking "outsiders" from accessing land, something that reinforced political regionalisation (Bazaara 1997: 69-73).

The one factor that more than any other complicated the national question on the eve of independence was the strong current of Baganda nationalism, itself fraught with strong internal tensions. Musisi (1999) analyses how the relationship between colonial and missionary intervention and changing local social relations stratified Buganda society and the local state along lines of religion, gender, class and ethnicity.

During the last few years of colonial rule, the Buganda aristocracy, landlords, traders and farmers, all from different perspectives, embraced the idea of self-rule, whereas urban based professionals and, evidently, non-Baganda migrant labourers and tenants preferred a unitary or federal solution for Buganda within Uganda. As already mentioned, a standoff between the colonial administration and the *Kabaka* over the future status of Buganda forced the latter into exile. His return two years later was part of an agreement that greatly expanded Buganda's autonomy within Uganda and which fuelled secessionist demands (Kasfir 1976: 121-122).

As independence approached, leading monarchists declared a ban on political forces in Buganda. Democratic Party (DP) did not heed the call, and participated in and won the 1961 elections that were boycotted by the *Lukiiko*. Finding itself isolated, the traditionalist elite at Mengo formed a political party, *Kabaka Yekka* (KY – "The King Alone"), with a populist base, and struck an alliance with UPC in order to pre-empt any political influence on DP's part. The unholy alliance between an ethnic monarchist movement and a "northern" trader based republican party shaped the outcome of the 1961 and 1962 constitutional negotiations. UPC supported KY's demands for a federal solution for Buganda and agreed to make the *Kabaka* of Buganda, Edward Mutesa, the president with ceremonial functions. In exchange, KY promised to block DP in Buganda and to back the formation of a UPC government.

Health and welfare under late colonialism

The political regulation of structural contradictions and demands from different social forces for welfare and recognition restructured and expanded the colonial state, in Uganda and elsewhere, in terms of bureaucratic and rudimentary welfare institutions. Demands for expanding welfare services were made on the state by different categories, such as the emerging middle and working classes and professionals.

The expansion of service provision was, however, modest and just like colonial state formation more broadly uneven and geographically concentrated to regions of capitalist expansion. The response to these demands must be seen in the light of the changing conditions for colonialism. Internationally, the balance of power had after World War II tilted against the older colonial powers, who sought to preserve domination and expand accumulation through a combination of coercion, social policy concessions and limited representation in response to the rapid emergence of radical nationalist movements.

What were the contours of the civil society that was shaped as a consequence of the colonial health regime? Religious institutions constituted an important section of civil society, and missions were initially closely connected to the colonial state. However, religious affiliation had from the inception of Christianity in Uganda in the 19th century carried significant political weight. Within the growing nationalist movement, religious divisions took on a political dynamic.

The important role of the churches in the provision of health and education as well as for social influence more broadly stratified the emergent political elite along religious lines and, from the 1950s onward, offered political clout to the political parties associated with the major denominations (Jørgensen 1981: 193-205; Mamdani 1976: 216-220; Mudoola 1993: 21-26). Alongside religious institutions, local governments were, particularly after the local government reform in 1949, given an important role in co-ordinating and providing primary services as part and parcel of the system of indirect rule, discussed in Chapter 3. Appointed chiefs were given the authority to impose norms of behaviour and exercise administration in most fields of society, including health and sanitation. Communal labour was made compulsory.

The colonial state also, but modestly, expanded its disbursements. Uganda had become the hub of medical research and education in East Africa. Already in the early 20th century, British specialists on tropical medicine were sent to Uganda to do research on sleeping sickness, and the evolving bureaucratic structures that were set off by their presence incorporated medical missionaries (White 1995: 1385). Makerere Medical School in Kampala gained a reputation as the regional outstanding centre for medical training and research, and although Ugandan doctors, predominantly Baganda, were rather few, they were still greater in number than African doctors in neighbouring countries (Iliffe 1998: 136).

Striving for recognition, these doctors had set up professional associations from the 1930s (Iliffe 1998: 92-94). The main one was Makerere Medical Graduates Association, which pushed for improved pay and working conditions, regularisation of status and the right to private practice. It developed a certain degree of political skill and clout and managed to enforce most of its demands (Iliffe 1998: 94, 101-102). With time, the agendas of the professional associations broadened beyond purely professional concerns. After World War II, they became influenced by and, through a few individuals, entered the rapidly growing radical nationalist movement (Iliffe 1998: 113-117).

The 1945 strike extended to workers in government hospitals (Iliffe 1998: 89-91); this was the first time that medical workers became involved in a broader political struggle. Collectively, doctors pushed for the Africanisation of the health sector; Africanisation was generally the main target for many groups. Individually, doctors became prominent in local and national politics – and more so than in neighbouring countries, due to the social composition of the political class. In Uganda the role of mass parties was marginal, and politicians were mainly recruited from the elite strata (Iliffe 1998: 113).

According to Iliffe, the doctors' relationship to the colonial state was deeply ambivalent – a familiar theme with regard to educated strata at around independence. On the one hand, the state protected the doctors as privileged elite; on the other, it blocked their further aspirations (Iliffe 1998: 91). This suggests that doctors were a rather insulated segment, divorced from most other social groups.

The ethos of professional elitism was partly reflected in their political outlook. Their point of entry into politics took place through party politics, whereas their professional association remained timid and primarily concerned with preserved status and Africanisation of the health sector. Unionism was on the whole an alien concept to doctors. This needs to be connected to colonial reforms that aimed at the dismantling of the radical wings of the nationalist movement and inclusion of the political elite into representative organs. These reforms stifled most sections of civil society. Others were effectively incorporated into state structures (Mamdani 1994: 522).

Professional bodies were set up for the purpose of regulating the sector. The Medical Practitioners and Dentists Board was a colonial creation, mainly concerned with issues of registration and disciplining of practitioners. The Uganda Medical Association (UMA), a professional interest group that was Africanised and changed its name from British Medical Association at independence, was on the whole not very efficient (Iliffe 1998: 124). However, the policy direction in the medical field was dominated by doctors in and out of politics and given a curative bias.

Professional relations were hierarchical, with status-based contempt for categories such as nurses and medical assistants (Iliffe 1998: 129). On the eve of independence, the Ugandan health regime was characterised by a medical model in terms of its policy content and incorporation by the state of professional interest groups in terms of the political regulation of the sector.

To summarise, late colonial reforms changed the colonial state and Ugandan society in a number of ways. They expanded the role of state intervention in the economy, they opened up for slightly more ambitious welfare reforms and they created institutions for limited political participation. However, these institutional changes took place within the guided continuity of the broader political economy framework.

Reforms were designed to promote a controlled transition to political independence. By cultivating close relations to the emerging middle-class and containing more radical social movements by supervision and control, the colonial state ensured that this transition would not involve economic or social transformation. This strategy decisively structured state-civil relations after independence.

A fragile coalition among the political elite inherited the colonial state with all its contradictions, as well as the underlying instability that colonial rule had produced in terms of economic underdevelopment, deep social cleavages, controversy over the national question and fragmented elite driven politics. As Bangura (1992: 58) argues with regard to Sub-Saharan Africa as a whole, in view of the wide range of social forces that made political demands on the colonial state:

> democratisation provided an institutional framework for reconciling the conflicting interests thrown up by the authoritarian colonial economy. It checked the absolutism of colonial rule by opening up space for popular participation in government and the rise of independent organisations. But the authoritarian character of the colonial economy prevented the growth of liberal democracy.

This contradictory relationship between formal political institutions and underlying social relations would in many countries resurface time and again after independence.

State and society after independence

In Uganda conflicts resurfaced almost immediately. Power struggles during late colonialism were articulated with a constitutional hybrid, which in a broader sense reflected the colonial trajectory of indirect rule and regional uneven development. As a way of accommodating the political elite in Buganda, the kingdom was given federal status.

Due to the weaker bargaining power of other regions the result was that the other southern kingdoms were given semi-federal status, and the northern parts were incorporated directly under the central government. Considerable autonomy was conceded to the Buganda *Lukiiko*. This body, effectively controlled by the Mengo elite, had powers of taxation, supplemented by federal grants, and managed Buganda's separate public administration, public services, court system and police force (Jørgensen 1981: 217; Mamdani 1976: 242). This leads to the overall conclusion that the constitution rested on a fragile coalition in which neither party trusted the other nor expected the coalition to last;

factional infighting among the bourgeoisie was articulated along ethno-regional lines in the struggle over state power.

The translation of economic competition into constitutionally guaranteed ethnicised political institutions, rooted in popular sentiments, raised the stakes and complicated matters enormously. Inherited colonial regional and social imbalances were thus reproduced as endemic political factionalism and instability, mediated by ethnic and religious modes of mobilisation and expressed as tensions over central-local relations, regional inequalities and racial segregation.

The government hence faced the challenges of reproducing and strengthening control of the state vis-à-vis a wide range of disparate social forces. That it had to do so within the given limits of a dependent economy, controlling an institutionally weak and territorially uneven state apparatus and relying on a fragile political base, severely restricted its policy choices. Therefore, the government initially avoided dramatic policy shifts and opted for a continuation of the development model of late colonialism.

What did that model consist in? The government continued to promote peasant based export production of cotton and coffee, regulated by centralised channels for processing and marketing, such as state controlled cooperatives and marketing boards. Export earnings were redirected into state promoted industries by way of partnerships between the state and international or Asian capital, expansion of infrastructure and public service and state administration. Industrialisation was modest but continuously growing in the post-war period. Asian capital was channelled from the agricultural export sector to manufacturing by various means of state intervention. Transnational capital entered into mining. There was also an expansion of Baganda entrepreneurs.

The growing economy and popular demands resulted in rising wages and an expanding domestic market (Loxley 1989: 68; Mamdani 1990 b: 454). Hence, it conformed to a fairly typical post-colonial state-regulated development model built on export earnings, foreign direct investments and foreign aid. State regulation of export production of cash, crops provided the resource base for import substitution industrialisation, infrastructure and social services as well as a growing private sector where international capital was invited to join partnership models under the auspices of Uganda Development Corporation. Foreign aid made possible expansion and redistribution of infrastructure and social services.

This economic model created considerable growth during the 1960s (5.8 per cent on average) but eventually proved to be unsustainable even in economic terms, as it remained dependent on world market prices for export commodities and as import substitution did not deepen or broaden industrialisation or enabled a qualitative change in agricultural production (Brett 1996: 313).

This was obviously not a strictly "economic" outcome but a profoundly political process, following from struggles over the control of the rural surplus (Bazaara 1997: 81). Hence, economic growth initially enabled the government to respond to popular demands for the fruits of independence, including the need to address regional and racial imbalances in what was a segregated economy. Strong popular demands were made for Africanisation. The government promoted rapid numerical Africanisation of the civil service and within parastatals and gradually opted for a more interventionist role in the economy.

Economic growth and expansion of social services and Africanisation built up the size and capacity of the state and bolstered political stability to a degree. But this expansion postponed rather than resolved conflicts, since the ways in which different social forces were rooted in the state at the same time undermined its capacity.

The UPC-KY coalition rested on an informal mutual non-interventionist agreement: the forces behind KY were to support the UPC government and would in exchange be guaranteed political monopoly and control over land in Buganda. The Baganda elite based their political clout on control over land and labour, and this mode of accumulation was safeguarded by the constitutional status of Buganda. However, built-in tensions between the central government and Baganda landlords over the control of Buganda's status and surplus were only temporarily contained.

UPC also faced demands in other directions from different constituencies, and hence, the weakness of central state institutions deepened rather than contained factional politics. Already from 1963, UPC started to make inroads into local councils in Buganda (Jørgensen 1981: 219).

At the level of national politics, the deteriorating alliance broke down over the issue of "the lost counties", two counties that had been seized from Bunyoro and allocated to Buganda by the British. Their contested regional status was decided in a referendum in October 1964, in which an overwhelming majority voted for returning to Bunyoro. The campaigns triggered a series of violent clashes, which heightened ethnic mobilisation at both elite and popular levels. Power relations in society had thus structured political alliances in ways that seriously destabilised the state.

With KY and DP marginalised after MPs from the two parties crossed the floor to UPC, political factions and conflicting demands were incorporated into an already fragmented ruling party. The radical wing of UPC was outmanoeuvred in 1964, and in early 1966 the conservative faction, with a base in Buganda including former KY politicians, challenged the government and the party centre over corruption among high-ranking politicians and army officers, with links to rebels in the Congo. Obote struck back, arrested his opponents and suspended the constitution. In its place he introduced a unitary interim constitution, which abrogated federal powers – a blow directed at Buganda.

Following the kingdom's refusal to accept the new constitution, the army attacked the *Kabaka*'s palace in May 1966, forced him into exile and declared a state of emergency in Buganda that would last until the coup in 1971. The following year, a new republican constitution abolished all traditional rulers and kingdoms (Jørgensen 1981: 227-231). The long standing economic and political confrontation between UPC and the Buganda elite was concluded with the elimination of the Buganda kingdom. The central government had resolved this political conflict with Buganda using military means.

Establishing political control by coercion came at a high cost, however. Not only had the government alienated Buganda as a whole, it also faced hostility among outmanoeuvred factions in UPC. During the following years the government sought to preserve control and reconstruct its political base through strategies of political and economic centralisation of power (Jørgensen 1981: 231-237; Mamdani 1976: 260-266). This has to be seen as crisis management to address a downward spiral of waning support, economic stagnation and political repression and militarisation. In the context of crisis the government tried to buy time.

One dimension of centralisation concerned territorial and administrative state structures. Arenas for local politics were fundamentally restructured with the 1966 and 1967 constitutions. The unit of local government was now the district, stripped of most of its powers in an attempt by the government to penetrate local political arenas.

Another way in which the government sought to establish control was by state intervention in the economy. After having secured political control, the government carved out a more ambitious role for the state. A basic contradiction from the government's point of view during the first years after independence had been that its political support base had aspirations contrary to the groups that contributed its fiscal revenue. In moving towards state involvement by introducing currency control and an export-import licensing system by strengthening existing and setting up new parastatals, the government attempted to weaken its economic dependence on Baganda and Asian merchant capital in order to gain some political autonomy.

The government had thus blocked the aspiring Baganda bourgeoisie both politically and economically. Instead, it promoted a state based African bourgeoisie through which it could retain and reward its core political base. The government also allied with Asian and foreign capital through joint ventures (Brett 1978: 275; Loxley 1989: 69).

While signs of economic weakness were present from the mid-1960s and had intensified political struggles, the late 1960s witnessed a looming crisis of accumulation. The government had built its strategy upon assumptions of increasing levels of foreign investment and foreign aid. Instead, it faced capital outflows due to falling world market prices and worsening terms of trade. In combination with stagnating productivity and expanding military expenditures, this eventually resulted in balance of payment deficits and inflation (Brett 1978: 276).

To reinvent its legitimacy and broaden its base, the government in 1968 launched a programme of economic nationalisation, built on "an appeal to republican, nationalist and populist sentiments" (Nabuguzi 1995: 194). The programme failed to take off. While not all parastatals were outright failures, some, and notably the National Trading Corporation, essentially became vehicles for corruption and patronage due to the government's reliance on support from the cadre controlling them (Brett 1978: 275).

Economic stagnation generated social unrest. The government increasingly had to rely on the rapidly growing military and on repressive modes of regulation in general. The militarisation of Ugandan politics hence occurred in the context of the incapacity of political rulers to enforce control through institutions and policies.

It is significant to note that militarisation needs to be understood as the extreme expression of authoritarian regulation of social conflicts in a wider sense. Such authoritarianism was rife during the late 1960s. The UPC government attempted to establish control over all social forces. State-civil society relations had not fundamentally changed with independence. On the contrary, it has been argued that the incorporation of autonomous social forces into the state nationalist project that was underway in the period preceding independence was consolidated afterwards. On this view, the Obote I government's interventions during the late 1960s reinforced colonial legislation and control mechanisms to undercut signs of autonomy. The parameters for autonomous public organisation or debate were effectively narrowed through the establishment of state-party controlled trade unions, women's organisations, religious institutions, and co-operative movements. The period leading up to the 1966 constitutional crisis had also to significant degrees been shaped by ethno-regional politicisation. Such expressions were thereafter stifled (Mamdani 1993: 542-543; Tidemand 1994: 47-51).

Economic stagnation and social unrest had by 1969 merged into a full-blown political crisis. In December an assassination attempt was made on President Obote, following which all other political parties were banned. The government now had its sole power base among select groups in a factionalised military and among security outfits.

Ethnicity was used as a tool to secure control over the internally fragmented military (Hansen 1977). Attempts to extend its nationalist policies so as to control the fluid situation proved futile, as these policies did not halt economic decay and political unrest. In January 1971 the government was overthrown in a military coup led by General Amin (Jørgensen 1981: 252-256). What has been suggested above about the main features of state formation under the Obote I government needs to be re-emphasised. The fragile nature of the government's political base revealed the inherent instability of the political economy. A legitimation crisis led to rapidly shifting alliances among a shrinking base and to destabilisation. Factionalist infighting among the political elite disrupted overall conditions for capitalist accumulation and cohesive state formation. The government's inability to accommodate conflicting interests into a territorially and institutionally weak state undercut the prospects for a long-term reproduction of state power.

Health and welfare from independence to the mid-1970s: Centralisation

At independence, demands were enormous. Uganda is normally described as having had one of the most well developed health care systems in Sub-Saharan Africa, with an impressive infrastructure of health care facilities and educated staff. However, as O'Manique (2004: 116-117) cautions, qualitative health care provision was largely curative and hospital based, and most people endured under conditions marred by poor sanitation and endemic diseases. During the first decade after independence, state-run health infrastructure, along with other social services, expanded rapidly, underpinned by a growing economy backed by coffee exports. Generally, governments of newly independent societies sought to balance the expansion of teaching hospitals and primary health care. The Ugandan health care system was initially based on a nationalised continuity of the colonial model, with emphasis on curative care in modern hospitals, at the expense of preventive and community based health care, although the expansion of facilities eventually created a rather dense network of rural primary health units (Nabuguzi 1995: 195-196; Okuonzi and Macrae 1995: 125). Ugandans enjoyed free and relatively accessible health care, and health indicators improved. Iliffe claims that the curative emphasis of the system can be explained with reference to a combination of a professional based policy regime, a well-established health infrastructure and strong local governments that sought to respond to demands for local hospitals (Iliffe 1998: 131). The sustainability of this welfare system was eventually undercut by the more deep-seated legacy of colonial rule, the structuring of social relations and political institutions (Nabuguzi 1995: 192).

How was the state structured in this field? Policy making under the Obote I government was shaped within conflicting logics. The dominant budget logic on the one hand, as recommended by the World Bank and followed until 1964, emphasised fiscal conservatism, and real expenditure on health fell by approximately 10 per cent from 1959 to 1964 (Iliffe 1998: 137). The domestic political logic on the other hand stressed expansion, as popular expectations on the fruits of independence created great political demands for services. In 1964 the government turned around and responded to these demands in order to consolidate its political support. In the context of a decentralised structure of government and local government control over primary health care and political challenges to the government from the opposition in most corners of the country, the central government built regional hospitals. Although this expansion was underpinned by a growing economy, it created underlying constraints on resources (Iliffe 1998: 139).

A defining feature of state formation in the early post-colonial years was the tense relationship between central and local governments. The political prestige and legitimacy associated with service delivery, or more broadly the development space according to Kanyinga (1995: 71) was significant, and struggles over it were fought out along different dimensions. Between 1962 and 1966, there was considerable leeway for local governments to shape policy. The central government was comparatively weak and needed to balance competing regionalised demands within a semi-federal constitutional order. The southern kingdoms had been given the power to levy taxes and make budgets. In addition, the concentration of missions to the southern parts of the country reinforced regional imbalances created during colonialism.

During the first years after independence, provision of public health care and social services was instrumental in creating local political power bases for individual politicians as well as for UPC as a party, in the face of the challenges in local political arenas that remained until the mid-1960s. UPC gradually took power over most local government assemblies outside Buganda (Jørgensen 1981: 222). In line with a general tendency towards a state led economy, planning and provision was centralised. Political relations between the central state and local government and society, and the relative importance of cultural and religious institutions at the latter level, were dramatically transformed during this period.

Traditional authorities and their welfare systems were dismantled by the creation of a republican unitary state in 1966-67. Federal regions were replaced by centrally controlled district authorities, through which the central government sought to impose tighter political control (Nabuguzi 1995: 196). The overthrow of the Obote government in 1971 did not immediately result in the collapse of state functions. On the contrary, social service expenditure continued to expand during the first years under the Amin regime, even though the quality of this expansion was gradually undercut by lack of maintenance and growing insecurity generally.

With regard to the formation of civil society in the health sector, the discussion needs to be situated in the wider context of state-civil society relations as they had unfolded through capitalist expansion and state intervention during the last decade of colonial rule. In Uganda, the combination of the pronounced reliance on indirect rule and an unevenly developing capitalism had set off a fragmented and relatively weak civil society in which religious and traditionalist ethnic forces tended to outweigh class-based ones. This was also a consequence of late colonial reforms, under which territorial and moderate elements of civil society and political parties actively disconnected its local as well as radical sections (Mamdani 1990a; 1994). Ethnic-monarchist organisations that straddled local government and civil society became very relevant, given local autonomy particularly in Buganda.

In the process of increased state control of economic life and restrictions imposed on civil society activity generally, the significance of mission hospitals was reduced. While missions and traditional authorities had been important providers of social services prior to independence and in the years following upon it, centralisation of political power in the mid-1960s reduced their role markedly. The tug-of-war between the state and religious institutions that followed did not completely marginalise the latter, who continued to offer services. However, they did so at a decidedly more modest level. In the education sector, the state had been even more successful in curbing the resistance of religious institutions when dismantling their education system in favour of a national secularised one (Nabuguzi 1995: 196).

The formation of civil society in relation to the making of the welfare system of the post-colonial state was not confined to modern and urban interest groups. In rural and lower income urban areas, parallel health systems co-existed without much effective institutional and legal integration. Apart from the public health system with its structures of health units down to the sub-county level, and the Anglican and Catholic hospitals, health units, private practitioners serviced rural and peri-urban areas.

The private service providers were profit-driven, and in remote areas they could offer quicker and more accessible services. On the other hand, the quality of service was poorer, and practitioners occasionally lacked formal qualifications. In most cases, private provision was beyond the realm of state supervision and was only regulated by market relations. In addition to these systems, traditional or community-based medicine had survived the introduction of modern medicine, and these practitioners, being equally available, functioned as the major providers in the rural areas.

In local arenas, which in substantial ways were deliberately disconnected from national politics, an array of groups for self-help emerged, based on kin, ethnicity, gender and class. The regulation of access to and quality of these systems, which overlapped with modern medicine as many providers started to offer a combination of indigenous and chemical medication, was based on both market and cultural imperatives. Overall, these systems were not coordinated, and overlapped only in informal ways (Mugaju 1999: 1-5).

The character of the health care system underwent changes in the 1960s. One source of change pertained to the relation between doctors and the state. Would health workers seek to create autonomy in relation to the independent state or steer close to it in order to make use of its close ties to the government? In Uganda, doctors were a privileged group, and a number of individual doctors were active and influential in politics.

Doctors had established control over the direction of policy content, and enjoyed a close alliance with the state. This alliance involved guarantees of relative privileges in terms of income and status in return for political loyalty, in line with general state corporatist tendencies (Iliffe 1998: 132). The doctors were thus part of the state-dependent intelligentsia. However, this alliance would come under strain as a consequence of the shrinking power base of the government. Political control over all social groups tightened rapidly from the mid-1960s. Economic decline did not halt the expansion of health infrastructure, but it did undercut the ability of the state to sustain the quality of services, including payment and other working conditions. Also, and in combination with overall social change, the relative status of doctors declined, and it became more difficult for those who so wished to uphold distinctions between different types of medical care. Nurses at Mulago went on strike in 1963, partly in protest against hierarchical work place relations in the sector (Iliffe 1998: 129).

Another change came from within medicine itself and consisted of the increased importance of chemotherapy, as drugs were made available to most sections of the population. The availability of drugs became "the central point of public criticism of medical systems" (Iliffe 98: 134). It also changed the balance between different practitioners towards "needle men", a category that included everyone from retired herbalists to untrained practitioners. The availability of drugs was connected to the decay of the state.

The weakened state eventually lost much of its supervisory capacity to enforce regulations, a process which undercut that part of the status distinction between formal and informal health systems which rested on availability of modern medication. Self-medication became all the more attractive because of "simplicity, convenience and privacy when compared with spiritual remedies" (Iliffe 1998: 135). What were the implications of this? Modern medicine was faced with rivals in competition for patients. On the other hand, it also offered previously unmatched opportunities for treatment.

A third type of change concerned transformations in the medical profession. At one level, medicine was no longer the only status-profession open to Africans. At another, doctors became the target for criticism when the quality of the health system declined. The profession itself became more internally divided along lines of status and political orientation. All of these factors interacted in transforming the underlying conditions for the health system (Iliffe 1998: 135).

To sum up the main features of the health policy regime during the first decade after independence, it was marked by expansion of social services as a response to the enormous expectations that political independence would also mean overcoming poverty, ignorance, and disease. This welfare state expansion was built on coffee exports. The Ugandan health system itself was to a great extent urban based, curative and professionalised in its orientation. In terms of political regulation, it was characterised by state incorporation of interest groups. Overall, Uganda followed a fairly typical pattern of state corporatism. From the perspective of the state, service delivery was, apart from the contribution that improved health and education of the population would make to development, an important means for creating political legitimation and power bases for nation building.

Until the abolition of the federal constitution in 1966, Ugandan political structures were decentralised, with local and federal governments and religious institutions providing primary social services. One dimension of this decentralisation centred on competition for political support among and within localities. Civil society in the health sector was composed of, on the one hand, Uganda Medical Association (UMA), and, on the other, providers linked to religious institutions and kingdoms. Strong regional and traditionalist, rather than modernist, power bases constituted the main political challenge to the central government. UMA had at an early stage been incorporated into state structures, partly a consequence of its limitations within elitist professionalism. Religious institutions were tolerated, but their activities were restricted. From the mid-1960s, there was a strong shift towards the centralisation of economic resources and political control. The government changed strategy from inclusion to repression in its attempts to regulate social forces. The health regime was characterised economically by growth and distribution; politically by a centralised state, narrowed scope for civil society and authoritarian forms of regulation and socially by the transformation of health care underneath the surface of hierarchical relations. The quantitative expansion of services continued during the first years of the Amin government but was gradually undercut by overall economic stagnation.

State and society 1971-1986

The Idi Amin coup in 1971 was initially received with euphoria by the many sections of society who had been hostile to Obote's government. Among the groups were traders who had opposed state intervention and, of strategic importance, most people in Buganda and in particular the bourgeoisie who hoped for improved economic opportunities. The coup had been carried out with the help from Israel and other foreign powers, notably Britain, quickly welcomed it (Furley 1989: 276-277).

The Amin government disbanded civilian political organisations and institutions. Lacking a firm social base, the government sought to consolidate its rule by embracing domestic and international constituencies through first right-wing and then nationalist populist measures (Jørgensen 1981: 270-273). One immediate source of political instability was the military itself. During the last few years of the Obote government, ethnic factionalism in the military was rampant. This pattern continued under the Amin government. Both control of the military and grievances and mutinies within it were mediated by ethnic and religious sectarianism. Many Acholi and Langi soldiers and civilians were massacred within the first six months after the coup. During the following years, other ethnic groups suffered the same fate (Jørgensen 1981: 303-306).

To extend support beyond select groups in the military, Amin conceded to demands for the "Africanisation" of the economy. Such demands came primarily from African traders, previously excluded and now ready to take on their Asian counterparts. In the context of a fragile political base, the government's response was the Economic War declaration – the expulsion in November 1972 of around 50,000 resident Asians without Ugandan citizenship. Their property, around 5,500 companies, was expropriated and redistributed through a Custodian Board to Amin loyalists in the military, in the bureaucracy and among the business community.

This proved to be fateful and restructured the Ugandan political economy in a lasting way. Relatively centralised economic structures initially provided the state with resources and institutional channels to control social forces. The weak economic bases of the bourgeois factions that gained access to economic opportunities made these groupings politically fragmented and dependent on the state (Mamdani 1976: 307-310). But the logic of the Economic War soon resulted in destabilisation. It "gave rise, in hot-house fashion, to a rapacious state-created, state-protected stratum of big proprietors, popularly known as the *mafutamingi*"[4] (Mamdani 1990b: 434) ."

The massive redistribution of wealth was erratic. While the state handed out resources to consolidate loyalty and control, widespread individual looting and mismanagement by beneficiaries undermined the state itself (Jørgensen 1981: 292-294; Mamdani 1976: 309-312). The long-term effects of this instant form of primitive accumulation through the state turned out to be recurrent redistributions of a stagnating economy, which sustained the short-term speculative nature of the economy. The structural contradictions were expressed as informalisation and militarisation of the state, economy and society. The dramatic

4 *Mafuta mingi* is a Kiswahili term, meaning abundant cooking fat or dripping fat, figuratively speaking "dripping with wealth."

restructuring that were the consequences of the Economic War took place in the context of domestic fiscal crisis, worsened by the stagnation of the world economy after the 1973 oil crisis.

The Amin government had, as described above, inherited the economic and political contradictions that had paved the way for the downfall of the Obote I government. These were now deepened. Its fragile political base and its reliance on the uncertain loyalty of the armed forces resulted in emphasis on short-term control. Real and perceived internal and external security threats generated dramatic increases in expenditure on coercive state institutions. This placed additional strains on the economy. Plunder undercut state institutions and the formal economy. Big proprietors evaded taxation due to political protection (Mamdani 1990b: 435).

To compensate for revenue decline, burdensome taxation and forced extractions of production were placed on the peasantry. Those who were able to made use of informal channels for export (Bazaara 1997: 129, 135). The consequence of this reshuffling of class structures by means of primitive accumulation was a fusion of state power, military might, economic wealth and ethnic affiliation in shifting and unstable patterns. While such tendencies had become visible during the later years of Obote's government, they took on unprecedented significance during Amin's rule (Saul 1976). In structural terms the economy was caught in a downward spiral of declining productivity, fiscal and balance of payment crises and rampant inflation, reinforced by political disarray and institutional collapse.

The later years of the Amin regime were characterised by the politics of survival at all levels. From the mid-1970s Uganda suffered from an acute economic crisis. The informal *magendo* economy had in effect become the real economy. Informal markets and smuggling evolved around the land, foreign exchange, import and export, coffee and public services. Patronage and plunder reinforced economic stagnation and further undercut state institutions. Insecurity was endemic. Violence and institutional decay led to the collapse of formal administration and the economy. With the erosion of both an economic and a social base, the state was reduced to an instrument for economic looting by its coercive apparatus (Mamdani 1984). Perpetuating decay only further undercut state capacity and reinforced violence and sectarianism.

As state power was exercised through erratic violence by both the army and other security organs, including the notorious State Research Bureau, political opposition took on a militarised form (Brett 1995: 140). Due to the brutal nature of repression, resistance was either based in exile, subterranean or occasional as in the case of religious institutions. The government's shrinking base in the military itself resulted in a mutiny in 1978. To divert grievances and to pre-empt a coup Amin ordered an attack on Tanzania. The Tanzanian army retaliated, and assisted by exiled Ugandan rebel forces overthrew the Amin government in April 1979.

The near collapse of the Ugandan state, in combination with the fact that the most important political groupings had been operating from exile, had left a power vacuum at the political centre. Efforts to spell out future directions under an 'umbrella' government under such conditions proved fruitless. Beneath the surface of national unity, political intrigue continued unabatedly. The major political forces mobilised clandestine support in the military to the effect that political "factionalism inside the dominant classes was almost mechanically reproduced inside the state apparatus" (Mamdani 1995: 11).

During the first year after the war, two brief politically weak administrations presided over the state, until the Military Commission took over power from the interim surrogate parliament in May 1980. Elections were held in December that year, with UPC, DP, Uganda Patriotic Movement (UPM) and the Conservative Party (CP), a resurrected KY, contesting. After what was widely perceived as a substantially rigged exercise, the Chairman of the Military Commission pronounced UPC to be the winner. The allegations of rigged elections were used by Yoweri Museveni's party UPM, regrouped first as Popular Resistance Army (PRA) and later National Resistance Army (NRA), as a justification to take up arms against Obote's new government in February 1981.

The rebellion soon escalated into a bloody guerrilla war. The government faced rebellions on several fronts. In West Nile, remnants of Idi Amin's army had taken up arms too. Buganda based groups, with ties to DP or Mengo, started urban guerrilla warfare. The main battlefield however was the "Luwero Triangle", the area northwest of Kampala, where the NRA was based. It sought support and protection from the civilian population there, and the Obote government responded with brutal repression. The military looted the countryside and harassed the population. Several hundreds of thousands were killed during the course of the war. Many were uprooted and resettled in what was officially termed "relief centres" – more adequately called concentration camps (Bazaara 1997: 163; Kasozi 1994: 180-185). As punishment for their perceived association with NRA, the government expelled the Banyarwanda community in 1982 (Mamdani 2001: 168). The Obote II government's legitimacy and power base was limited from the outset and diminished further as economic stagnation set in.

The government responded to the inherited economic crisis by agreeing upon the terms of economic reform required by the IFIs in exchange for financial support. The terms of that reform programme were typical of adjustment programmes in Africa: exchange rate adaptation, devaluation, reduced state control of prices and import licenses, among other conditionalities (Loxley 1989: 73). On the surface, the reform programme appeared to have reasonably favourable macroeconomic aspects up to the second half of 1984. The government could register GDP growth and reduced inflation rates and budget deficits. However, the programme failed to address structural imbalances, which were reinforced by for instance political manipulation of the exchange rate regime to the benefit of top politicians and civil servants (Mugyenyi 1991: 64-65).

What conclusions can we draw about the political economy of the adjustment programme? Fundamentally, the application of the programme generated political dynamics that the government could not control, similar to the cycles of stagnation, adjustment and protest common in other African countries during the 1980s. As just referred to, state based or state connected capitalists made quick and big money from trade and foreign exchange manipulations. The UPC politicians and their business associates replaced the *mafutamingi* strata, which had emerged during the Amin years. The fact that this group profited heavily from the liberalisation programme underlines the continuity of the Ugandan political economy after the economic war. Factions of capital with weak autonomous financial and organisational bases were dependent on access to state power for their immediate reproduction based on speculative accumulation, mainly in trade.

The losers from the economic reforms, on the other hand, were politically estranged. The transformation of a state regulated economy to a market regulated system involved privatisation measures that would remove advantages for the government's political base within the state apparatus. The alienation of this constituency further undermined the limited support the government enjoyed.

Various groups, whose purchasing power had been reduced by adjustment measures, made demands on the weakened state, which responded by raising salaries for workers and soldiers in order to cultivate short-term support. Pressure also forced the government to recruit 20,000 civil servants between 1982 and 1985 (Nabuguzi 1994: 11). However, IFI conditionalities and budget ceilings blocked workers in service sectors from bargaining for raised salaries (Banugire 1989: 99). These groups resorted to exit strategies within and beyond their official workplaces, something that reinforced the institutional deterioration of the official economy and strengthened structures of the *magendo* variant.

In rural areas, a heavy burden of adjustment was placed on the peasantry, as state services in terms of health and education as well as agricultural extension services were cut. Insecurity with regard to both land tenure and the political situation more broadly prompted peasants to withdraw their incomes from the state, whose agricultural revenue hence diminished. Instead, politicians and capitalists with political protection made use of their control over the UPC-ified local state and bought land for purposes of protecting themselves against inflation (Bazaara 1997: 164-173).

During 1984, intensification of the civil war undermined macro-economic achievements through a rapid increase in military spending. These different sources of expenditure triggered fiscal deficit and inflation. The withdrawal of IFIs in 1985 was a severe blow to the possibilities of the Obote II state to reproduce itself. To understand the wider social and political dynamics of the situation, it should be noted that the economy had shrunk by more than 20 per cent between 1970 and 1985. By 1986, inflation ran at 169 per cent and the shilling had depreciated dramatically (Brett 1996: 314-315).

By late 1984, the UPC government had been pushed to the wall. Rebellions raged on different fronts. Ethnic factionalism was rife in the military. Most segments of society had withdrawn from formal welfare institutions into informal coping strategies. A stagnant economy was overburdened by military expenditure. Just like during the last years of the Obote I government, a crisis of accumulation was transmitted into a political crisis, which culminated in a military coup. In July 1985, a military junta of mainly Acholi officers took over power. It could not however establish control. Kampala was divided into zones, each controlled by a faction of the military or by a militia. In spite of a peace agreement in Nairobi in December, the NRA continued with its armed struggle, and captured state power on 26 January 1986.

Health and welfare 1975-1986: State breakdown

During the first years of Amin's rule, the government expanded the infrastructure of hospitals and health centres. It was not until the mid-1970s, however, that economy collapsed as a result of Amin's "economic war", the main features of which have already been discussed above. Economic activity moved from official structures to informal networks. The decline of the health sector was an integral part of this wider deterioration of state institutions. Main features of this decline included informalisation and *de facto* privatisation, related to widespread corruption and mismanagement in the public sector, along with serious demoralisation of the work force.

It is estimated that real per capita income declined by over 6 per cent annually between 1973 and 1980 and that the real expenditure per capita of the Ministry of Health fell by 85 per cent from 1972 to 1982 (Iliffe 1998: 146-147). Health indicators fell rapidly from the mid-1970s. Working conditions became untenable. The physical infrastructure collapsed. Hospitals and health centres lacked even the most basic equipment and drugs. Water and power supply was erratic.

As inflation skyrocketed, health workers' salaries declined in real terms, and in 1984 they amounted to less than 10 per cent of their 1971 value (Nabuguzi 1995: 198); for instance, the real value of interns' salaries fell from $244 in 1972 to $4.80 in 1987 (Iliffe 1998: 166). Medical workers were forced to resort to a combination of coping strategies. Some set up parallel private clinics; these were by official declaration shut down by the government between 1974 and 1978, but operated illicitly (Iliffe

1998: 148). Others sought employment in mission hospitals. Still others derived a parallel income from farming or private business. Finally, a large proportion of the educated medical and administrative staff fled the country. Among those who stayed, coping strategies resulted in absenteeism and widespread diversion of public drugs and equipment for private purposes (Iliffe 1998: 147-148; Nabuguzi 1995: 198-200; Mugaju 1999: 7-12).

After the fall of the Amin regime in 1979, a Sub-committee for the Rehabilitation of Medical Services was set up. The result of its discussions was to prioritise rehabilitation of the system and to expand primary health care. Rehabilitation first faced the challenge of drug supply. The government lacked foreign exchange to buy drugs. A big proportion of the available drugs was either stolen or got spoilt. It was also difficult to encourage staff to rejoin the sector, as they were severely underpaid (Iliffe 1998: 151). The rehabilitation of primary health care was heavily dependent on international NGOs.

This decline continued until the mid-1980s. The Obote II government's strategy for rehabilitating the economy rested on a structural adjustment programme negotiated with the IMF in 1981. The outbreak of armed rebellion the same year however swallowed most of the state's resources, and political conflict and social dislocation prevented economic recovery.

The formal structures of government social service provision turned into an empty shell. This took place at the same time as health needs escalated, due to civil war and population displacements as well as the outbreak of the HIV/AIDS epidemic. The effects of the epidemic were further aggravated by the war and its consequences in terms of the breakdown of social structures. Large parts of the population turned to non-state institutions, such as traditional medicine, which was cheap and accessible; private services from professional providers for the well off and quacks for everyone else; and self-treatment. Faith healing emerged as a new feature from the late 1970s (Mugaju 1999: 13-14).

The consequence was clear and rapid stratification of access to and quality of health services, which had implications beyond the sector. Bazaara argues that the combined effects of inflation, devaluation and cuts on subsidies for social welfare led to land struggles, as tenants sought to sell off fallow land to finance fees for education or medical treatment, but they were blocked by landlords from doing so (Bazaara 1997: 179).

Systematic support from foreign NGOs came to a halt, except for emergency relief intervention. As formal structures underwent decay and declined, the backbone of the health care system now consisted of the wide spectrum of non-state actors. Mechanisms for co-ordinating and regulating social welfare were by and large fragmented and eroded and increasingly dependent on external intervention. The latter consisted in large parts of vertical programmes of, for example, immunisation. Necessary as these programmes were, they nevertheless contributed to the fragmentation of the delivery system and the erosion of regulatory capacity (Nabuguzi 1995: 200; Okuonzi and Macrae 1995: 125-126).

In the 1960s, the parallel provision systems survived on the margins of the expanding public health system, but their relevance was undercut by the expansion of free public health care. The period following the mid-1970s, on the contrary, was primarily characterised by the dissolution of effective formal regulation of health. The state disintegrated, civil society withdrew, and mediating channels for regulating the relationship between them were informalised. As a consequence of the crisis of the public health system, informal modes of provision and regulatory regimes regained their significance. These systems lacked formal supervision and were characterised by market or cultural modes of control. The lack of formal regulation did not exclude overlap between formal and informal systems. However, their interpenetration was itself unregulated and unintended. Both health workers and patients moved across systems. The structural effects of the fragmentation of health systems manifested themselves in the erosion of state capacity to supply and regulate services, exit strategies of health workers from the formal systems and stratification of access and quality. The legacy of this period still shapes the health system, as underlying structures were destabilised. The repercussions of that destabilisation, in terms of unregulated forms of provision and interpenetration between formal and informal modes of provision, would continue to affect the public health provision system as well as institutional reforms of it.

Conclusions

This chapter has brought out defining features of state-society relations in Uganda, critical among which is a widely recognised pattern of increasingly elite driven, fragmented, unstable and militarised politics. Subsequent governments inherited and passed on a "development problem", in terms of a colonially distorted political economy and in turn destabilised this order further, until public institutions collapsed in the mid-1980s. With regard to health care, social groups responded to institutional decline in a diversity of ways, which resulted in a very complex constellation of provision systems and health regimes.

How are these features to be explained? Various structural constraints to democracy were connected in a specific sequence. Regionally uneven capitalist development and underdevelopment, reflected in and reinforced by indirect rule and traditionalist ideology and organisational forms, resulted in weak class based organisation, political fragmentation along ethno-regional lines and elite-driven politics, consciously disconnected from popular influence by the converging interests of the colonial power and the emerging local political elite. The weak and fragmented ruling class lacked an independent financial base and was dependent on state power for economic accumulation; this led to intense struggles over the control of fragile state institutions, which further weakened them. In the process, alternative social bases of political power were subjected to close state control. Inherent tensions between fiscal and social bases of state power resulted in contradictory and eventually self-defeating policies by state rulers.

In order to trace the roots of post-colonial state formation, it is worthwhile to return to and expand on the of state-society relations offered by Mamdani on the composition of the nationalist movement, its internal contradictions and the variety of forms of state interventions and state-society relations this set off (1976: 148-49). Democratic forces in society were weak and internally disconnected. In rural areas, extra-economic forms of compulsions gave the regulation of the peasant economy an authoritarian bent. While the fused power of the local state often was less than absolute, it was still a significant factor in shaping the social basis of authoritarian politics. I will return to the specific histories of different strands of civil society in chapters 4 and 5.

In its 10-point programme, the NRM stated its intent to overcome precisely those features of Ugandan society discussed here: fragmented, or "sectarian", elitist and militarised politics in a dependent economy – a political order that reproduced authoritarian social relations at the level of the state. The following chapter discusses political and economic reforms during the first years of the NRM in power.

Chapter 4

The Transformations of State, Society and Health Regimes, 1986-1992: Regulating Instability

The disintegration of state institutions discussed in the previous chapter had disastrous consequences for service provision and the health status of Ugandans. As political structures were rebuilt from the mid-1980s, welfare systems were also slowly rehabilitated, but under conditions rather different from those which had sustained them in the immediate post-independence period. The collapse of the economy left scarce resources for state reconstruction, especially in the context of the economic liberalisation programme. The years of turmoil had further generated institutional fragmentation.

Most sections of the population had been forced to develop repertoires of coping strategies in a context of a plurality of health systems, ranging from public health to formal and informal private clinics, self-medication and traditional medicine. Informal modes of provision took on permanent existence alongside formal ones. This legacy of institutional pluralisation and complexity continues to have effects on the character of the state and civil society in this field, on strategies to reform and regulate the Ugandan health system and on the resulting patterns of access, equity and quality of health care.

This chapter analyses health sector developments from 1986 to the early 1990s in relation to the historical transformation of state and civil society. These changes will be analysed from the theoretical perspective developed in Chapter 2. This perspective, to repeat, sets out from the notion of a regulatory regime that encompasses rules and practices and suggests that this regime is developed within a matrix of general political relations, economic and development policies and sector specific challenges. The overall argument further posits that the capacity for social forces to democratise the state hinges on relations of domination within both state and civil society and among international actors. State-civil society relations are shaped by parallel and overlapping logics. All states are forced to handle contextually specific requirements of resource extraction and political control by a combination of coercion and consent. In contemporary Uganda, state formation occurs within a liberalised and donor dependent economy, and the government has sought to accommodate and balance a liberalisation strategy for resource extraction, demands for services that create legitimation while preserving control over civil society.

In exploring the changes in the political regulation of the sector, the focal point will be on how structures of interest representation and conflict regulation have been created by the interplay between state strategies and demands from groups in society. What has characterised the framework of interest representation and conflict regulation that these relations have generated?

The argument further proposes that the formation of state and civil society in this field is shaped by a combination of the historical legacy of the sector, overall economic policies and their relation to health and social policies and overall state-civil society relations. There is an interplay between the capacity of the state and its social base. In Uganda towards the late 1980s, structures for health care were dilapidated, and the government had just embarked upon a strategy for state reconstruction from a vulnerable political position, within weak state institutions and with very limited resources. This would naturally shape the contours of the demands that different political forces would make on priorities for the rehabilitation of health care and for social services in general.

State-society relations 1986-1992

State institutions and formal mechanisms for regulating state-society relations had almost collapsed in 1986. How did NRM seek to relate to the political expressions of different social forces during its first years in power? How did these forces attempt to root themselves in the state? This section addresses these issues by reconstructing the main features of state formation in order to provide the broader context within which health sector developments took place. As many observers (e.g. Kasfir 2000: 63; Khadiagala 1995; Mamdani 1996: 216) point out, NRM had a very narrow political base upon taking power.

The main challenge that the new government faced was to establish control in the face of threats to fundamental statehood – that is, to secure a minimum of political and economic stabilisation in terms of security and resource extraction within a very volatile political context. The measures taken included the defeat of insurgency groups; the incorporation of rebel leaders and politicians of old parties; the setting up of structures for local governments in order to establish an effective presence in rural

Uganda; and reviving the economy. The following sections will discuss these aspects of state formation in that order. The main argument is that while each of these strategies was relatively successful in the sense of lending the state some degree of legitimation and revenue, it was a very uneven process, whereby unresolved contradictions of the past that affected for example the role of the military in politics were reproduced in slightly new ways. The rather different trajectories that characterised the different arms of the state led to an incoherent composition of the state, where some institutions dominated in certain regions (such as the coercive organs in the north) while being rather insignificant in others (the central and western regions).

Security and military control

To restate a truism, an absolutely central preoccupation for any state is to establish sovereign control over the territory it claims to govern and to assure the compliance of the population within that territory. Rather than to protect its citizens, the Ugandan state had turned into the very source of violence against them.

In this respect, the challenges the NRM faced after taking over power were to restore effective law and order and to fend off threats by armed rebellions. Ever since 1986, the NRM government has faced armed uprisings, each of which, in the words of Kayunga, "represents a different and rather contradictory phase in Uganda's political history" (2000: 118). With the victory of the NRA, military power had suddenly shifted to the south.

This structural shift created a military and political vacuum and ultimately a profound social crisis in northern Uganda, as well as fertile ground for armed resistance against the new government. Between the Okello-Lutwa coup of July 1985 and NRM takeover in January 1986, Kampala was divided into zones, each to varying degrees controlled by different armies or paramilitary units.

After January 1986, these groups either fled or surrendered. The NRA followed the former government army north, and within a few months it managed to capture the northern parts of the country without encountering any great resistance. Sections of the former government army, however, had withdrawn to bases in south Sudan to regroup, and in August 1986 the Uganda People's Democratic Army (UPDA) launched an attack on NRA barracks in northern Uganda (Kayunga 2000: 112).

Initially, UPDA enjoyed popular backing in Gulu and Kitgum, its main area of operation, in spite of not being rooted among the population. Most observers explain this relative support by referring to extensive human rights violations carried out by the NRA (Finnström 2003: 104-105; Kayunga 2000: 112). As the NRA proceeded to the north, it included elements of other rebel groups in order to expand its forces. Among these groups was the Federalist Democratic Movement of Uganda (FEDEMU), a Baganda rebel force that had fought the Obote II government but entered an agreement with the Okello government following the coup. After the fall of Okello, FEDEMU was incorporated into NRA's 35 Battalion. Although human rights abuses were widespread, the FEDEMU elements, stationed in Kitgum, have been singled out to have been particularly brutal. Plunder, torture and rape became the common means for humiliating the local population. In August 1986, as retaliation for the UPDA attack, they massacred some 40 civilians in Namokora, Kitgum, home of Tito Okello (Doom and Vlassenroot 1999: 15; Lamwaka 2002). This generated resentment against the NRA from the local population and triggered a vicious circle of deep mistrust with strong ethnic overtones. The NRA, finding itself in a hostile terrain without the presence of its political leadership, resorted to coercive means, which made the local population more receptive to UPDA and its version of events: that the new government was determined to wreak havoc on the Acholi in revenge for the Luwero war (Asowa-Okwe 1997: 99). These narratives were embedded in broader discourses emphasising "north-south" or "Bantu-Nilotic" divisions, themselves reinforced by colonial and post-colonial structural underdevelopment – c.f. the discussion by Finnström (2003: Chapter 3).

Its relative support notwithstanding, the UPDA soon exhausted its military capacity. In June 1988, a peace agreement was signed, after which the bulk of the UPDA forces were incorporated into the NRA. A new force stepped into the vacuum, the Holy Spirit Movement (HSM) of Alice Auma Lakwena, formed in 1986. Whereas UPDA had been a conventional military organisation with the aim of recapturing state power, the peasant based HSM and its armed wing Holy Spirit Mobile Forces (HSMF) embodied a combination of armed and spiritual resistance (Finnström 2003: 110), and its message of purification and redemption found resonance in crisis stricken Acholi.

The HSM rapidly mobilised support and, joined by some UPDA soldiers, HSMF marched towards Kampala. These forces were finally defeated near Iganga (eastern Uganda) in November 1987. Alice Lakwena's father, Severino Lukoya, made an attempt to reactivate the Holy Spirit Movement. This proved to be a short interlude, and by 1989 the remnants of HSMF, along with a faction of UPDA, had been incorporated into what later was to be known as Lord's Resistance Army (LRA), under the command of Joseph Kony.

The war between the LRA and the government raged until 2006 and was characterised by exceptional brutality against the civilian population in Acholi. While this cruelty was mainly on the part of LRA, it is important to point out that the UPDF has been implicated in abuse of the civilian population as well (Amnesty International 1999; Human Rights Watch 2005b). The conflict has ebbed and flowed and shifted character over the years.

After the peace agreement in 1988 with the UPDA, the NRA initiated major operations against the remaining rebels. The counter-insurgency campaign set out from a scorched-earth strategy. The rebels were to be denied access to food and other resources. Civilians were evacuated into camps, food stocks destroyed and domestic animals were killed (Human Rights Watch 2004a: 4; Lamwaka 2000).

This strategy was repeated in 1991, after the government had launched "Operation North" aimed at putting an end to the insurgency. Acholi and Lango were sealed off for four months. While it weakened the LRA, Operation North soon became notorious for its brutal methods and further alienated the civilian population from the NRA. Civilians and local political leaders accused of being rebel collaborators were arrested and tortured (Doom and Vlassenroot 1999: 23; Refugee Law Project 2004: 6). At the same time, most civilians in the region denounced the LRA's brutality as they found themselves caught between the two warring parties.

Popular resentment against the military was also a consequence of the latter's failure to protect communities. Cattle raiders from Karamoja took advantage of the general instability in the region, and in between 1987 and 1988 they looted cattle from Acholi, Lango and Teso regions on an unprecedented scale. In all these regions, this bred widespread bitterness against the government army, which was broadly considered to be guilty of complicity, both by turning a blind eye to Karamojong raids and by active cattle thefts (Finnström 2003: 105).

Cattle rustling was also the major reason for the series of rebellions in Teso from 1987, the major one being the one spearheaded by the Uganda People's Army (UPA), another faction of the former government army and containing elements of the former Special Branch of the police.[5] The Karamojong raids were preceded by a government decision to disarm local militias, something which created bitterness in Teso as it was seen as a deliberate strategy to leave them vulnerable to rustling.

[5] The other rebel movements were Force Obote Back Again (FOBA) and the Ninth October Movement (NOM)

The region was a UPC stronghold, and subsequently became subject to a heavy handed counter-insurgency, which in turn was exploited by UPC activists (Kayunga 2000: 112). The Teso insurgencies ended in 1992 by a combination of military defeats and political negotiations whereby sections of the rebel groups were incorporated into the army and government

Other regions in the north also experienced rebellions. In West Nile, the Uganda National Rescue Front (UNRF), a legacy of the national army under Amin, joined the new government, but splinter groups carried on armed resistance well into the 1990s.[6] Lango region was also the site for insurgencies in the late 1980s.

What were the implications for state formation? It is argued here that the common denominator is that issues of security have fundamentally shaped the way the state is rooted in local society. The important starting point for such a discussion is that the war in Uganda did not end in 1986 – it shifted battlegrounds. While there is some truth to the statements in conventional overviews of Ugandan politics that the NRM government brought security and peace to large parts of the country (Mutibwa 1992), this transformation has coexisted with the dark undercurrent of continuous violence. And from a northern regional perspective, this violence has not just been the undercurrent of politics but its very core.

The dramatically shifting regional experiences should not come as a great surprise; at a basic level, they point to the difficulties involved in overcoming politically derived underdevelopment and the relations of domination this has given rise to. State-society relations, the composition, capacity and legitimacy of the state and the political content of formal political institutions can in most societies be expected to vary considerably between regions, due to the balance of power within society and the capacity of different social forces to reproduce themselves in the state.

---

6 For instance West Nile Bank Front (WNBF) and UNRF II.

Regional imbalances have been taken to an extreme degree in Uganda; security has by and large been established in, from the government's perspective, politically significant regions. Security has been one of the most important assets for government legitimacy in the south, central, and western regions and a resource frequently utilised in election campaigns. Following war or insurgencies, the NRM government has also been able to replace old political loyalties with new structures in different regions, whether by design or default.

The post-colonial Ugandan state is rooted in trajectories of violent and authoritarian regulation of social relations, and while in some regions it has succeeded in containing violence by incorporating insurgents into political structures, in other regions political differences have been militarised by rebels and the state alike.

Political inclusion

At the level of national politics, the NRM addressed the challenges to authority that stemmed from its own weakness in the context of contending political forces by including leaders of older political parties and rebel movements (on the condition that these put down their arms). Some were given cabinet posts whereas others were integrated into the army and their commanders were given higher military ranks.

This accommodation was in the process presented as "broad based" politics for purposes of peace and national unity, a deliberate antidote to political sectarianism (Kasfir 2000:65). Notable among incorporated rebel groups were, after their surrender in 1988, leaders and rank-and-file of Uganda People's Democratic Movement (UPDM), and remnants of the old government army. The most important political group that entered into an alliance against UPC, their common enemy, was DP; sections of DP had gone to the bush to fight the Obote II government in 1981 under the banner of Uganda Freedom Movement (UFM).

DP was rather discredited after having joined the Okello military government and was grateful for the opportunity to rehabilitate its reputation through an alliance. The NRM-DP alliance rested on a gentlemen's agreement that the ban on party politics was to be lifted after a transition period. This consensus was disrupted in 1988, when the NRM proposed that the reintroduction of party politics should be postponed and only after having been subject to a referendum (Kasfir 1991:254). Thus, the NRM was, out of strategic necessity, buying loyalty and time.

The ideological rationale for a broad-based government became compromised, as pragmatic considerations of inclusion undercut the long-term political direction that this institutional arrangement was set to promote. While the inclusion of leaders and soldiers of rebel armies created stability at the level of elite politics, this type of pacts often failed (as in the north and most evidently in Acholi) to address underlying structural problems of underdevelopment and political exclusion.

Having come to power by force and with a narrow social and regional base, the NRM faced the challenge of extending control and securing legitimacy as a civilian government with at least some degree of democratic credentials. It developed one key resource for those purposes in its framework for popular democracy, the proposed alternative to the "sectarian" multi-party politics of the past.

The content of such a popular alternative to politics has not remained fixed or precise apart from its resistance to parties, and Kasfir (1999; 2000) has traced the shifting meaning and use attached to the concepts of "no-party" or "Movement" democracy. Initially, the transition to more democratic rule centred on Resistance Councils (RCs), a five-tier system of elected committees first set up during the guerrilla war in NRA controlled areas. RCs (later LCs) are subject to a vast body of literature (e.g. Ddungu 1994; Kasfir 1991, 1999; Mamdani 1994, 1996; Ottemoeller 1996; Tidemand 1994).

During the course of the war, RCs had been functional in providing popular support for the armed struggle and in securing the NRA protection and resources such as food and water, intelligence and recruitment from local communities in the areas – most importantly the Luwero triangle – where the war took place. In addition to this, RCs were useful vehicles for the mobilisation and politicisation of communities by the NRA. The uneven and turbulent making of RCs essentially reflected the conjunctures of the war. Intermittently, in NRA controlled territory during periods of security, they were openly extended into organs for self-organisation and dispute settlement among these communities (Ddungu 1994: 376-377; Kasfir 2005; Tidemand 1995 b: 228).

The popularity of these structures during the war led the NRM to energetically extend the RC system (five-tiered – village to district) to the entire country. The institutionalisation of RCs is generally considered as having been the NRM's most important democratic reform during its first years in power. The system dismantled the structures of the colonial and post-colonial local state, where judicial, legislative and executive power was concentrated in the institution of the chief, and replaced this mode of authority with elected committees.

The committees furthermore granted representation to "social minorities" such as women and youth (Kasfir 1991: 252-254; Mamdani 1996: 209, 215). From the NRM point of view, the extension of the RC system was a considerable strategic achievement, as it expanded NRM's influence and democratic credibility throughout the territory, albeit unevenly, and indirectly incorporated administrative functions of local society into the structures of the central state. The democratic gains of the system were presented in contrast to colonial and post-colonial systems of non-accountable forms of power in general and to "bourgeois sectarian elite politics" of DP and UPC in particular. Needless to say, the legitimacy of the system was regionally uneven. The population in most parts of eastern and northern Uganda viewed the new government with suspicion – some as a military regime with a southwestern base. Furthermore, within a few months armed uprisings emerged in these regions which made the effective introduction of civilian political forms more complicated (Kasfir 1991: 260).

Whereas, predictably, participatory democracy worked best at the village level, indirect elections from that level to councils and committees higher up in the system deliberately disconnected local participation from broader representation. The government rejected a proposition by the 1987 Commission of Inquiry into the Local Government System for direct elections at all levels with the claim that this would reinforce sectarianism (Kasfir 1999: 207). This should be seen as a government strategy to retain control by reforming the institutional forms of local politics while at the same time ensuring that contending political forces at the national level could not forge links with rural areas.

The latter possibility was real. As a new political force, the NRM was unsure of its political support. By 1988, DP candidates had done well in various RC elections. According to Kasfir (1991: 255), they held a majority of seats in two thirds of the district councils and most probably enjoyed solid support at lower levels too. Real political power was, however, and particularly up to the 1989 elections, concentrated in the National Resistance Council (NRC) – the parliament – or more precisely its Central Committee dominated by the inner circle of NRM leaders.

The introduction of RCs was initially done without legal anchoring. The unclear legal status of RCs in the context of their co-existence with old administrative structures created uncertainty regarding their mandate and function. It was not until May 1987 that legislation concerning the RC system was formulated. This legislation spelt out that RCs were placed under the minister of local government who was given power to dissolve individual councils. This suggested the limits to which RCs could influence the central state.

Another measure of control was the introduction of District Administrators (DAs) – presidential appointees, who were given power to intervene in council affairs (Tidemand 1995b: 228-29). Along with legal regulations came administrative functions, such as tax collection and road building. Kasfir claims that these duties over the following years, in combination with biting poverty and the fact that lower level committee work was supposed to be voluntary, made RCs less popular. The balance shifted from deliberation to administration (2000: 67).

To what extent did the RC system during the early years of its existence result in the transformation of the local state? Assessments have stressed the different strands inherent in the system. The debate in the late 1980s over the relative importance of democratic, bureaucratic or sectarian (that is, pro-NRM) features of the system pointed to the constraints to local participation when disconnected from direct representation (Mamdani 1988).

The contradictions between these tendencies came to the surface through numerous conflicts between RCs and DAs and through the legislation that imposed limits on the mandate on RCs. Mamdani (1996: 209, 215) claims that the democratisation of the institutional forms of village politics was accompanied by social conservatism, which tended to reproduce social inequalities inside reformed political institutions. To precipitate the conclusions of this study, political monopoly would eventually undercut the relative democratic gains as well.

Until 1989 only local government elections had been held under the new government. With in a short notice, local and national elections were called in February that year. Ugandans would, through the RC system, elect a parliament, the National Resistance Council (NRC), which would in the process be extended from 38 to 279 members. The political landscape at the time was fluid and, from the government's point of view, very uncertain. The NRM had allocated itself (and had been granted, in a gentlemen's agreement with DP and other minor parties to the coalition) a four-year transition period, which would come to an end in early 1990.

With only one year to go, the NRM had been fighting two civil wars; the economy was only on the verge to becoming stabilised; and DP, the temporary NRM ally in the broad based government, was the stronger political force on the ground. At the same time, NRM legitimacy had to a large extent rest on promises to restore democracy as part of the "fundamental change" it claimed to install. Hence, it needed to build political bases in local and national political arenas with preserved or enhanced democratic legitimacy (Kasfir 1991: 254). How could the NRM test its strength without risking its hold onto power?

For the 1989 elections, it addressed this dilemma by shaping the rules of the elections to its advantage. It assured itself of substantial presence in the NRC by granting the 38 appointed MPs extended terms without elections, and it gave the military and the president powers to appoint 30 more. In addition, it instituted the National Executive Committee (NEC) within parliament and ensured that Museveni would remain president regardless of the parliamentary majority. There was also a ban on discussing issues, as opposed to the merits of individual candidates.

The government sought to justify this move as a means to prevent sectarianism. It is also far from clear that the NRM would have been able to compete successfully with the old parties in competitive representative politics. Again, NRM's claims to democratic legitimacy were based on a doctrine of popular democracy, as opposed to the exclusive and divisive politics of the past. Furthermore, while local elections were direct, elections to the NRC were indirect hence, conducted through electoral colleges. This made it very clear that the NRM would accept the influence of RCs at the local level only, or in the words of Ottemoeller, "much of the populist energy that had been directed into the RC system in hope of reaching or influencing the state was, by 1989, either absorbed into the NRM ruling hierarchy, or directed into the growing body of NRM critics" (1996: 51).

Local structures for representation were thus disconnected from the direct influence on national politics and were instead enmeshed in local struggles over influence with organs of the central state. While opposition groups contested the restrictions, even more damaging to the legitimacy of the NRM was the fact that the NRC soon after the elections extended the provisional NRM government and ban on parties by another five years. This mistrust was to shape the politics of constitution making in the early to mid-1990s.

State-civil society relations

How were state-civil society relations affected by the change of government? Most sections of civil society, already restricted in their operations from the late 1960s, were severely weakened during the years of turmoil, partly as a result of state repression, partly following from the informalisation of both the state and of economic structures. In spite of strict state supervision, some social movements, including trade unions (Barya 1991: 36-38) had protested against both declining living conditions and political autocracy during the Obote II years, but typically in a sporadic fashion in the face of repression. Nor had civil society organisations played an active or instrumental part in the change of government. The legacy of repression and informalisation shaped the contours of the emerging relations between NRM and civil society. What were the main features of these relations?

The relative relaxation of state control laid the foundations for a quantitative revival of civil society groups. Another source of this was the weakened capacity of the state to fulfil its tasks. During the years of crisis, NGOs and CBOs entered into this vacuum in rural and poor urban areas, creating networks that turned into more permanent organisations in a stabilised political context. At the same time, economic hardship and internal wrangling had seriously weakened the previous backbone of civil society, cooperatives and trade unions. Informalisation had critically undercut the preconditions for collective organisation and representation because of individualised coping strategies by members of these organisations and the subsequent fragmentation of associational life. This trend continued to limit their strength up to the early 1990s. Economic liberalisation undercut the viability of cooperatives, and trade unions suffered from stagnation and retrenchment. Strikes and protests during the late 1980s were largely temporary outbursts that failed to impact on this overall trend (Kjær 2001: 155). Ironically, at a time when the legislative framework had become more benevolent for autonomous organisation, unions were too weakened to be able to take advantage of this opportunity. The overall effect was a recomposition of civil society, and in extension of state-civil society relations, towards the predominance of NGOs. Sections of the middle class responded to retrenchment and insufficient earnings by entering or creating NGOs, among other things.[7]

This shift was underpinned and reinforced by a third aspect of the structural transformation of civil society, on its way since the early 1980s, namely its "donorisation". This was expressed in terms of both an influx of international NGOs and of

[7] Technically speaking, trade unions are of course non-governmental organisations. However, I keep to an established distinction in both popular and academic Ugandan discourse that singles out NGOs from for example unions and religious institutions.

deepened donor dependence among Ugandan ones (Hearn 1999). Such dependence obviously increased the probabilities of the disconnection of NGOs from the constituencies they claimed to represent, weakening their roots in society and muting their political voice. This connects to a fourth dimension of state-civil society relations during this period and beyond: the gradual "apoliticisation" of most civil society groups – that is, their tendency to adapt to calls for loyalty by strategies of keeping a distance from controversial issues.[8]

This had a number of causes. In addition to recomposition and donor dependence, it needs to be recognised that the NRM enjoyed the support of many interest groups that had developed by and large in tandem with the state. This occurred in a number of ways. Some groups developed cordial relations to the state relatively voluntarily, others were co-opted, and still others were set up by members of the political elite for purposes of business, or in order to channel political support and block more autonomous and critical counterparts. In order to impose control over NGOs, the state instituted legal and administrative instruments through the 1989 NGO Registration Act. The Act opened up for the creation of a National Board of NGOs, which along with other departments was supposed to enforce the regulation of NGO activities (Barya 2000b: 18-22). However, the relatively weak state refrained from exercising direct repression or excessive control of civil society (which it lacked capacity to implement anyway) – that recognised the limits of its own freedom of movement in the context of the overall monopolisation of political space.

On an everyday level, the immediate expression of inclusion was the fact that RCs had become the main channel for interest representation. Obviously, social forces continued to influence

---

[8] I owe the observation that this tendency is better captured by the term apoliticisation to denote an active choice under constraints rather than depoliticisation as demobilisation from above to Joe Oloka-Onyango.

the state, but not only or even mainly in a collective organised form. Various informal structures remained important. Ruling class politics was articulated as competing state based networks, whereas its middle and working class equivalents were concerned with coping rather than accumulation activities. Similar struggles were played out in RCs, where local configurations of party-political affiliation and social power continued to exercise informal influence.

Economic reforms

The economy was in deep crisis. Rebuilding the political infrastructure would require the simultaneous reconstruction of a stabilised and growing economic base. How did the NRM deal with the informalisation of economic life? How did it seek to rebuild the state and root itself in society through the economy? Out of ideological conviction, and conscious of popular disillusionment with the adjustment experience of the Obote II government, the NRM initially embraced a socialist-nationalist economic programme.

The main components of this programme involved a Produce Marketing Board (PMB) monopoly on export of crops and PMB organised barter trade for essential import goods, as international creditors viewed the new socialist-leaning government in Uganda with suspicion; this feeling was reciprocated by the NRM. The government fixed the exchange rate well below the parallel market rate. The results were devastating, and structural problems magnified: the inflation rate was 147 per cent during the second half of 1986, the budget deficit grew and GDP declined by 2 per cent. Faced with threats of economic collapse in a context of military insurgencies, the government sought accommodation with the IFIs (Mugyenyi 1991: 70).

The Economic Recovery Programme (ERP) was launched in May 1987. In an exchange for a Structural Adjustment Facility loan from the IMF, the government committed itself to stabilise the economy through budgetary and monetary policies such as devaluation and fiscal reforms for increased levels of revenue, privatisation and liberalisation of trade and foreign exchange regulations – measures that were aimed at bringing under control balance-of-payment and budget deficits and inflation, and at stimulating growth. In return, it was given considerable financial rewards. On average, Uganda received $500 million annually between 1987 and 1992 through the IMF, the World Bank and bilateral donors (Mwenda and Tangri 2005: 452).

The main components of the ERP centred on stabilisation, and the programme did not address issues of adjustment through institutional reform. During its first five years the reform programme promoted rapid growth, albeit from a very low level, but did not attend to poverty or inequality. As Brett (1996: 342) wrote, "a social security crisis continues to exist in Uganda with a substantial proportion of the population living in extreme poverty".

Poverty and inequality were not mainly the result of adjustment but had much deeper roots. Nevertheless, and of particular importance for this study, the preoccupation with security and macro-economic stability and rapid growth, together with the lack of concern with institutional reform or equity measures, meant that the social sectors were left to be shaped by institutionalised informalisation and fragmentation. Public health care, although receiving slightly more resources than before, remained under-funded and poorly regulated. As regards other dimensions of economic recovery, industrial development was modest during the first years of the ERP and blocked by structural problems in terms of low levels of education, infrastructural constraints and weak institutional capacity. Also, inflation and budget deficits were not controlled as prescribed by policy. This followed partly from exogenous shocks, and partly from contradictory implementation (Brett 1996: 318-325).

The process of implementation was mediated by political struggles on different levels. In one sense, it was an ongoing negotiation process between the structuralist position of the Uganda government and the monetarist view of the IMF – for details, see Mugyenyi (1991: 69-76). It is argued here that such struggles within the state grew in importance with NRM's increasing control over contending political forces. The turnaround of economic policy direction followed upon, but did not mark the end to, the tug-of-war between the two on issues of devaluation, exchange rates and privatisation. Ideologically, the shift to liberalisation was a bitter pill for the NRM to swallow, and hence, justified as a tactical retreat in order to achieve a strategic aim. In a very difficult situation it nevertheless found it necessary to give up a great deal of control over economic policy making in exchange for resources and recognition, in order to guarantee a revenue basis for stabilisation and recovery, and for its own political survival, as well. While obviously being the weaker party, the Uganda government was able to create some scope for influence. As Brett points out, it was "a complex and ambiguous process of negotiation in which compliance could only be achieved where the former [IFIs] were able to offer real incentives and create effective links with local political and administrative actors" (1996: 340).

However, this overall shift and its contradictory implementation cannot be understood merely with reference to contestation between the NRM and the IMF. It more fundamentally reflected struggles for control between contending political forces in Ugandan society and within the Ugandan state. The new government confronted hostility within the bureaucracy, as many of the cadre of the latter had DP or, to a lesser extent, UPC inclinations. While the overall policy direction of economic reform was designed to promote capitalist development through liberalisation, contestation between factions of the bourgeoisie over this process and its outcome was fierce. The extension to a broad based government after 1986 reproduced contradictory perspectives on economic policy within the bureaucracy as well as within the NRM itself.

The NRM included groupings with different ideological inclinations, such as Lule's Uganda Freedom Movement (UFM) with ties to Baganda landlords. According to Loxley (1989: 82), the latter supported continued state intervention, for purposes of safeguarding extra-economic control over agricultural production. Powerful sections within and outside the state had vested interests in the preservation of an economic model that enabled them to use their political position for economic purposes. Others with a different social base sought to achieve a similar goal through liberalisation. This struggle between adherents of state control and proponents of liberalisation resulted in a haphazard implementation of the ERP during its first five years and, according to observers (Loxley 1989: 80-82; Mugyenyi 1991: 62), in the paralysis of decision making in the context of weak institutions.

Still, the basic thrust of reforms was towards liberalisation, undercutting the position of the salaried middle-class and workers. Retrenchment from parastatals, withdrawn subsidies of social services and rising living expenses generated protests and strikes among urban groups during the late 1980s and early 1990s. The struggles centred on the consequences of reforms, mainly in terms of wages, but they also included demands for improved working conditions. Wildcat strikes erupted in most parastatals. Students and university lecturers went on strike in 1989, protesting against withdrawn subsidies and the introduction of user charges.

Nurses and interns went on strike in Jinja in 1989 and in Mulago in 1990. Market traders protested against rising fees. In both urban and rural areas there were tax riots (Iliffe 1998: 166; Nabuguzi 1994: 16). Even so, the political impact of the strikes and protests was modest. The trade union movement was internally fragmented and its capacity for collective action had been severely weakened by informalisation and liberalisation. The state could afford to ignore its demands or to offer workers benefits in lay-off package deals.

There were also other reasons for the relatively limited influence of the protests, which can be helpfully outlined with comparative reference to Ghana (see Hutchful 1996: 154-155). There, just as in Uganda, the economic crisis had evolved over many years and preceded the adjustment programme. This created a consensus around the need for some kind of reforms, which to a greater degree than in most other African countries came to be associated with economic and institutional recovery rather than decay. This recovery was propped up by large inflows of foreign aid and some level of protection of social services. Also, in both Ghana and Uganda the longstanding crisis had entrenched overlapping formal and informal survival strategies for both urban and rural constituencies. Such exit rather than voice options both offered some protection against shocks in the formal economy and fragmented the scope for consistent collective action.

In a similar vein, Nabuguzi argues (1994:15-16) that liberalisation involved a recomposition of the political base of the government. The reduced role of the state in the economy offered fewer direct possibilities for accumulation through the state. The middle and working classes that lost their employment had to find new ways of coping. The government sought to reduce its political dependence on these urban constituencies and actively attempted to bypass them by rooting its institutions and policies among rural groups. Meanwhile, the pro-liberalisation forces within the NRM crafted an alliance with a section of the bourgeoisie that had been excluded from influence during the Obote II years and who found an opening in the liberalised economy. The same process also created business opportunities for the NRM's own followers.

The consequences of the reform programme, including its impact on state-society relations, were uncertain and fragile during its first five years. The overall direction of the liberalisation programme spelt out a reduced role for the state in the economy. At the same time, the Ugandan state had been dramatically weakened over the previous decade and a half. The shrinking of the state would have to be preceded by its reconstruction, and the struggles over the direction of this reconstruction resulted in uneven implementation. However uncertain the process was, it nevertheless contained the seeds for future developments. The growing dependence on foreign aid in the reconstruction process anchored the state's resource base among transnational institutions. Hence, the gradual transformation of the Ugandan economic policy regime towards a model of aid-driven liberalised capitalist development would eventually also involve the reconstruction of the state as regulatory instance, which then made it possible for state rulers to reconfigure political alliances and solidify political bases among the peasantry and upcoming sections of capital. Nabuguzi (1994: 18) proposes that what was at stake was:

> less the withdrawal of the state from certain economic activities, which *de facto* had taken place, but the rehabilitation of the state apparatus [...] Structural Adjustment for Uganda was therefore a relegitimating exercise intended to reroot the state and reconsolidate its institutions.

This section has described the diverse ways in which the NRM government sought to strengthen its position and to root the state in society and how different social forces tried to root themselves in the state, from 1986 up to the early 1990s. The priority of security and macro-economic stabilisation made social policy a marginal concern. The character of the health regime, the subject of the next section, needs to be understood in the context of these priorities and as shaped by the lingering legacy of informalisation and fragmentation of the health care system itself.

State reconstruction in the health sector: 1986-1992

Following the takeover of state power by the NRM in 1986, state institutions were slowly reconstructed. What strategies did the government pursue to rebuild health infrastructure in terms of policy content and political regulation of interest representation? What was the degree of purpose and coherence of these strategies? How did different sections of civil society respond to this? Did the reconstruction of the state involve its democratisation, or was this process shaped by lingering authoritarianism? This section explores the defining features of social policymaking and the evolving health regime during the first years of NRM rule. It starts with an overview of the physical and institutional legacies of state decay and then addresses different attempts to reconstruct the health system in that context as well as responses to those reconstruction strategies.

Legacies of decay

Physical and institutional infrastructure deteriorated in mutually reinforcing ways during the 1970s and 1980s. While the construction of physical structures for health provision actually continued during this period, institutional capacity to manage and provide services eroded (Dodge 1987). There was no foreign exchange with which to purchase drugs. Large budget deficits halted salary payments, the real value of which was being eroded by inflation anyway. In 1986, health sector salaries ranged from $2 to $12 per month (Iliffe 1998: 151). Other figures also indicate the level of deterioration that lingered on after some relative stabilisation. In the 1986/87 fiscal year, public spending on health amounted to 0.1 per cent of GDP, equivalent to 4 per cent of public expenditure (Macrae *et al* 1996: 1097). During the period of decay, the health status of Ugandans worsened due to poverty and civil war, which contributed to the spread of tuberculosis, measles and HIV/AIDS (Macrae *et al* 1996: 1097).

Hospitals and health units were in a desperate state. Five hospitals had been demolished in the 1979 war. In Mulago Hospital, Kampala, the main hospital in Uganda, a small number of operating theatres were restored during the early 1980s. There was however no consistently functioning water supply, electricity or sewerage system (Iliffe 1998: 151). In rural areas, conditions were even worse, as insecurity made these areas difficult to access. Overcrowded and understaffed units lacked even basic equipment and consumables such as beds, bandages and drugs (Mugaju 1999: 5); this was still the situation some years after 1986, according to the findings of the Health Policy Review Commission appointed in 1987 (Iliffe 1998: 154).

Resources allocated to health care and the use to which they were put was conditioned by the informalisation of Uganda's political economy in general. Medical staff had to rely on parallel coping strategies to make ends meet. This in turn worsened conditions in health units as staff demanded informal fees from patients or stole drugs in order to sell them in their private practices. People with access to large-scale distribution made vast amounts of money out of illegal procurement of medication. The government lacked the capacity to intervene and enforce regulations (Iliffe 1998: 153; Mugaju 1999: 11). At the national level, infrastructural decay both promoted and was reproduced by the narrow scope and limited incentives for planning and policy making, as well as by low qualitative and quantitative levels of medical training (Iliffe 1998: 152). Civil servants were paid very poor wages, and units set up for purposes of policy formulation lacked even basic utilities such as pens and paper (Macrae *et al* 1996: 1097). Institutional and physical deterioration were connected in a vicious circle that resulted in the near collapse of the public health system. Following below is a detailed examination of the nature of these connections and the different provision systems that had emerged as a response to state collapse.

It is important to recall that the emergence of a fragmented and informalised health system did not only reflect the wider pattern of an informalised political economy linked to uneven capitalist development. It was also in some respects a long-standing feature of health service delivery. While the collapse of the regulatory capacity of the state was more distinctly linked to violent destabilisation and civil war, institutional pluralism was a central feature of social service provision during colonialism, as demonstrated above. What state decay brought about was "incremental privatisation" (Birungi *et al* 2001: 80) of the public health system, with "licensed or unlicensed private clinics, drug shops, home providers and increased self-medication" (Birungi *et al* 2001: 80).

Provision systems had their own distinct trajectories, and the state exercised different forms and degree of regulation vis-à-vis each of them. These systems were and are not separate entities but interlinked in different and mainly informally regulated ways. That is, while unlicensed practices were unregulated by statutory frameworks, they were structured by informal modes of regulation based on market or social principles. This also means that the historical legacy of regional differentiation of service coverage, in favour of the central region, was eventually accompanied by social stratification of access to quality health services within regions.

It became very difficult to disentangle formal from informal aspects of the health delivery system. State services were to a considerable degree informally privatised through under-the-table payments for supposedly free services and reliance on private practitioners; formally employed by the government. On the other hand, private services were in crucial ways regulated by the state, although with difficulty; the interpenetration between private and public, and between formal and informal aspects of both sectors in terms of funds, staff and drugs was extensive (Nabuguzi 1995: 205). The public sector was informalised, with high degrees of straddling between public and private services in terms of "free circulation of staff, drugs and medical knowledge between them" (Birungi *et al* 2001: 82).

In spite of this institutional complexity and interconnectedness, it is useful to outline commonly made distinctions between different forms of provision. This will also help to clarify the nature of their interlinkage. The main distinction is drawn – by both analysts and policy makers and hence taking on an institutional existence – between public and private sectors. The latter is normally sub-divided into private-not-for-profit, which refers to NGOs and missions, and private-for-profit. With reference to not-for-profit providers, missions have historically had an important role in service provision, whereas the phenomenon of other foreign NGOs is more recent. The private-for-profit category is further divided into formal and informal providers. In short, formal providers are licensed and registered, while informal providers are not but operate anyway, and "are appreciated and legitimised by communities" (Birungi *et al* 2001: 82). This category includes private clinics and drug shops, home providers and traditional healers. There are also a great number of community based organisations (CBOs), some of which are registered as NGOs. A study conducted in the 1990s found that 79 per cent of the curative care in Uganda is provided by various private actors (Hutchinson 1999: 10). All of these providers and the provision networks they set off are to varying degrees incorporated into political projects to secure legitimacy and authority.

Religious institutions have historically constituted the main type of private-not-for-profit health provision. The first religious health unit in Uganda, Mengo Hospital, was set up in 1897. During the first decades of the twentieth century both Catholic and Protestant churches created networks of health centres and hospitals, which constituted the backbone of the colonial health care system. In the 1950s, the Catholic and Anglican churches established medical bureaux which were given the responsibility to access and distribute resources from the colonial government and from donors abroad, and to coordinate the functions of their growing networks.

The colonial state had slowly started to expand welfare services during the 1950s. As an institutional guarantee, and due to financial limitations, the religious bodies felt the need to create a working relationship with the state. The colonial and then the independent state agreed to support church based health care and education with grants through ad hoc arrangements. The government offered subsidies to support a certain number of staff. This arrangement, however, was constantly subject to potential deactivation (interviews, Executive Secretary UCMB, 7 February 2005; Head of Administration, UPMB, 9 February 2005).

As demonstrated above, state-church relations around education became tense after independence, with the state seeking to rein in faith-based education and eventually nationalising the sector. While mission health units continued to operate, state subsidies diminished during the 1960s and 1970s. During the Amin government, most providers experienced difficulties due to general unrest. One exception was Islamic institutions, which flourished as encouraged by Amin and sponsored by aid from the Arab world (Nabuguzi 1995: 200). As the quality of public health care declined during the 1970s, more people turned to religious providers. In order to counter the dwindling state subsidies in a context of increased demand for services, religious providers had to rely on a combination of external donations and user fees. They also set up the Joint Medical Stores in 1979 (Reinikka and Svensson 2003: 8). Thanks to continued external support and user fees, health provision of churches and missions managed to retain some degree of quality during the late 1970s and the first half of the 1980s. Following the decay of the public sector, religious providers remained a central pillar of health service delivery throughout the 1980s. Their financial resources were however deeply affected by the 1987 devaluation.

Although missions can technically be described as non-governmental organisations, they occupy a special place among health providing NGOs by way of the historical continuity and geographic extension of their service networks. Nevertheless, other forms of NGOs, both national and foreign, also offer services. Single- or multi-purpose welfare arrangements have long been a main source of provision in areas poorly covered by state infrastructure.

Following the decay of state institutions from the 1970s onward, they took on greater importance. During this period, general insecurity made it very difficult for formal Kampala based organisations to operate. Most forms of health provision mushroomed as responses to state contraction and reflected more or less regular and continuous forms of coping or accumulation strategies of providers and patients. In this context, their emergence did not mean the manifestation of a discrete and independent sub-sector. Staff moved between public, private and NGO based clinics or laboratories in ways that further undercut already weak statutory regulatory capacity. Most service providers in Uganda are local entities, CBOs, more or less formalised groups, often not registered, and not systematically studied. Nabuguzi (1995: 200) proposes that there are indications that community initiatives could have taken over malfunctioning or abandoned state health facilities but emphasises that this cannot be verified.

The combination of weak state structures and humanitarian emergencies in the wake of civil unrest opened the door for international relief organisations and programmes. This was during a period when donors started to shape the Ugandan political economy more broadly through macro-economic stabilisation programmes. NGOs such as AMREF and Save the Children entered as key actors in the Ugandan health system from the early 1980s. Such organisations had been around since the 1950s, but their extensive and systematic presence occurred during this latter period. International NGOs, typically operating on project basis, was one expression of the internationalisation of the health care system (Okuonzi and Macrae 1995: 125).

Another was the intervention by UNICEF programmes on vaccinations, in response to the famine in Karamoja, northeastern Uganda, in the early 1980s. While these programmes were badly needed in themselves, their combined systemic legacy was one of fragmentation and international dependence of the health care system. In spite of the fact that vertical programmes and disconnected projects were intended to sustain an integrated policy for primary health care as spelt out in the Obote II government's Revised Recovery Programme, neither of which were systematically incorporated into economic policy making (Okuonzi and Macrae 1995: 126). This was largely because defence and security expenses dwarfed other sectors and because the destabilised political situation further eroded coordination capacity of long-term strategies.

As indicated above, informal modes of health provision filled the vacuum left by state contraction. Due to inadequate salaries, medical workers in the public sector retreated to coping strategies, such as the setting up of formal or informal private practices or drug shops run by themselves or by proxy staff. They used their position in the public health system by charging informal fees or diverting drugs (Van der Heijden and Jitta 1993). Others sought employment with NGOs or left the country. Doctors had run licensed private practices since the 1940s, the majority of which were run by Asians (Iliffe 1998: 83-84; 121).

Around independence, following the expansion of medical services, all categories of medical workers established practices (not all of which were licensed) in both urban and rural areas. According to Mugaju (1999: 2), many medical workers exploited their employment in the public or NGO sectors to divert resources to their part-time private practices. Such units were flexible to instant service demands and often constituted the only option. However, many of these providers lacked both formal and real competence, and the state had no capacity to supervise their operation in remote areas.

In the late 1960s, Obote's government made moves towards nationalising health services the way it had done with education (Iliffe 1998: 140). As long as the public health system functioned relatively well, informal private care was a fairly marginal phenomenon. But as government services deteriorated and falling wages were paid infrequently if at all, both patients and medical workers turned to private health care for survival strategies. The informal private-for-profit sector exploded beyond control, as acknowledged by the 1987 Health Policy Review Commission (Whyte 1991: 130).

The informal interpenetration between public, NGO and private provision in terms of staff, medication and resources mirrored the overall informalisation of the Ugandan economy. This in turn set the stage for further withdrawal of patients by way of self-medication with often illegally sold drugs or syringes. According to Whyte (1991: 141-143), the essential meaning of self-medication was the introduction of pharmaceutics into a tradition of family care, alongside or replacing herbal medicine.

The blurring of distinctions between formal and informal modes of health provision was clearly visible in the field of drugs. Drugs found their way into the informal market through smuggling, or theft from government health facilities, and could be found in private clinics, markets, on the street or in villages. In the 1980s, the Medical Council tried to enforce existing legislation but lacked the power to do so, since it was believed that medical workers would then leave the country (Iliffe 1997: 153). In a case study from eastern Uganda in the late 1980s, Whyte (1991) argues that there was no shortage of drugs as such – they were available from different sources such as the Central Government Medical Stores, the Joint Medical Stores of the Christian Medical Bureau, from multilateral or bilateral donor programmes, or entered the country as contraband (1991: 131-132). They were however dispersed at different stages of the commercial chain due to breakdown of the supervisory capacity.

Traditional medicine, which also flourished (Obbo 1996) was, on the other hand, met with greater understanding, and the government made sporadic attempts to incorporate this diverse field of provision into the national health system. There was no rivalry as such between practitioners of traditional and regulated modern medicine; more so between traditional practitioners and extralegal skilled or untrained medical workers who set up clinics in rural areas, especially since both categories combined traditional and modern methods of treatment and targeted similar constituencies. Traditional healers had formed an association, *Uganda n'Eddagala Lyayo*, in the early 1980s, but it appears to have suffered from infighting and inefficiency (Iliffe 1998: 164-165).

The above examples have sought to illustrate how state capacity eroded and how different sections of Ugandan society and external forces responded to this gradually evolving informal privatisation. It is important to note, first, that these different responses had separate historical roots and operated on different levels and scales, and, second, that although they were connected, they were so by default rather than by design. Some reflections on this will be offered here as a basis for the next section.

The contours of different civil society groups and the demands they make are best understood with reference to their history and the context they operate in. The common denominator for the different health providers discussed above (and for most expressions of civil society in Uganda at the time) is that they emerged as coping strategies that sought to escape from the militarised and, in terms of welfare, incapacitated state, rather than as interest groups seeking to impose their demands on it. Most civil society organisations were also severely weakened during the years of turmoil. Or in other words, the main tendency of state-civil society relations from the mid-1970s to the mid-1980s was the weakening of both state and civil society and the disengagement by the latter from the former, as opposed to the expansion of both under capitalist growth. This shall be understood not as distinct shift but as a tendency with undercurrents.

Linkages remained between the state, providers and citizens, and these linkages were instrumental in shaping the reconfiguration of institutions for welfare provision. These institutions consisted of a plurality of health regimes, which combined formal and informal modes of regulation in rather fluid ways. Medical workers' income from private practices subsidised and sustained their public sector work, albeit at low-intensive levels. Hence, conditions for public health care were transformed by parallel health regimes that had been created by coping strategies. These features impacted on the capacity of civil society groups to engage the state even after relative security was established in the central and southern parts of the country after 1986. It also had implications for the inclination and capacity of the state to regulate them, and for the different means chosen for doing so.

As the above description has made it clear, lack of resources undercut regulatory capacity. Systematic policy formulation and implementation were replaced by ad hoc declarations that were only erratically enforced. Deepening informalisation of public health resulted in straddling between formal and informal modes of provision within a policy vacuum. The rather different character of various forms of non-state provision presented shifting challenges for regulation. Not only did fragmentation and institutionalised informalisation undercut the regulatory capacity of the state, it also complicated the possibilities for collective action and bargaining for civil society organisations. Unsurprisingly, these patterns remained pervasive for some time after 1986.

Strategies and responses in the late 1980s

What strategies were chosen by the NRM to respond to the crisis of the health system? During the first half of the 1980s, health policy had been aimed at institutional rehabilitation, supplemented by the introduction of primary health care (PHC) in line with the WHO guidelines as spelt out at the 1978 Alma Ata meeting. These two strands of policy were the outcome of a compromise between priorities as envisaged by politicians and technocrats during the interregnum regimes of 1979-80 (Iliffe 1998: 151). PHC was declared the foundation for the national health strategy in the early 1980s; however, lack of resources and institutional decay in the context of the civil war meant that none of these priorities could be met during the first half of the 1980s, in spite of the available foreign aid. Importantly, the policy shift towards PHC, albeit merely rhetorical, opened the doors for donor funded vertical programmes, something that continued to shape formal health provision structures well into the 1990s.

In 1987, a Health Policy Review Commission was appointed. The commission came up with the proposal of returning to the standards of the 1960s by drastically increasing public expenditure and reintroducing user fees, among other things. Rehabilitation would be supplemented by expansion of primary health care, hence retaining the policy direction of earlier governments. The rehabilitation plan would come at an estimated cost of $203 million. Given the lack of resources, most observers, including UNICEF, had recommended that the government needed to choose between these aims (Macrae *et al* 1998: 1098). As a result of weak coordinating capacity in the Ministry of Health and the low priority given to the sector, it was not until 1991 that the recommendations of the commission were turned into policy (Okuonzi and Macrae 1995: 126).

Meanwhile, health services continued to be provided, although in a policy vacuum. The programme for restored primary health care was slowly and unevenly implemented. To a large extent it consisted of vertical national programmes, which did not necessarily support the reconstruction of a rehabilitated infrastructure for community health care. Multilateral and bilateral donors responded to requests from the new government and reactivated existing programmes or set up new ones.

In view of the weak capacity of the state, most donors sought to bypass it with vertical programmes or projects and in the process created still more parallel structures. Some of the main programmes, such as the – twice revived since the early 1970s – UNEPI (Uganda National Expanded Programme on Immunisation) and the Essential Drugs Management Programme, and the newly established World Bank First Health Project, were set up as distinct bodies and projects managed by separate units within the Ministry of Health (de Torrenté and Mwesigye 1999: 11-12). Much as the services that these programmes provided were desperately needed, the combined systemic effects of the relative dominance of vertical programmes, carrying only weak links to formal statutory structures, were to reinforce fragmentation and donor dependence. That is, while rehabilitation aims were met to some degree, these programmes "did not provide the basis for a functioning health system" (Macrae *et al* 1996: 1100).

The fact that the government opted for policies which could be predicted to promote institutional rehabilitation as opposed to create foundations for primary health care is explained by Macrae *et al* with reference to the political capital invested in symbolic normalisation and visible restoration (Macrae *et al* 1996: 1100). In addition, the government was dependent on all external resources that were offered, especially the low priority given to social sector rehabilitation. On their part, donors were keen to assist with resources to meet vast needs, without attaching conditionalities for their use.

When the recommendations from the Health Policy Review Commission finally turned into the National Health Plan in 1991, the plan was rejected by key donors, and in particular the World Bank, for exceeding available resources. In its place, the government, under the influence of donors, presented a more austere three-year health plan in 1992. While the 1991 National Health Plan elaborated ambitious objectives for expanding and reorganising services over the following decade, the revised version consisted of a minimal package that centred on consolidating rather than expanding services, and on reorienting the system towards PHC (Okuonzi and Macrae 1995: 126-127). World Bank supervision in revising the plan was, according to Okuonzi and Macrae (1995: 127), the starting point for the Bank's leading role in shaping Uganda's health policy. It may be suggested that the government took the opportunity to abandon a state interventionist programme, especially since donors seemed prepared to fund the health sector.

A central feature of state reform, which affected the health sector, was the thoroughgoing civil service reform, initiated in 1990. The aim was to rationalise the sector so as to improve performance. Totally, the civil service was reduced from 320,000 in 1992 to 160,000 in 1995 (Robinson 2006: 16). Central line ministries were to be reduced to "a core of planners, policy makers, advisors and inspectors" (Okuonzi and Lubanga 1995: 9). Retrenchment went hand in hand with a rise in salaries. Nevertheless, low remuneration remained a major problem (Okuonzi and Lubanga 1995: 9). No exact figures exist with regard to health, but there is no reason to believe that the changes were less than dramatic. Investigations had revealed that more than 2,000 ghost workers were on the payrolls of the Ministry of Health (Iliffe 1998: 159).

In spite of the intentions of the reforms, working conditions of medical workers remained poor. The Uganda Medical Association approached the President in the late 1980s with a list of complaints about low pay – the real value of an intern's salary in 1987 equalled $ 4.80 a month (Iliffe: 1998: 166) – and lack of housing, transport facilities and working conditions broadly. Nurses and interns went on strike in Jinja in 1989 and in Mulago in 1990. The government's response was lukewarm, probably because the bargaining position of medical workers was very weak, reinforced by their tendency to use the exit responses to problems. Such exit strategies included the setting up of legal or extra-legal practices, demanding under-the-table payments from patients or going into exile. Hence, the state could rely on creeping privatisation as a strategy to channel the frustration of medical workers.

The very limited resource base thus blocked the rehabilitation of state institutions. The state relied on assistance from international NGOs, bilateral donors and the UN system for resources. Even though, as Brett (1996: 336) shows, health and education expenditures increased in both absolute and proportional terms after 1989, this was insufficient for reversing developments. In recognition of increased inequality, the World Bank in 1990 initiated the Programme to Alleviate Poverty and the Social Costs of Adjustments (PAPSCA), without either much resources or impact.

Health workers' low salaries led to informal charges. With the stated aim to curb this tendency, or perhaps more correctly adapting policy to reality, the National Task Force for Health Financing recommended the introduction of user charges. In 1990, a Bill was tabled in Parliament, but the proposal was met with stiff resistance, not least from the President, and was thrown out. Nevertheless, a legal loophole enabled district authorities to raise user charges if they found it necessary to do so, meaning that national formal policy and local practices again diverged. Domestic political forces did not decide the issue, however. In 1992, the World Bank resolved the standoff by attaching user charges as conditionality for the release of a new loan (Okuonzi and Macrae 1996: 128).

To conclude this section, the main determinants of state-society relations in health during this period were the legacies of informalisation and fragmentation of rather different health regimes as a consequence of civil war and state decay. Or more precisely: as a result of immediate challenges to state authority in the context of structural constraints imposed by lack of resources as well as physical and institutional capacity, the government made security and economic stabilisation its priorities. The state lacked sufficient resources and capacity to enforce regulations or to set up new institutional frameworks or policies. Most expressions of civil society, or of public demands more broadly, were politically weak. To an extent, informal structures functioned as social safety nets. Loyalty or exit options where far more common than voice strategies (Hirschman 1970). Given these constraints, the health sector was shaped by its specific institutional legacy of fragmented health provision systems.

Locally, RCs were given responsibility for some aspects of local service delivery, such as the building of health units. They were on the whole not, however, incorporated into donor programmes (Macrae *et al* 1996: 1100). The trajectory of informal privatisation shaped both state and civil society and reinforced the disconnection between them. Domestic civil society groups had typically emerged as responses to state collapse, as opposed to being expressions of political struggles, something which of course shaped the parameters of their response. Furthermore, there were vast unregulated fields of informal welfare regimes. This shaped both liberalisation and control efforts: economic liberalisation was used as a strategy to adapt to the very structures of informal privatisation that continued to undercut control efforts.

Formal privatisation attempted to address some consequences of informal privatisation, but this was by its very nature complicated by the interpenetration of formal and informal modes of provisioning. The state aspired to reclaim control over these providers by balancing its aims of extended privatisation of services and preserved political control. This also sought to ensure support among the population through the political legitimacy that is invested in capacity to directly or indirectly meet demands for services. The reshaping of state-civil society relations generated new sources for the creation of state legitimacy. In order to rein in groups in civil society, legal and administrative means of supervision were employed, such as the 1989 NGO Registration Act and the introduction of a National Board of NGOs.

With regard to welfare provision, however, the NGOs and CBOs that had proliferated in the new context of relatively relaxed control were not disturbed as long as they remained politically loyal to the new government. Effective government control over (especially foreign) non-governmental actors, and even more so, over informal health systems, was uneven. It proved very difficult for the state to regulate the many forms of informal exit strategies, something which reinforced the fragmented landscape of fluid and very often informal welfare regimes. This underlines the shortcomings in terms of institutional incoherence and resulting problems of efficiency, quality, access and sustainability of services.

Conclusions: State formation until the early 1990s

One immediate observation that follows from this account concerns the uneven and unstable nature of this political project. Given the depth and complexity of the crisis, this is hardly surprising. The NRM came to power with a very limited political base and inherited a state with only rudimentary institutional structures and a minimal resource base. Internationally the new government was received with suspicion. It was thus equipped with rather limited means to sustain legitimacy and control.

The uneven process of consolidation stemmed from the very different challenges to the state in different regions and the uneven inclination and capacity by the state to meet these. This followed both from the balance of power among social forces seeking accommodation, recognition or protection in the new political order and from the institutional incoherence of the state. Hence, the question to which extent the new government managed to root the state in society needs to be disaggregated with regard to differences among state organs and is linked to the question of how competing social forces managed to root themselves in the different apparatuses of the state. Another aspect of the uncertainty that marked this period was that many institutions and arrangements, including the informal agreement on what was stated to be a temporary ban on party-political activities, were proclaimed to be transitory, awaiting the writing of the new constitution.

For the purpose of consolidating its hold on power, the NRM had to broaden its political base by incorporating other sections of the political elite. The negotiations and concessions necessarily involved reproduction of political conflicts inside the state. In the case of economic reform. This resulted in rapid pendulum swings, which in a broader perspective were reflective of a lack of direction, due to a deadlock between contradictory interests.

The chief means by which NRM countered the political ambitions of competing elites was to stifle them at the national level by placing a ban on party-political activities and to by-pass them in rural areas by building political institutions at the local level. During these early years, RCs were the real innovation with regard to the form and content of the state, but RCs too were characterised by instability. Their effective democratic content was circumvented by restrictions on political freedoms, and defined by the local balance of forces entering into them.

In order to balance strategies for accumulation and legitimation, the government undertook different measures. However, instruments such as military coercion and political concessions were occasionally difficult to reconcile and contributed to deepened institutional incoherence and regional cleavages. While it is correct to say that this period saw the state regain some of its relevance and a degree of purpose and coherence to its activities in the south and the west, and that improved security and political incorporation were the main assets in this process, coercion, exclusion and the lack of security was the main experience and the defining feature of state-society relations in the north.

The conclusion that state formation is a contradictory process with undercurrents and unintended consequences may seem self-evident. Nevertheless, it is fundamental to understand that state formation in Uganda since 1986 has been characterised by regional and institutional disconnections at different levels, which appear to reproduce rather than to overcome the historical trajectory of underdevelopment and regional conflicts. Many issues remained in the balance in the early 1990s. The next chapter discusses the relative stabilisation of these fragile processes and the consolidation of state power that characterised the mid- and late 1990s Uganda.

Chapter 5

State and Civil Society since the Early 1990s: Consolidating the "Governance State"

Recently, the Ugandan showcase model of transformation under the NRM government (as presented by e.g. Brett (1998); Nsibambi (1998b); Collier and Reinikka (2001); Villadsen and Lubanga (1996)) has been challenged by a more critical interpretation of current Ugandan politics. According to the latter perspective, the incumbent government initially brought about successful change, but its performance then declined, and contemporary politics has relapsed into being no less patronage ridden than under earlier governments (Barkan 2005; Robinson 2006; Tripp 2004).

While the critical literature identifies certain problematic aspects of Uganda's politics, and by and large offers a valid description of recent changes, it fails to explain the sources of these changes. For instance, it too neatly demarcates subsequent phases of "success" and "decline" and links these phases primarily to the will of political leaders (and more precisely President Museveni) to reform and the lack of such will, respectively.

In order to explain the sources of change, I propose that while dramatic manifestations of political authoritarianism are more recent, its causes can be traced to the political economy of the earlier years of the NRM in power and to the way dominant political coalitions shaped the state. Continuity and change with regard to state-society relations have coexisted all along, and they continue to be highly unevenly articulated with regions and state institutions. Economic liberalisation has indeed transformed Uganda's political economy, but in ways mediated by the specific Ugandan context and not necessarily as described in the "success story" literature. It is therefore important to analyse the interplay between continuity and change and between democratisation and authoritarianism, as manifested in the state.

The underlying argument of the proposition offered above is straightforward. As has been argued throughout this book, all capitalist states are formed within some given constraints of inherited contradictions of capitalist development and by the competing political forces and the demands these contradictions generate. In their attempts to accommodate and balance requirements of resource extraction and political control, ruling groups combine coercion and consent.

Clean breaks with the past or "fundamental changes", as has been suggested with reference to the case of post-1986 Uganda, are rare indeed. The partial changes that do occur are rooted in and limited by historical trajectories. In order to bring out strands of continuity and change with regard to authoritarian and democratic politics, this chapter the contradictions underlying the reconstruction of the Ugandan state from the early 1990s.

Such contradictory features have been manifest also in the health sector. While Uganda has been hailed internationally as a role model in fighting HIV/AIDS, most other basic health indicators, such as high maternal and infant mortality rates, point in other directions and suggest that the health system suffers from enormous challenges and shortcomings, many of which have deep roots.

The present day problems within the health sector range from factors at the macro level, including unclear and shifting policies and bureaucratic management, to more context-bound effects, such as inconsistent implementation, and the shortage of medication, equipment and staff. General problems in health centres and hospitals include absenteeism, moonlighting and brain drain, both to abroad and to other sectors (Brett 1996; de Torrenté and Mwesigye 1999; Iliffe 1998).

In this chapter, I examine the character of welfare and health regimes as these have evolved since the early 1990s. This is undertaken in tandem with a critical of the overall development of the state, and I discuss the role of health and welfare in rendering stability and creating legitimacy to the Ugandan state. The first two sections offer an overview of the general changes in state-civil society relations and public sector and health sector reforms, respectively, including the role of donors in shaping the state. Following on, the third and fourth sections interrogate the case studies of two health regimes, in terms of state-civil society relations in the health sector through the examples of the medical workers union and NGO oriented incorporation into "participatory" poverty reduction policy processes.

I conclude that while formal openings for participation have been created, those political spaces were in the process depoliticised and detached from issues of influence and power in the case of NGOs – or not utilised because of weakened bargaining power in the case of Uganda Medical Workers Union

(UMWU). The weakness of democratic forces at the level of civil society politics also contributes to explaining the prevalence and expansion of ruling class excesses in terms of corruption and repression, and illustrates the continuity of authoritarianism.

Formation of state and civil society from the early 1990s

For the purpose of interrogating the issues of how external and internal political forces have tried to shape the state, this section will explore important changes that have shaped state and civil society from the early 1990s and in what ways these changes were rooted in history. To what extent did the changes impact on the prospects for democratising the state?

The following two sections argue that the broad tendencies of state formation during this period are best captured by notions of consolidation, expansion and restructuring. I will discuss these tendencies through an overview of recent political change that runs from the politics of constitution making and the consolidation of Movement structures, over state reform and privatisation to emergent contradictions of the current development model and challenges to the NRM state.

Constitutional politics

Already during the bush war, the NRM asserted its intentions to write a new constitution once in power. In 1989, the Uganda Constitutional Commission (UCC) began its work but it did not submit its report and a draft constitution until late 1992 (Amaza 1998: 174-175). The most contentious stance of the report and the draft, and one that would shape political debate in Uganda over the following decade, was the recommendation by the UCC that Uganda should continue with a "movement" political system as distinct from both a one-party or a multiparty one. This recommendation had followed upon the president's clearly stated wish to ban political parties permanently with the argument that parties inevitably exacerbated sectarian divisions in a pre-capitalist society and hence created destructive instability (Khadiagala 1995: 40).

The movement system was conceptualised as one in which political participation would be ensured through representation based on what was called individual merit within an overall system of unity and accommodation. According to the UCC proposal, the system would operate for five years, after which the question of political system would be subject to a referendum (Oloka-Onyango 2000: 47). Unsurprisingly, this proposal encountered strong resistance from DP and UPC, who saw it as a contravention of the earlier agreement to limit the restrictions to political party activities to a transition period.

For that and other reasons, mainly the connected issues of federalism vs decentralisation,[9] land, and the role of traditional rulers in the wake of the 1993 restoration of the Buganda kingship, the 1994 elections to the Constituent Assembly (CA) that was to debate the proposal, became intensely politicised – "[e]thnic, religious, regional and party factors combined to make the CA elections a referendum to some degree on the NRM government" (Kasfir 1995: 150).

Whereas delegates were formally elected to the CA on an individual basis, the combination of the above identities, and mainly party vs Movement affiliation, implicitly underwrote their campaigns. The contentious issues came to dominate the CA debates in ways shaped by political loyalty to either camp (Oloka-Onyango 2000: 45-53). The result of the CA debates as included in the 1995 Constitution were to reinforce the UCC proposals, which also happened to be the positions taken by the government – namely to constitutionally safeguard the movement

[9] Federalism and decentralisation are evidently logically compatible forms of devolution. The contentious issue revolved around Buganda's specific demand for a federal status which would also involve a more pronounced political role for the kingdom.

system for a period of five years after which a referendum would take place to choose between decentralisation over federalism as a model for deconcentrating the political system.

A significant observation is that the victory for the government line on these fundamental points resulted in a political turning point: the decisive consolidation of NRM control over state power and the political arena at large, something which was further reinforced by the 1996 presidential and parliamentary elections, in which president Museveni and candidates associated with the NRM won by large margins. Donors endorsed the constitutional change and electoral victory of NRM, and thus confirmed the Movement hegemony, political parties were contained and most civil society organisations chose to adapt to the prevailing order. These major political achievements signified a high point of convergence between successful legitimation and control for the NRM government.

At the same time, consolidated control had come at the cost of cracking political coalitions. The legal establishment of the Movement system had in the process revealed the growing disjuncture between its theory and its practice – "the CA deliberations and outcome were the last nail in the coffin of 'broad-basedness' which was ostensibly the fundamental criterion on which the NRM had been constructed" (Oloka-Onyango 2000: 52).

Not only had the multipartyist resistance pointed to the limits to consensual accommodation, opposition to the Movement line among a number of important NRM insiders but had also revealed emergent conflicts within the NRM itself, conflicts that would result in deep splits and eventually in the formation of the Forum for Democratic Change (FDC) party a decade later. And as the NRM was closely interwoven with state structures, such political conflicts would throughout the period of no-party politics be played out as conflicts within the state. With hindsight, sharpened political conflicts in the aftermath of the constitution making process can be said to have foreboded the increasing use of control and coercion, at the expense of waning consent. Political consolidation also interplayed with changes in economic and institutional reforms, to which we now turn.

Economic liberalisation and state reform

Contrary to countries such as Kenya, where the main aim of adjustment programmes was to dismantle the role and resources of the state, Uganda, along with a few other cases such as Ghana (Hutchful 2002), experienced the simultaneous moves towards economic liberalisation and a growing scope for a reconstructed state.

This observation immediately calls for qualifications, as the Ugandan state was and remains extremely dependent on external revenue for its existence and consequently found itself operating within the fairly narrow restrictions for macro-economic policy set by the IFIs. Nevertheless, the means exercised by the latter in enforcing their aims in Uganda gradually came to centre on incentives and rewards rather than on strict conditionalities and punishment (Harrison 2004). This was mutually reinforcing: the more Uganda lived up to donors' expectations, the greater the rewards in terms of financial and ideological support – and consequently the greater the scope for domestic political control.

What explains the scope for state reconstruction? One reason was that the Ugandan state was so run down by decades of destabilisation and informalisation that it was obvious already from the late 1980s that it needed to be rebuilt. Another reason suggested above was Uganda's relatively successful macro-economic performance. Uganda's strict adherence to the adjustment agenda, particularly from the early 1990s, made donors place trust in the government's inclination and capacity for reform. This coincided with a gradual shift in emphasis in the international development orthodoxy towards a greater reliance on a relatively downsized and reconstructed state, capable of promoting capitalist development by creating an enabling environment through enhanced institutional stability.

In the words of Brock *et al* (2002), "policy communities" emerged around a common ideology and shared interests among key donors, political leaders and top technocrats in core ministries. Still another way to say this is that vertical constellations of political forces straddled external-internal divides as important aspects of Uganda's state formation were incorporated into a "transnational state" in the making (Robinson 2004). This occurred through discursive affinity, governance methods and techniques and donor-funded expatriate advisory personnel in the state (Harrison 2004: 89-90).

This mutual accommodation between donors and the Ugandan state evolved over time. As shown in Chapter 3, it was not until the early 1990s that the pro-liberalisers in the government got the upper hand in the internal NRM struggle over reform directions. From then on neo-liberal orthodoxy reigned supreme among the top cadre within the powerful Ministry of Finance, Planning and Economic Development (MoFPED), a super ministry created in 1992 after the sacking of a Minister of Finance that was hostile to the collaboration with the IFIs. Another ministry that towered above the rest thanks to its allotted coordinating role during the early days of institutional reforms was the restructured Ministry of Public Service (MoPS) (Harrison 2004: 85-87). Together, the two ministries bore the operative responsibility for the restructuring of the state.

One of the first major administrative reforms to be carried out was the Civil Service Reform Programme in 1990, the aims of which were to drastically restructure the civil service. The main components of the programme, at that stage closely linked to the adjustment logic, included the downsizing of ministries and retrenchment of staff, primarily of "ghost workers", and included donor funded redundancy packages. Other aspects that came to the fore during later phases of the programme included personnel management and pay reform (Kjær 2001: 15).

The reform programme had the strong backing of the president and could therefore be rapidly implemented with the result that the size of the bureaucracy was reduced dramatically. During the same time, the military was downsized from 90,000 to 50,000 in a demobilisation programme (Robinson 2006: 16). Retrenchment met only limited resistance, much due to donor funded redundancy packages (Brett 1995) and the fact that most civil servants had already developed coping strategies during previous decades.

Success as measured in relation to the stated aims helped to earn the Uganda government the status of a good adjuster (Harrison 2004: 85). It was amply rewarded. From the late 1980s until 1996, foreign development aid averaged $500 million annually and $800 million after 1996. Donors and creditors have for years contributed a vast percentage of the state budget: around 50 per cent, or roughly 10 per cent of GDP, and 80 per cent of development expenditures throughout the 1990s (Tangri and Mwenda 2006: 122, notes 4 and 5). The fiscal basis of the state was radically transformed.

After having downsized the central bureaucracy, the government embarked upon a decentralisation programme, which gradually re-expanded the size and cost of public administration. Several factors shaped decentralisation reforms, including the introduction of Resistance Councils during the guerrilla war. The legislative framework for decentralisation was provided by the Local Government Statute of 1993, the 1995 Constitution and the Local Government Act of 1997, which converted RCs into LCs.

Apart from wishing to restore state capacity, another dimension of this shift was more directly political, as the NRM leaders were known not to have trusted the political inclination of inherited civil servants and were happy to replace these with more reliable cadre in the local state (Kjær 2001: 138-139), functionaries who could furthermore anchor the NRM among local communities. Both civil service and decentralisation reforms were policy arenas where donor and government interests coincided.

The inflow of development aid contributed to rapid economic growth from the early 1990s. Annual growth rates averaged 7.4 per cent between 1994 and 1998 (Harrison 2004: 75) and were combined with one-digit inflation rates. International investment added to economic expansion. Asian industrialists and business people, expelled by Amin, were asked to come back, and their confiscated properties were returned to them. A boom in coffee prices during the mid-1990s contributed to rising real income and falling poverty numbers throughout the 1990s. Together, all of this added to real improvement among a large section of the population in central and western Uganda, something that was translated into political support for the NRM.

One aspect of economic reforms that contributed to the opposite, meaning undercutting support for the NRM, was the privatisation programme, initiated in the early 1990s but not fully operating until a few years later – that is, when the NRM had already secured a higher degree of political control and insulation from opposition politics. By and large, privatisation of parastatals was manipulated by and for an emerging section of the ruling class with political connections, and it seriously tainted the reputation of the government (Tangri and Mwenda 2001; World Bank 1998). When discontent over this was linked to other grievances, latent political conflicts were set in motion.

The NRM state consolidated and challenged

The Movement system was promulgated into law with the contentious 1997 Movement Act, the details of which made it very difficult to separate the Movement from the formal structures of the state, or indeed from the informal structures of the NRM. The state-funded Movement did not only embody monopolised institutional forms for all political activity at every level, but it also incorporated various entities of the state and in society, such as the military, the police, trade unions and private sector bodies (Oloka-Onyango 2000: 56-57). To its critics, the Movement had been turned into a one-party system in all but name (Human Rights Watch 1999).

The constitution had spelt out that the issue of the choice of political systems would after five years be subject to a referendum, due to take place in 2000. This referendum was highly contested and was boycotted by most proponents for the restoration of multiparty politics, not only on the grounds that it was bound to be conducted under unfree and unfair conditions, but also that the referendum, according to the opposition, was fundamentally illegitimate in itself. The referendum, boycotted by the opposition, was held in June 2000 with a low turnout (51 per cent), with the outcome that the Movement system remained in place.

The reinforcement of the Movement corresponded to an increased control over other political forces. After many years of struggles in parliament, the likewise controversial Political Parties and Organisations Act of 2002 effectively restricted all activities normally associated with political parties. Civil society organisations were also subject to tightened control. The 1989 NGO Registration Statute has recently been replaced by the Non-Governmental Organisations Registration (Amendment) Bill, 2006, to the effect that the NGO registration board would include representatives of security organisations. These pieces of legislation obviously strengthened the Movement system's formal and the NRM organisation's real political monopoly. But the fact that they were deemed necessary in the first place pointed to growing resistance and deepening, but for some time contained, cleavages emerging also within the NRM itself.

What did such resistance consist in? In brief, it revolved around related accusations of corruption, political repression and militarism. As was mentioned above, the privatisation process which had gained momentum from the mid-1990s was tainted by scams that involved individuals with their base in politics, the military or business, and all of them with close links to the very inner circles of power, including crucially the president and his family. The fact that action was rarely taken against high level corruption even when this was clearly documented bred allegations of nepotism and impunity (Tangri and Mwenda 2006).

The corrupt practices on behalf of networks within the politics, military and business nexus alluded to above were deeply entrenched in commercial activities that related to the military, including influence peddling in procurement processes, the creation of "ghost soldiers" and the plundering of gold and other natural resources in the DRC; the latter was also linked to complicity in grave human rights abuses. The interventions of UPDF in DRC was also highly contested in itself (Human Rights Watch 2001b; Nganda and Tumusiime 2005; Tangri and Mwenda 2003; United Nations 2002).

To critics, such manifestations of primitive accumulation not only confirmed that the military still played an uttermost important role in Uganda's politics but also that accumulation of economic wealth and political power were closely linked. In this light, restrictions on political freedoms and the entrenchment of authoritarianism were seen as means of containing resistance to both corruption and militarism. Due to shrinking political space, resistance was only sporadically voiced and did not result in any major organised opposition throughout the 1990s.

The decisive change came with the challenge by Kizza Besigye to president Museveni in the 2001 presidential elections. For the first time, Museveni was challenged by an insider who shared his most important power bases – the NRM, the military and the southwestern region – and who represented growing discontent among these bases. The strong support for Besigye caught the government by surprise and shock, and it responded with coercion.

The 2001 elections were the most violent in Uganda's history. Opposition candidates were intimidated and beaten up by the military and (extra-)constitutional security organs; some were even killed. Besigye appealed against the election result, and the Supreme Court ruled that while the elections had not been free and fair, there had not been sufficient rigging to significantly affect the outcome. Besigye left for exile in South Africa, but violent persecution of political opponents accelerated (Human Rights Watch 2004a).

The government's coercive response needs to be understood in a context where sufficient political loyalty could no longer be cultivated or maintained. Nevertheless, both the 2000 referendum and the 2001 elections made it clear that the Movement system had outlived its usefulness for purposes of containing political dissent. It became increasingly difficult to accommodate the political contradictions that emerged from the specific model of donor driven capitalist development resting on economic liberalisation and political repression.

In order to pre-empt a real rift within the NRM and the military, the government in March 2003 decided to open up for political pluralism in implicit exchange for a change in the constitution to the effect that presidential term limits would be abolished so as to allow Museveni to run for the presidency again. Another aspect of the government's change of mind was its need to reinvent its legitimacy vis-à-vis donors, who showed clear signs of running out of patience. Political liberalisation was therefore essentially a strategy for reconsolidating political control.

The period between 2003 and the 2006 presidential and parliamentary elections were marked by a rather bitterly contested and narrow political transition process to multi-party politics, marred by violence. Opposition politicians and supporters were frequently harassed. During the same time, a change occurred in the relations between Uganda and important donors, some of whom now publicly voiced their concern and occasionally froze or cut levels of development aid.

Corruption and rising levels of administrative public expenditure was one concern and human rights violations was another. Also, the northern war – unlike before – was consistently discussed as a national and regional problem by both domestic and international actors. NRM's and Museveni's international reputation deteriorated dramatically, but as yet, donors appeared to remain committed to Uganda. Besigye returned to Uganda in October 2005 in order to be elected party leader for FDC, only to be arrested and charged with treason and rape a few weeks later.

The February 2006 elections resembled the preceding ones in that president Museveni and the NRM emerged victorious amid accusations of rigging. Again, responding to the opposition's appeal of the Electoral Commission's verdict, the Supreme Court ruled that elections were not free and fair. However, although not unfair to the extent that the outcome was significantly affected. In 2011, Museveni and the NRM won again, this time by wider margins. The NRM continues to rule Uganda, but it now presides over deep structural problems, including stagnating agricultural and industrial production. It remains to be seen how the configuration of economic and political contradictions will affect the conditions for future reproduction of state capacity and legitimation.

State and health sector reforms from the early 1990s

The effects of civil service reforms on the health sector were discussed in Chapter 3. The Ministry of Public Service coordinated and executed a restructuring of the Ministry of Health that involved the retrenchment of a great number of employees. The emphasis on downsizing and cost-cutting was very much an integral part of the adjustment logic of the late 1980s. In 1997, a second Public Service Reform Programme commenced with a focus on Results Oriented Management. It contained grand intentions to restructure the institutional set-up of the public sector through the introduction of "more than 100 proposed autonomous and semi-autonomous bodies". It seems, however, that these plans never really took off (Harrison 2004: 86; Therkildsen 2006: 65).

Privatisation and decentralisation are in their programmatic variants components of the New Public Management agenda with its emphasis on management decentralisation and the introduction of market mechanisms in the public sector, such as results oriented management and outsourcing of service delivery. These technical models are underpinned by a broader reformulation of the role of the state.

The concluding recommendations of the revised 1993 National Health Plan were that the state's role in the health sector should be redefined from provider to policy maker and supervisor of various private providers, and to the extent that the state did provide services, it ought to concentrate on preventive care (Okuonzi and Macrae 1995). As should be clear from previous chapters, however, privatisation in the health sector was neither only nor straightforwardly imposed by donor or government policy. It was an incremental and informal process with multiple roots and manifestations.

Deliberate and designed privatisation was just as much an adaptation to the reality of longstanding informal privatisation as it was a distinct expression of certain policy preferences. Formal privatisation in the health sector thus took on many forms, from the recognition of different kinds of private providers, over outsourcing of public agencies and activities to specific reforms such as the introduction of user charges, as described in chapter 3. We shall return to the issue of private providers below.

Decentralisation of health services began in earnest in 1993 as a component of the overall decentralisation programme. Both the general institutional framework of the state and different policy sectors have been decentralised with reference to fairly typical justifications, such as bringing government closer to the people, and improving accountability and efficiency in service provision, or phrased as "improved service delivery … more efficient and effective …performance-based … responsive … [and] … transparent" by the Ministry of Public Service in 1993 as quoted in Therkildsen (2006: 56). Just as with privatisation, decentralisation reforms have multiple roots in the Ugandan political economy, with consequences that go far beyond the stated intentions of reforms, as will be discussed in chapter 5.

As political decentralisation of the RC system merged with administrative decentralisation from 1993 and did so in a broader context where political consensus started to give way to polarisation over the issues of pluralism, the identity of local governments gradually changed from an early official emphasis on popular democracy to more typical references to improved service delivery.

Administrative decentralisation transferred power over decision making, including budgets, to local governments. As the role of central government ministries was at the same time transformed to a function of policy development and supervision as discussed above, the main responsibility for planning and executing service delivery shifted from ministries to local governments. Districts and sub-counties were in turn given responsibilities to collect and, to some extent, utilise revenue (Jeppsson 2002: 2054-2055).

Just as with many other institutional reforms carried out in Uganda during the 1990s, decentralisation soon displayed contradictions between policy and reality and was subject to change (Jeppsson and Kisubi 2001). A fundamental problem proved to be the not uncommon mismatch between authority and resources. For most districts, local revenue was very modest indeed. Following the need to supplement local resources and the ambition to counter possible inequalities between central and peripheral districts, the pendulum swung back from an experiment with fiscal decentralisation towards a greater degree of conditional grants from the central government (Francis and James 2003: 327, 334).

Yet another reform aimed at bringing the central government back and superseded the vision of far-reaching decentralisation. As the idea that development aid should be coordinated and channelled through the central government gained in strength among donors during the late 1990s, codified in the so called "Paris Agenda", financial aid increasingly came by way of direct budget support, or "basket funding". The thrust of this reasoning was to bypass the fragmentary effects of project support and instead via a Sector-Wide Approach (SWAp) strengthen central government capacities for budget control and for sector policy formulation and in extension also its political "ownership".

SWAp was introduced in Uganda's health sector in 1998. It soon gained momentum and strengthened the role of the Ministry of Health and the central government in general vis-à-vis both local governments and NGOs (Jeppsson 2002: 2055-

2056). This process was in turn linked to the advent of Poverty Reduction Strategies (PRS), which will be addressed below.

Health regimes: State-civil society relations in the health sector

How did the state attempt to root itself in society with regard to health policy making? How did various interest groups seek to link up with the state? To what extent did they create access to the state and scope for shaping its capacity through policy choices and political regulation of interest representation? This section the cases of Uganda Medical Workers Union (UMWU) and NGO participation in the Poverty Eradication Action Plan (PEAP), in that order. The experiences are situated within the contexts of state-labour and state-NGO relations. It needs to be recalled that these examples are only two of a great many state-(civil) society relations in the health sector. Other health regimes, particularly in remote areas, are as indicated not primarily regulated by the state but by economic and cultural relations.

State-labour relations

In this section I address the constitutive features of state-labour relations in Uganda, historically and under the present NRM government. The trade union movement in Uganda has historically been comparatively weak. The relative lack of impact by the Ugandan labour movement has been explained with reference to limited industrialisation and a numerically small working class, internal shortcomings of labour organisations and the general features of state control over autonomous interest groups introduced by the colonial government and taken over by post-colonial legislation (Mamdani 1994).

Trade unions were subject to supervision and regulation and largely failed to resist these arrangements. Colonial legislation served the purpose of nipping any potential radical politicisation of unionism in the bud. Correspondingly, the colonial state cultivated a conservative leadership (Barya 1991: 32; Mamdani 1994: 529). According to Mudoola, "The ICFTU and the British unions devised an indoctrination programme for the trade unions through training leadership abroad" (Mudoola 1993: 62).

Another aspect that weakened the labour movement was the strong localist and traditionalist tendencies that fragmented the nationalist movement. Ethnic and religious bases of politicisation overtook other forms of social mobilisation. As we have seen, central-local struggles continued unabated after independence. Much political mobilisation outside ethnic and religious organisations tended to feed into the latter due to their perceived clout. This tendency divided trade unions along ethnic and religious lines (Mudoola 1993: 61). Divisions were further encouraged and manipulated by factions within the UPC government after independence:

> tensions within the ruling party spilled over into the trade union movement and vice versa and as a result a socio-political force which could have sustained constitutionalism became a victim of party-political considerations and factional politics (Mudoola 1993: 61).

These conflicts were further deepened by trade unions' dependence on outside financial support and subsequent factional East-West affiliations during the cold war (Barya 1991: 41-43).

Seeking to contain workers' militancy and the alliance between the radical Uganda Federation of Labour (UFL) and the left wing of UPC and its Youth League in the context of mounting political tensions, the government introduced a new legal framework in the mid-1960s (Barya 1991: 16-17). However splits and mergers continued. The swiftly changing national political landscape sealed the fate of autonomous trade unionism. Following the 1966 crisis and the subsequent concentration of power, parliament passed the 1970 Trade Union Bill, which restricted unionism to the Uganda Labour Congress (ULC), closely controlled by the government (Mudoola 1993: 72).

The Amin government substituted this Act with the 1976 Trade Union Decree and replaced the ULC with the National Organisation of Trade Unions (NOTU) (Barya 1991: 20-24). On paper, this was a more liberal law than the 1970 Trade Union Act, but general political repression and economic collapse made it next to impossible to make use of its openings (Barya 1991: 34-35; Mamdani 1994). The effects of Amin's "economic war" in the early 1970s, and the expulsion of Asian industrialists dealt a blow to industrial production, the public sector and the entire national economy and, consequently also to the trade union movement. In spite of being undermined by structural changes and internal weaknesses, workers still resisted oppression and divide-and-rule interventions under the Obote II government (Barya 1991: 36-38).

The Uganda trade union movement has experienced hardships during the post-1986 period. Although labour legislation has been liberalised under the NRM government and the ban on all public sector workers to unionise has been lifted, the trade union movement suffers both from its legacy of internal divisions and a weak centre and from the impact of economic liberalisation. The 1993 Trade Union Law and the provisions on workers' freedom of association in the 1995 Constitution have created legal space for trade union activities, but trade unions have largely been unable to make use of this space. Not only has privatisation and public sector retrenchment weakened unions' bargaining power through eroded membership and financial basis of NOTU and its affiliated unions. Repression was added to marginalisation.

The government has come out firmly against the unions' protests in favour of international investors. Workers' demands are met with reluctance or outright hostility from employers. Formal rights, where these exist, are often neglected in practice by employers (Barya 2001). It may also be suggested that decentralisation has further hampered the unions' capacity to organise and bargain.

In line with its general strategy of incorporating interest groups, NRM in 1989 proposed to introduce workers' representation in parliament, to be recruited from a newly created body. The creation of a new organisation was resisted successfully by NOTU, which sought to preserve its autonomy and unity. Similarly, the government in 1993 attempted to divide workers along blue-collar and white-collar lines; again, this was resisted (Barya 2001: 50).

Workers were granted parliamentary representation, elected from different trade unions within NOTU. Barya (2001: 34-40) discusses the contradictory effects of such representation. On the one hand, workers' MPs have been able to call attention to workers' interests and more specifically to counter, at least to some degree, the effects of privatisation and retrenchment, for instance through the Workers' Compensations Act of 2000.

On the other hand, parliamentary representation lacks, according to Barya, an "organic link with the trade union movement, apart from their individual unions [...] Thus, there are no rules as to whom they should report or account, with whom or when and how they should consult" (Barya 2001: 37). In other words, with severed or only weakly institutionalised links to the constituencies they represent, workers' MPs have been vulnerable to political manipulation and co-optation, particularly in the sixth (1996-2001) and seventh (2001-2006) parliaments when they were torn between the principles of representing the constituency of all workers in Uganda and acting in line with the politics of "individual merit".

The case of Uganda Medical Workers Union

Prior to the 1993 Trade Union Law, public servants were not allowed to organise themselves in unions. "White-collar" employees' interests were supposed to be channelled through professional associations. In the health sector, the professional association, the Uganda Medical Association (UMA), was assumed to function as a gatekeeper up- and downward.

While established as a professional association to cater for the interests of its members, it was also expected to regulate the field, broadly according to the wishes of the Ministry of Health. The Minister of Health, for instance, claimed in 1996 that "the medical command structure is like in the army" (*New Vision*, 15 July 1996). In the context of a fragmented and partly unregulated health system, UMA had little capacity to exercise such a function.

Fragmentation and informalisation of the health system into overlapping fields of provisioning, lack of resources in the public health sector and social change had over a generation drastically undermined the position of medical workers in terms of both income and status; their professional identity was confused when inherited ideals were so very far detached from reality (Iliffe 1998: 165).

Uganda Medical Workers Union was formed in 1992 by a number of doctors and nurses in Entebbe Hospital, as a response to the long-standing decline of working conditions and salaries of medical staff (interview, Apollo Nyangasi, UMWU National Chairman, 25 February 2003). It was registered with the Registrar of Trade Unions in 1994, following the legal opening for public sector unionisation (Barya 2001: 16-20). Nevertheless, public sector unionism was a new concept that initially was not received with much enthusiasm from the Ministry of Health.

After some fruitless efforts to get official recognition and to be heard, the union "saw no other way than going for industrial action" (interview, Apollo Nyangasi, 25 February 2003). Repeated strikes had taken place in different hospitals throughout the country since the late 1980s, but so far only by paramedics, interns and nurses. Doctors, it was assumed, were above such irregular behaviour.

Strikes in 1994 had led to promises of pay rise, but when the government failed to meet its promises, UMWU called for a strike in September 1995. The strike, which lasted for two weeks and spread all over the country, was declared illegal by the minister of Health, and was finally disrupted by force after intervention by the president (*New Vision*, 28 September 1995). After patients had died due to the absence of care, public opinion started to turn against the striking workers. Three union leaders were arrested, and others were transferred (*Monitor*, 25-27 September 1995). Victimisation of activists continued, according to the union (interview, Apollo Nyangasi, 25 February 2003).

Negotiations were promised over the union's demands for improved working conditions, higher salaries and legal recognition, but none was fully met The main concession from the government came in form of lunch allowances. A committee drafted an agreement on future negotiations. This report was later shelved (interview, Apollo Nyangasi, 25 February 2003). A legacy of the strike and a reflection of the general government attitude towards labour rights is that the government has still not signed any recognition agreement with the union. The absence of institutional mechanisms for bargaining or for settling of disputes evidently undercuts union sustainability in a fundamental sense – "for each round of negotiations there are ad hoc committees … our asset is mainly threats of industrial action. It's a set-back" (interview, Apollo Nyangasi, 31 January 2003). Even more seriously, there is the continuous risk that the union can be outlawed.

There have indeed been threats of further strikes, and one was carried out in 1997. The government has mostly managed to negotiate temporary solutions through promises of meeting demands, promises which have never been kept. The absence of a negotiation machinery takes a lot of time and resources from what should be routine work, such as recruiting members. It has also prevented UMWU from engaging consistently in policy discussions around the public sector reform and reforms of labour law.

Decentralisation has further complicated the union's work. Out of approximately 10,000 members, only about 4,000 pay their membership fees regularly. In some districts, UMWU has managed to negotiate check-off agreements with district administrations, where employers deduct union fees at source, but this depends to a high degree on the location of the district and on personal relations and leadership skills on both sides (interview, Apollo Nyangasi, 25 February 2005).

The conflict between UMWU and the government also highlights a certain degree of incoherence internal to the state, deliberately utilised or not, as demands for recognition and pay are recognised by one ministry, the Ministry of Health, but blocked by another, the Ministry of Public Service. Just as in the case of many other interest groups, this has caused UMWU to turn directly to President Museveni, something which may be useful as a one-off tactical move but which creates ties of dependence that undermines sustainable strategic achievements.

UMWU's struggles have not only been fought with the government. It has also been deeply involved in leadership wrangles within NOTU, conflicts that eventually led to the breakaway from NOTU in 2003 by five unions, including UMWU, and to the formation of an alternative national centre, the Confederation of Free Trade Unions in Uganda (COFTU) the following year.

The Secretary General of UMWU, Sam Lyomoki is also the Workers' MP, and aspired for the leadership of NOTU. Elections were prohibited for a long time following a government ban on trade unions holding general meetings. When elections took place in October 2003, the circumstances were controversial which then led to the break off from NOTU. The Commissioner for Labour in the Ministry of Labour refused to recognise another trade union centre with reference to the 1976 Trade Union Decree. UMWU claimed that the decree had been overtaken by provisions in the 1995 constitution, filed a petition to the government over the matter and took the Ministry of Labour to court, a court case they won in early 2004. There have been attempts to start a splinter union, Uganda Nurses and Midwives Union, but according to UMWU Nyangasi (interview 17 February 2004), this was a failed attempt by NOTU to block COFTU from taking off.

Rivalry within NOTU and later between the two aspiring trade union centres has to a great extent paralysed the trade union movement and could easily be used by the government in divide-and-rule politics. The fact that Sam Lyomoki is an MP and harbours personal political ambitions cuts both ways for UMWU, pointing to the general problem with workers' MPs as stressed by Barya (2001). Like all other MPs, Lyomoki found himself dragged into the succession debate after 2003, and like most others, he came out favouring the lifting of term limits. Likewise, Nyangasi in 2005 declared his ambition to stand for parliament in Tororo on an NRM ticket, but lost, something that may have undercut his position in relation to both NRM and medical workers. These examples in turn illustrate the weakness and vulnerability of most trade unions in relation to the government.

State-NGO relations

Some reasons for the virtual explosion of NGOs since 1986 were outlined in Chapter 3. The contours of the NGO landscape has both general and specifically Uganda dimensions, relating to the way in which coping strategies are situated within the contexts of economic liberalisation and state reform, as well as to the strategies to accommodate sections of civil society in order to consolidate political control.

While development NGOs have been warmly welcomed by the NRM government, their operations, whenever these tend to extend their prescribed mandate, are carefully controlled. This reflects the fact that the government badly needs NGOs in the field of social development for service delivery and the indirect legitimacy they bring. Beyond this, however, the government has been rather suspicious of autonomous activities, and has employed a variety of control measures, such as registration and monitoring, co-optation, dissolution, reorganisation and, in some cases, repression and imprisonment. There was more room for manoeuvre in late 1980s and the early 1990s, when the institutional, financial and political state capacity was weaker. As stated in the NGO Registration Statute, NGOs need to register with the NGO Registration Board, and for their continued existence they need to conform to the broad NRM agenda (Dicklich 1998: 98-106).

For many years, a new NGO Bill with proposed new restrictions was debated in parliament. It was seen as a threat to any ambitions for autonomous advocacy. The building of such capacity was seen by many donors as a major political advance during later years of no-party politics (De Coninck 2004: 62-67). The Bill was finally turned into the NGO Registration Act in April 2006 to the effect that NGOs would be vetted by security organs before registration.

The government's continued reliance on NGOs for welfare provision gave the bigger and wealthier among them a bargaining position vis-à-vis ministries or local governments. Some humanitarian agencies have recently spoken out against the military approach to the northern war. The incorporation of sections of civil society into policy making results in new institutional configurations, sometimes with unintended consequences. There are different policy processes going on. Before turning to the overarching and emblematic process, the Poverty Eradication Action Plan (PEAP), I will summarise recent attempts made in creating a coherent framework for various providers in the health sector.

Integration of private providers

Attempts at incorporating private providers into the health sector started in 1987 with the recommendations in that direction by the Health Policy Review Commission. The process did not gain momentum until 1996, however, when a health sector NGO panel was appointed for the purposes of establishing means of collaboration between the government and the private sector and reviewing regulations and legislation relating to this (Birungi *et al* 2001: 82). Because of the plethora of health regimes and actors, the complicated relations between some of them and the illicit operations of others, integration has proven difficult to achieve.

Debates around inclusion revolved around definitional issues, such as the meaning of "integration" or "collaboration", or "public" and "private" (Birungi *et al* 2001: 83). Apart from the fact that legitimacy is invested for different actors in the promotion of a certain set of vocabulary at the expense of another, the discussion clearly illustrated how blurred boundaries between private and public still characterise Uganda's health system.

A number of actors were left out or were only marginally included into this process. After health care, along with other policy areas, had been decentralised with the 1993 Decentralisation Act, responsibility for implementing policy rested with the districts. In spite of this, district leadership was only marginally consulted. Likewise, representatives of private informal providers were left out, as were representatives of health workers. Under the current institutional framework, the Ministry of Health is responsible for "formulation, co-ordination and implementation of national health policy" (Birungi *et al*

2001: 84) while the different arms of the district health teams are expected to ensure implementation within the public sector. The private sector by and large remains separate. Hence, several interests are not represented in the policy making process (2001: 84).

The government is supposed to regulate the sector in its entirety. Laws and regulations that concern private sector activity have been updated. Furthermore, professional bills were passed in 1996. Among them are the Uganda Nurses and Midwives Bill and the Allied Health Workers Professional Bill, which established councils that were assigned the task of regulating and supervising professionals. The latter are to be registered with their respective councils and are expected to be members of a professional association. Some of these regulations have been seen as ineffective, poorly enforced and out of touch with the real structures of private health provision (Birungi *et al* 2001: 84).

The National Drug Authority and its local extensions, the District Drug Inspectors, suffer from low capacity and limited resources and are largely unable to carry out their mandate. Illegal and expired drugs continue to flood the market. Most private practitioners remain unlicensed but enjoy popular acceptance as they provide services that would otherwise not be there at all. Different health regimes continue to co-exist and cut across. When it comes to health, the state largely lacks the capacity to organise and regulate sector policies.

The Poverty Eradication Action Plan: "Governance corporatism"

A more accommodating attitude towards interest groups was expressed during the revision of the Poverty Eradication Action Plan (PEAP), where select civil society groups were included in a consultation process, in line with governance ideals of civil society participation in policy processes. The "policy spaces" (Brock *et al* 2002) or, with another turn of phrase, fora for "governance corporatism" with tripartite configurations of state, donors and certain NGOs, are continuously in the making in different arenas such as PEAP working groups, sector working groups and even consultative group meetings. As opposed to corporatist arrangements in a strict sense, these policy spaces are largely informal and activated and deactivated rather unpredictably.

Without institutionalised structures for bargaining and interest representation, there is little continuity and consistency in government policy making. Priorities between access, quality, equity and sustainability are implemented on a basis of ad hoc intervention. The extent to which participation can be translated into influence on even fairly uncontroversial issues is limited (Brock *et al* 2002: 41).

As stated above, the discourse emanating from the World Bank had shifted in focus towards the mid-1990s, partly as a response to the failure of the orthodox monetarist structural adjustment programmes and the extensive critique that they had been subjected to. In place of its image as a haven for hard-core number crunchers and cost-cutters, the World Bank now wanted to present itself as the chief agent of poverty reduction, with policies that critically involved and were to be implemented by a reconstructed state, capable of institutionalising capitalist development supported by social services, civil society participation and popular empowerment.

The World Bank's need to reinvent itself internationally coincided with a similar imperative for the NRM to renew its foundations of external legitimacy. The government realised that in order to create leeway for consolidating its own political control and creating sufficient state capacity to build institutions and promote policies for doing so, it needed to deliver growth and social services in return (Brock *et al* 2002: 3). To some extent it had already done so, and it was able to draw on the substantial achievements in order to create a credible narrative of poverty reduction. This was statistically supported by the 2001 poverty headcount, which showed that the number of Ugandans living in poverty (in accordance with the conventional, if arbitrary, definition of living on less than the equivalent of $1 per day) had decreased from 56 per cent in 1992 to 35 per cent in 2000 (Brock *et al* 2002: 35); although this number rose again to 42 per cent by 2003.

The government elaborated a framework for dealing with poverty reduction very much similar to the World Bank's conception. The government had established some vaguely "pro-poor" policies, notably the introduction of Universal Primary Education (UPE) in 1997, offering Uganda as a suitable testing ground for new policy directions. It is important to note that the production of an approved Poverty Reduction Strategy, as the World Bank version called it, was a precondition for becoming a beneficiary of the Highly Indebted Poor Countries (HIPC) debt relief programme; in 2000, Uganda became the first country to qualify (Harrison 2004: 40).

The first PEAP was launched in 1997, and it has since been the overarching framework for steering policy making. It is based on four "pillars"– creating an enabling environment for economic growth, ensuring good governance and security, promoting the ability of the poor to raise their incomes, and improving the quality of life of the poor. The programmes are to be implemented by local governments that utilise resources from the Poverty Action Fund (PAF), in turn derived from funding available through HIPC and from other sources (Francis and James 2003).

Largely through the pressures of key donors in combination with a core of actors in the government, there has been a broadening of the range of "legitimate actors" in the "policy making community". A number of civil society actors, separate or jointly, have been invited into processes. New channels and arenas for influence have been created, and the policy making process is more complex than before.

The change towards an extension of arenas and the creation of new mediating mechanisms was an institutional innovation with regards to state formation as well as an attempt by the government at producing a new foundation of legitimacy in relation to demands made by both donors and civil society for extended participation. At first glance, these new arenas might seem to obscure to the distinctions between state, civil society and donors through the involvement of parallel entities of "participating stakeholder" identities.

The internal hierarchy of these stakeholders, however, in relation to decision making, as opposed to participation, was never put into question. It should be noted, though, that at least according to the overall statement speaking for civil society groups, participation for the PEAP revision was more substantial and effective than for the first PEAP (Uganda NGO Forum 2003). On the other hand, another civil society voice, that of Uganda Debt Network, claimed that both poverty reduction policies and participation and influence in the PEAP process were undercut by the superior role ascribed to macro-economic stability (Chekwoti 2003).

The actual impact of PEAP on budget allocations is not apparent. According to one study (Opolot and Kibikyo 2003: 12-19), health care remained underfunded. Certainly, the health sector has received a significant increase in funding since the late 1990s from 6.4 to 9.6 per cent of the state budget (Opolot and Kibikyo 2003: 12-13). In the early 1990s, public expenditure on health was roughly $2 per person per annum. In 2000, this amount had increased to $4 (O'Manique 2004: 139). This is still much below the set targets for achieving the Millenium Goals, and reflects the bargaining power of the Ministry of Finance vis-à-vis the Ministry of Health in the struggles between the two, where the former, backed by IMF claimed that increased funding to social sectors would destabilise exchange rates and trigger inflation.

It is very difficult to evaluate the effects of any specific set of policies, as so many other factors impact on complex outcomes. A striking feature, however, is that poverty levels actually increased since 2000 and that inter and intraregional inequalities deepened. Infant mortality levels increased from 81 to 88 deaths per 1,000 births between 1995 and 2000, with under-five mortality at 152 per 1,000 live births as compared to 147 in 1995. Maternal mortality rates, on the other hand, fell from 527 to 505 per 100,000 births between 1995 and 2000 (Opolot and Kibikyo 2003: 17-18). These are aggregated numbers, with the situation in the north being much worse than elsewhere.

Conclusions

From the early 1990s, Uganda's political economy experienced a phase of political consolidation, economic growth and institutional expansion. These three interrelated dimensions had domestic roots but were all critically underwritten by the inflow of vast amounts of foreign aid. The state underwent expansion and a degree of qualitative improvement in terms of capacity. Health and social welfare was, compared to the immediately preceding period, to a greater degree incorporated into the reconstruction project.

Social sector budgets expanded, infrastructure was being rebuilt and policy making was improved and integrated into the macro-economic framework. During this period, the international governance agenda shifted emphasis from market based solutions and participation of civil society to the embeddedness of these features in "poverty reduction strategies" channelled through and promoted by the state. This was symbolically articulated in the 1997 World Development Report (World Bank 1997). The shift in discourse and policy interplayed with the revival and reconstruction of the Ugandan state in more ways than one and was to influence the shape it took.

A growing inflow of donor resources was increasingly channelled by way of budget support, and the state exercised its poverty reduction policies through a deconcentrated state apparatus. The role of civil society in Uganda was reformulated from relatively free entities in the context of a weak state into loyal but participating partners in poverty reduction policy making (De Coninck 2004). The different meanings of this were demonstrated in the case studies.

In the case of UMWU, it has been shown how such partnership and participation was effectively blocked by the state. Along with organised labour in general, it was not allowed to voice demands that transcended civil society engagement and workers' interests as defined by the government. The harsh way in which it was dealt with by the state underscored the disjuncture between a relatively liberal legal framework and an authoritarian practice.

Other sections of civil society, however, were welcomed and incorporated into informal structures of policy making in a way that was in agreement with the aspirations expressed in donors' "poverty reduction policy" frameworks. A selection of groups was brought onboard national and transnational fora for decision making. Even so, one needs to distinguish between the scope that was opened up for participation on the one hand, and actual influence on the other. This distinction gradually came to preoccupy many groups, as the potential concern that participation would not translate into influence proved to be a real one. This is not to say that civil society input came to nothing; only that there were and are clear limits to the kind and extent of their engagement.

Certain issues, such as the overall direction of economic policy making, political pluralism, high level corruption or the role of the military could clearly not be discussed even though they quite obviously impacted upon poverty reduction policies, and very few policies at all could be discussed in a challenging manner. The changing composition and orientation of the state has been a consequence of the need to respond to structural change. The move towards civil society participation was partly a means for the government to broaden its social base at a time when domestic political coalitions started to crack and partly used to pre-empt donor demands for democracy; thus, it amounted to a strategy to reconstruct legitimacy and fitted into the broader NRM project to reshape state-society relations along the lines of technocratic governance; the guided incorporation of civil society into the policy arenas of poverty reduction policy.

To summarise, donors supplied resources and legitimacy to the revived Ugandan state in exchange for its continued implementation of the agreed macro-economic framework and its sustained governance performance in relation to externally defined criteria. The shift in the fiscal base of the state carried evident political implications. Rather than negotiating power sharing as concessions for resources and loyalty with domestic constituencies only, the elite factions in control of the Ugandan state could now bypass some of the political struggles that this would inevitably have involved and instead bargain with external providers of resources and contain their domestic allies. At the same time as expenses for public administration in the local state grew, financial and in extension political autonomy of local governments was undercut.

From 33 districts in 1986 and 45 in 1995, districts have multiplied around each election, and the current number is well over 100. "Financial transfers to districts increased from 11 per cent of the budget to 34 per cent in 2003" (Mwenda and Tangri 2005: 458), and conditional grants amounted to over 80 per cent of the transfers by 2001 (Francis and James 2003: 329), with the result that local government structures have expanded dramatically at the same time the central government financial and political control over it has increased. During the same period, the civil service was rebuilt in the central state, albeit in the new and semi-autonomous sense of creating enclaves in the state through a vast number of secretariats, autonomous bodies and commissions.

With the delivery of security and services, the government was able to consolidate its political base among the rural population in the central and southwestern regions. Externally, it was able to enjoy a widespread legitimacy among key donors, at least until recently. It was able to trade the execution of economic and institutional reforms in their prescribed variants in exchange for vast resources and the tacit acceptance of political monopoly. Resources could be channelled into the stabilisation and expansion of the state as well as the securing of political control.

Nevertheless, this phase of stable mutual accommodation proved to be short-lived. During the late 1990s, state credibility was gradually undercut among both domestic and foreign constituencies following the intervention in the DRC, rising levels of corruption and political intolerance; among donors, though, it tended to result in to ambivalence rather than in sustained critique. As the coalitions of the early and mid-1990s were disintegrating, the government sought to reconstruct alliances among ethno-regional voting blocs, interest groups and donors in different political arenas of policy making and electoral politics.

Underlying tendencies at the level of accumulation, visible in the privatisation process, tended to undermine the logic of capacity building and coherence as factional infighting among the ruling class penetrated the state. Resistance to the latter tendency was weak due both to the internal weaknesses of democratic forces, political monopoly at the level of state power and the silent complicity of external actors.

There were several logics of state formation at work which shaped the health sector. After 1986, the sector suffered from immense dilapidation in terms of institutional decay and economic bankruptcy. In rehabilitating the sector, the government had to balance the need for the improved provision of services while at the same time exercising control over different providers and their potential constituencies in the context of economic crisis and privatisation. Social service delivery remains an important source of legitimacy for both central and local governments, even if actually provided by NGOs. Social sector improvements have also been a significant component of poverty reduction, vital, again, for regime legitimacy in the eyes of both domestic constituencies and key donors. But "poverty reduction success" is a relative and abstract phrase which is not necessarily experienced on the ground.

In the context of health, some sections of (civil) society operate within a vast informal and unregulated field – or a field regulated on the basis of the logic of the market. This means that the regulatory capacity of the state is highly uneven. There is incorporation and a politicisation of a section of civil society alongside repression of interest representation in certain health regimes and state disengagement, continuing informalisation and lack of effective regulation in others. These regimes are furthermore connected, and both fragmentation and straddling complicate representation and regulation.

The legacy of informalisation and fragmentation of economic and social relations continue to reinforce a disconnection between state and society on many levels. This makes it complicated for the state to regulate some health regimes, and difficult for civil society to engage the state. While the state has successfully incorporated selected actors in national level policy making, whole areas of provision, especially in rural areas, have been institutionalised informally outside the formal health system but interacting with it. In both informal and formal welfare regimes, the reality may for most people more often than not be one of health units without trained staff or adequate medicine.

Even so, social forces have not to any great extent been able to make effective demands on the state for social service expenditure. On the whole, there is limited politicisation of health as an object or arena for political struggles. Civil society groups have weak capacity and propensity to make real use of their increased but conditional access to the policy process. Although the broadening of the policy arena, which may eventually change the character of the welfare arms of the state, has been premised on a higher degree of accountability, it continues to be ambiguous in democratic terms.

State-civil society relations continue to be characterised by fundamental disconnections between policy and politics and between participation and interest representation. The state has created structures for participation – but not for interest representation and debate on policy issues. While service delivery is essential for political support, security concerns as conventionally defined have remained the main priority to this day. Underlying the institutional transformation of the state is a legacy of continued authoritarianism. Contrary to the intentions of the demobilisation programme, defence expenditures have gone up dramatically, from "$88 million in 1996, over $155 million in 2003, and $196-203 million in 2004" (Mwenda and Tangri 2005: 456).

On several occasions, including in 2004, the budget for health and other social sectors was cut in favour of increased defence spending. The tension between inclusion and repression or between governance and militarism has resulted in increased fragmentation of the Ugandan state with regard to its administrative and infrastructural capacities. This displays continuous incoherence within the state apparatus and disconnection between its organs and levels.

The changes in the health sector described in this chapter exemplify how such complex recomposition of state and civil society has unfolded. It illustrates how the regulation of political space through a combination of accommodation, marginalisation and repression has been shaped by the interplay between international intervention and the political forces on the ground. Taken together, these configurations explain the pattern of guided inclusion and parallel authoritarianism and the relatively limited scope for civil society organisations to shape the state.

Chapter 6

The Decentralised State: Regulating the Politics of Local Development

As preceding chapters have shown, one of the main ways in which the NRM has sought to establish control for itself and the state at large in society is by creating institutions for local government. Decentralisation has been a cornerstone of contemporary state reform in both its generic governance sense and specifically in the Ugandan setting.

Strong claims have been made about the transformative role of RCs and LCs in reshaping the local state in democratic and developmental directions. New openings for exercising citizenship is said to have been created (Brett 1998; Regan 1998; Wunsch and Ottemoeller 2004). This chapter examines local level implications for state-civil society relations, as these are played out in Masaka District.

The first section of the chapter addresses different theoretical and empirical dimensions of local politics in a comparative perspective. The second section outlines the historical development of local politics in Uganda, Buganda and Masaka area in particular. Following this, the third section discusses the different cases of the study, that is the different manifestations of state-civil society relations in Masaka. The fourth section then approaches the same topic from a different angle, in that it discusses local politics and its various institutions as seen from the community perspective. The chapter concludes with a discussion of contemporary local politics in Masaka and in Uganda at large.

Local politics: Sources and content

During the last two decades, most countries in the developing world have experienced a turn from centralised development strategies towards localised politics. A great many factors have contributed to a variety of expressions of "localisation of politics" (Harris *et al* 2004: 2), ranging from formal decentralisation by design to informal localisation by default.

Conspicuous among these features are downsizing of the central state capacity, privatisation, decentralisation reforms and political liberalisation. As a result political forces at the local level have also been reactivated. Localisation has affected both economic relations and political institutions. The main features of the current development orthodoxy comprise neo-liberal economic policies contained within decentralised political structures. State reforms were introduced in order to provide the institutional corrective to market reforms. To relate to Harrison's discussion of the "governance state", decentralisation has been a key element in this "project to embed neoliberalism" (Harrison 2004: 5).

Disparate sets of powers and responsibilities have been decentralised to institutions at district and sub-county levels.

This has been echoed by a discursive emphasis on community participation in local arenas, where development is supposedly more efficient and democratic. The typical official rationale for decentralisation centres on the assumption that such a policy brings government closer to the people, as the catchphrase goes, and thus enhances bureaucratic and political accountability and responsiveness – especially if combined, as is often the case, with participatory development strategies with the stated aim of empowering local communities.

Great expectations initially underpinned this most recent wave of decentralisation. The combination of a decentralised state, civil society and market forces was frequently claimed to constitute a key mechanism for giving ordinary citizens political voice and to bring about improved service delivery and increased political and administrative accountability, especially when contrasted to what was seen as failed centralised states (Harriss *et al* 2004: 3; Wunsch and Olowu 1995: 1-22). Just like with civil society (White 1994), enthusiasm for the emancipatory potential of local politics was voiced by neo-liberal, radical and culturalist proponents alike (Crook and Manor 1998: 1-2).

The obvious possibility that power relations may remain unaffected or might even be strengthened in local arenas was frequently concealed. In real terms, decentralisation often remained underwritten by centralised control. The mixed experience of actual practice has tempered such optimism, and most observers are now aware that not only are reforms often hampered by lack of resources and capacity that tends to result in re-centralisation, but also that decentralisation is an inherently political process:

> Political variables are key, not because of variations in formal structure or broad regime types, or technical failure of implementation, but because decentralisation is essentially about the distribution of power and resources, both amongst

> different levels and territorial areas of the state, and amongst different interests in their relationship to ruling elites (Crook and Sverrisson 2003: 2).

According to Mamdani (1996), as most people in rural Africa are deprived of/or lack the capacity to exercise real citizenship, decentralisation has essentially meant bringing oppression rather than accountability closer to the people. Local political arenas have in this view essentially been marked by authoritarianism as opposed to popular democracy. Another strand in the literature emphasises the pervasive structures of patron-client relations, which are predicted to be revived as local political arenas are reactivated (Chabal and Daloz 1999: 105).

Regardless of the democratic potential of the reforms, local political units have re-emerged as arenas for struggles over access to resources. This tends to expand the scope for political actors at that level, along with their connections up- and downward. Such connections are, with regard to social policy, critically related to semi-privatised development, which occasionally turns into an important source for accumulating money or influence, based on religion, ethnicity and political affiliation. All these forms of local politics are in turn shaped by the nature of central-local relations, through the ways in which central governments seek to deal with local elites and to penetrate communities for political support and by responses from different sections of local society.

The recent turn to local politics has thus different and contradictory sources, all deeply embedded in power struggles. What are the implications of decentralisation and localised politics and development for the relations between state and civil society? What is the scope for different social forces to open up political space? In order to address these issues, it is suggested here that the necessary starting point is the need to approach local politics in a historically and conceptually broader sense than the preoccupation with contemporary institutional reforms allows for. To broaden it conceptually means including all forms of social power in local arenas. The argument advanced here follows the general framework of the study and sets out from the premise that the realisation of decentralisation and local governance reforms will be shaped by struggles among different social forces, including efforts by ruling elites to structure their relations to the former.

Decentralisation is undertaken for a variety of political purposes, which create variation in the structuring of local politics also within countries. Central authorities seek to maintain or transform local political arenas and the mechanisms that mediate between the central state, the local state and local society. These processes are likely to include pressures from or concessions to interest groups or regional or local elites that seek to establish or consolidate their own political influence (Boone 2003).

Local social forces and politicians struggle to negotiate the terms of such a transformation. Institutional reforms take root in different ways due to interventions and contestation by competing actors both central and local, who seek to create access and influence through channels of external connections and local control. Local political arenas are thus shaped in the interface between different forms of externally intervening political projects and the range of aspirations and responses that emerge locally.

For these reasons, local level politics is never distinctly "local" Just as institutional reforms are interwoven with political struggles over access to resources in terms of property, power, prestige or hegemony, local politics is fundamentally connected to wider arenas. This holds for both temporary networks linking local elites to national centres of power and for the way central governments historically have structured their relations to different regions – a most significant factor in the case of Masaka and Buganda. There are furthermore several levels within the local, from the village to the district, where issues are negotiated. At all of these levels, civil society organisations are often centres of power in themselves through the way they seek to mobilise and represent different sections of society (cf. Kassimir 2001).

It has been argued that in the absence of the effective presence of the state in parts of rural Africa, elite sections of local society have taken on the function as public authority (Bierschenk and Olivier de Sardan 1997). This observation needs to be qualified. State formation and public authority is not merely complicated by the absence of the state or of formal structures of regulation that create room for informal modes of rule. More crucially, formal and informal regulatory structures are closely connected, both in immediate local arenas, and in the ways in which these arenas are linked to actors and institutions at higher levels of politics. Lund (2001: 40) makes the observation that "the local" is not necessarily distinctly local in a spatial sense but functions, equally importantly, as a point of reference for different forces invoking authority. Hence, local politics is constituted both by the ways in which central governments have institutionalised their relations to different sections of local society in the form of direct intervention or mediated by middlemen and by the specific trajectories shaped by the constellation of forces that emerge among the latter.

A few assumptions can be made about civil society in local politics with reference to the above remarks. First, local interest groups or networks are likely to be deeply embedded in local society, making state-civil society relations less distinctly sector-specific than at the national level. Secondly, these groups typically seek to claim authority in different arenas simultaneously, basing their influence on overlapping forms of social distinctions, including class, gender, ethnicity, and religion. Thirdly, they can be expected to do so by combining formal and informal resources and regulatory mechanisms. The relations involved cannot be sufficiently captured by references to "patronage", as such claims may also relate to issues of equality or democracy.

What factors are central to explain variations in the political content of decentralisation? In a review essay on the topic, Crook and Sverrisson (2001: 2, 4) identify as critical the nature of political relations between central and local governments, "which derives principally from the character of and power bases of the ruling elite and their relationships with local elites" in combination with "the configuration of local ... economic, social and political structures". Power relations in local society underlie central-local relations, both in terms of shaping political resources and interests of local elites and the strategies developed by central governments to deal with these elites.

Similarly, Boone offers a detailed argument (2003: Chapter 2) about the relations between local political struggles, based on communal structures, relations of property and production and institutional mechanisms for reproducing social authority and the shifting scope that this offers for central government intervention. It enables her to elaborate a framework that is both general enough for encompassing common parameters (central governments seek to tax and govern society, and they are capable of doing so according to the balance of power in the local society) and sufficiently specific to capture local variations. Ideal typically, according to Boone, the outcome is either power sharing, usurpation, administrative occupation or other models of engagement that are less based on the integration of the various levels.

The argument predicts regional diversity according to central-local relations and configurations of power in the local state and society, following from cleavages, such as class, gender and ethnicity, as well as their organised expressions in civil society. These perspectives are broadly in line with the overall theoretical framework of this study: that state formation can be explained by the relation between the balance of power in civil (and in this case local) society and within the state as mediated by various political institutional arrangements. I will return to these arguments in the final section of this chapter.

What may be hypothesised about local politics in the period of the study? Are there any specific traits that distinguish this period from earlier rounds of local politics? The period is characterised by economic, and in some cases political, liberalisation and the resurrection of civil society organisations. As demonstrated in previous chapters, Ugandan society underwent relative stabilisation from the late 1980s.

Yet, at the level of civil society macro-economic stabilisation corresponded to a stabilisation of already relatively fluid informal coping strategies, rendering the latter greater predictability without necessarily formalising them. As will be explored in the case of Masaka, the combination of a plural and fluid society and a re-established state resulted in evolving but tentative and internally incoherent formal and informal regulatory mechanisms that went hand in hand with continued overall fragmentation. At the level of national politics and civil society, this social pluralism evolved within a context of consolidation of political control and deliberate separation of social organisation from political power. What happened in the local political arena?

Local politics in Uganda

As has been made clear throughout the foregoing chapters, tense and unresolved relations between the central government and regional power centres, notably Buganda, have fundamentally shaped territorial and institutional dimensions of state- and nationhood. A brief rehearsal of changes in central-local relations is therefore called for in order to bring out how the many layers of local politics, as discussed in the previous section, interplay with contemporary institutional reforms.

The history of local politics in Uganda is the subject of a vast literature (see among others Burke 1964; Richards 1982; Sathyamurthy 1982; and Tidemand 1994 for an overview), as is that of Buganda (see e.g. Apter 1997; Fallers 1964; Karlström 1999; Mafeje 1977; Tidemand 1994). Two dimensions are seen as relevant in the present context, both of which underline the arguments about the multiple sources and the fundamentally political character of local institutions and the very shifting manifestations of local political arenas.

One dimension concerns Uganda's dramatically uneven regional development, following from a combination of colonially imposed economic underdevelopment and indirect rule, reinforced by elite-controlled politics after independence. This type of leadership in turn rested on regional power bases and was prone to manipulate ethnic and regional identities that entrenched centrifugal tendencies in the central state.

The other dimension is related to the informalisation of the economy and society that expanded dramatically during the years of institutional decay, which set off a multiplicity of informal institutions, mainly local in scope. These entities have interacted in complex ways with formal institutions of local politics.

The first dimension resulted in struggles between central and regional centres of power at the level of national politics that were temporarily settled with the independence constitution and its hybrid status with regard to regional autonomy. The struggles took off again immediately after independence and resulted in the constitutional crisis of 1966. The new order that followed ushered in centralisation in terms of real control over district councils, as codified in the Local Government Act of 1967.

Frequent changes in institutional regulation of local government throughout the entire post-independence period and the great swings between centralisation and decentralisation points to the importance paid by all state rulers to controlling local arenas. Simultaneously, the rapid decay of formal institutions from the early 1970s – the dimension of informalisation – resulted in "great discrepancy in post-independent Uganda between the formal rules and actual performance of local administration" (Tidemand 1994: 56).

The history of RCs and LCs was outlined in chapters 3 and 4. What was suggested there, and can be concluded here, is that the move towards localised politics stems from very different sources. They included official claims about the introduction of popular democracy through RCs, and later rephrased in terms of more efficient service delivery and better governance through LCs.

The move was also prompted by the unannounced interests of the government to ensure local political control through the same institutions. The process is confounded by traditionalist aspirations for ethnic communities, kingdoms or not, and by the informalisation of economic and social life generating informal networks. There is now an extensive body of valuable literature that analyses political change in the present-day Uganda from different regionally specific and historically anchored perspectives – see for West Nile, Leopold (2005); for Acholi, Finnström (2003); for Karamoja, Knighton (2003); for Teso, Jones (2005); and for the case of Jinja, Byerley (2005).

Taken together, these contributions illustrate the general theoretical perspective on the multiple sources and institutional complexity of local politics outlined above. They put into perspective the rather narrow policy-oriented strand that has characterised part of the literature and which has centred on successful decentralisation reforms (Regan 1998; Villadsen 1996). The latter writings have been a great deal more influential in shaping the orthodox narrative about contemporary Uganda, and in extension the policy that guides it.

Buganda in contemporary Uganda

The relations between the central government and Buganda have been both delicate and decisive for the politics of the country. The central government has had to contain potentially explosive politicisation of ethnic sentiments in the central region also after the kingdom was dismantled after the 1966 crisis.

The Luwero war added further animosity towards Obote and UPC among most Baganda, something that the NRM took advantage of. The loyalty that developed between the NRM and the Baganda is said to have been strengthened by promises from the NRM leadership to reinstall the kingdom after taking power. The truth of such claims remains controversial in view of the government's continuous reluctance to concede too much autonomy to the kingdom. For discussions of this issue, see Kasfir (2005: 282-283) and Oloka-Onyango (1997: 176-177).

NRM was initially hostile to politicised ethnicity in general and "feudal kingdoms" in particular due to a set of factors, including the complicated history of political ethnicity in general, the radical leftist outlook of leading NRM politicians during the 1980s and considerations of difficult future alliances, as most of high-ranking NRM politicians and NRA officers came from western Uganda. In a move to accommodate Buganda and ensure the continued political support of the Baganda, however, NRM turned around, and in 1993 kingdoms were restored on the premise that they would restrict themselves to the purely "cultural" domain (Karlström 1999).

The distinction between culture and politics proved as difficult to uphold as the one between politics and development. This was made very clear when, in the wake of the restoration of the Buganda kingdom, a project for reinstalling the Ankole kingship was cut short by the government (Doornbos and Mwesigye 1995). As the home region of President Museveni and most other NRM leaders and senior army officers, Ankole was not expected to develop into an alternative centre of power, particularly not one with intra-ethnic divisive potential.

When it became clear to politicians from Buganda kingdom that their autonomy remained circumscribed, their relation to the central government turned sour. The kingdom set up its own government structures and continues to push for federalism and real autonomy by way of silent negotiations and expansion of its repertoire of activities in, for instance, arenas of development work and media. The push was combined with regular public demonstrations and far-reaching demands. That Buganda would not succumb to the requirements of restricting kingship to the sphere of culture, was made clear during the discussions preceding the making of the new constitution in 1995, where a highly contested political issue concerned the unit of local government.

Decentralisation advocates came up against proponents for the restoration of federalism as the extension of ceremonial kingship. The government insisted on decentralisation – one objective of which, according to Crook (2001), was to fragment and undercut important sub-national political entities, and in particular Buganda region.

The monarchist movement contained a militant wing, but a small armed monarchist rebel group fighting for federalism that emerged in rural Buganda in the mid-1990s does not seem to have been long-lived (Karlström 1999: 113). The government remained dependent on the political support of the Baganda and has sought by various stick-and-carrot measures to contain their discontent, although with increasing difficulty.

Buganda has not been entirely domesticated as a stable government power base or a united electoral bloc. More broadly, politicised ethnicity has in spite of NRM's stated aims refused to go away. Most other ethnic groups articulate dissatisfaction about what is perceived as the western ethnic base and bias of the state, and the civil war in the north has dramatically deepened regional cleavages that complicate the resolution of the national question.

Masaka in Buganda

Masaka District is located roughly 100 km southwest of Kampala, and is part of Buganda region. According to the 2002 census, the population of Masaka Town stands at 61,300 and that of the district at 767,759. The majority of its inhabitants are Catholic Baganda, with religious (Protestant and Muslim) and ethnic (mainly Banyarwanda and Banyankore) minorities. The bulk of the population are small-holding peasants.

The district has been subdivided throughout the years, as Kooki, Ssesse and Ssembabule counties which became Rakai (1974), Kalangala (1989) and Ssembabule (1997) districts. Historically, the area has been one of the most important food producing regions in the country, coffee being the main cash crop. Following the slump of world market coffee prices and overall decline of agricultural production, however, the former bread basket of Kampala has sunk into stagnation (Bazaara 1997: Chapter 5). This is reinforced by lack of investments and employment facilities in Masaka town and other urban centres. Poverty is further deepened by the severe impact of HIV/AIDS in the region.

The history of Buddu county (the older name for the territory covering today's Masaka) has impacted deeply upon contemporary Masaka. From 1888 to 1892, religious conflicts between Muslims, Protestants and Catholics shook Buganda. These conflicts were settled by the gun as the Protestants called upon Britain for military assistance. As consolation for defeat, the Catholics were given Buddu County by the British as a base in which to resettle and rebuild their religious community. Political and commercial ties with the rest of Buganda remained, but with Protestant hegemony established, "Catholics migrated towards Buddu in a 'trek' and there emerged an exile mentality, which became a distinguishing feature of the Buddu church" (Waliggo 2002: 66).

Masaka town was created as a trading centre in the late 19th century by Asians, who just like in colonial Uganda at large came to dominate trade and the manufacturing sector. The local political economy revolved around agriculture, coffee in particular. A great number of coffee barons in Kampala of the 1950s and 60s made their wealth from large farms in the district. A domestic bourgeoisie emerged, and proximity to Kampala created conditions for the expansion of agri-business, including a prominent role for the Coffee Marketing Board and cooperative unions ever since late colonialism (Kasita 2005; Mamdani 1976: 203). Many workers on farms and in the agri-industry were migrant labourers, significantly from Rwanda, Burundi and Kigezi, who came to settle there and assumed Baganda names – Buganda is much less ethnically homogeneous than what is often thought.

Compared to its relative economic prosperity, Masaka suffered from political marginalisation. Under colonial rule, Catholics were socially and politically subordinated to Protestants in Buganda in general, and to the so-called Mengo establishment that dominated the institutions of kingship and the court in particular. The Anglican Church was systematically incorporated into the colonial project, at the expense of Catholic leaders. With independence closing in and the political temperature rising, religious rivalry translated into political struggles.

The Democratic Party was founded in 1956 with a Catholic base as a distinct response to perceived religious discrimination and a way to counter Protestant hegemony. Although DP was successful in Kigezi, Ankole and West Nile (Jørgensen 1981: 199), it was and remains seen as an essentially Catholic Baganda affair with particularly strong backing in Masaka. When Mengo boycotted the pre-independence parliamentary elections in 1961, DP candidates won unopposed and formed the majority.

In response, the Mengo elite formed the Kabaka Yekka party and allied with UPC in order to shut out DP from power; and this bloc won the 1962 elections (see Chapter 2). Soon after independence, most DP members of parliament crossed the floor to UPC, something which turned Masaka into a political backwater. This tendency was further pronounced after 1967, with the imposition of UPC one-party rule and the dissolution of all kingdoms. According to one observer, as this coincided with the era of the coffee boom, local elites took a hands-off stand towards politics, and political leadership in the region stagnated. Obote had appointed a stratum of local leaders and civil servants, and during Amin's rule, there was no political activity at all. This created a political gap that lasted until 1979 (interview, John Kawanga, 22 February 2004).

The overall decay of the economy under the Amin government was accompanied by falling coffee prices and smuggling of coffee. Informal *magendo* structures permeated the public sector and destroyed the private sector. Asian farmers, traders and industrialists were expelled in 1972. Rampant smuggling undercut long-term investments. The hardest blow, however, came during the 1979 war between Uganda and Tanzania, when Masaka town was almost completely destroyed. Another fundamental shock came with the HIV/AIDS epidemic from the early 1980s onwards, with Rakai and Masaka being its epicentre in Uganda.

After the ousting of Amin, DP was revived nationally and was widely seen to have been robbed of the victory in the 1980 elections. Under the new UPC government, Masaka was marginalized both politically and economically during the first half of the 1980s. Its role as a DP stronghold has been difficult to translate into direct political clout, as DP has invariably been on the losing side in national politics. During the NRA guerrilla war, Masaka was viewed by the government as potential rebel territory. Development was blocked and economic recovery after the destruction in the 1979 war was prevented. The population experienced a great deal of political pressure and harassment from the central government until 1986. The last six months before NRM took over power, that is during the Okello Lutwa regime, Masaka was cut off from Kampala, with the NRA being in control of the area.

After 1986 RCs were set up, and people were made to renounce their political partisanship. The NRM took advantage of the power vacuum and managed to wipe out DP structures by incorporating its most important leaders, including the former speaker of parliament, Edward Ssekandi and former minister of finance, Gerald Ssendaula. Another important factor that reshaped the local political landscape was the strong support from the Catholic Church for the NRA during the guerrilla war and for NRM in post-1986 politics.

At the district level, appointed leaders remained until the 1989 elections, when some old guard DP politicians were voted into RCs. NRM domination over local politics during the late 1980s later became a more complicated game, as DP opted out of the broad-based government and the NRM's relations with both Mengo and the Catholic Church became strained from the mid-1990s. Even though political parties were dormant under the no-party system and the geographically peripheral and less developed parts of rural Masaka in particular tend to be pro-NRM, DP still exercised some degree of informal influence at most levels of the political system. In Masaka town and other urban areas, there was widespread support for DP, and in the two most recent parliamentary elections, in 2001 and 2006, the party's John Kawanga has been elected MP for Masaka town.

None of the NRM leaders that hail from Masaka have made it into the very innermost circle of national power politics, and while they have enjoyed influence within their constituencies, none has been able to control local politics in a strongman fashion, as has been the case elsewhere. There is a balance between different centres of power, including MPs and the district council and its then chairman, Vincent Ssempijja, in the running of Masaka politics. Ssempijja has for a long time been popular and perceived as an efficient politician with a base among most communities. During the 2006 elections, however, local and national NRM politicians challenged him through the ad hoc Masaka Election Task Force, accusing him of corruption.

Just like all districts, Masaka is economically dependent on revenue from the central government. Precisely for that reason, informal connections through MPs or cabinet ministers are very important for local development. Political dependence on the centre is marked but not entirely straightforward. The district leadership may take on importance in its own right, for example in struggles between the central government and Mengo for local support, where local leadership is given the opportunity to exercise some degree of bargaining power in either direction. Central-local relations further interact with intra-local relations of domination. There are important constituencies that the district political leadership needs to accommodate, critical among which are representatives of the Buganda kingdom, the Catholic Church and landlords, not seldom in combination. Landlords and kingdom representatives were outraged by the 1998 Land Act that sought to diminish their influence, something which led to extreme tensions.

Religion and ethnicity are political factors, as everywhere, and Masaka is not as homogeneous in those terms as conventional wisdom would have it. But nor are these dimensions constant objects or arenas for political struggles the way they are in other places such as in neighbouring Ssembabule, a hotbed of politicised ethnicity. The attitude towards Buganda nationalism is ambivalent. The contemporary revival of the kingdom seems to be met with caution by many Masaka Catholics, and many claimed to embrace the cultural dimensions of monarchism, being hesitant towards its political undercurrents.

Likewise, the Catholic Church is an influential social institution, both on a general level of spiritual guidance and as a political player in a more mundane sense (Kassimir 1998). Religious leaders, especially at lower levels such as parish priests, are prominent brokers or mobilisers of political support, making them targets for appeals and pressure from politicians. Popular gossip consequently concerns their political orientation, but most observers agreed that the importance of the established churches have declined, in marked contrast to the rising influence of evangelical churches. Local state-church relations are characterised by mutual accommodation. Local government leaders cannot afford to be seen to openly confront the church, in view of the risks this would involve in terms of lost support. Likewise, the church will not risk being seen as a vehicle of partisan dissent.

Recurrent narratives centre on Masaka's economic stagnation and peripheral role in contemporary national politics in a way which shows some resemblance to discourses of decline in the former industrial hub of Jinja (Byerley 2005). Popular interpretation puts much of the blame for the economic decline on the central government and its perceived deliberate efforts to promote the western region – the home area of President Museveni – at the expense of Buganda. Masaka town has never quite recovered after the destruction in 1979. The collapse of cooperative unions after the civil war in the 1980s undercut the structures of the commercial economy in the district and left unemployment and poverty in its wake (Kasita 2005). Even if there are signs of recovery, they are modest.

There is a looming agricultural crisis, caused by land fragmentation and erosion, overpopulation and falling coffee prices after a brief mid-1990s boom. This has resulted in rural poverty, reinforced by lack of investments and employment opportunities in urban areas; the local market is therefore very limited. By way of agriculture, most people are small holding peasants. Although many benefited from the coffee boom, this was temporary. In terms of business, most main investors are active or retired politicians or civil servants based in Kampala, with contacts in or having returned to Masaka either doing business in town or as big coffee farmers. Most business people in Masaka intervene in formal politics by proxy if at all. Many back political campaigns, and are rewarded afterwards.

At village level, community based organisations are promoted by district development officials, religious institutions and foreign NGOs. They include burial societies and income generating groups such as rotating savings schemes that operate on the basis of age and gender and other identities. The sustainability of these development efforts is evidently highly dependent on external funding through linkages to local government and NGOs.

The kingdom seeks to establish itself through different manifestations, for instance in the activities run by Buganda Cultural and Development Foundation (BUCADEF), its development organisation. The kingdom has set up structures of representation parallel to, and in reality overlapping with, the Local Council system. Religious institutions are prevalent. Apart from the Catholic church, a great number of evangelical churches have succeeded in establishing themselves in rural areas and attract an ever-increasing number of followers.

International NGOs are also a typical feature of local civil society. Most of these stay for a limited period only. Given the extremely weak fiscal base of local government, there is constant competition between districts for attracting them. At the time of fieldwork, World Vision was by far the biggest and most important development NGO in Masaka District. With regard to health care, Masaka displays the general institutional complexity outlined in previous chapters. The public health care system co-exists with semi-private or private forms. Alongside the government hospital, there are two Catholic mission hospitals, long well established, which also run mobile home units. Both government and NGOs train community health workers. Self-help groups dealing with the impact of HIV/AIDS is a common feature. There are also a number of private clinics and pharmacies.

State-civil society relations in the health sector

The involvement and engagement of civil society organisations with local government and communities depends on their point of entry and their social base in local society. There are evidently great differences in that respect, which two of the cases studied here amply illustrate. The stagnation of Masaka region along with the dramatic transformation of political arenas has made the population politically dependent and vulnerable.

DP loyalists have gone underground following the imposition of NRM hegemony. In the context of a region in decline, with a rampant HIV/AIDS epidemic, and in the absence of viable state structures, World Vision, an evangelical Christian organisation, has established itself as a provider of social welfare with claims to spiritual empowerment. The scope for action and the legitimacy it enjoys are shaped by the vast resources at its disposal and its possible effectiveness as a development organisation, having entered into the heartland of Uganda's Catholicism and the silent competition with the welfare arms of the Catholic church and the religious factionalism this is likely to set off.

BUCADEF embodies the developmental efforts of the Buganda kingdom to reclaim a foothold among the peasantry and to regain relevance at community level. Its presence needs to be understood in the context of changing economic, social and political conditions for ethnic monarchism. The historical depth of monarchism and cultural mobilisation embodies a series of tensions, including the role of Masaka in Buganda, the unresolved issue of decentralisation or federalism, the fear of potential ethnic exclusion and the nature of citizenship in Uganda. I will return to World Vision and BUCADEF but, before doing so, however, I will briefly summarise the experiences of UMWU in Masaka.

UMWU: Public sector unionism under strain

The potential strength that the union can claim in terms of its nation wide scope and large membership base is undermined by decentralisation reforms, which impose parallel arenas for negotiations, beyond the scope of the union's present capacity. The capacity of UMWU differs markedly between districts due to qualities and attitudes of the district leadership. Unsurprisingly, national leaders see decentralisation as a means to fragment the union's capacity and would prefer a re-centralisation of certain policy making mechanisms. Furthermore, the union has not made any inroads in the private or NGO run hospitals and health units (interview, Apollo Nyangasi, 25 February 2003).

The Masaka Hospital branch of the union was set up in 1995, and in 2003 it had reportedly around 200 members, or between 75 to 80 per cent of the staff. There are contradictory views on the history of the union, though. When I was there in 2000, I was told that a union branch had just been established, after UMWU representatives had been in Masaka to introduce the idea, but that the union was not yet effectively operating.

The Union branch of Masaka was said by national union leaders to be weak, primarily due to ineffective leadership (interview, Apollo Nyangasi, 25 February 2003), and has failed to become included into decision making structures at the district level. While the members claimed that they have not experienced interference from the hospital administration with regard to the holding of meetings, the leadership had failed to expand the scope of union activities.

The main achievements of the union were seen to be, here as elsewhere, salary increases and secured lunch allowances. However, the salary was still viewed as insufficient, and members continued to voice demands for housing and travel allowances. They were paid Ush12,000 (at the time roughly US$7) a month, plus housing allowances. Asked how they cope, they mentioned their *shambas*, or small plots for cultivation (group discussion with nurses, Masaka Hospital 18 October 2000).

The main problem was understaffing with resulting overworked staff. They would want insurance, including life insurance and subsidised or even free medical treatment in case something happened, such as AIDS or Ebola (an Ebola epidemic was raging in northern Uganda as we spoke). Transport costs to and from work, increased salaries and improved accommodation were other demands mentioned, as was the wish that the government would pay children's school fees. They had not approached the hospital management with their demands directly, as their assumption was that the management would merely refer to higher levels of authority (that is the Ministry of Health) – the level they identified as causing the major problems of the sector. Although they were undecided on the issue of whether decentralisation has made any change, findings from elsewhere suggest that working conditions for health workers have if anything deteriorated under decentralisation (Kyaddondo and Whyte 2003).

Working conditions in the public sector have set off multiple coping strategies. In Masaka, quite a few of the members double as private practitioners. The union has intervened to ameliorate living conditions of its members through temporary income generating projects, such as small-scale farming and agricultural credit. So far, however, it has not succeeded to establish itself as an effective vehicle for interest representation in relation to the structures of decision making. It has yet to prove its consistent relevance in defending the interests of members, as suggested by my discussion with the union leadership and a group of health workers (Masaka Hospital, 4 March 2003). Our discussion followed upon a major staff meeting a few weeks earlier, where UMWU's national leadership had come to address workers, and where the local leadership was taken to task for past failures.

One issue that runs through any discussion of health workers is the accusation of low "staff morale" and allegations of wide spread corrupt practices among staff. This is a legacy of the collapse of the public sector, and a key issue in the debate about the problems of health delivery for patients, staff and officials alike; although addressed from different perspectives. Although health workers themselves denied first-hand knowledge of any such practices, other sources claimed that they were very common. The union's insistence on the connection between workers' rights and service delivery is countered by officials at the ministry, district or hospital level, who claim that underpayment is but a partial explanation to this behaviour. The debate in this arena for conflict reveals the overall difficulty of UMWU to impress its understandings of the problems of the sector and its solutions on its counterparts.

World Vision: Community empowerment through faith

World Vision is an international evangelical Christian development and relief NGO, founded and based in the United States. According to Bornstein (2002: 6), a few years back it was operating in 95 countries, and it is one of the biggest NGOs in Uganda where it is active in many districts. Most of its rather extensive economic resources come from private donations. World Vision has operated in Uganda since 1986 and in Masaka since 1990, where its presence was originally motivated by the devastating effects of HIV/AIDS.

At the time of fieldwork, World Vision ran three Area Developing Programmes in the district and planned to expand their geographical and thematic scope. World Vision is engaged in rural development in a general sense; their activities cover most sectors and range from constructing health units, schools and boreholes, to providing agricultural credit and running programmes for sponsorship of the education of orphans. In the health sector, their main contributions include the construction of health units, making overviews of sanitation and immunisation needs as well as providing training for community health workers. These mundane activities are informed by an evangelical perspective on development that seeks to achieve transformative worldly and spiritual empowerment through participatory community activities.

World Vision is the main foreign NGO in the district, and their staff and local politicians alike are very aware of their crucial role for social service delivery in Masaka. As compared to indigenous NGOs, World Vision enjoys relative affluence. In more ways than one, World Vision functions as a link to the outside world through the contacts and material resources it introduces into the local settings where they operate.

World Vision is included in decision making fora at the district level, where as one member of the District Health Management Team put it, "their views are not binding, but persuasive" (interview, 10 February 2003). Politicians, technocrats and NGO staff emphasised that the relationship is marked by close co-operation and confirmed by written memoranda of understanding. All parties expressed a desire to expand the role of World Vision in community development. World Vision staff were keen to point out that they engage in development, not politics. This distinction is a key feature of NRM ideology and is upheld as regulatory principle for the NGO sector through threats of more restrictive legislation and self-imposed practice. Local political leaders and NGOs seek mutual accommodation, as long as this follows the logic of relations at the level of national politics. While state supervision of NGOs is weaker at the local level, the latter have learnt to keep within the limits imposed by the strict legislation.

When asked about the line between politics and development, a World Vision staff admitted that it is indeed difficult to uphold such a distinction and that power relations significantly shape implementation of development projects, including their own. "Most politicians try to influence resource allocation" (interview, World Vision Programme Co-ordinator (WVPC), 11 March 2003).

Conflicts over access to resources are acted out on the basis of ethnicity, political orientation and above all, gender. Allegedly, it was not uncommon that loans given to women were taken by their husbands (interview, WVPC, 20 November 2000). The organisation tries to deal with political hijacking by exercising influence "by example, not by input" in order not to directly challenge the political leadership. "We don't wish to destabilise our smooth working conditions" (interview, WVPC, 20 March 2003).

Another strategy is to make use of religious leaders for indirect mobilisation of communities against for example "corrupt practices" in general and certain individual politicians in particular (interview, WVPC, 20 March 2003). In this context, it may be added that World Vision as a way to address the spiritual dimension of community empowerment, has a stated aim to "empower churches of all denominations" (interview, WVPC, 11 March 2003).

Where churches run programmes, World Vision adds resources. Although downplayed by World Vision representatives, such an intervention may of course add on to the competition between the Catholic, Anglican and the Pentecostal churches. Representatives of World Vision were aware of the tensions between its own ideas of harmonious community empowerment and the local power politics that might prevent the former to occur. To avoid such difficult issues they preferred to emphasise lack of education and enlightened cultural habits, rather than structural conflicts over property and power (cf. Kelsall and Mercer 2003).

Thus, World Vision avoids being seen as a political actor in the public arena. They do, however, play an important role in the context of state formation. World Vision and similar NGOs are central in anchoring state reform and semi-privatised development. The creation of an "enabling environment" for NGOs is a key dimension of the formation of state capacity and legitimacy in the field of social welfare, as well as a boost for local politicians who predictably try to capitalise on the presence of foreign NGOs. Most NGOs develop institutional structures for community participation, which in combination with the resources they command turn them into influential entities at the local level. World Vision has set up development committees at village, parish and sub-county levels. A fundamental problem from a democratic perspective is that although most NGOs wrestle with problems of accountability, the institutional links for participation that are created between NGOs and community are subordinated to structures of bureaucratic accountability that come with external funding.

BUCADEF: The power of the kingdom?

Buganda Cultural and Development Foundation (BUCADEF) is in itself a rather modest NGO. BUCADEF is also, however, a component in a political project for restructuring the relations between state, nation and ethnicity. In their own words, they are the "grass root development arm of the Buganda Government" (interview, BUCADEF Chief Executive Officer, 23 November 2000). As was discussed in previous chapters, the Buganda kingdom's demands that cultural monarchism should be expanded to include federalism and real autonomy have so far been denied, but federalism remains the object of ongoing political struggles between Mengo and the central government. Hence, the sources of BUCADEF's capacity to articulate concerns and mobilise support are defined not only by its relations to the local state, but also by relations between the Buganda kingdom and the central government ultimately, also to local governments in Buganda region.

BUCADEF was established in 1995 for the purpose of improving the standard of living of the people of Buganda kingdom. It is involved in a wide range of welfare provisioning and income generating projects and run a number of programmes in health, education and water sectors "using culture as a device for mobilisation and development" (interview, BUCADEF Chief Executive Officer, 23 November 2000). In the field of health they are engaged in preventive health care, such as nutrition projects, early childhood development, the protection of water sources, and in immunisation programmes on behalf of the district.

The programme for nutrition and early childhood development is a government one, or to be precise, the money comes from the World Bank through the Government of Uganda to UNACOH (Uganda National Association of Community Occupational Health), and the project is contracted out to BUCADEF for implementation. The first phase of the programme had run for five years and would expire in June 2003, but was expected to be renewed (interview BUCADEF District Co-ordinator, 17 February 2003).

Their main source of funding is the kingdom, although BUCADEF has also managed to attract resources from donors, including USAID, DFID, and the World Bank. For the kingdom, there are limits set to resource mobilisation. Under the 1995 Constitution, the kingdom is not allowed to tax the population within its territory. However, the kingdom has assets, parallel bodies for commercial activities including an investment company that owns most of the shares in several companies, including a travel agency, a newspaper and a commercial radio station.

Ordinary and not-so-ordinary Baganda in Uganda and among the Diaspora also contribute in cash or kind. The kingdom's symbolic manifestations, most prominently the 1993 coronation and the 1999 royal wedding, offer fertile ground for donations of different scale, from modest to extravagant. Even though contributions from ordinary Baganda display an element of commitment and popular support for the kingdom, there is certainly also a degree of social pressure, which gives it the character of informal taxation (Englebert 2002: 355). Furthermore, most sub-counties within Buganda make annual donations to the kingdom "for cultural purposes" (interview, BUCADEF Chief Executive Officer, 28 February 2003). District administrations make contributions by covering travel expenses, staging workshops and functioning as brokers for consultancy work. Influential politicians of the kingdom are of course also prominent brokers in the competition for contracts.

BUCADEF operates in most of the districts of the kingdom and has activities in roughly 50 out of the 168 sub-counties in the kingdom. In Masaka, the representative expressed a strong desire to become more consistently included into frameworks for social service provision. Previously, BUCADEF, just like most organisations of the kingdom was met with great suspicion from central and local governments and seen as a way to undermine NRM structures and hegemony. On the whole, however, they claim that they no longer encounter mistrust from government officials but are regarded as a legitimate actor. They seek and get recognition and accommodation; in fact, they "don't want to be independent, since then if something happened, the government couldn't help us. The government is always there" (interview, BUCADEF District Co-ordinator, 17 February 2003).

Just like World Vision, BUCADEF stress their non-political nature, in accordance with the overall terms of agreement between NRM and the Buganda kingdom. However, the presupposed distinction between "politics" and "development" does not mean that political considerations are absent, only that they are not talked about in public. BUCADEF's discourse on development is embedded within a wider narrative that centres on monarchist conceptions of Buganda's rightful but neglected place and the need to revive a selection of traditions in order to create development in the region. Present-day decay of law and order, the status of health and education, the low level of economic productivity and communal spirit are contrasted against a "golden age" characterised by "hard work and honesty" and "family and village cohesion" (Buganda Kingdom, 1999: 5) upheld by the structures of the kingdom. These interpretations are also tied to carefully presented demands for federalism (Kingdom of Buganda 2001).

The contours and orientation of any civil society organisation is shaped within the broader political context within which it has emerged. While the capacity of BUCADEF to turn into a prominent development actor is limited, the importance of the kingdom in regional and national political arenas gives all its institutions a special role and obliges all political leaders to deal with them. As regards state-ethnic relations, the transformation of the relations between the NRM and the Buganda kingdom from confrontation to compromise has created its own channels of mediation. The government is careful not to provoke ethnic sentiments that could be used by the kingdom politicians for withdrawing their support for the regime.

The political importance of BUCADEF should be seen from the perspective of politicised ethnicity at large and the contestation over ways of organising political participation and public life. The political project of the kingdom involves the restructuring of state-society relations that is potentially far-reaching. While BUCADEF and kingdom representatives claim to work for all the people in Buganda, this formulation points to unresolved fundamental issues about the precise meaning of political citizenship in Buganda as a political entity, as ethnic minorities perceive it as a threat. Again, as with so many other issues in the local development arena, rumours abound. While BUCADEF stresses the harmonious relations between Baganda and other ethnic communities, other informants told stories about distinct hierarchies among their beneficiaries on precisely that basis.

The claim that everyone is welcome so long as they are willing to join hands for the development of Buganda is a far from innocent statement, and in fact a controversial basis for social organisation in Buganda. There is also a class dimension to this, related to social struggles around land and ultimately to struggles over the political meaning of ethnicity and monarchism. BUCADEF has in a sense been a vehicle for the kingdom to attempt to bypass such struggles. However, even though a project of the Buganda kingdom, when confronted with conflicts on the ground, BUCADEF inevitably incorporates contradictory tendencies.

Monarchism has historically been the focus of struggles between the Mengo establishment and the peasantry, and between different strata of the latter (Mamdani 1995: 120). This has created complications of its own. BUCADEF cannot bring up the question of land in any direction. On the one hand, to criticise the government's land policy would be to go beyond their "developmental" mandate. On the other hand, to criticise landlords would also be difficult, even when these are seen as to be standing in the way of community development.

Many observers agreed on the view that BUCADEF's capacity as a development organisation was limited not only by lack of resources, but also by the role of the kingdom at large. The cultural ties between the kingdom and the population in Buganda have weakened with intermarriages, general individualisation and economic links which have been severed with the selling of land. People seemed to view the kingdom pragmatically, welcome its cultural expressions, make use of its development facilities but keep a distance to its political project as long as it looks weak. It should be noted that these are impressions on my part, and not based on systematic research, but it is a perception supported by other studies (Englebert 2002). The tensions and ambiguities about what Buganda is and should be still go to the core of the formation of the Ugandan state.

Community, civil society and state

The above discussion has centred on the relationship between civil society groups and the state. How do both relate to the constituencies which they claim to represent in society at large? There are simultaneous mechanisms at work. While civil society groups themselves seek recognition or a permanent role in the institutional framework, political actors employ them as vehicles for penetrating into the communities. To the latter, they may be viewed as intervening institutions that offer access or protection. The capacity to make use of such resources is of course unevenly distributed among sections of the community. Because of the financial weakness of most local government institutions, they cannot orchestrate the distribution of development projects at will.

Local governments depend on others, especially foreign NGOs. Although this dependence runs both ways as NGOs need a good working relationship with local politicians to be able to stay in the district. This evidently puts restrictions on their propensity to speak out against corruption or other forms of abuse of power. There is often stiff competition between different administrative units over the geographic allocation of NGOs. This competition is played out along lines of political orientation (expressed in terms of "movementists vs multipartyists" during the period of field work, when the movement system was still operational), religious affiliation, ethnic belonging or a combination of these.

Fieldwork for this study has been conducted in different parts of Masaka. However, a more detailed study of village politics was undertaken in Bulando village, Buwunga Sub-county, a division of Bukoto East County, located in the eastern part of Masaka. Buwunga is roughly five miles southeast of Masaka town towards Lake Victoria. It was previously a prosperous area, but its fortunes were reversed along with the rest of Masaka. Kasozi (1994: 235-239) provides an account of the changes from the early 1960s to the mid-1980s.

Good health and education facilities were left to decay after looting had destroyed the health centres during the 1979 war. The local economy stagnated with the demise of the cooperatives, since bananas and coffee grown by the local population had been sold to the cooperative union. It is now considered a relatively poor part of the district, due to the low fertility of the soil and the dependence on a single cash crop, that is, coffee. Poverty is also self-reinforcing through low levels of education.

Politically, Buwunga is rather similar to other rural areas of the district. NRM candidates generally win elections, although by far lesser margins than is the case in other regions with stronger degrees of NRM hegemony. It is difficult to establish how solid and genuine this support is. In the past, the NRM had structures for mobilisation around campaigns, whereas opposition candidates were not allowed to run officially for instance on a DP ticket. I was told by several informants that during the presidential elections in 2001, armed soldiers surrounded polling stations in different areas of Bukoto East Constituency to ensure that people voted "correctly". This was however denied by local politicians and local government officials. In view of the well-documented violent character of the election campaign generally (Human Rights Watch 2001a), the information can certainly not be dismissed. Political opposition has been repressed for some time.

Foundation for Human Rights reported in its annual report for 1999 that the local population of Buwunga (and other places) was coerced into taking part in pro-government *chaka mchaka* "political education" courses; in some cases, non-compliance resulted in torture (Foundation for Human Rights Initiative 1999). Vincent Kimera of DP, later FDC and the main campaigner for Kizza Besigye in Masaka in 2001, was elected to represent Bukoto East in the sixth parliament (1996-2001) but was forced to stand down due to lack of academic qualifications. In both 2001 and 2006, John Alintuma Nsambu was elected on an NRM ticket, and President Museveni received a majority of the votes in the Presidential elections.[10]

Bulando village has around 200 inhabitants, and is rather remotely located, with roads that get washed away in the rainy season. Most inhabitants are small-holding farmers, and several are tenants. Coffee has for a long time been the main cash crop. There is a primary school in the village, but the nearest public health centre, Mazinga, is located a mile and a half away in a neighbouring parish. The closest hospital is a highly regarded Catholic Mission one, Kitovu Hospital. Its quality services would normally be beyond the financial scope for most villagers. When talking about the main problems with regard to health, the discussions understandably soon entered into broader debates on poverty and its causes, since all the interlinked problems that were discussed had their common roots in poverty. Health problems were seen as mainly caused by poor sanitation, such as the absence of latrines, and lack of protected water sources, and specific diseases are invariably worsened by malnutrition. The biggest health problem for the community as a whole is the

[10] In 2001, Museveni received 157 out of 235 votes cast at Bulando polling station, while his closest challenger Kizza Besigye got 0 votes! On the other hand, the DP linked candidate Francis Bwengye is – incredibly – claimed to have got in Bulando 121 of the 373 votes he received in the whole of Masaka District; most likely, there was an administrative confusion of votes (Source: Electoral Commission, Masaka District).

large number of AIDS orphans. Some of them are taken care of by their grandparents. Others, however, have nobody to cater for them.

Mothers have not been helped by the shortcomings of public health care, as experienced by the people of Bulando. While most participants welcomed the construction of new health centres, many lamented the shortage of medication and trained staff there. That this is a widespread problem, is confirmed by the Uganda Poverty Status Report 2003 (Republic of Uganda 2003: 147). Health centres were also said to lack electricity for storing medication when available. A common experience was that instead of being given drugs in the health centre or hospital, patients were directed to a certain pharmacy to buy them. This fed into a wider narrative of corruption in the sector, and that services that were supposed to be free had to be paid for – sometimes so much that it was impossible, or that one had to sell land or cattle to pay the fee. People linked this and what was seen as poor quality services to the low pay for health workers, and remarked that the government was aware but did not care.

In Bulando, informants viewed World Vision, along with a few other NGOs as the most efficient vehicle for development efforts. World Vision had been in Buwunga since 1989 (group discussion, 7 November 2000), and has mainly dealt with assisting orphans with education and supporting the construction of schools, health units and safe water sources. Other NGOs operating in the area include Kitovu Mobile; the mobile health care unit of Kitovu Hospital, Masaka Diocese Development Organisation (MADDO), Uganda Women's Effort to Save Orphans (UWESO), and the Swedish VI agroforestry programme. Over time, these NGOs have set up their own development committees, although some people were dissatisfied with having been left out, claiming that leading persons only distributed information and resources among themselves (group discussion, 25 February 2004).

BUCADEF was disregarded due to its perceived incapacity to deliver. Many people had bought the kingdom's certificate, but neither BUCADEF nor any other kingdom representatives had made consistent efforts to promote development, according to informants (group discussions: 3 March, 5 March, and 7 March 2003).

Some people were members of community based organisations (CBOs). There were groups for women, youth, animal keepers and crop growers, such as the Bulando United Farmer's Association (BUFA), *munno mukabi* ("a friend in need") and the *Tusitukiremu* women's group. These multi-purpose groups raise funds for specific occasions such as funerals, provide credit or offer links to bigger NGOs. A general observation was that it proved difficult to sustain their activities due to lack of funds (group discussions: 5 March 2003, and 7 March 2003). It was suggested by the men that women's groups were relatively more consistent and more efficient (group discussion, 5 March 2003). Connections to an outside resource provider are essential.

Everybody was very aware of the resources accompanying the presence of a major NGO. Access to World Vision or similar entities, or to higher levels of the local government system was in turn viewed as altogether dependent on a "good leader", generally perceived as one with contacts to such organisations or to someone capable of serving as a broker. Again, there is constant competition within and between villages, parishes and sub-counties around the access to services and resources. The ability of leaders to create access is in turn based on ethnic, religious or political belonging, and their capacity to secure protection and strike alliances on that basis. Differences between leaders in this respect can be substantial. While the LC I Chairman in Bulando was generally seen as lacking in connections, his counterpart in another village, a big farmer and businessman, was perceived as more efficient in securing funds.

Local councils were generally looked upon by people in Bulando as effective for settling intra-communal disputes, but by and large as inefficient in securing development, although the population squarely placed responsibility for creating development and providing and regulating services on the government. Views diverged on the activities of the village council, with some complaining that meetings were few and far between and distribution of information imperfect (group discussion, 7 March 2003). Likewise, views differed on the usefulness of the Parish Development Committee, with some defending its work whereas others did not even know of its existence.

A common view, not confined to Bulando, describes local government structures at higher levels as "thoroughly corrupt". While this may be a common popular view of politicians anywhere, widespread graft within the local government system has been amply documented (Centre for Basic Research 2005), although LCs occasionally fulfil a scapegoat function in this respect. Stories about district and sub-county corruption are frequently used by politicians at higher levels to conceal the ways in which local fraud is systematically connected to large-scale corruption encouraged or orchestrated from the national level. Complaints about difficulties in making the sub-county council hand down the allocated grants to parishes and villages were also rife. Most development interventions come with institutional structures, frequently in the form of development or user committees at different levels.

In the study area, government development committees were only rarely operating due to lack of funding. It was often remarked that "it is very difficult to mobilise people for voluntary work", and many local politicians and administrators pointed to an increasing monetisation of public life, not least during election campaigns. In cases where NGOs were able to offer some basic incentive, committees would occasionally take off (field notes,

2003). Likewise, there was unanimous disappointment with the two successive MPs (group discussions: 7 November 2000, 3 March 2003, 5 March 2003, and 7 March 2003) who both were said to have come with empty promises only to forget them – "we can only see him on posters" (group discussion, 5 March 2003).

It is obviously impossible to generalise from one village, but it is plausible to assume that this reflects a more general tendency. The limits to popular participation have been underlined, for instance, by de Torrenté and Mwesigye (1999) in their study on health sector provision in other parts of Uganda and more recently by Golooba-Mutebi (2005). The proliferation of development committees at community level echoes the general trend towards pluralised development structures. This may be read as the creation of parallel structures of local government.

In principle, institutional ambiguity could result in parallel sources of democratic interest representation. While it should not be ruled out that some genuine capacity building may take root in the future, for the time being the dominant overall effect seems to be a institutional and political fragmentation. These parallel structures also turn into sites for intense elite competition over peasant loyalties and collective identities, a burning issue especially in the context of the struggle over federalism. This competitive dimension is likely to become more pronounced with the advent of multi-party politics.

A striking feature of the interviews and discussions that were held during fieldwork was the deep cleavage between the "official story" of local politics and the picture that most people gave on conditions of anonymity. While this hardly is unique to Masaka or Uganda, the discrepancy was so marked that a few reflections on it are required. For obvious reasons, I am unable to reference the observations in a stringent way. Nonetheless, it forms an important part of my understanding of local politics during the period of fieldwork.

While the official perspective, offered not only by politicians but by most categories spoken to, predictably emphasised bottom-up planning, popular participation and collaboration between all involved stakeholders striving for deepened empowerment, the version told on conditions of anonymity instead involved themes such as top-down control, politicised rivalry, corruption and influence peddling by elites and political apathy among "ordinary citizens". While nobody would of course officially claim that conditions were perfect, obstacles were portrayed as temporary and as emphatically not rooted in power structures. What is interesting about this is perhaps not so much the fact that there existed a separation of narratives, but rather the lengths to which most people were willing to go in order to avoid "sensitive" issues, since the observation that politics is shaped by power would appear to be perfectly commonplace.[11] An alternative picture of local politics emerging from these accounts would read something like this:

Relations between different kinds of actors in society and in the state are characterised by separation and conflicts to a greater extent than by collaboration and synergy. This seems also to hold true for actors at different levels and acting in different capacities in the political system, such as politicians and technocrats. Between levels, relations are typically hierarchical, and there is limited scope for influence, especially from the community or village level upward. It is very difficult, for instance, for LC I leaders to approach sub-county councillors, let alone challenge them, due to the overall distribution of power, resources, status and hegemony. Both structural, such as levels of education and income, and conjunctural factors, such as political subordination and loyalty, tend to reproduce these hierarchical relations.

---

[11] Another possible interpretation is that local government officials were uncertain of what was really going on in local communities but would not want to be seen to admit this, which is why they held on to the official version. I owe this argument to Frederick Golooba-Mutebi.

There is no iron law governing this, although the leadership factor often makes a crucial difference with regard to access to and distribution of resources among communities. However, political structures are weak, and to a great extent dependent on functioning personal relations. Access is perceived as sustained through a combination of formal and informal attributes, including religious affiliation, ethnic belonging, age, gender, class and political orientation; in short: all markers of social differentiation interact.

According to one observer, "politicisation of community development is all-pervasive..." There is much uncertainty and mutual suspicion, reinforced by persistent interpretations of political orientation, religious affiliation and ethnic identity:

> regardless of whether there is any clear substantial basis of the rumours or not. Interpretations arise directly or by association from questions like 'where does this money come from?' If a political leader suspects the involvement of 'opposition politics' in relation to a government programme or an NGO, close supervision will follow (interview notes, February 2004).

According to another observer:

> there is ongoing politicisation around resource allocation from councillors or MPs, who will seek to capitalise on development. If the local leaders have good connections the area prospers – if not, it will perish. District and sub-county leaders will first need to cash in during their time in office. Only after that can they extend to the general good.

The umbrella organisation, Masaka NGO Forum had failed to take off because it was regarded as a vehicle for control on behalf of local political leaders. The typical pattern of political relationship between leaders and citizens is one of top-down manipulation, not bottom-up engagement. Social relations are hierarchical, most of civil society at the local level is weak in terms of finances and general capacity, which:

> puts both civil society and communities in a position of dependence, and communities are dependent upon both politicians and NGOs. Both formal and real mechanisms of accountability are weak, and structural obstacles are enormous (interview notes, February 2004).

Another local observer had this to say:

> local patrons and brokers dictate the needs of the community. Big farmers operate as credit providers. Local capitalists are too busy to engage in politics on a personal basis, but interfere by pulling the strings. Normally, there is at least one in a village – a landlord, a businessman or a prominent farmer – who is the credit provider. These people use their influence in a number of ways, by allocating loans, events, resources from condoms to health units, in order to control the election of chiefs and councillors… Of course this distorts the allocation of public resources and public interests (interview notes, November 2000).

According to one account:

> there is a widespread popular interpretation of what is going on… Leaders and ordinary people at lower levels lack education, skills, experience and power to challenge those above them, be it NGOs or LCs. In order to get whatever little they can, they cannot afford to make trouble or to be seen as 'antigovernment'. To a great extent LCs have developed into Movement structures, and where there are opposition leaders these tend to keep a low profile.

One sub-county suffers from having a chairman from the multiparty camp and he is therefore sidelined by the district leadership (interview notes, February 2004).

In the experience of several informants, political connections are activated around elections. Loyal and effective mobilisers are rewarded, others are removed. Economic contributions during campaigns are rewarded after them by, for example, allocation of contracts. Perceived opposition activists or sympathisers are harassed, co-opted or even beaten up; dissent, however modest in scope or reach, is cut short. Development initiatives normally benefit people according to hierarchy; political and economic favours need to be paid back, and future political support cultivated. No group or community can afford to be seen as critical – risking to be cut off from networks of access. Everybody knows the implicit limits. This creates a political culture of silence – or whispering – reinforced by the ever growing disjuncture between official rhetoric claiming "all encompassing" politics and the reality manifested as nepotism, corruption and arrogance.

The fact that local political discourse during the time of field work to such a large extent was characterised by assumptions, whispers and rumours necessitates a methodological remark. Evidently it is impossible to verify the truth of all stories, in particular, those about corruption and nepotism that circulated in the popular rumour mill. All storytellers can be assumed to have their own reasons for wanting to influence my understanding. The relatively unambiguous direction of the stories, in combination with documentation from elsewhere (Centre for Basic Research 2005), gives them a degree of credibility, though. Moreover and regardless of the precise truth of the accusations, they reflect popular political consciousness and as such a parallel and influential version of what is going on. Furthermore, the subdued mode of discourse is significant in its own right, as it points to the constraints imposed on open critique by no-party politics.

This evidently also affects civil society. There is an ongoing competition between NGOs where both political and religious identities are tacitly assumed. Most CBOs are seen as having direct or indirect religious and political attachments, in terms of funding and coverage. All civil society organisations need to navigate in this politicised terrain, by its nature informal and implicit. Some groups disengage, while others are even created by or are loyal to leading politicians. As suggested by Brock *et al* (2002), there may well be a certain dialectic to this, as such connections may function as channels of interest representation for poor communities. In contexts where democratic institutions for representation of social and political concerns are weak, parallel forms of participation and representation are bound to emerge. Informal linkages may be used both for imposing subordination and control and for breaking out of such relations of domination.

What can be said about the composition of the local state? Whereas the welfare arms of the state suffer from underfunding and weak capacity especially at lower levels, the coercive arms of the state and their extensions in society exercise effective presence when, for instance, the need arises to silence opposition activists and sympathisers, especially around the time of elections. This means that the capacity and legitimacy of the local welfare state is dependent upon NGOs, who by their nature refrain from becoming involved in local policy making for fear of becoming dragged into local politics. This opens up for regulated or unregulated parallel arenas of provision and welfare regimes.

Parallel systems are reinforced by NGO Development Committees and continued illegal health provision as the latter undercuts the capacity and legitimacy of the state. According to one observer, "it is difficult for the local government to compete with NGOs, especially international ones that come with a lot of resources and get things done." Government provides contacts, but for "development" people turn to NGOs. "Many government programmes are associated with failure. People prefer "NGOs" and "PRA" [Participatory Rural Appraisal, an acronym which is now firmly integrated into community discourse.][12] Their work is very useful", but the overall unintended consequences are that

[12] However, "PRA" has since ca 2003 assumed another and much less attractive meaning altogether, namely "Peoples Redemption Army", a rebel group normally referred to as "shadowy". The government links it to the political opposition party FDC, who in return claims that PRA is a government creation set up in order to frame its leaders.

it tends to fragment the institutional landscape and "alienate communities from the government" (interview notes, February 2004). However, the other side of the coin is that the local state also capitalises on the presence of NGOs, and regains some legitimacy by proxy.

Locally, health services are a potential political asset, although they are rarely given priority in budgets (Jeppsson 2001). Politicisation may occur in several ways. One way concerns the political capital always linked to social service delivery, either as offered by the government directly or by its "creating an enabling environment" for donors and NGOs. Another dimension concerns the competition for money and allocation of infrastructure that comes with competition for this political asset within localities. When asked about general conflicts relating to the health sector context, several informants mentioned the use of condoms (and contraceptives generally) as a sensitive issue, due to the influence of the Catholic church in the region. Catholic leaders have actively agitated against contraceptives and family planning, issues that are on the agendas of most development organisations. It is not uncommon that priests use prayer services and fundraising occasions to attack this practice and people associated with it, notably certain NGOs and the District Health Development Team.

Another source of conflict is the theft of drugs and the prescription of sub-standard drugs, associated with but not limited to the practice of "fake doctors", yet another typical feature of the informal health regime. Unqualified personnel open up clinics in rural areas and pose as doctors. This is said to be a problem mainly in remote rural areas where the population is unaware, poor and dependent or lack other options; urban people are more aware and have a greater selection of clinics to turn to. It is very difficult for the local government to close down clinics if it cannot offer the local population an alternative health unit. This constitutes a dilemma for authorities. Enforcing regulations would play into the hands of local politicians who would like to capitalise on popular discontent. As a result, authorities sometimes refrain from confronting quacks for fear of losing popularity. Informal welfare regimes linger on and interpenetrate with formal ones.

What explains the weak capacity of democratic forces in civil society and among communities to participate in and influence policy making and policies? Popular democratic influence is blocked by a number of factors. Financial resources are scarce, placing restrictions upon policy choices. Social relations are hierarchical, especially with regard to age, class, gender and level of education. It is very difficult for people at lower levels to effectively challenge their superordinates.

Local communities are often left at the mercy of higher levels of government and NGOs through patrons and brokers. Lack of politicisation from below is however not merely a matter of cultural-structural obstacles such as poverty, ignorance and illiteracy. Nor is it mainly due to lack of fiscal decentralisation and real scope for decision making alone, although the above factors are significant and contributing. Rather, those factors are better understood within the context of political power relations between centre and locality, between the local state and critical sections of civil society and between local elites and critical sections of local society.

The main reason is the strong degree of NRM control and authoritarian hierarchical relations more broadly on the local level. There is a general pattern of seeking to fit in and lying low, for different reasons. In the case of BUCADEF, the general scope for articulating political demands has waxed and waned with the fortunes of the Buganda kingdom in its negotiations with the central government, but it is also restricted by its own weak position vis-à-vis the local government. World Vision, just like all other foreign NGOs, enjoys a privileged role in development on the condition that it does not engage in politics.

Some conceptual remarks on (de-)politicisation might be in order at this point. The concept is elusive and relative: what is politicisation in one sense may be depoliticisation in another. This is all the more relevant to recognise since the concept is at work on different levels, as analytical category and as used in practice in a particular social formation, and as such an object of study. In Uganda as in other societies, politics has specific connotations. Under the no-party system, one of these was that (competitive) politics equalled "sectarianism" and should be condemned.

At the same time as the official discourse has separated "development" from "politics" as two distinct types of activities, and with the specific recommendation that civil society organisations should concentrate on the former and stay away from the latter, it has reconnected development to politics in another, implicit yet more specific sense, namely through the suggested link between successful development and government politics. But as can be seen in the quotes above, people use politicisation to refer to how certain political groups have hijacked development projects. What can be concluded, however, is that intensified politicisation in the sense of ideological control from above has, as is common, corresponded to the closure of independent politicisation of discourse and organisation from below. Development has largely become politicised in a specific and narrow sense, with an ideology and institutional forms that stress direct community participation at the expense of interest representation and influence within or beyond the political system.

The rather "numb" feeling about the cases examined here should be understood to reflect precisely the reality of a political culture of silence at the time of my fieldwork. Political control, fear and limited outlet for dissent subsumed social conflicts, blocked the scope for independent politicisation and sent political struggles to a shadowy terrain of rumours. Due to the inherent legal-political contradictions of the Movement system, opposition remained muted. Political repression has been on the increase in Uganda since 1996 and exploded during the 2001 campaigns. Campaigns were characterised by intimidation, in Masaka, DP's John Kawanga was prevented from campaigning in Buwunga and Bukoto South. The military travelled around at night, threatening people with terrible consequences if they did not vote for Museveni and NRM. Contrary to perspectives that emphasise poverty and ignorance, patronage politics and lack of effective citizenship must be understood in relation to hegemony and the exercise of power through the political system.

Conclusions: Community, state and civil society in local political arenas

This chapter has brought out the complexities of local politics. Struggles over power shift, even from one parish to another. Complexity necessitates a detailed level of analysis, yet there is a general pattern within which local complexity is played out. Relations of domination between different social forces have both central-local and intra-local dimensions. Central-local relations impact on both the local state and civil society. In the case of Masaka politics, they include specific features as the role of Buganda in national politics; the legacy of Catholicism and DP; past wealth built on coffee production; civil war and local decline; the HIV/AIDS epidemic; and the rise of the NRM to domination in local political arenas.

The leaders of local governments need for their political survival to consider both the central government and local power holders. They have to manoeuvre in relation to a variety of political interests that are mobilised, by for instance landlords, the Buganda kingdom, the Catholic church and donors. The scope for any of the groups studied to influence policy directions is shifting but overall relatively limited. Capacity for influence has been explained in this chapter with reference to the balance of power between social forces and state interventions to regulate them but also by factors internal to the organisations and to the local settings.

Uganda Medical Workers' Union is hampered by the political weakness of a fragmented trade union movement nationally and the absence of collective bargaining mechanisms, due to state hostility to labour rights and a strategy geared towards marginalising and repressing trade unions. Therefore, their perspectives on problems and solutions of the sector are effectively sidelined. Locally, the union has suffered from weak leadership. The weak financial base of BUCADEF and the restrictions on the political dimensions of cultural institutions place constraints on their effective presence as a development institution. World Vision primarily establishes itself as an efficient development provider but not as an actor on the political scene. To some degree, these incapacities, from the point of view of political demand-making can be related to other reasons as to why NGOs refrain from advocacy work or politicisation. The very nature of their involvement in scattered projects, limited in time, creates a temporary and spatially restricted ad hoc form of engagement, with weak roots in local society. The internal structures for decision making of NGOs tend to impede effective democratic demands by their beneficiaries – who are often addressed in precisely such terms. Furthermore, NGO discourses on poverty and development, with their focus on a narrow interpretation of community empowerment and participatory development tend to stress symptoms rather than underlying causes of problems, which reproduces a neglect of power relations.

It would be misleading, however, to assess the relations between these groups and local politics of development singularly against a model of pressure groups. At least in the cases of BUCADEF and World Vision, it is probably more interesting to point to the parallel forms of demand-making and mobilisation that they generate. At the same time as these groups to different degrees lack the ability or will to function as pressure groups, they nonetheless constitute alternative sources of social power, who can offer protection and access to different constituencies. At the very local level, to the sub-county and below, World Vision tends to assume the role of the actual welfare arms of the state. BUCADEF is able to draw on the prestige of the kingdom to ensure cultural loyalty in exchange for development programmes.

The kingdom has reintroduced a competing version of how to structure relations between state and society and has expanded some political space for itself, even though that space is regularly narrowed every time the kingdom tries to push the central government for further autonomy. The presence of these organisations in local society furthermore offers points of access to the political system or ways of by-passing it, mainly for more influential sections of the population.

Due to their shifting capacity for interest representation, the organisations under study encounter the state in very different ways. UMWU's weak leadership in the context of ad hoc forms for negotiations has left the union in a vacuum. Issues that should normally be subject to systematic negotiation are now handed down (or not, as is sometimes the case) from the hospital administration or achieved on an individual basis. BUCADEF is accommodated into structures for community development, also in an ad hoc manner, through the interconnectedness between public, private and NGO forms of provision.

At lower levels, they tend to negotiate their presence with the local state in an everyday informal mode. World Vision has been taken on board in important forums of decision making at the level of the district council but prefer to exercise influence in silent ways. Structures of regulation are overlapping and shifting, and these groups target different regulatory mechanisms. Many issues remain beyond democratic control because they are handled by institutions that are privatised and informalised. At lower levels and with regard to welfare functions, it is certainly possible to speak of a degree of "civil societisation of the state" (Gibbon 2001: 833).

Different impulses, such as the preoccupation of the central government to seek control over local politics, NGOs' concern with projects of participation, donors' desire to promote governance, calls for the activation of "traditional institutions" by the Buganda kingdom and responses to all these merge in shaping local politics.

Institutional configuration is shaped by interest groups, networks and relations of rural authority. At the local level, regular interest groups are often characterised by uneven presence and weak capacity. Local arenas are distinguished by a plurality of overlapping institutional mechanisms, more or less formal or hierarchical, in terms of for example elite networks and coping strategies in the case of poorer social strata.

These mechanisms of control are often based on class, gender, (sub-)ethnic and religious identities. They all have differing capacities to create access to state institutions, due to local power relations, linkages to national or transnational centres of power. Occasionally these entities claim public authority in competition with or in the absence of the state. The links between formal and informal sources and forms of authority run in both directions. While formal rules may not regulate all aspects of public life, they do affect informal rules and activities. All actors need to relate to the imposition of formal rules, which serve to provide a system of reference for control. This has been illustrated above by discussing the different historical trajectories of these organisations and the extent to which they are able to relate to "sedimentary institutions" (Bierschenk and Olivier de Sardan 1997; Jones 2005: 31-33) that is to say the layered institutional configuration as shaped by past experiences and contemporary examples.

BUCADEF, World Vision, UMWU and other groups have very different kinds of resources to draw upon when seeking to root themselves in the local state and among communities. For World Vision, the lack of social roots is compensated by an abundance of economic resources. As much as BUCADEF represents an institution with cultural prestige, it has been unable to translate this into political influence because of the limited developmental resources on offer. Yet, conditions for local politics in Masaka are strongly shaped by the role of Buganda in national politics, but also by the relations between Masaka and Mengo.

There is considerable overlap between the district administration and NGOs in the field of community development in terms of staffing, funding and implementing. Close contacts tend to spill over into interdependence between sections of the political elite and the NGO leadership, each with their own networks linking them to national centres of power. Does this mean that the current development model tends to reproduce political patronage rather than promotion of democratic citizenship and interest representation? Francis and James (2003) suggest that the "technocratic mode" of design clashes with the "patronage mode" inherent in local politics.

While patronage politics certainly is pervasive, such an interpretation needs to be qualified. Their argument gives much weight to latent structures of undemocratic cultures marked by corruption and clientilism, without discussing the broader political processes that they are part of. It has been argued here that the low degree of autonomy and capacity of interest groups to push for inclusion into decision making processes needs to be understood with reference to the balance of power between social forces, the propensity of the state to engage with some of these forces and exclude others, and the structures of the state.

The contradiction that Francis and James call attention to does not merely follow from a technocratic design that is captured and perverted by some vague "patronage mode" that prevails locally. That contradiction is more persuasively understood in relation to political strategies of the central government to forge alliances with social forces and to the balance of power in society that shapes the responses to these strategies. Fundamental restrictions are imposed on the local leadership, just as they are on local governments all over the country; by dependence on revenue from the central government and by the strong political influence of NRM over LCs, particularly before, but also after the introduction of multi-party politics.

The practice of decentralisation is thus hampered by extremely weak fiscal bases and low levels of capacity, which tend to reinforce re-concentration over decentralised administrative units by way of their reliance on funding and staffing from central line ministries and donors (Francis and James 2003: 327, 334). In contrast to official claims of democratic accountability to the local electorate, there is financial and bureaucratic accountability to donors and the central government. The same contrast between the participation rhetoric and a hierarchical reality can be observed between the different levels of the local government system.

At lower levels of the political system, state-society relations become even more blurred and state institutions are only weakly and unevenly exercising routine functions of taxation, administration and welfare provision. Links between different levels of local government, in particular between LC I and LC III, are often disconnected. The scope for influencing higher instances of decision making is for most sections of society very limited indeed. This does not exclude the possibility of occasional individual or collective economic "empowerment", but it points to the limits to systematic political engagement set by a combination of factors, including poverty, hierarchical social relations and political closure in terms of the absence of both (until recently) political parties and of organisationally and politically viable civil society groups. Local politics is strongly shaped by the interpenetration of social, economic and political inequality.

Turning finally to the more general theoretical theme of state formation at the local level, a few reflections will be offered. The inquiry into what constitutes "the state" locally sets out from the observation that central states in developing countries lack capacity to penetrate society in order to extract resources, engage in nation building or to enforce legislation and administration evenly, consistently and reliably all over their territories in the terms of Michael Mann (1986), they lack "infrastructural power". This theme has been subject to a vast literature. It has been suggested that the state is unable to capture society (Hydén 1983) or that it finds itself captured by local strongmen in society (Migdal 1988). Without going into detail about the relative merits of these contributions, I have drawn on arguments that stress the mutual constitution of state and society (Boone 2003; Migdal *et al* 1995) in terms of how state institutions are rooted in and reshape power relations in society.

State-centred views with their focus on the capacity of states to penetrate and control society tend to gloss over the ways in which state intervention is fought out at different levels, as well as the different and occasionally competing trajectories of state-society relations. State institutions, especially at the local level, are not only shaped by political calculations of the regime and the presence of a state-party machinery to ensure for instance electoral loyalty. They are also formed by the social forces that try to mould them.

In the present conjuncture, the downsizing and decentralisation of state structures have opened up space for political forces at the local level. From the perspective of communities, these forces intervene as parallel political entities with specific resources to offer. International NGOs may undertake the role as welfare providers; ethno-regional movements may operate with tacit claims to statehood, to draw on two of the cases discussed here. This has been interpreted as the creation of a continuum of governmental apparatuses (Ferguson 1998: 58) or "degrees or kinds of ´state-ness'" (Doornbos 2000: 255), where the redrawing of lines between "state" and "society" is crucial for the construction of authority (Hansen and Stepputat 2001). Where the lines should be drawn keeps being contested in all political arenas.

In this chapter, specific features of a specific context are highlighted. But local contexts are not unique, of course; they share some common denominators. In the concluding chapter, I shall broaden the of local politics by placing this close-up picture of the local complexities in Masaka in a wider Ugandan perspective.

Chapter 7

Conclusions: Militarism and "Governance" in Uganda

This book has sought to theorise and explain how different state-society relations create or block conditions for democratising the state. It has situated the in the context of the apparent contradictions that characterise Ugandan state formation and state-society relations, from militarism to "governance" participation. The problem has been addressed from a theoretical perspective that stresses the interplay between relations of domination in society, within the state, and among transnational political forces. More specifically, it has analysed state-society relations as they have unfolded through modes of conflict regulation and interest representation in the health sector.

The book set out from the argument that the exercise of state power is mediated – accepted, modified, resisted, and so on – at the level of society and that the character of the state in terms of its capacity and propensity for, among other things, development and democracy is shaped by its internal composition and its roots among different domestic and external constituencies. When holders of state power seek to consolidate control, they do this in relation to demands from domestic social forces and external pressure and within structural constraints such as inherited institutions and revenue base. What does this mean in contemporary Uganda?

This book argues that the present government has, from an initial position of weakness, consistently sought to secure control in response to demands from contending political forces and to challenges of promoting accumulation and revenue, and consolidating legitimation and a political base, but that political pressure has over time led to tactical moves that result in contradictions between different aims such as economic growth and political support. Measures taken to sustain political support, such as for example creating new districts, has increased public expenses and gradually slowed down growth, as opposed to an earlier phase of reforms when political support and growth complemented each other to a greater degree.

Although achievements of economic recovery and institutional reconstruction are real, they are institutionally and regionally uneven and fragile due to inherent contradictions reproduced by underlying structural weaknesses maturing into political conflicts. The scope and political content of these reforms are heavily shaped by a historical legacy of authoritarian structures, as well as by the contemporary reinforcement of part of them through the way state rulers have based their power among forces in society that have tended to block rather than to promote democracy. While this argument should not be taken to imply structuralist determinism, it does underline the structured conditions for political agency and institutional reforms. In this chapter, I will start by summarising the findings before situating them in a broader context in the second part.

The analysis in Chapter 2 explored the historical roots of the institutional landscape of a multiplicity of welfare regimes. Prior to independence, health services were mainly provided by missions, and modes of provision and regulation evolved in tandem with the fragmenting structures of indirect rule and regional underdevelopment under the colonial state. Only under late colonialism did the state make modest efforts to expand basic welfare services to Africans. After independence, the Ugandan state responded to popular demands by embarking upon ambitious programmes to expand health and education services. Underpinned by export earnings from coffee, facilities rapidly expanded in numbers and improved in quality.

While Uganda in a broad sense followed a rather typical model of state-directed economic policy making and incorporation of interest groups, strong regional power centres provided a countervailing tendency, regional and national welfare systems coexisted, albeit ever more uneasily. While religious institutions continued providing health services, this took place under the close supervision of the state.

The state-centred orientation of the political economy became much more pronounced after the constitutional crisis in 1966 and the resulting centralisation of state power, including economic policy making. Public health and welfare services suffered badly from political and particularly economic destabilisation from the late 1960s onward, and so did the health status of Ugandans. Public infrastructure and regulatory capacity fell apart.

The disintegration of the economy, society and state left a legacy of fragmented and increasingly informal welfare regimes that survived under great stress. Military rule, interim governments and civil war destabilised society in fundamental ways and in the context of institutional decay, insecurity and political repression, most people had no option but to take recourse to a combination of coping strategies. The legacy of connections between formal and informal, public and private health regimes continues to shape policy making and institutional reform in subtle ways. This was very visible during the period covered in Chapter 3, from 1986 to the early 1990s.

When the NRM government embarked upon economic liberalisation reforms, it did so from a position of political weakness and implemented policies through very weak state institutions. These years were characterised by intense political struggles over NRM's future direction and armed rebellions that challenged the state. Priority was consequently given to macro-economic stabilisation and security. While NRM faced little effective resistance from a delegitimised and damaged political opposition, greater threats came from armed rebellions. State control and legitimation revolved around the politics of security or its absence.

As the NRM had both its institutional origins and its claims to control over state power rooted in armed struggle, restoration of security was and for a long time remained a chief political asset in the south for the new government. In the north and east, on the contrary, insurgencies continued to destabilise society. In both southern and northern Uganda, but for different reasons, health and social welfare in general continued to be neglected during the late 1980s.

Although a great number of civil society organisations were set up, reactivated and or entered from abroad during this period, most were concerned with immediate issues of coping and survival. With only an embryonic health policy, there was rather little to influence in a systematic fashion. As a consequence, informal welfare regimes remained highly important for most strata of Ugandan society and continued to dominate the institutional landscape and influence the workings also of formal public health care.

From the early 1990s the Ugandan state was restructured and expanded. As compared to countries where adjustment programmes straightforwardly demanded downsizing of the public sector, the Ugandan case was more contradictory. Certainly, Uganda government enforced thoroughgoing retrenchment in the civil service and set out on a military demobilisation programme. But only a few years later, the bureaucracy had grown again, although now in the local state, and the military was being expanded once more.

This change cannot be understood unless related to processes of regime consolidation and state-society relations. By the mid-1990s, the NRM had established firm control over the political landscape. Its achievements in stabilising the state and reviving the economy had consolidated its control over much of Ugandan society as well as award it with legitimacy among donors, who also provided a growing revenue base for the state.

Political control was sealed through the enactment of the new constitution in 1995 and the presidential and parliamentary elections the following year. This offered the government a much stronger degree of autonomy vis-à-vis both domestic and foreign constituencies. Accordingly, it could thereafter embark upon political projects more independently. The more closely the government seemed to be following advice from key donors with respect to macro-economic and formal governance reforms, the further it could stray from donors when it came to more hard-core political issues such as political pluralism and the role of the military in politics. Or more precisely, while donors may have developed concerns about corruption and NRM monopoly over state power (and the US did try to push for multi-party politics during the Constituent Assembly debates in 1995) they have by and large given priority to economic policy compliance.

Chapter 4 tells the story of this period of consolidation, expansion and restructuring of the Ugandan state. State legitimation exercises were carried out in relation to both domestic and foreign members of the broad-based coalition, who in turn from different directions sought to reshape the political orientation of the state. From the mid-1990s, the government was encouraged by donors to include civil society in policymaking. Due to increased institutional, financial and political state capacity, it could do so from a position of strength and extend its invitation of participation to a selection of favoured civil society actors who had emerged in close connection to the NRM state.

With regard to the social base of the state, including the broad contours of state-civil society relations, the contradictions between the encouragement of partnership and participation, and the closure of interest representation and influence at the level of politics may be qualified in several ways. Participation is limited to certain actors, and the more resources they command, the greater their influence, obviously. The scope for participation varies between ministries, with "soft ministries" being more accommodating. It is restricted to certain issues, while the most important issues and institutions were safely insulated from popular demand making. These contradictions were given their legal expression in the NGO Registration Act of 2006. The structural contradictions that has underpinned the conditions of NGO operations in terms of a threatening restrictive legal framework but accommodative practices render them vulnerable to political manipulation, and have been reproduced in the discourses and practices of individual NGOs, who wrestle with contradictory logics of "partnership" with the state and "advocacy" against it, or of both implementing and monitoring government policies.

What then can be said of the character of the local state and civil society as analysed in Chapter 5? At one level local politics is indeed very composite and structured by specific historical legacies that result in subtle or sometimes even marked differences in levels of development and political orientation even within a single district. At the same time, most local governments are strikingly similar with regard to their relation to the central government in that they are financially unviable and correspondingly politically vulnerable – particularly so, of course, under no-party rule.

Differences between districts largely follow regional historical fortunes, in individual cases bolstered by the influence of national political heavyweights where these are prominent sons (in most cases) or daughters of the soil. The latter impact may of course lead to occasional clashes between intervening kingmakers, normally operating in the national political arena, and local strongmen, but this is invariably preferred by the local population as compared to the absence of such power.

The extraverted character of the local state at the district level stems from other external influences apart from the reliance on revenue from the central government and political dependence on NRM under no-party rule. One such influence follows from the weak resource base and is manifested in the form of resourceful NGOs such as World Vision or other donor programmes. These are taken onboard as unacknowledged but effective partners in local government, rather like the central government has embraced donors at the national level. At the same time, NGOs need to navigate in the local political context, downplay any political pretensions that they might have and relate to local power brokers such as politicians, religious leaders and landlords.

In the case of Masaka, yet another external influence comes from Mengo and the leadership of the Buganda kingdom. BUCADEF has been analysed as a tool for the kingdom to reclaim its relevance in developmental terms, and the district leadership needs to relate to the shifting terrain of Mengo-central government relations. While the district leadership evidently is not keen to surrender power to Mengo, they still have to tread carefully in view of the popular support that the *Kabaka* enjoys. Hence, within rather severe financial and political restrictions for the local state, local politics is fluid and shaped by delicate considerations of informal influence.

The findings from Bulando suggest that the state is weakly and unevenly institutionalised at lower levels of the system, or the level where the majority of Ugandans lead their lives. Again, this is not to suggest that the state is absent; it is to say that different state institutions are present and absent to varying degrees from the point of view of local communities. This differs between regions, over time and in relation to different sections of society.

In the case investigated at the level of village politics, the welfare institutions of the state were notable by their relative absence, or more precisely: by their distant presence. But this needs to be qualified. There is a primary school in Bulando (although the standard of education there may be subject to debate, as with most government primary schools in poor rural areas); but the health centre is far away, as are the sub-county headquarters, not least in terms of political accessibility. A village elite has access to the state at the sub-county level or beyond. Although not always manifestly present, the state impacts on local politics in some ways, even if indirectly and by unintended consequences.

The findings point to the existence of parallel political arenas at the local level. Following Jones (2005: 33-34), one may speak of a *bricolage* of sedimentary state and non-state institutions, which to shifting degrees are embedded in local society. For some groups and individuals, local informal arenas replace state ones, while for others these arenas function as vehicles for connecting to the state at other levels, directly or through power brokers. Foreign NGOs, in themselves power brokers, may be superficially rooted even where they energetically develop their own institutions such as development committees. One cannot, however, exclude the possibility that in the absence of immediate state welfare and development intervention, these groups may, by combining material and spiritual resources as in the case of World Vision, gain relevance and quickly embed themselves by becoming indispensable.

A decade ago, Bazaara (1997) wrote about a looming agrarian crisis even in previously relatively prosperous Masaka. In the context of rapid population growth, land fragmentation and erosion, coffee wilt, unresolved land tenure relations, small local markets and limited employment opportunities, one may foresee the continued relevance of NGOs and the permanent institutionalisation of ad hoc forms of governance structures that characterise current local health regimes.

The above summary may unsurprisingly be interpreted as a critique of governance expectations. Does it then render support to a neo-patrimonialist reading of Ugandan politics, recently articulated by critics of the NRM government (Barkan 2005; Mwenda and Tangri 2005)? Not really. The influence of vast patronage networks, informal and corrupt modes of primitive accumulation among complexes of business people, politicians and high ranking military, and an arbitrary and increasingly personalised rule over key state institutions by the president need to be related to relations of domination in society as these impact on state formation, as well as to countervailing tendencies in order to explain the scope for and content of patronage politics. As such, these features may be thought of as both causes and symptoms of authoritarian rule. I shall seek to illustrate this from the health sector.

In August 2005, what came to be known as the Global Fund scandal was made public. It was disclosed that billions of shillings, donated by the Geneva based Global Fund to fight AIDS, tuberculosis and malaria, had been diverted in an extensive series of scams and that government ministers had been involved. Popular fury exploded against what was seen as vultures stealing from the sick and dying patients. The three ministers in the Ministry of Health had to leave the government after the 2006 elections in a move widely interpreted as a concession to donors and the public. Evidently, this was a temporary and largely symbolic sacrifice that may not have affected the deeper structures of corruption or the regulation of the health sector.

The Global Fund saga nevertheless typifies the decay of state institutions and points to underlying problems. Parallel to struggles between ministries over resource allocation and policy priorities, informal political and economic struggles over accumulation and influence are manifested within state institutions. Alongside its routine functions, the state is an arena for intense factional infighting among the ruling class, and the main reason that countervailing forces within for example, the Ministry of Health are too weak to resist is that certain groups are seen to have been given leeway "from above". But again, such patronage politics needs to be understood in a broader context of presidential authoritarianism and suppression of opposition politics at party political and civil society levels. I shall elaborate this argument in the remaining part of this concluding chapter.

* * *

This study set out with a broad aim: to examine the formation of state-civil society relations in contemporary Uganda in order to illustrate and explain the scope and capacity of different social forces to create access to/and democratise the state. It has sought to address that topic empirically through detailed case studies of state-civil society relations in the health sector, at the level of national politics as well as locally in Masaka District. Now, it is obvious that the case studies undertaken cannot in themselves provide sufficient evidence in order to either corroborate or refute any specific theory about the nature of the contemporary Ugandan state.

First of all, the findings need to be related to conclusions from other studies. But that is not enough. Empirical conclusions are also essentially dependent on the plausibility of theoretical arguments. This final section seeks to situate the findings in precisely such empirical and theoretical contexts in order to relate the parts to the whole, so as to enable a return to the wider scope that motivated the study. It draws on existing empirical work on other instances of state-civil society relations, in other political arenas and regions, as well as on theoretical reflections of a more general kind.

In order to bring out what is specific and what is typical about the cases studied here, I will start by going beyond the civil society cases in this study and compare them to other instances and then move on to discuss how state-society relations have evolved in other parts of Uganda. After that I seek to situate health politics on the broader canvas of state formation in Uganda in its entirety, including the transnational dimensions of this. Finally, I shall make a brief comparison between Uganda and a few other countries.

Civil society beyond the present cases

How typical are the cases analysed in this study? Judging from a number of studies (Dicklich 1998; Bazaara and Nyago 1999; Oloka-Onyango and Barya 1997), the conclusions offered above about the characteristics of the groups under study correspond to a typical pattern. When the NRM came to power, it accommodated all political forces, placed a ban on political parties and incorporated civil society. Civil society was structurally weak and eroded after years of turmoil. The NRM chose to promote and elevate development NGOs and subordinate unions and cooperatives, a strategy that was both consistent with neo-liberal economic policy and an integral part of its project to institutionalise a political base.

The persistent weakness of Ugandan civil society can be attributed to structural factors, such as weak middle and working classes and strong landlords, reinforced by political conjunctural ones, such as the incorporation of many civil society organisations into a project for state reconstruction under a restrictive political framework. This has created a pluralised but apolitical NGO based civil society, oriented towards "community development". Appeals to popular participation are prominent in shaping civil society strategies for rural development more broadly.

The preoccupation with "ownership" and "empowerment" has its ideological roots in the governance model and shares important aspects of its technocratic governance populism with both the NRM doctrine and the World Bank's notions of "poverty reduction" (Harrison 2004: Chapter 6). Workshop and consultancy modes of engagement with constituencies dominate. Cross-cutting forms of public-private-NGO "partnership models" in social policy making seem, not only in Uganda, to lead to institutional fragmentation of decision making arrangements and ad hoc forms of regulation and frequently also to informalised structures of access to these arrangements – in contrast to expectations of synergy effects (Bangura 2001).

One common denominator of the institutional linkages created between state and society under the present government in Uganda is that they encourage community participation within given limits but disconnect participation from interest representation around issues of power and influence in society. This is further reproduced by an official discourse that stresses the separation of "development" from "politics" and appeals to populist notions of "the poor" and restrains the articulation of different perspectives on more general aspects of economic policy making or politics at large.

The participatory tendency may transform the modes of regulation of interest representation within certain policy arenas, as more actors are accommodated into them. However, this is not without its own complications of accountability. Furthermore, it is contingent upon particularistic as opposed to institutionalised relations and overdetermined by the main tendency towards authoritarianism. When interest representation around political issues, both within the political system and through civil society organisations, is effectively blocked, most people are likely to turn to any channel of influence that may serve their interests. Another aspect of this, as observed by Tidemand (1995a: 44) is that with different degrees of privatisation of social services, there is less to control through political institutions.

In shaping its relations with society, the NRM government opted for a general strategy of inclusion. This had different dimensions. One was the close regulation of the parameters for the operation of interest groups in order to ensure that no undue "political" activities were undertaken. The suppression of most politicised social movements, the successful incorporation of some (the women's movement being the most prominent example – see Goetz 2002, Tripp 2000), and the emergence of NGOs and community based organisations through a combination of local coping strategies and structures of foreign development aid has resulted in a recomposition of civil society (Dicklich 1998).

When it comes to issues of real influence, the NRM state has sought to bypass urban based, middle class political elites with bases in civil society and parties and instead forge direct links with the peasantry through LCs. With a crude distinction, the NRM state is populist, with the President's frequent claims to rule on behalf of the masses (Carbone 2005), rather than corporatist, although it displays characteristics of both modes of incorporation.

Gibbon (1992) suggests that some general characteristics of contemporary "post-developmentalist" civil society are the predominance of groups with a narrow social base; a mode of internal decision making based on hierarchy rather than membership; and ad hoc forms of engaging the state. Such propositions rhyme with the findings of this and other empirical studies on Uganda's civil society. Overall, while this kind of civil society may pluralise social life and in some cases expand the parameters for participation, it only rarely strengthens the interests of subordinate groups (cf. Huber *et al* 1997: 328).

At this point, it is useful to elaborate on the discussion by Huber *et al* of the "contradictions of contemporary democracy" (1997). More precisely, what is relevant for the purposes of this discussion is their contention that while decisive structural features, including the nature of class relations, state structures and international relations which frame the current conjuncture are conducive for promoting formal democracy, these features at the same time tend to obstruct social and participatory dimensions of democracy.

This apparent contradiction bears clear resemblance to the arguments and conclusions of this study. In the Ugandan context, the paradox has in this study been formulated as the coexistence of technocratic governance and militarism. To draw on Huber *et al*'s argument, these seemingly diverging tendencies are really linked and constitute different expressions of authoritarianism. They are not linked in a seamless way, though, but connected as different expressions of how political groupings have over time sought to shape the form and content of the state. This is best explained by the balance of power between different political forces in society, within the state and at the transnational level.

In Uganda, the apparent paradox was during the no-party system in place until 2006 not manifested as contradictions between formal and deepened democracy but as tensions between controlled participation in LCs and for selected civil society organisations in poverty reduction arenas, on the one hand, and marginalisation and militarised repression of outright opposition politics, on the other. However, since the introduction of multiparty politics in 2006 the contradictions between fragile democratic forms and institutions and authoritarian social relations and political practices that tend to undercut the former have been deep and wide.

The guided and restricted participation for accommodating social forces and the suppression of more radical ones is also, incidentally, reminiscent of the political reforms of late colonialism. The structural shift in the fiscal base of the state towards an economy and a state built on foreign aid has meant that external political interests in the shape of governments, transnational organisations, private capital and international civil society are more influential in shaping the state than are domestic social forces. Also, this has occurred in far from straightforward ways, as many of these external interests have had different agendas and have also in different ways been appropriated by the government as their interventions have increasingly been internalised into the state apparatus.

Yet another dimension of this discussion is that much social life is not easily captured by references to state and civil society. Informalisation continues to shape patterns of coping strategies with regard to health provision and beyond. State presence is highly shifting in extent and quality. Much provision of health and access to it, such as formal and informal markets for medication and treatment, remains very weakly and unevenly regulated. In spite of stated aims of partnership between public and private forms of provision, the health regimes, particularly – but not only – in remote areas, remain shaped by coping strategies, resulting in institutional fragmentation and stratification of access and influence.

Local politics beyond Masaka

As was demonstrated in the previous chapter, local politics in Masaka has many roots. State-society relations are shaped by overlapping trajectories, including the role of Buganda within

Uganda, the role of Catholicism and DP within Buganda as well as in national politics, the political economy of migrant labour and coffee production and the destruction of Masaka town during the 1979 war and its continued decay after that. Compared to the other parts of the country, however, the central region has enjoyed stability and relative prosperity since 1986.

The theoretical argument proposes that local political arenas are shaped by a combination of intra-local and global-national-local power relations and that regional differences run deep. The admittedly fluid patterns of public authority found in local arenas are shaped in relation to systematic attempts at structuring central-local relations. The scope for local claims to authority is closely connected to the uneven ways in which such structuration is played out. Before exploring further how such differences may shape state-society relations, I will make a few comments on the common denominators within which local variations currently occur.

The organisation of local government along lines of "popular participation" has been an important dimension of restructuring state-society relations. In the early 1990s, RCs were transformed by more general decentralisation reforms, placing the responsibility for service delivery with Local Councils. On one level, the dynamics of local politics is shifting between districts due to variations in local government capacity and the specific local social structure. More broadly, however, a main tendency has been a gradual shift in the political identity of local government institutions from democratic organs to state-party vehicles for local mobilisation.

While LCs transformed the role of local government in a more democratic and participatory direction by replacing the institution of the chief with elected councils, the limits to both participation and democracy have gradually become more visible. Political participation was until 2005 exclusively confined to LCs, within the Movement system. The real as opposed to formal autonomy of LCs is circumscribed by their generally low capacity for revenue extraction, leading to dependence on centrally allocated and to a large extent earmarked financial grants. This reinforces the political vulnerability of these economically unviable local entities. It would be an exaggeration to describe LCs as through and through politicised by the government. They have harboured contradictory tendencies, being at the same time arenas for popular participation, NRM structures and local state forms.

Regional support for suppressed opposition forces places limits on the imposition of NRM hegemony, consistent and effective political penetration from above is typically more limited at lower levels. As the tensions inherent in the construction of the NRM itself grew, however, the political and bureaucratic dimensions of LCs gradually came to overshadow the democratic one. By and large, LCs have over time become increasingly significant in cultivating a rural political base for the regime. This rural electoral support has in turn been instrumental in securing NRM victories in the presidential and parliamentary elections of 1996, 2001, 2006 and 2011.

Again, though, there are important regional and local variations on this theme. Many peripheral areas – not only in the north, but also in the east, the far west and within the central region, such as Kalangala District to the immediate east of Masaka, in Lake Victoria – are weakly integrated into the national economy and state structures. This similary has many causes, including historically rooted underdevelopment in combination with contemporary political marginalisation. Even in core areas, effective state presence is geographically, temporally and institutionally uneven. For the sake of clarity, I shall illustrate this point through a discussion of health and state-society relations in, until very recently, war-torn Acholi region. This will also make it possible to interrogate state formation through the medium of health and social policy regimes in a more elaborate way.

As discussed in previous chapters, a series of rebellions broke out in northern Uganda after 1986, following the disintegration of former national armies and in the context of a political vacuum and a deep social crisis in the north. Another way of viewing these rebellions is that the civil war never ended in 1986 – it just shifted battleground from central to northern Uganda (Finnström 2003: 100-109). Acholi region was then plagued by civil war until 2006.

The LRA rebellion started in the late 1980s and has ebbed and flowed in intensity. After a government offensive in the early 1990s, peace settlements were close in 1994 but failed at the last minute. The war took off again and entered a more brutal phase as LRA targeted civilians on a massive scale, killed adults and abducted children. In 1996, the UPDF started to move civilians into what it called "protected villages". The process of displacement is disputed. The government has always claimed that people moved voluntarily but critics have charged that much displacement was forced and that property was looted.

In 2002 the UPDF set out on its "Iron Fist" offensive, which pushed the LRA into southern Sudan but which also triggered abduction of children on an unprecedented scale. The number of people living in camps doubled to 800,000, ordered there by a UPDF ultimatum pronounced that anyone moving outside the camps would be treated as a potential rebel. In 2006 it was estimated that 1.6 million people – among whom 95 per cent were of the Acholi population – were internally displaced in Acholi, Lango and Teso regions (Rodriguez 2006). Conditions in the camps were appalling. Security, supposedly the reason for their establishment in the first place, was erratic; poverty was rampant. HIV/AIDS rates by far exceed the rest of the country. People died in the hundreds – or according to one report, even more than one thousand (WHO 2005) – every week from easily curable diseases (Medecins Sans Frontieres 2004; Nannyonjo 2005). Many lived in the camps for years, seeing no future (CSOPNU 2004; Finnström 2003: Chapter 5).

While local governments in other parts of Uganda have functioned as vehicles for control and inclusion, state-society relations in Acholi are characterised by their dramatic distortion, epitomised by forced displacement into camps that has created a lasting state of emergency and in effect informal apartheid with regard to basic citizenship entitlements. Hence, state-society relations mean something completely different in Acholi than in Masaka. The welfare arms of the state have by and large withdrawn and have for some time been replaced by a great many entities. Education and health is provided by local and international NGOs and UNICEF, while is food by the World Food Programme. Much as humanitarian aid is indispensable for the survival of the population in the camps, it is unwittingly caught in the logic of militarised politics (Finnström 2006).

The composition of the Ugandan state in Acholi is heavily tilted towards its coercive institutions. Much of civil society, apart from religious organisations, has entered from outside and remains unable to address the humanitarian catastrophe in political terms, partly because they are the very same groups that also constitute part of the informal welfare institutions of the state, disconnected from and subordinate to its repressive organs. This has also resulted in the subordination of local civil society and the separation between state and society, as the state denies local society possibilities of expression and organisation.

It is easy to note that the Uganda government denied its citizens in the camps their basic human rights in terms of protection, health and education. With regard to the protection of its citizens, the state failed completely in Acholi. In order to go beyond this observation, however, one needs to apply a political of how this humanitarian disaster came to be and why political demands related to it have been neglected for so long. One such analysis offers the perspective that for a long time, the state consciously dismantled all forms of political opposition stemming from parties or social movements among the Acholi population by discrediting such expressions as "rebel collaboration", leaving the bulk of the population in a vacuum between the government and the rebels by condemning them to a life in camps (Branch 2005; Finnström 2003).

The Ugandan experience of state formation points among other things to ad hoc regulation of welfare functions and the simultaneous systematic creation of new institutional linkages that deepens the political presence of the central state in local arenas, but with deep regional differences. While fragmentation of state authority is more visible in relation to welfare functions of the state, its coercive apparatus is often marked by its heavy presence. of local level complexity needs to be connected to patterns of central-local relations, including the way central power holders have developed specific strategies in order to deal with regional power configurations in combination with the capacity of these social forces to respond to intervention. This requires that both state and civil society are analytically disaggregated, and that the "presence" or "absence" of the state is discussed in relation to the balance of forces in society.

State formation beyond health: Militarism and governance

While noting that the situation in the camps certainly is extreme, it is absolutely fundamental not to treat the north as a residual category or as an exception. On the contrary, it is necessary to incorporate the northern war fully into the understanding of the Ugandan state in order to expand and revise that understanding. The particular composition of state and civil society in Acholi region, and the role of militarism and coercion there, requires a discussion of the overall direction of institutional and territorial dimensions of state formation in Uganda and how power relations have structured the selective generation of development and consent as well as underdevelopment and coercion.

In this section, I attempt to elaborate the theoretical argument about state-society relations, not only within the contexts drawn upon in the previous sections, but also in relation to central dimensions of state formation in Uganda more broadly. The incumbent government has consistently sought to re-root the state in society in response to challenges of consolidating control and of cultivating a political base. While achievements of economic recovery and institutional rehabilitation are real, they remain not only fragile due to inherent contradictions reproduced by underlying structural weaknesses, but also one-sided and shallow due to the absence of democratic processes. What follows is an attempt to answer the main research question, the scope for social forces to democratise the state, although in its reverse form. What has been the nature of the social relations that have blocked democracy and formed the basis of authoritarian rule?

A fundamental assumption with regard to the argument developed below is that state formation is a very complex process and needs to be understood in its historical context. While, as in the Ugandan case, donors might be more concerned with institutional reforms of for example the bureaucracy, often with a technicist bias, such reforms are always enmeshed in political struggles with economic, social and institutional roots over state power and its social foundations. This proposition is coupled to the argument that state formation as ongoing process is bound to consist of parallel, overlapping and contradictory tendencies, more or less coherently regulated.

Contradictions arise partly from the different interests of ruling elites, other social forces and external actors, which are all reflected in the state. They are also due to the fact that state managers seek to gain control, frequently within institutions marked by relatively low regulatory capacity, through often contradictory strategies for accumulation and legitimation. For similar perspectives on state formation, see Hutchful (2002) on Ghana, and Mustapha (2002 a) on Nigeria. That said, the following remarks about "main tendencies" need to be understood as crude shorthand, which implicitly encompasses these complexities and countervailing trends.

In 2005, the government succeeded in amending the constitution to the effect that presidential term limits were abolished. This highly contentious issue had become a central part of a transition process which also involved legalising political party activity, so that the February 2006 presidential and parliamentary elections were held under a multi-party dispensation. Although not quite as violent as the 2001 elections, the process as a whole was marred by state-inspired coercion, in addition to other forms of rigging. Furthermore, throughout the entire period from 2001, there has been a marked tendency towards the militarisation of the entire state apparatus. How can such intensified repression be situated within the long-term process of state-society relations? The following explanation revolves around an account of shifting forms of government control, and in particular its changing accumulation and legitimation strategies, following from struggles and cracking and emerging coalitions within and outside the state.

When NRM took over power, it was a relatively weak political grouping, prompting it to strike alliances with stronger political and social forces in order to consolidate its rule. In so doing, it inevitably incorporated contradictory political tendencies. Among those tendencies were not only representatives of established power centres such as the DP and the Catholic church. The government also accommodated factions of capital which had acquired their wealth through state-based speculative accumulation during the era of economic instability, the groups popularly known as *mafutamingi*. Even though liberalisation and inflow of development aid gradually transformed the revenue base of the state and dismantled parastatals as direct source of instant wealth creation, old and new groups of *mafutamingi* entrenched themselves in the state (Mamdani 1990b: 434-435). As Mamdani pointed out in the mid-1990s:

> there is today a contradiction between the democratising tendency in our politics and the anti-democratic thrust of the economic policy contained in the IMF programme. Given time, that contradiction is bound to surface (1995: 78).

It is a principal contention of this book that this is precisely what happened and that the anti-democratic tendency superseded the democratic one in the political arena, too, through the ways in which speculative economic elites entrenched themselves in the state, and the government correspondingly suppressed autonomous organisation that seemed to offer resistance to those and similar forces.

The structure of the state was transformed during the period of NRM consolidation from the early 1990s, with civil service, privatisation and decentralisation reforms. Another shift, strongly promoted by key donors, was to concentrate domestic influence on decision making around economic policy to an insulated group of elite technocrats in the Ministry of Finance, Planning and Economic Development (MoFPED) – a "poverty reduction Dream Team" (Brock *et al* 2002: 7), or a transnationalised elite (Robinson 2001). The relative success of the economic reform programme stabilised a ruling coalition of leading donors, pro-liberalisers in the government and economic beneficiaries of the reforms, underpinned by broader political support for NRM from rural constituencies in southern Uganda and many sections of civil society.

The policy framework promoted by this coalition centred on macro-economic growth, low inflation, state reforms and, implicitly attached, an acceptance from donors for a continued political monopolistic control – a dominant model which in this study has been called a technocratic governance regime, under which the ruling coalition was successful in coordinating and ensuring its economic and social reproduction.

NRM reformulated its legitimation efforts, based on governance success, and directed it towards new local and foreign constituencies. References to radical nationalism and popular democracy were unsurprisingly used more infrequently, but were not altogether abandoned as they occasionally proved instrumental in justifying restrictions on political pluralism. NRM consolidated its political control with the 1995 Constitution in which, to repeat, the temporary no-party agreement was codified into a Movement system in spite of protests by parties and moderate NRM leaders, through the 1996 elections. This froze the political landscape, and with hindsight made it appear more stable than it probably was. Seeds of discontent and renewed opposition were sown during this period. The tensions between different political interests, not least within NRM and the state apparatus, some of whom were sidelined politically and economically by this model, grew stronger during this period, and would eventually burst open.

During this period, privatisation of parastatals took off, something which dramatically raised the stakes in the struggles for resources and state power and generated factionalism. Many cases of privatisation were tainted by gross corruption, repeatedly involving friends and family of the President or other top politicians in collusion with international capital (Human Rights Watch 2005a; Tangri and Mwenda 2001). States are always shaped by contestation between different interests, including strategies of ruling class factions to develop both their own institutions and those of the state in order to ensure the reproduction of both in relation to resistance from subordinate forces. When subordinate political forces are weakened and alienated, the state runs the risk of being reduced to nothing more than an arena for intense factional infighting among the ruling class. Something of the sort that happened in Uganda in the late 1960s and during the 1970s. So what about the Movement state? Was there greater scope for various social forces to impose discipline on ruling elites?

Undeniable achievements had been made compared to earlier governments. Economic liberalisation removed some forms of parasitic economic monopolism, and during a coffee boom in the early to mid-1990s this generated some degree of surplus in coffee producing areas. A "peace dividend" brought about favourable conditions for building and construction, the developing of infrastructure and economic expansion generally, which created a new middle class with vested interests in the new order. The local state had been democratised to some degree. Bureaucratic, coercive and welfare institutions of the state were relatively stabilised, which reinforced their capacity and legitimacy, not least in the eyes of the beneficiaries of the economic boom.

This seems to support an argument of uneven, contradictory but still transformative capitalist development. The argument advanced here, however, is that those achievements have proven to be unsustainable as they were undercut by another destabilising tendency through the social relations that underpinned state reconstruction and economic liberalisation. The basis of resource extraction and distribution, control and legitimacy of the NRM government has rested upon a conjuncturally specific and therefore unstable configuration of regionally distinct rural political loyalty, donor support and military control. One aspect of this is the persistent weakness of the domestic bourgeoisie and the growing factionalism and corruption discussed above.

Informalisation is longstanding and has been profound in reshaping the economy, society and state. It impacted on the formation and economic activities of all classes, but in different ways. Hence, the fractured organised expressions of class interests are structurally rooted in this kind of political economy. This is linked to the fragmented and incoherent nature of the state and its institutions that are sites of struggles between different social forces. Precisely because of the unstable and unsettled nature of state formation in Uganda and subsequent weak institutionalisation, such struggles are often magnified rather than contained by state institutions themselves, at least during conjunctures when contradictions evolve into political stress and crisis.

Related to this is the insulation of technocrats and the suppression of alternative perspectives on economic policy making. Evidently, there is a crucial difference between forms of accumulation based on plunder and forms conforming to legality, however narrow the scope of such legality. The common denominator of these manifestations of control over economic policy making, however, is that they were protected and promoted by restrictions on the political freedoms that would have enabled protests from either competing elites or from subordinate groups. The argument advanced by Rueschemeyer *et al* (1992: 64-66) is that in order for democracy to be promoted and consolidated, the state must be relatively autonomous from dominant classes and, correspondingly, sufficiently embedded among subordinate ones. In Uganda, the opposite has been the case.

Following an initial period of intense political struggles, to a large extent within the state, over the direction of economic policy making and political pluralism, a more antidemocratic faction of the NRM, supported by key donors and linked to sections of speculative capital gradually gained control from the early 1990s and demobilised political opponents within the broad-based coalition. Dominant political groups with their power bases in the military entrenched themselves in the state and offered privileged access to their economic protégés. This led to cracking coalitions and fall-outs within the state-party, during the 1990s expressed as internal and informal struggles over the direction of NRM, but from the late 1990s fought out in the open, resulting in new lobby groups and eventually in political parties.

As resistance to corruption and political authoritarianism grew in strength, the government responded with increasingly repressive measures. In addition, dominant classes loyal to the government have been divided among themselves. Parallel informal modes of primitive accumulation created tensions between different factions of capital, and those tensions constitute another source of the current political instability and authoritarianism.

It is hence necessary to understand authoritarianism in a broader sense than as militarisation. Therefore, it is worth restating Bangura's argument about the social basis of authoritarianism presented in the first chapter: "the basis for authoritarian rule should be located primarily at the level of material relations, that is, it expresses a particular resolution of contradictions in particular forms of accumulation" (1992: 46). In line with that argument, it is claimed here that capitalist development in Uganda is still uneven and weak, epitomised by the implications of the 1972 expulsion of the Asian capitalist class and the expropriation of their property by Amin loyalists. While distinct changes have taken place in terms of the sources of accumulation, the ethno-regional composition of the bourgeoisie and the precise nature of their relationship with the state, there is continuity with regard to the importance of the interconnection between business, state power and military influence.

In Uganda, the military is for historical reasons institutionally significant, and firmly entrenched in the political economy, including illicit and illegitimate forms of accumulation (Vlassenroot and Raeymaekers 2004: 400-410). One significant feature in the Ugandan case is that the present army, just like the ruling party, has its origins in a rebel group. This has created inseparable ties between the political and the military leadership, and in extension also to business groups. The rise of strong state, business or military networks means that military coercion constitutes both a source of accumulation of wealth and power and a means of repression of protests against it.

This double bind, the military as source of political and economic control, is what explains the recurrent militarisation of the state as a response to real and perceived opposition even within the state itself, as has been the case with militarised threats by the executive against the judiciary. This should not be understood in the reductionist sense that the military functions as the inevitable tool for the ruling class. There are many sources of the important role of the military in Uganda's politics. The historical institutionalisation of the military in the Ugandan state corresponds to the weakness of class based organisations and is what gives authoritarianism in Uganda its characteristic as militarised form.

The weakness of different factions of the bourgeoisie and middle and working class based organisations has enabled informal and clandestine connections in the business, military or politics nexus of primitive accumulation alluded to above (Tangri and Mwenda 2003; United Nations 2002). Again, this is not the only kind of accumulation that takes place, and it certainly does not justify the catch-all notion of a "criminalisation of the state" (Bayart *et al* 1999; see Mustapha 2002b for a critique). But it is a crucial dimension of capitalist development in contemporary Uganda, and it goes a long way to explain the violent response by the state to accusations of political authoritarianism linked to military corruption made by former allies and insiders in the context of a shrinking and fragmenting power base and reluctance of foreign supporters. Or to phrase it according to the theoretical argument of this study, struggles between political forces in society and within the state have transformed the institutional structure and the political orientation of the state in a militaristic direction.

This feature is also central in explaining the precarious balance between authoritarian and constitutionalised modes of regulation of social relations in Uganda generally. To return to Bangura's argument, its second part runs: "the dynamics of authoritarian rule and struggles for democratisation develop at the level of civil society" (Bangura 1992: 46). Just as primitive accumulation points to the convergence between militarism and capitalism, its manifestations have provided a focal point for converging resistance against it.

For a long time, the NRM government managed to merge economic liberalisation and authoritarian political control without encountering serious resistance. The inherent contradictions between institutional reforms and contentious social processes that were mentioned in the introduction of this book were eventually reproduced and manifested, however, in both state and society. Protests were waged against corruption and repression at both popular and elite levels of politics and led to cracks in the NRM and in the military, then eventually to the decision in 2005 following a referendum to legalise political party activity. Since this change goes to the heart of the research problem, the democratisation of the state, it is pertinent to elaborate on it.

The opening up for multi-party politics was an undeniable victory for the proponents of political pluralism. To some extent it was a consequence of pressure from the political opposition within the no-party system, and donors' impatience also had significant impact. Still, it is important not to overestimate the democratic content of this change. It was very clear already from 2003, when the proposal was first made that the NRM's official position was, that the turn-around was not due to a change of heart regarding the merits of political democracy, but a tactical manoeuvre in order to control internal dissent. Also, formal political pluralism has been undercut by intensified repression.

I propose that the move to multi-party politics is best understood in terms of crisis management as opposed to being a genuine conversion to political pluralism – the NRM leadership sought in one stroke to reinvent its legitimacy among donors, to pre-empt dangerous divisions within the military and to soften criticism against its bid to abolish constitutional restrictions on presidential terms. And it is power relations between political constellations within and outside the state that explain why formal democracy has so far been subverted by increasing militarisation.

While most donors now appeared to realise that many claims about governance success were incomplete at best, different donor agencies pursued their own political agendas, resulting in a shift from a rather stable coalition to a more fluid pattern of multiple games between the Ugandan state and its foreign partners. For instance, the President suddenly changed his position on fighting HIV/AIDS in a direction that would supposedly please the US religious right influencing politics on reproduction under the incumbent administration. He also came out quickly in support for the war in Iraq in 2003 and later claimed that he had been "mislead" into doing so.

In structural terms, though, it can be argued that contradictions of accumulation, through an intensification of factional infighting entrenched in the state, resulted in that patterns of primitive accumulation within the business, military and politician nexus discussed above undercut the logic of technocratic liberalism, a shift which enabled corruption and ultimately reinforced political authoritarianism. The very form of primitive accumulation the *mafutamingi* thrived upon, promoted and, in the face of growing opposition, even necessitated a shift in authoritarian political regulation from stabilising control, to destabilising coercion. For these reasons, and also because of the significant institutional role of the military in Ugandan politics, coercion as a mode of control gradually grew in importance also in the previously peaceful south.

It should be noted, though, that the government's revealed militaristic approach to political opposition is nothing new. Already in 1993, it responded to political rallies by rather insignificant groups within DP and CP with a display of police and military force and threats of "dead bodies" (Barya 2000a: 10). Just as the northern war has merged with the "war on terrorism", so has politics in Uganda more broadly. In its attempt to pin down the opposition, the government seeks to suggest links between the leading opposition party Forum for Democratic Change (FDC), an elusive rebel group called People's Redemption Army (PRA), LRA, and (implicitly) Rwanda. The leader of FDC, Kizza Besigye, was charged with treason for those very reasons.

Such measures have intensified the securitisation of politics. Over the last few years, this has been manifested by the militarisation of other state institutions – military control over the police force, representatives of security forces on the NGO Registration Board, the Media Centre and the tug-of-war over supremacy between the High Court and the General Court Martial which has lead to dramatic stand-offs. These events have sharply brought to a head the wider struggles between political forces outside and within the state over the political orientation of the latter: over whether the supreme locus of state power in Uganda is the constitution or the armed forces.

Special military forces surrounded the high court when the latter, following a ruling by the constitutional court, was about to release a number of persons, accused of being members of PRA and of being guilty of high treason, on bail in November 2005 and again in February 2007. Hence, contrary to claims of reconstruction and stabilisation of the Ugandan state, that tendency is countered and undercut by a continuous pattern of violent destabilisation; most clearly in the north, but more recently, as shown here, extended to Uganda as a whole as the government has become less concerned with covering its iron fist with a velvet glove.

Rather than viewing the northern experience as an exception, one needs instead to incorporate it into an overall understanding of the Ugandan state and to recognise the inbuilt contradictions of different strands of control as foundations for state power in the way they very unevenly articulated between regions. This unevenness has been so pronounced that it has been possible to present it as a dualism rather than as two related parts of one political whole: regional divisions have become engrained to the point that they are seen as a fact of life. In effect, contemporary state formation tends to reproduce colonially created patterns of regional underdevelopment, inasmuch as the state takes on completely different shapes in the north and the south.

The post-1986 period witnessed a reconstruction of the state, almost from the ashes, by a political organisation that started out as an army. In its early phase, the politics of the NRM state fundamentally revolved around enforcing control and security, however much the content of those concepts differed among regions. In both symbolic and very concrete (defeating and incorporating rebel groups) terms, the initial legitimacy and control of the NRM in its relevant constituencies – the regional political bases – rested on the military and the security it had brought to the civilian population by politicising armed conflicts.

Correspondingly, forms of resistance and the repression of that resistance outside those areas were articulated in terms of militarised politics. The very dimensions that rooted the state in society in southern Uganda took on their inverted manifestation in the north and alienated the state from society. In the north, the military repressed rather than guaranteed security and blocked as opposed to promoting the exercise of citizenship. In Acholi, the purely coercive aspect of the state has outweighed all others, and its heavy presence has also provided avenues and protection for shady forms of accumulation. Throughout Uganda, security has remained the central dimension of politics under the NRM but in ways rooted in and reproducing colonial and postcolonial divide-and-rule politics in a way that is reminiscent of Mamdani's (1996) argument about the bifurcated state.

Uganda in the international political economy

The Ugandan political economy is incorporated into the world economy from a subordinate position and in an uneven and distorted ways. Within that context, the NRM government has strengthened its political autonomy throughout Uganda's career from the international basket case to showcase and successful African adjuster. The particular coalition of external and internal political forces at work has had a vested interest in rebuilding and consolidating a "governance state" (Harrison 2004). But what is the nature of the political order that promotes such a coalition? If imperialism is taken to mean something like global "structures of inequality reproduced through a capitalist system of both political and economic power" (Bracking and Harrison 2003: 7), then it appears clear that it makes sense to locate subordinate groups in Uganda at the receiving end of imperialist relations that are also inherently expansionist and potentially violent. But one needs to recognise the historically specific forms that imperialism takes on under global capitalism, beyond mere nation-state rivalry (Halliday 2002: 86-88).

In the Ugandan case, the relatively successful application of economic and institutional reforms elevated it to a "success story" status and created considerable leeway within the framework of imperialist domination for the local ruling class in shaping domestic politics through rebuilt state structures. It is argued in this study, however, that the constellation of forces upholding the state has reproduced regional disjuncture and institutional fragmentation and incoherence, as well as authoritarianism at the level of state power.

The changing structuration of the state and the shifting expressions of authoritarian rule are explained by a relatively stable coalition that emerged during a critical formative period in the 1990s between the domestic and external forces with converging interests in upholding authoritarian political regulation of uneven accumulation and by the weak and fragmented nature of resistance to authoritarian domination. With growing resistance to authoritarianism and corruption and related struggles between political forces in society and in the state, the government has increasingly responded by militarising the state in its entirety.

External intervention impacts on state formation in critical but ambiguous ways. Pronounced international presence has translated into complex configurations of power as opposed to straightforward external political influence. Faithful application of economic adjustment programmes has given greater, not less, political autonomy to state managers, in relation to both other domestic and external constituencies. The acceptance by the government to surrender sovereignty of decision making over the framework for economic policy has given it sufficient autonomy and resources to control of state power.

The high degree of dependence upon external revenue and the concessions made in relation to foreign providers of resources resulted in the Ugandan state having as much of an external constituency as a domestic one. A convergence of interests emerged between the NRM, sections of domestic and international capital and donors around constructing a state capable of attracting investments, promoting capitalist development and protecting these interests through relatively stable bureaucratic and coercive state institutions in the endemically volatile Great Lakes region.

But also more routine modes of accumulation point to an imperialist logic. Tensions that follow from capitalist development are not reducible to conflicts within the ruling class. The state sides with foreign and domestic capital against labour and is an active part in subordinating labour to the expressed or imagined demands of capital. The labour regime is a clear case of repressive "Bloody Taylorism". Defending the decision to fire around 200 striking female workers who protested against poor working conditions, including abuse, in a foreign-owned textile factory outside Kampala in 2003. The President was reported to have said that "I sacked those girls because of indiscipline and their action would have scared off investors who had plans of setting up business here. They would have thought that the labour force in Uganda is undisciplined" (Monitor, 040322), and "Anybody who encourages strikes among our workers at the very moment we are struggling to break into external markets is worse [enemy] than Kony [the LRA leader]" (Monitor 031029-031105). Labour organisations have been too weak and disorganised to offer consistent resistance. The very recent changes in labour legislation in favour of labour are an exception, but are also to a considerable degree, concessions from the government to demands from US labour organisations that Uganda, if not respecting labour rights, would lose its preferential African Growth and Opportunity Act (AGOA) status.

There are links between militarism and labour. In Gulu district, one former LRA leader who was granted amnesty (and then immediately turned into a vocal government supporter) was made manager of a government funded farm, on which rescued LRA abductees work under slave-like conditions, including torture (Nyakairu 2006). The wider theoretical point that this admittedly anecdotal evidence is intended to illustrate is that institutionalised regulation of social relations of, for instance, labour and militarism are different but related modes of authoritarian control with common roots in the domestic social relations and foreign intervention that underpin state power.

The other dimension of imperialism concerns the geopolitical intervention in the post September 11 context of "the war against terrorism". It is evident that local, national, regional and global power relations have interplayed throughout the northern war. From the mid-1990s, it expanded into a regional proxy war. The LRA was backed by the Sudanese government in response to Uganda's support of SPLA rebels in southern Sudan. Recently, LRA forces moved into north-eastern DRC. All along, important donors have either accepted or supported Uganda's military strategy, even though demands have been made to complement this with negotiations. In the wake of the September 11 attacks, Uganda quickly adopted an anti-terrorism law, and the LRA was put on the US list of international terrorist organisations.

Already in the late 1990s, the US incorporated Uganda, along with Ethiopia and Eritrea, as a front-line state in the "war against terrorism" to contain the Islamist government in the Sudan. Another form of international intervention, supposedly a counterweight to militarism, comes by way of humanitarian aid. Much as this is indispensable for the survival of the population in the camps, it can be argued that humanitarian aid is caught in the logic of militarised politics (Finnström 2006).

A third expression of external intervention comes from the International Criminal Court (ICC), which on requests by the Ugandan government has issued arrest warrants for LRA leaders. Foreign intervention as such does not create but enters into and reinforces militarisation of politics and local and regional conflicts. Neither governance nor violence is exclusively shaped by imperialism. However, domestic and regional political struggles are fought in a context at least partly defined by external intervention.

With the turn to economic liberalisation, donors and creditors were brought on as informal but crucial members of the broad-based coalition. Transnational power structures hence impact heavily on policy making in contemporary Uganda. This does not mean that the state and the interests represented by it are essentially dependent on whims of external actors. In one sense, the state is obviously financially dependent on donors and creditors for sustaining even its most basic functions. There is rather limited leeway for departing significantly from the economic framework preferred by these external actors; something the government is quick to point out when the need arises to divert protests over unpopular reforms.

On the other hand, there is also room for bargaining. Straddling between external and internal interests runs both ways, as the restructuring of the state has involved the incorporation of its highest echelons into what Robinson (2004) calls a transnational state, something which also, though more indirectly, is the case with the state apparatus in different policy arenas and at all levels.

The often-proclaimed special status of Uganda as a success story has created a two-way dependence between donors and the government, with different implications for different donors. Not only has international intervention influenced the resource base, but also the institutional composition and the legitimacy of the state in relation to domestic constituencies. The historical trajectory of Uganda's politics has also in turn impacted significantly upon external intervention there, not in spite of, but precisely because of ambitious external involvement by way of investing vast financial resources and legitimacy.

Supporters of economic reform find it hard to admit that their policies may not be very deeply rooted in economic transformation and that they actually in certain ways reinforce corruption. Others, with regional security interests, are willing to accept slow and democratically ambiguous political reforms in exchange for loyalty and perceived stability (de Torrenté 2001: 43-51). In recent years, relations between donors and the Ugandan government, seemingly increasingly unwilling to perform according to the script of good governance, have turned sour. However, within the ongoing "war on terror", most recently in Somalia, Uganda is likely to reaffirm itself as a reliable ally of the West, something that is likely to continue feeding into a domestic logic of militarised politics.

State and civil society beyond Uganda

The Ugandan experience may finally be illuminated by some brief notes on state-society relations in a few select comparative cases. In Kenya, the historical trajectory of state-society relations led to a different turn. In the early 1990s, a coalition made up by a section of the Kikuyu bourgeoisie in central region cut off from state patronage networks under the Moi government, and increasingly vocal groups in civil society consisting of professional associations, human rights groups and religious institutions protested against an authoritarian and corrupt government with waning support presiding over a decaying economy.

The opposition was given support by key donors who put pressure on the Kenya Africa National Union (KANU) government. A decade-long process culminated in a change of government in 2002; it is increasingly clear, however, that this has not transformed the deeper structures of Kenyan politics. In Uganda, state legitimacy has been stronger; democratic forces in civil society are weaker and transnational power structures have so far seemed to favour political status quo. In Tanzania, political liberalisation was introduced by the ruling Chama Cha Mapinduzi (CCM) party itself, fully aware that in view of the practical absence of political opposition it could steer a guided democratisation process. Even though party political opposition was long frozen in Uganda, political cleavages run much deeper than in Tanzania, and the NRM government cannot as easily as CCM control the political transition uncontested.

A third case, in many ways similar to Uganda, is Ghana, where a faction of the military led by Jerry Rawlings took power in 1981. Ghana's experience during the 1980s, when a military government shifted from radical rhetoric to textbook economic liberalisation with large degrees of external involvement coupled with "no-party democracy" and rural populism as social and ideological base preceded Uganda's by a few years. Why did Ghana return to stable, albeit shallow, multi-party democracy in the early 1990s?

Returning to the explanatory framework, one difference between Ghana and Uganda is the composition and orientation of civil society and the way this corresponded to state structures. In Ghana, a relatively autonomous and active civil society, based on both ethnic and class identities, had emerged during colonialism. While its democratic expressions were largely suppressed during the 1980s, a coalition, in many respects similar to the Kenyan one (though with a bigger role for labour and a smaller for religious institutions) managed to impress its demands for political pluralism on the government (Akwetey 1994). This process found the support of donors.

In Uganda by contrast, civil society, especially its nationwide and class based expressions, is much weaker. The political aspirations of Buganda were contained through the alliance with the government. By contrast, the People's National Defence Council (PNDC) government in Ghana never managed to contain the economic and cultural elites in southern Ghana, including the important Ashanti region, as loyal regional power brokers.

Urban-based civil society groups in Uganda are only weakly connected to rural areas, where the NRM has developed a dense institutional network of mobilisation through the LC system. Correspondingly, the role of the military in the structures of the state is more marked in Uganda, leading to severe repression of political dissent. External actors, finally, for different reasons accepted the government's rationale for the ban on parties.

While the World Bank and IMF as well as certain bilateral donors had become dependent on Uganda as "success story" and were unwilling to "destabilise" the reconstruction, the US came to embrace the NRM government largely for geo-political reasons.

Along with other governments of "the new breed", Ethiopia and Rwanda, Uganda assured the US of its regional indispensability in containing the Islamist government in Sudan. Both the US and UK have supplied military aid to the Uganda as a reward for its loyalty in "the war against terrorism". While enthusiasm among western governments with "new breed" leaders faded some time ago, ties to them remain in place.

In contrast to cases which since the 1980s have experienced phases of economic stagnation and adjustment and political authoritarianism followed by popular protest and political liberalisation, Uganda's trajectory after 1986 led to the convergence of economic liberalisation and reconstruction by way of economic growth, enforcement of law and order and institutional rehabilitation in the southern parts of the country. Weak internal political (as opposed to armed) resistance to reforms and strong external support for them reinforced government legitimation efforts and political support in these areas.

The political and economic contradictions inherent in the NRM model of authoritarian political control over economic liberalisation for capitalist development and state formation became manifest only from the mid-1990s. From the 1996 elections, it was evident that the government increasingly used coercion to assert its control. The tensions that could no longer be accommodated by the Movement model included growing economic inequalities, fuelled by corruption with perceived ethno-regional connotations linked to military involvement and enhanced control – in increasingly striking contrast to the official discourse of national development, inclusive participation, demilitarisation and political liberalisation.

These tendencies have manifested the underlying instability of the reconstruction of the Ugandan state. Institutional reforms have proven fragile, as they have been only weakly anchored among social forces with the capacity and inclination to defend hard-won gains – and this is how the empirical cases of this study help to illustrate the general argument, which in turn provides the context within which to understand these particular experiences.

The contradictions summarised above, were only episodically and weakly politicised during the 1990s. The interconnection between the balance of power in society, state structures and transnational power structures prevented the emergence of sustained political opposition until the late 1990s, when grievances were systematically linked and articulated by dissatisfied insiders in a way that in the run up to the 2001 elections threatened to divide the core power bases of the government: the NRM as its political organisation, the military as its initially silent, but increasingly vocal back-up, and the western region as its key voting block.

The process of liberalisation that was initiated from above in 2003 and translated into multi-party politics in 2005 has been heavily shaped by this legacy insofar as there is striking incongruity between formal expansion and real stagnation and occasional contraction of the democratic space. This is an immediate continuation of the 2001 election campaigns, where the military and a range of more or less regular security apparatuses were instrumental in influencing the outcome. More profoundly, though, the contemporary re-militarisation of Uganda's politics points to a continuation of long-standing and deep-rooted authoritarianism that has fundamentally shaped the construction of Uganda as a state and society in ways this book has sought to explore, empirically as well as theoretically.

References

Ajulu, Rok 2001, "Uganda's Flawed Presidential Election Bodes Ill for the Future." *Global Dialogue* 6(2), pp. 20-22.

Akwetey, Emmanuel 1994, *Trade Unions and Democratisation: A Comparative Study of Zambia and Ghana*. Stockholm: Department of Political Science, Stockholm University.

Allen, Tim 2006, "AIDS and evidence: Interrogating some Ugandan myths." *Journal of Biosocial Science*, 38(1), pp. 7-28.

Allen, Tim and Suzette Heald 2004, "HIV/AIDS Policy in Africa: What has worked in Uganda and what has failed in Botswana?" *Journal of International Development*, 16, pp. 1141-1154.

Amaza, Ondoga ori 1998, *Museveni's Long March from Guerrilla to Statesman*. Kampala: Fountain Publishers.

Amnesty International 1999, "Uganda: Breaking the Circle: Protecting human rights in the northern war zone." AFR 59/01/99.

Apter, David E. 1997 [1961], *The Political Kingdom in Uganda: A Study in Bureaucratic Nationalism* (revised third edition). London: Frank Cass.

Asowa-Okwe, Charles 1997, "Insurgency and the Challenges of Social Reconstruction in Northern and North-East Uganda, 1986-1996" in Fredriksen, Bodil Folke and Fiona Wilson (eds.), *Livelihood, Identity and Instability*. Copenhagen: Centre for Development Research, pp. 95-111.

Bangura, Yusuf 1992, "Authoritarian Rule and Democracy in Africa: A Theoretical Discourse" in Gibbon *et al* (eds.), pp. 39-82.

Bangura, Yusuf 1994, "Economic Restructuring, Coping Strategies and Social Change: Implications for Institutional Development in Africa." *Development and Change* 25(4), pp. 785-827.

Bangura, Yusuf 2001, "New Directions in State Reform: Implications for Civil Society in Africa" in Beckman *et al* (eds.), pp. 250-282.

Banugire, Firimooni R. 1989, "Employment, Incomes, Basic Needs and Structural Adjustment Policy in Uganda, 1980-87," in Onimode, Bade (ed.) *The IMF, the World Bank and the African Debt, Volume 2, The Social and Political Impact*. London: Zed, pp. 95-110.

Barkan, Joel 2005, "Uganda: An African '*Success*' Past its Prime?" Paper presented at the Woodrow Wilson International Centre for Scholars, June 2, 2005.

Barya, John-Jean 1991, "Workers and the Law in Uganda." CBR Working Paper No. 17. Kampala: Centre for Basic Research.

Barya, John-Jean 2000a, "The Making of Uganda's 1995 Constitution: Achieving Consensus by Law?" CBR Working Paper No. 57. Kampala: Centre for Basic Research.

Barya, John-Jean 2000b, "The State of Civil Society in Uganda: An of the Legal and Politico-Economic Aspects." CBR Working Paper No. 58. Kampala: Centre for Basic Research.

Barya, John-Jean 2001, "Trade Unions and the Struggle for Associational Space in Uganda: The 1993 Trade Union Law and Article 40 of the Constitution." CBR Working Paper No. 63. Kampala: Centre for Basic Research.

Bates, Robert 1981, *Markets and States in Tropical Africa: The Political Basis of Agricultural Policies.* Berkeley: University of California Press.

Bayart, Jean-Francois 1993, *The State in Africa: The Politics of the Belly.* London: Longman.

Bayart, Jean-Francois, Stephen Ellis and Béatrice Hibou 1999, *The Criminalisation of the State in Africa.* London: James Currey.

Bazaara, Nyangabyaki 1997, :Agrarian Politics, Crisis and Reformism in Uganda 1962-1996." Unpublished Ph.D. Thesis, Queen's University, Kingston Ontario.

Bazaara, Nyangabyaki and Kintu Nyago 1999, "Civil Society, Empowerment and Poverty Reduction," mimeo, Centre for Basic Research, Kampala.

Beckman, Björn 1993, "The Liberation of Civil Society: Neo-Liberal Ideology and Political Theory." *Review of African Political Economy* 58, pp. 349-369.

Beckman, Björn and Attahiru Jega 1995, "Scholars and Democratic Politics in Nigeria," *Review of African Political Economy* 64, pp. 167-181.

Beckman, Björn, Eva Hansson and Anders Sjögren (eds) 2001, *Civil Society and Authoritarianism in the Third World.* Stockholm: PODSU.

Berman, Bruce 1990, *Control and Crisis in Colonial Kenya: The Dialectic of Domination.* London: James Currey.

Berman, Bruce and John Lonsdale 1992, *Unhappy Valley: Conflict in Kenya and Africa.* London: James Currey.

Bierschenk, Thomas and Jean-Pierre Olivier de Sardan 1997, "Local Powers and a Distant State in Rural Central African Republic." *Journal of Modern African Studies,* 35 (3), pp. 441-468.

Birungi, Harriet, Frank Mugisha, Xavier Nsabagasini, Sam Okuonzi and Anders Jeppsson 2001, "The policy on public-private mix in the Ugandan health sector: Catching up with reality." *Health Policy and Planning,* 16(suppl.2), pp. 80-87.

Boone, Catherine 2003, *Political Topographies of the African State. Territorial Authority and Institutional Choice.* Cambridge: Cambridge University Press.

Bornstein, Erica 2002, "Developing Faith: Theologies of Economic Development in Zimbabwe." *Journal of Religion in Africa* 32(1), pp. 4-31.

Bracking, Sarah and Graham Harrison 2003, "Africa, Imperialism & New Forms of Accumulation." *Review of African Political Economy,* 95, pp. 5-10.

Branch, Adam 2005, "Neither Peace Nor Justice: Political Violence and the Peasantry in Northern Uganda, 1986-1998." *African Studies Quarterly,* 8(2), pp. 1-31, [online] URL: http://web.africa.ufl.edu/asq/v8/v8i2a1.htm

Bratton, Michael 1989, "Beyond the State. Civil Society and Associational Life in Africa." *World Politics,* 41 (3), pp. 407-430.

Bratton, Michael and Nicolas van de Walle 1997, *Democratic Experiments in Africa. Regime Transitions in Comparative Perspective.* Cambridge: Cambridge University Press.

Brett, E A 1978, "Relations of Production, the State and the Ugandan Crisis." *West African Journal of Sociology and Political Science,* 1(3), pp. 249-284.

Brett, E A 1995, "Neutralising the Use of Force in Uganda: The Role of the Military in Politics." *Journal of Modern African Studies,* 33 (1), pp. 129-152.

Brett, E A 1996, "Uganda" in Engberg-Pedersen, P. *et al.* pp. 311-346.

Brett, E A 1998, "Responding to poverty in Uganda: structures, policies and prospects." *Journal of International Affairs,* 52(1), pp. 313-337.

Brock, Karen, Rosemary McGee and Richard Ssewakiryanga 2002, *Poverty Knowledge and Policy Processes: A case study of Ugandan national poverty reduction policy.* Brighton: IDS Research Report 53.

Buganda Kingdom 1999, "The Royal Development Drive (2000-2005)" (http://www.buganda.com/develop.htm)

Burke, Fred G. 1964, *Local Government and Politics in Uganda.* New York: Syracuse University Press.

Byerley, Andrew 2005, "Becoming Jinja. The Production of Space and Making of Place in an African Industrial Town." Stockholm: Department of Human Geography, Stockholm University.

Carbone, Giovanni M. 2005, *Populism Visits Africa: The Case of Yoweri Museveni and No-Party Democracy in Uganda.* Crisis States Working Paper Series 1, No. 73, Crisis States Research Centre, London School of Economics.

Centre for Basic Research 2005, "The Impact of Political Corruption on Resource Allocation and Service Delivery in Local Governments in Uganda." Kampala: Centre for Basic Research.

Chabal, Patrick and Jean-Pascal Daloz 1999, *Africa Works: Disorder as Political Instrument.* London: James Currey.

Chazan, Naomi 1992, "Africa's Democratic Challenge." *World Policy Review* 9(2), pp. 279-307.

Chekwoti, Caiphas 2003, "Uganda's Poverty Reduction Strategy Papers and Resource Allocation to the Health Sector." Kampala: Uganda Debt Network.

Collier, Paul and Ritva Reinikka 2001, "Reconstruction and Liberalisation: An Overview" in Reinikka, Ritva and Paul Collier (eds) *Uganda's Recovery: The Role of Farms, Firms and Government.* Kampala: Fountain Publishers, pp. 15-47.

Comaroff, John L. and Jean Comaroff (eds) 1999, *Civil Society and the Political Imagination: Critical Perspectives.* Chicago: University of Chicago Press.

Cooper, Frederick 1996, *Decolonisation and African Society: The Labor Question in French and British Africa.* Cambridge: Cambridge University Press.

Cowen, Michael and Robert Shenton 1996, *Doctrines of Development.* London: Routledge.

Crook, Richard 2001, "Strengthening Democratic Governance in Conflict Torn Societies: Civic Organisations, Democratic Effectiveness and Political Conflict." IDS Working Paper 129. Brighton, Sussex: Institute for Development Studies.

Crook, Richard and James Manor 1998, *Democracy and Decentralisation in South Asia and West Africa. Participation, Accountability and Performance.* Cambridge: Cambridge University Press.

Crook, Richard and Alan Sverrisson 2001, "Decentralisation and Poverty Alleviation in Developing Countries: A Comparative or, is West Bengal Unique?" Working Paper 130, Brighton: IDS.

CSOPNU 2004, "Nowhere to Hide. Humanitarian Protection Threats in Northern Uganda." Kampala: Civil Society Organisations for Peace in Northern Uganda.

Ddungu, Expedit 1994, "Popular Forms and the Question of Democracy: The Case of Resistance Councils in Uganda" in Mamdani, Mahmood and Joe Oloka-Onyango (eds) pp. 365-404.

De Coninck, John 2004, "The State, Civil Society and Development Policy in Uganda: Where are we coming from?" in Brock, Karen, Rosemary McGee and John Gaventa (eds) *Unpacking Policy: Knowledge, Actors and Spaces in Poverty Reduction in Uganda and Nigeria*. Kampala: Fountain Publishers, pp. 51-73.

De Torrenté, Nicholas 2001, "Pragmatism as Principle? The Donors and 'no-party democracy' in Uganda" in Musisi, Nakanyike B. and Cole P. Dodge (eds.) *Transformations in Uganda*. Kampala: Makerere Institute of Social Research, pp. 35-56.

De Torrenté, Nicholas and Frederick Mwesigye 1999, "The Evolving Role of the State, Donors and NGOs Providing Services in a Liberal Environment: Some Insights from Uganda." CBR Occasional Paper No. 2. Kampala: Centre for Basic Research.

Diamond, Larry 1994, "Rethinking Civil Society. Toward Democratic Consolidation." *Journal of Democracy*, 5(2), pp. 4-17.

Dicklich, Susan 1998, *The Elusive Promise of NGOs in Africa: Lessons from Uganda*. Basingstoke: Macmillan.

Dijkstra, A. Geske and Jan Kees van Donge 2001, "What Does the 'Show Case' Show? Evidence of and Lessons from Adjustment in Uganda." *World Development* 29(5), pp. 841-863.

Dodge, Cole P. 1987, "Rehabilitation or Redefinition of Health Services" in Wiebe, Paul D. and Cole P. Dodge (eds) *Beyond Crisis: Development Issues in Uganda*. Kampala: Makerere Institute for Social Research, pp. 101-112.

Doom, Ruddy and Koen Vlassenroot 1999, "Kony's Message: A New *Koine*? The Lord's Resistance Army in Northern Uganda." *African Affairs*, 98, pp. 5-36.

Doornbos, Martin 2000, *Institutionalising Development Policies and Resource Strategies in Eastern Africa and India: Developing Winners and Losers*. London: Macmillan.

Doornbos, Martin 2001, "'Good Governance': The Rise and Decline of a Policy Metaphor?" *Journal of Development Studies*, 37(6), pp. 93-108.

Doornbos, Martin and Frederick Mwesigye 1995, "The new politics of kingmaking" in Hansen and Twaddle (eds.) 1995, pp. 61-77.

Ekeh, Peter 1975, "Colonialism and the Two Publics in Africa. A Theoretical Statement." *Comparative Studies in Society and History,* 17(2), pp. 91-112.

Ellis, Frank and Godfrey Bahiigwa 2003, "Livelihoods and Rural Poverty Reduction in Uganda." *World Development,* 31(6), pp. 997-1013.

Engberg-Pedersen, Lars and Neil Webster 2002, "Introduction to Political Space" in Webster, Neil and Lars Engberg-Pedersen (eds) *In the Name of the Poor: Contesting Political Space for Poverty Reduction.* London: Zed, pp. 1-29.

Engberg-Pedersen, Poul, Peter Gibbon, Phil Raikes and Lars Udsholt (eds) 1996, *Limits of Adjustment in Africa*. London: James Currey.

Englebert, Pierre 2002, "Born-again Buganda or the Limits of Traditional Resurgence in Africa." *Journal of Modern Africa Studies* 40 (5), pp. 345-368.

Fallers, Lloyd A. (ed.) 1964, *The King's Men: Leadership and Status in Buganda on the Eve of Independence.* London, New York, Nairobi: Oxford University Press.

Feierman, Steven 1985, "Struggles for Control: The Social Roots of Health and Healing in Modern Africa." i *African Studies Review,* 28(2/3), pp. 73-147.

Ferguson, James 1998, "Transnational Topographies of Power: Beyond 'the state' and 'Civil Society' in the Study of African Politics" in Marcussen, H. Secher and Signe Arnfred (eds) *Concepts and Metaphors: Ideologies, Narratives and Myths in Development Discourse.* Occasional Paper no. 19. International Development Studies, Roskilde University, pp. 45-71.

Finnström, Sverker 2003, *Living With Bad Surroundings: War and Existential Uncertainty in Acholiland, Northern Uganda.* Uppsala: Uppsala Studies in Cultural Anthropology, PhD Thesis.

Finnström, Sverker 2006, "Survival in war-torn Uganda." *Anthropology Today,* 22 (2), pp. 12-15.

Fowler, Alan 1991, "The Role of NGOs in Changing State-Society Relations: Perspectives from Eastern and Southern Africa." *Development Policy Review,* 9(1), pp. 53-84.

Foundation for Human Rights Initiative 1999, "The Human Rights Reporter 1999." Kampala: FHRI.

Francis, Paul and Robert James 2003, "Balancing Rural Poverty Reduction and Citizen Participation: The Contradictions of Uganda's Decentralisation Program." *World Development,* 31(2), pp. 325-337.

Furley, Oliver 1989, "Britain and Uganda From Amin to Museveni: Blind Eye Diplomacy", in Rupesinghe (ed.), pp. 275-294.

Geertz, Clifford 1973 [1963], "The Integrative Revolution: Primordial Sentiments and Civic Politics in the New States" in *The Interpretation of Cultures.* New York: Basic Books, pp. 255-310.

Gibbon, Peter 1992, "Understanding Social Change in Contemporary Africa." Paper to a workshop on "The State, Structural Adjustment and Changing Social and Political Relations in Africa." Scandinavian Institute of African Studies, Uppsala.

Gibbon, Peter 1996, "Some Reflections on 'Civil Society' and Political Change" in Rudebeck, Lars and Olle Törnquist (eds) *Democratisation in the Third World. Concrete Cases in Comparative and Theoretical Perspective.* Uppsala: Seminar for Development Studies, pp. 21-50.

Gibbon, Peter 2001, "Civil Society, Locality and Globalisation in Rural Tanzania: A Forty-Year Perspective." *Development and Change*, 32(5), pp. 819-844.

Gibbon, Peter, Yusuf Bangura and Arve Ofstad (eds) 1992, *Authoritarianism, Democracy and Adjustment: The Politics of Economic Reform in Africa.* Uppsala: Nordiska Afrikainstitutet.

Goetz, Anne Marie 2002, "No Shortcuts to Power: Constraints on Women's Political Effectiveness in Uganda." *Journal of Modern Africa Studies*, 40(4), pp. 549-575.

Golooba-Mutebi, Frederick 2005, "When Popular Participation Won't Improve Service Provision: Primary Health Care in Uganda." *Development Policy Review* 23(2), pp. 165-182.

Gough, Ian 2004, "Welfare regimes in development context: A global and regional analysis" in Gough, Ian and Geof Wood (eds) *Insecurity and Welfare Regimes in Asia, Africa and Latin America.* Cambridge: Cambridge University Press, pp. 15-48.

Grugel, Jean 2002, *Democratisation: A Critical Introduction.* Basingstoke: Palgrave.

Halliday, Fred 2002, "The Pertinence of Imperialism" in Rupert, Mark and Hazel Smith (eds) *Historical Materialism and Globalisation.* London: Routledge, pp. 75-89.

Hansen, Holger Bernt 1977, *Ethnicity and Military Rule in Uganda.* Uppsala: Scandinavian Institute of African Studies.

Hansen, Holger Bernt and Michael Twaddle (eds) 1991, *Changing Uganda.* London: James Currey.

Hansen, Holger Bernt and Michael Twaddle (eds) 1995, *From Chaos to Order: The Politics of Constitution-Making in Uganda*. Kampala: Fountain Publishers.

Hansen, Thomas Blom and Finn Stepputat 2001, "Introduction. States of Imagination" in Hansen, Thomas Blom and Finn Stepputat (eds) *States of Imagination: Ethnographic Explorations of the Postcolonial State*. Durham: Duke University Press, pp. 1-38.

Harbeson, John 1994, "Civil Society and Political Renaissance in Africa" in Harbeson, John, Donald Rothchild and Naomi Chazan (eds) *Civil Society and the State in Africa*. Boulder: Lynne Rienner, pp. 1-29.

Harrison, Graham 2001, "Post-Conditionality Politics and Administrative Reform: Reflections on the Cases of Uganda and Tanzania." *Development and Change*, 32 (4), pp. 657-679.

Harrison, Graham 2004, *The World Bank and Africa: The Construction of Governance States.* London: Routledge.

Harriss, John 2002, *Depoliticising Development*. London: Anthem Press.

Harriss, John, Kristian Stokke and Olle Törnquist (eds) 2004, *Politicising Democracy*. Basingstoke: Palgrave.

Haynes, Jeff 2001, "'Limited' Democracy in Ghana and Uganda. What is Most Important to International Actors: Stability or Political Freedom?" *Journal of Contemporary African Studies,* 19(2), pp. 183-204.

Hearn, Julie 1999, "Foreign Political Aid, Democratisation, and Civil Society in Uganda in the 1990s." CBR Working Paper No. 53. Kampala: Centre for Basic Research.

Hirschman, Albert 1970, *Exit, Voice and Loyalty. Responses to Decline in Firms, Organisations and States*. Cambridge: Harvard University Press.

Huber, Evelyne, Dietrich Rueschemeyer and John D. Stephens 1997, "The Paradoxes of Contemporary Democracy. Formal, Participatory and Social Dimensions." *Comparative Politics*, 29(3), pp. 323-342.

Human Rights Watch 1999, *Hostile to Democracy.* New York: Human Rights Watch.

Human Rights Watch 2001a, *Uganda: Not A Level Playing Field – Government Violations in the Lead-Up to the Election*. New York: Human Rights Watch.

Human Rights Watch 2001b, *Uganda in Eastern DRC: Fuelling Political and Ethnic Strife.* New York: Human Rights Watch.

Human Rights Watch 2004a, *State of Pain: Torture in Uganda*. New York: Human Rights Watch.

Human Rights Watch 2004b, *Northern Uganda: Understanding and Solving the Conflict.* Nairobi/Brussels: ICG Africa Report No. 77.

Human Rights Watch 2005a, *The Curse of Gold.* New York: Human Rights Watch.

Human Rights Watch 2005b, *Uprooted and Forgotten: Impunity and Human Rights Abuses in Northern Uganda.* New York: Human Rights Watch.

Hutchful, Eboe 1996, "Ghana" in Engberg-Pedersen *et al* (eds.), pp. 143-214.

Hutchful, Eboe 2002, "The Fall and Rise of the State in Ghana" in Samatar and Samatar (eds.) pp. 101-130.

Hutchinson, Paul 1999, *Health Care in Uganda. Selected Issues.* Washington D.C: The World Bank.

Hydén, Göran 1983, *No Short-Cuts to Progress: African Development Management in Perspective.* Berkeley: University of California Press.

Hydén, Göran 1992, "Governance and the Study of Politics" in Hydén, Göran and Michael Bratton (eds) *Governance and Politics in Africa.* Boulder: Lynne Rienner, pp. 1-26.

Hydén, Göran 1995, "Bringing Voluntarism Back In" in Semboja and Therkildsen (eds.), pp. 35-50.

Iliffe, John 1998, *East African Doctors: A History of a Modern Profession.* Cambridge: Cambridge University Press.

International Crisis Group 2000, *Scramble for the Congo: Anatomy of an Ugly War.* Nairobi/Brussels: ICG Africa Report No. 26.

International Court of Justice 2005, "Case Concerning Armed Activities on the Territory of the Congo (Democratic Republic of Congo v. Uganda)." Judgement of 19 December 2005, ICJ, The Hague.

Jeppsson, Anders 2001, "Financial priorities under decentralisation in Uganda." *Health Policy and Planning* 16(2), pp. 187-192.

Jeppsson, Anders 2002, "SWAp dynamics in a decentralised context: Experiences from Uganda." *Social Science and Medicine*, 55, pp. 2053-2060.

Jeppsson, Anders and Mohammad Kisubi 2001, "Decentralisation of the Health Sector in Uganda: Formal or Informal Partnership with Users?" *Uganda Health Bulletin*, 7(1), pp. 9-12.

Jessop, Bob 2002, *The Future of the Capitalist State.* Cambridge: Polity Press.

Jones, Benjamin 2005, *Institutional Landscapes at the Margins of the State.* PhD thesis, Development Studies Institute, London School of Economics.

Jørgensen, Jan Jelmert 1981, *Uganda. A Modern History.* London: Croom Helm.

Joseph, Richard 1997, "Democratisation in Africa After 1989: Comparative and Theoretical Perspectives." *Comparative Politics,* 29 (3), pp. 363-382.

Kanyinga, Karuti 1995, "The Changing Development Space in Kenya. Socio-Political Change and Voluntary Development Activities" in Gibbon, Peter (ed.) *Markets, Civil Society and Democracy in Kenya.* Uppsala: Nordiska Afrikainstitutet, pp. 69-120.

Karlström, Mikael 1999, "Civil Society and Its Presuppositions: Lessons from Uganda" in Comaroff and Comaroff (eds), pp. 104-123.

Kasfir, Nelson 1976, *The Shrinking Political Arena: Participation and Ethnicity in African Politics with a case study of Uganda.* Berkeley: University of California Press.

Kasfir, Nelson 1991, "The Ugandan elections of 1989: Power, populism and democratisation" in Hansen and Twaddle (eds) 1991, pp. 247-278.

Kasfir, Nelson 1995, "Uganda Politics and the Constituent Assembly Elections of March 1994" in Hansen and Twaddle (eds) 1995, pp. 148-179.

Kasfir, Nelson (ed.) 1998, *Civil Society and Democracy in Africa: Critical Perspectives.* London: Frank Cass.

Kasfir, Nelson 1999, "'No-Party' Democracy in Uganda" in Diamond, Larry and Marc F. Plattner (eds) *Democratisation in Africa*, Baltimore: Johns Hopkins University Press, pp. 201-215.

Kasfir, Nelson 2000, "'Movement' Democracy, Legitimacy and Power in Uganda" in Mugaju, Justus and Oloka-Onyango (eds), pp. 60-78.

Kasfir, Nelson 2005, "Guerrillas and Civilian Participation: The National Resistance Army in Uganda, 1981-86." *Journal of Modern African Studies* 43 (2), pp. 271-296.

Kasita, Ibrahim 2005, "Masaka: Fallen town that struggles on," *Sunday Monitor,* 10 April 2005.

Kasozi, A.B.K. 1994, *The Social Origins of Violence in Uganda.* Kampala: Fountain Publishers.

Kassimir, Ronald 1998, "The Social Power of Religious Organisation and Civil Society: The Catholic Church in Uganda" in Kasfir (ed.), pp. 54-83.

Kassimir, Ronald 2001, "Producing Local Politics: Governance, Representation and Non-State Organisations in Africa" in Callaghy, Thomas, Ronald Kassimir and Robert Latham (eds) *Intervention and Transnationalism in Africa: Global-Local Networks of Power.* Cambridge: Cambridge University Press, pp. 93-112.

Kayunga, Sallie Simba 2000, "The Impact of Armed Opposition on the Movement System in Uganda" in Mugaju and Oloka-Onyango (eds), pp. 109-126.

Kelsall, Tim and Claire Mercer 2003, "Empowering People? World Vision & 'Transformatory Development' in Tanzania." *Review of African Political Economy*, 96, pp. 293-304.

Khadiagala, Gilbert 1995, "State Collapse and Reconstruction in Uganda" in Zartman (ed.), pp. 33-47.

Kingdom of Buganda 2001, "Major Issues to be presented to the Uganda Constitutional Review Commission." Kampala: Bulange, Mengo.

Kiondo, Andrew 1995, "When the State Withdraws: Local Development, Politics and Liberalisation in Tanzania" in Gibbon, Peter (ed.) *Liberalised Development in Tanzania*. Uppsala: Nordiska Afrikainstitutet, pp. 109-176.

Kisubi, Mohammad and Justus Mugaju (eds) 1999, *Rural Health Providers in south-west Uganda*. Kampala: Fountain Publishers.

Kjær, Mette 2001, *The Politics of Civil Service Reform: A Comparative of Uganda and Tanzania in the 1990s*. Aarhus: Politica. PhD Dissertation, Political Science, Aarhus University.

Knighton, Ben 2003, "The State as Raider Among the Karamojong: 'Where There Are No Guns, They Use the Threat of Guns'." *Africa*, 73(3), pp. 427-455.

Kyaddondo, David and Susan Reynolds Whyte 2003, "Working in a decentralised system: A threat to health workers' respect and survival in Uganda." *International Journal of Health Planning and Management*, 18, pp. 329-342.

Lamwaka, Caroline 2002, "The Peace Process in Northern Uganda 1986-1990" in Lucima Okello (ed.) *Protracted Conflict, Elusive Peace: Initiatives to End the Violence in Northern Uganda*. Conciliation Resources and Kacoke Madit. Available online: http://www.cr.org/accord/uganda/accord11

Leftwich, Adrian 1993, "Governance, Democracy and Development in the Third World." *Third World Quarterly*, 14(3), pp. 605-624.

Leopold, Mark 2005, *Inside West Nile: Violence, History and Representation on an African Frontier*. London: James Currey.

Loxley, John 1989, "The IMF, the World Bank and Reconstruction in Uganda" in Campbell, Bonnie K. and John Loxley (eds.), *Structural Adjustment in Africa*. London: Macmillan, pp. 67-91.

Luckham, Robin 1994, "The Military, Militarisation and Democratisation in Africa: A Survey of Literature and Issues." *Africa Studies Review,* 37(2), pp. 13-75.

Lund, Christian 2001, "Politics in a Sahelian Town: Dori and the Art of Alliance" in Benjaminsen, Tor and Christian Lund (eds) *Politics, Property and Production in the West African Sahel.* Understanding Natural Resource Management, Uppsala: Nordic Africa Institute, pp. 40-56.

Lwanga-Lunyiigo, Samwiri 1989, "The Colonial Roots of Internal Conflict" in Rupesinghe (ed.), pp. 24-43.

Lyons, Maryinez 1996, "Foreign Bodies: The History of Labour Migration as a Threat to Public Health in Uganda" in Nugent, Paul and A.I Asiwaju (eds) *African Boundaries. Barriers, Conduits and Opportunities*. London: Pinter, pp. 131-144.

Macrae, Joanna, Anthony B. Zwi and Lucy Gilson 1996, "A Triple Burden for Health Sector Reform: Post-Conflict Rehabilitation in Uganda." *Social Science and Medicine,* 42(7), pp. 1095-1108.

Mafeje, Archie 1977, "The Legitimacy of the Uganda Government in Buganda" in Cliffe, Lionel, James S. Coleman and Martin Doornbos (eds) *Government and Rural Development in East Africa: Essays on political penetration,* pp. 99-116.

Mahoney, James 2003, "Knowledge Accumulation in Comparative Historical Research: The Case of Democracy and Authoritarianism" in Mahoney and Rueschemeyer (eds.), pp. 131-174.

Mahoney, James and Dietrich Rueschemeyer (eds) 2003, *Comparative Historical in the Social Sciences*. Cambridge: Cambridge University Press.

Mamdani, Mahmood 1976, *Politics and Class Formation in Uganda*. Kampala: Fountain Publishers.

Mamdani, Mahmood 1984, *Imperialism and Fascism in Uganda*. Trenton: Africa World Press.

Mamdani, Mahmood 1987, "Extreme but not Exceptional: Towards an of the Agrarian Question in Uganda." *The Journal of Peasant Studies,* 14(2), pp. 191-223.

Mamdani, Mahmood 1988, "Uganda in Transition: Two Years of the NRA/ NRM." *Third World Quarterly,* 10(3), pp. 1155-1181.

Mamdani, Mahmood 1990a, "State and Civil Society in Contemporary Africa: Reconceptualising the Birth of State Nationalism and the Defeat of Popular Movements." *Africa Development,* XV (3/4), pp. 47-70.

Mamdani, Mahmood 1990b, "Uganda: Contradictions of the IMF Programme and Perspective." *Development and Change,* 21(3), pp. 427-467.

Mamdani, Mahmood 1994, "Pluralism and the Right of Association" in Mamdani and Oloka-Onyango (eds), pp. 519-563.

Mamdani, Mahmood 1995, "And Fire Does Not Always Beget Ash: Critical Reflections on the NRM." Kampala: Monitor Publications.

Mamdani, Mahmood 1996, *Citizen and Subject: Contemporary Africa and the Legacy of Late Colonialism.* Princeton: Princeton University Press.

Mamdani, Mahmood 2001, *When Victims Become Killers: Colonialism, Nativism, and the Genocide in Rwanda.* Princeton: Princeton University Press.

Mamdani, Mahmood and Joe Oloka-Onyango (eds) 1994, *Uganda: Studies in Living Conditions, Popular Movements and Constitutionalism.* Vienna: JEP.

Mamdani, Mahmood and Ernest Wamba-dia-Wamba (eds) 1995, *African Studies in Social Movements and Democracy.* Dakar: CODESRIA.

Mann, Michael 1986, "The Autonomous Power of the State: Its Origins, Mechanisms and Results" in Hall, John A. (ed.) *States in History*. Oxford: Basil Blackwell, pp. 109-136.

Medecins Sans Frontieres 2004, *Life in Northern Uganda: All Shades of Grief and Fear.*

Migdal, Joel 1988, *Strong Societies and Weak States. State-Society Relations and State Capabilities in the Third World.* Princeton: Princeton University Press.

Migdal, Joel, Atul Kohli and Vivienne Shue (eds) 1995, *State Power and Social Forces: Domination and Transformation in the Third World.* Cambridge: Cambridge University Press.

Mkandawire, Thandika 1995, "Fiscal Structure, State Contraction and Political Responses in Africa" in Mkandawire, Thandika and Adebayo Olukoshi (eds) *Between Liberalisation and Oppression. The Politics of Structural Adjustment in Africa.* Dakar: CODESRIA, pp. 20-51.

Mkandawire, Thandika 2001, "Thinking about Developmental States in Africa." *Cambridge Journal of Economics,* 25(3), pp. 289-313.

Mkandawire, Thandika 2004, "Social Policy in a Development Context: Introduction" in Mkandawire, Thandika (ed.) *Social Policy in a Development Context.* Basingstoke: Palgrave, pp. 1-33.

Mohan, Giles and Kristian Stokke 2000, "Participatory Development and Empowerment: The Dangers of Localism." *Third World Quarterly*, 21 (2), pp. 247-268.

Mohan, Giles, Ed Brown, Bob Milward and Alfred Zack-Williams 2000, *Structural Adjustment: Theory, Practice and Impacts*. London: Routledge.

Moore, Mick 1993, "Introduction." *IDS Bulletin*, 24(1), pp. 1-6.

Mudoola, Dan 1993, *Religion, Ethnicity and Politics in Uganda*. Kampala: Fountain Publishers.

Mugaju, Justus 1999, "On the Eve of the Rural Health Programme" in Kisubi and Mugaju (eds), pp. 1-15.

Mugaju, Justus and Joe Oloka-Onyango (eds) 2000, *No-Party Democracy in Uganda: Myths and Realities*. Kampala: Fountain Publishers.

Mugyenyi, Joshua 1991, "IMF Conditionality and Structural Adjustment Under the National Resistance Movement" in Hansen and Twaddle (eds.), pp. 61-77.

Musisi, Nakanyike B., 1999, "Morality as Identity: the Missionary Moral Agenda in Buganda, 1877-1945." *The Journal of Religious History* 23(1), pp. 51-74.

Mustapha, Abdul Raufu 1992, "Structural Adjustment and Multiple Modes of Livelihood in Nigeria" in Gibbon *et al* (eds), pp. 188-216.

Mustapha, Abdul Raufu 2002a, "Coping with Diversity: The Nigerian State in Historical Perspective" in Samatar and Samatar (eds.), pp. 149-176.

Mustapha, Abdul Raufu, 2002b, "States, Predation and Violence: Reconceptualising Political Action and Political Communities in Africa." Paper presented at 10th CODESRIA General Assembly, Kampala, 8-12 December.

Mutibwa, Phares 1992, *Uganda Since Independence: A Story of Unfulfilled Hopes*. Kampala: Fountain Publishers.

Mwenda, Andrew M. and Roger Tangri 2005, "Patronage Politics, Donor Reforms, and Regime Consolidation in Uganda." *African Affairs*, 104, pp. 449-467.

Nabuguzi, Emmanuel 1994, "Structural Adjustment and the Informal Economy in Uganda." Working Paper 94:4, Centre For Development Research, Copenhagen.

Nabuguzi, Emmanuel 1995, "Popular Initiatives in Service Provision in Uganda" in Semboja and Therkildsen (eds), pp. 192-208.

Nannyonjo, Justine 2005, "Conflicts, Poverty and Human Development in Northern Uganda." Research Paper no. 2005/47. Helsinki: UNU-WIDER.

Nganda, Ssemujju Ibrahim and James Tumusiime 2005, *Weekly Observer*, May 26, "Shs 600bn lost to army ghosts" http://www.ugandaobserver.com/today/news/news200505261.php

Nsibambi, Apollo 1998a, "Conclusion" in Nsibambi (ed.), pp. 140-144.

Nsibambi, Apollo (ed.) 1998b, *Decentralisation and Civil Society in Uganda. The Quest for Good Governance.* Kampala: Fountain Publishers.

Nyakairu, Frank 2006, "From captivity to slavery," *Daily Monitor,* March 31, http://www.monitor.co.ug/socpol/socpol03311.php

Obbo, Christine 1996, "Healing, Cultural Fundamentalism and Syncretism in Buganda." *Africa,* 66(2) pp. 183-201.

Okuonzi, Sam Agatre and Francis X. Lubanga 1995, *Decentralisation and Health Systems Change in Uganda.* Mimeo, Ministry of Health, Kampala.

Okuonzi, Sam Agatre and Joanna Macrae 1995, "Whose Policy is it Anyway? International and National Influences on Health Policy Development in Uganda." *Health Policy and Planning,* 10(2), pp. 122-132.

Oloka-Onyango, Joe 1991, "The National Resistance Movement, 'Grassroots Democracy' and Dictatorship in Uganda" in Cohen, Robin and Harry Goulbourne (eds) *Democracy and Socialism in Africa*, Boulder: Westview Press, pp. 125-141.

Oloka-Onyango, Joe 1997, "The Question of Buganda in Contemporary Ugandan Politics." *Journal of Contemporary African Studies*, 15(2), pp. 173-189.

Oloka-Onyango, Joe 2000, "New Wine or New Bottles? Movement Politics and One-Partyism in Uganda" in Mugaju and Oloka-Onyango (eds.), pp. 40-59.

Oloka-Onyango, Joe 2004, "'New Breed' Leadership, Conflict and Reconstruction in the Great Lakes Region of Africa: A Sociopolitical Biography of Uganda's Yoweri Kaguta Museveni." *Africa Today,* 50(3), pp. 29-52.

Oloka-Onyango, Joe and John-Jean Barya 1997, "Civil Society and the Political Economy of Foreign Aid in Uganda." *Democratisation*, 4(2), pp. 113-138.

O'Manique, Colleen 2004, *Neoliberalism and AIDS Crisis in Sub-Saharan Africa: Globalisation's Pandemic.* Basingstoke: Palgrave.

Opolot, Samson James 2002, "Migrants and Public Health in Uganda: From 'Pathogens' to Agents of Public Health Care Development" in Opolot, Samson James (ed.) *Building Healthy Cities: Improving the Health of*

Urban Migrants and the Urban Poor in Africa. Washington: Woodrow Wilson International Centre for Scholars, pp. 83-98.

Opolot, Samson and David L. Kibikyo 2003, "Uganda's Health Delivery Trends – A Critical Assessment of the Impact of the PEAP/PRSPs: 1997-2003." Paper presented to the Panos Eastern Africa Workshop on *Budgetary Allocations to the Health Sector with the Implementation of the PRSP*, Nairobi, August 5-7 2003.

Ottemoeller, Dan 1996, "Institutionalisation and Democratisation: The Case of the Ugandan Resistance Councils." Unpublished PhD Dissertation, University of Florida.

Packard, Randall 1997, "Visions of Postwar Health and Development and Their Impact on Public Health Interventions in the Developing World" in Cooper, Frederick and Randall Packard (eds.) *International Development and the Social Sciences.* Berkeley: University of California Press, pp. 93-115.

Parkhurst, Justin O. 2005, "The Response to HIV/AIDS and the Construction of National Legitimacy: Lessons from Uganda." *Development and Change*, 36(3), pp. 571-590.

Piron, Laure-Hélène and Andy Norton 2004, "Politics and the PRSP Approach: Uganda Case Study." Working Paper 240, Overseas Development Institute, London.

Potter, David 2000, "Democratisation, 'Good Governance' and Development" in Allen, Tim and Alan Thomas (eds.) *Poverty and Development into the 21st Century*. Oxford: Open University Press, pp. 365-382.

Potter, David, David Goldblatt, Margareth Kiloh and Paul Lewis (eds) 1997, *Democratisation*. Cambridge: Polity Press.

Refugee Law Project 2004, "Behind The Violence: Causes, Consequences and the Search for Solutions to the War in Northern Uganda." Kampala: Refugee Law Project Working Paper No. 11.

Regan, Antony 1998, "Decentralisation Policy: Reshaping State and Society" in Hansen, Holger Bernt and Michael Twaddle (eds) *Developing Uganda*, London: James Currey, pp. 159-175.

Reinikka, Ritva and Jacob Svensson 2003, "Working for God? Evaluating service delivery of religious not-for-profit health care providers in Uganda." Policy Research Working Paper Series 3058. Washington DC: The World Bank.

Republic of Uganda 2003, "Uganda Poverty Status Report, 2003." Kampala: Ministry of Finance, Planning and Economic Development.

Richards, Audrey 1982, "Changing Local Government Policy, 1950-1970" in Robertson, A. F. (ed.) *Uganda's First Republic: Chiefs, Administrators and Politicians, 1967-1971*. Cambridge University, Cambridge African Monographs No. 1, pp. 20-36.

Robinson, Mark 2006, "The Political Economy of Governance Reforms in Uganda." IDS Discussion Paper 386. Brighton: IDS.

Robinson, William I. 2001, "Social Theory and Globalisation: The Rise of a Transnational State." *Theory and Society* 30, pp. 157-200.

Robinson, William I. 2004, *A Theory of Global Capitalism. Production, Class, and State in a Transnational World*. Baltimore: Johns Hopkins University Press.

Rodriguez, Carlos 2006, "Genesis of IDP camps," *Weekly Observer*, April 27, http://www.ugandaobserver.com/new/features/spec/spec200604272.php

Rudebeck, Lars 1990, "The Effects of Structural Adjustment in Kandjadja, Guinea-Bissau." *Review of African Political Economy*, 49, pp. 34-51.

Rueschemeyer, Dietrich, Evelyne Huber Stephens and John D. Stephens 1992, *Capitalist Development and Democracy*. Cambridge: Polity Press.

Rueschemeyer, Dietrich 2003, "Can One or a Few Cases Yield Theoretical Gains?" in Mahoney and Rueschemeyer (eds), pp. 305-336.

Rupesinghe, Kumar (ed.) 1989, *Conflict Resolution in Uganda*. London: James Currey.

Samatar, Abdi Ismail and Ahmed I. Samatar (eds) 2002, *The African State. Reconsiderations*. Portsmouth, NH Heinemann.

Sandbrook, Richard 2000, *Closing the Circle: Democratisation and Development in Africa*. London: Zed.

Sathyamurthy, T.H. 1982, *Central-Local Relations: The Case of Uganda*. University of Manchester: Manchester Papers on Development.

Saul, John S 2004 [1976], "The Unsteady State: Uganda, Obote and General Amin", *Review of Political Economy*, reprinted in Mohan, Giles and Tunde Zack-Williams (eds) *The Politics of Transition in Africa*. Sheffield: ROAPE, pp. 17-31.

Semboja, Joseph and Ole Therkildsen (eds) 1995, *Service Provision under Stress in East Africa*. London: James Currey.

Sjögren, Anders 2001, "State, Civil Society and Democratisation: Theoretical Debates Past and Present" in Beckman *et al* (eds), pp. 21-48.

Summers, Carol 1991, "Intimate Colonialism: The Imperial Production of Reproduction in Uganda, 1907-1925." *Signs: Journal of Women in Culture and Society*, 16(4), pp. 787-807.

Tangri, Roger and Andrew Mwenda 2001, "Corruption and Cronyism in Uganda's Privatisation in the 1990s." *African Affairs*, 100, pp. 117-133.

Tangri, Roger and Andrew Mwenda 2003, "Military Corruption and Ugandan Politics Since the Late 1990s." *Review of African Political Economy*, 98, pp. 539-552.

Tangri, Roger and Andrew Mwenda 2006, "Politics, donors and the ineffectiveness of anti-corruption institutions in Uganda." *Journal of Modern African Studies*, 44(1), pp. 101-124.

Therkildsen, Ole and Joseph Semboja 1995, "A New Look at Service Provision in East Africa" in Semboja and Therkildsen (eds.) pp. 1-34.

Therkildsen, Ole 2006 "Elusive Public Sector Reform in East and Southern Africa", in Bangura, Yusuf and George A. Larbi (eds) *Public Sector Reform in Developing Countries. Capacity Challenges to Improve Services.* Basingstoke: Palgrave, pp. 53-81.

Tidemand, Per 1994, *The Resistance Councils in Uganda. A Study of Rural Politics and Popular Democracy in Africa.* Unpublished PhD thesis, Department of International Development Studies, Roskilde University.

Tidemand, Per 1995a, "New Local State Forms and 'Popular Participation' in Buganda, Uganda", in Kanyinga, Karuti, Andrew S.Z. Kiondo and Per Tidemand, *The New Local Level Politics in East Africa. Studies on Uganda, Tanzania and Kenya.* Uppsala: Nordiska Afrikainstitutet, pp. 22-49.

Tidemand, Per 1995b, "Popular Versus State Provision of Local Justice" in Semboja and Therkildsen, pp. 223-237.

Tripp, Aili Mari 2000, *Women and Politics in Uganda*. Oxford: James Currey.

Tripp, Aili Mari 2004, "The Changing Face of Authoritarianism in Africa: The Case of Uganda." *Africa Today*, pp. 3-26.

Uganda NGO Forum 2003, *In Search of Consensus of a New Development Path: A Synthesis of Civil Society Views into the 2003 PEAP Revision Process.* Kampala: Uganda National NGO Forum.

United Nations 2002, "Final Report of the Panel of Experts on the Illegal Exploitation of Natural Resources and Other Forms of Wealth of the Democratic Republic of Congo," October 16, 2002 (S/2002/1146).

Uzoigwe, G.N. 1982, "The Agreement States and the Making of Uganda: I. Buganda" in Uzoigwe, G.N. (ed.) *Uganda: The Dilemma of Nationhood.* New York: NOK Publishers International, pp. 57-93.

Van der Heijden, Thomas and Jessica Jitta 1993, "Economic Survival Strategies of Health Workers in Uganda." Kampala: Child Health and Development Centre, Makerere University.

Villadsen, Søren 1996, "Decentralisation of Governance" in Villadsen and Lubanga (eds) pp. 60-78.

Villadsen, Søren and Francis Lubanga (eds) 1996, *Democratic Decentralisation in Uganda.* Kampala: Fountain Publishers.

Vlassenroot, Koen and Timothy Raeymaekers 2004, "The Politics of Rebellion and Intervention in Ituri: the Emergence of a New Political Complex?" *African Affairs* 103, pp. 385-412.

Waliggo, John Mary 2002, "The Bugandan Christian Revolution: The Catholic Church in Buddu, 1879-1896" in Maxwell, David and Ingrid Lawrie (eds) *Christianity and the African Imagination. Essays in Honour of Adrian Hastings.* Leiden: Brill.

White, Gordon 1994, "Civil Society, Democratisation and Development (I). Clearing the Analytical Ground." *Democratisation* 1(3), pp. 375-390.

White, Luise 1995, "'They Could Make Their Victims Dull': Genders and Genres, Fantasies and Cures in Colonial Southern Uganda." *American Historical Review*, 100(5), pp. 1379-1402.

WHO 2005, "Health and mortality survey among internally displaced persons in Gulu, Kitgum and Pader districts, northern Uganda." Geneva/Kampala/Gulu: World Health Organisation.

Whyte, Susan Reynolds 1991, "Medicines and Self-Help: the Privatisation of Health Care in Eastern Uganda" in Hansen and Twaddle (eds) 1991, pp. 130-148.

Williams, David and Tom Young 1994, "Governance, the World Bank and Liberal Theory." *Political Studies*, 42 (1), pp. 84-100.

World Bank 1992, *Governance and Development.* Washington DC: The World Bank.

World Bank 1997, *The State in a Changing World.* Washington DC: The World Bank.

World Bank 1998, *Recommendations for Strengthening the Anti-Corruption Program, Uganda. Poverty Reduction and Social Development, Africa Region*, PRSD Anti-corruption Series No. 1. Washington DC: The World Bank.

Wunsch, James S. and Dan Ottemoeller 2004, 'Uganda: Multiple Levels of Local Governance" in Olowu, Dele and James S. Wunsch (eds) *Local Governance in Africa: The Challenges of Democratic Decentralisation.* Boulder: Lynne Rienner, pp. 181-209.

Young, Crawford 1994, *The African Colonial State in Comparative Perspective*. New Haven: Yale University Press.

Zartman, William (ed.) 1995, *Collapsed States: The Disintegration and Restoration of Legitimate Authority*. Boulder: Lynne Rienner.

Newspapers

The Monitor (later renamed *The Daily Monitor*, also *Sunday Monitor*) Kampala daily, http://www.monitor.co.ug

The New Vision, (also *Sunday Vision*) Kampala daily, http://www.newvision.co.ug

The Weekly Observer, (later renamed *The Observer*) Kampala Weekly, http://www.observer.ug

Interviews

Babirye, Robinah, World Vision Programme Co-ordinator (Kampala, 20 November 2000).

Giusti, Daniele, Executive Secretary, Uganda Catholic Medical Bureau (Kampala, 7 February 2005).

Kasozi, Joseph, Masaka District Secretary for Health (Masaka, 10 February 2003).

Kawanga, John, (DP), MP Masaka Town, (Masaka 22 February 2004).

Kyambadde, Faustin, BUCADEF District Co-ordinator (Masaka, 17 February 2003).

Kyewalabye, David Male, BUCADEF Chief Executive Officer (Kampala, 23 November 2000).

Mutibwa, Martin, World Vision Programme Co-ordinator (Masaka, 11 March 2003).

Mutibwa, Martin, World Vision Programme Co-ordinator (Masaka, 20 March 2003).

Nyangasi, Apollo, National Chairman, Uganda Medical Workers Union, (Kampala, 31 January 2003).

Nyangasi, Apollo, National Chairman, Uganda Medical Workers Union, (Kampala, 25 February 2003).

Nyangasi, Apollo, National Chairman, Uganda Medical Workers Union, (Kampala, 17 February 2004).

Sabiiti, Benon, Head of Administration, Uganda Protestant Medical Bureau (Kampala, 9 February 2005).

Serwadda, Badru, BUCADEF Acting Chief Executive Officer (Kampala, 28 February 2003).

Group Discussions

Nurses, Masaka Hospital (Masaka, 18 October 2000).

Union leadership and health workers, Masaka Hospital (Masaka, 4 March 2003).

In Bulando

Men, 7 November 2000.

Women, 7 November 2000.

Women, 3 March 2003.

Men, 5 March 2003.

Youth, 7 March 2003.

General, 25 February 2004.

Index